Principles of Ec

F. Stuart Chapin, III
Gary P. Kofinas
Carl Folke

Editors

Principles of Ecosystem Stewardship

Resilience-Based Natural
Resource Management
in a Changing World

Illustrated by Melissa C. Chapin

 Springer

Editors

F. Stuart Chapin, III
Institute of Arctic Biology
University of Alaska Fairbanks
Fairbanks, AK 99775, USA
terry.chapin@uaf.edu

Gary P. Kofinas
School of Natural Resources
    and Agricultural Sciences
    and Institute of Arctic Biology
University of Alaska Fairbanks
Fairbanks, AK 99775, USA
gary.kofinas@uaf.edu

Carl Folke
Stockholm Resilience Centre
Stockholm University
SE-106 91 Stockholm, Sweden
and Beijer Institute of Ecological
    Economics
Royal Swedish Academy of Sciences
PO Box 50005, SE 104 05 Stockholm
Sweden
carl.folke@beijer.kva.se

ISBN 978-1-4899-9650-3   ISBN 978-0-387-73033-2 (eBook)
DOI 10.1007/978-0-387-73033-2

Printed on acid-free paper

springer.com

Preface

The world is undergoing unprecedented changes in many of the factors that determine its fundamental properties and their influence on society. These changes include climate; the chemical composition of the atmosphere; the demands of a growing human population for food and fiber; and the mobility of organisms, industrial products, cultural perspectives, and information flows. The magnitude and widespread nature of these changes pose serious challenges in managing the ecosystem services on which society depends. Moreover, many of these changes are strongly influenced by human activities, so future patterns of change will continue to be influenced by society's choices and governance.

The purpose of this book is to provide a new framework for natural resource management—a framework based on stewardship of ecosystems for human well-being in a world dominated by uncertainty and change. The goal of ecosystem stewardship is to respond to and shape change in social-ecological systems in order to sustain the supply and opportunities for use of ecosystem services by society. The book links recent advances in the theory of resilience, sustainability, and vulnerability with practical issues of ecosystem management and governance. The book is aimed at advanced undergraduates and beginning graduate students of natural resource management as well as professional managers, community leaders, and policy makers with backgrounds in a wide array of disciplines, including ecology, policy studies, economics, sociology, and anthropology.

The first part of the book presents a conceptual framework for understanding the fundamental interactions and processes in social-ecological systems—systems in which people interact with their physical and biological environment. We explain how these systems respond to variability and change and discuss many of the ecological, economic, cultural, and institutional processes that contribute to these dynamics, enabling society to respond to and shape change. In the second section we apply this theory to specific types of social-ecological systems, showing how people adaptively manage resources and ecosystem services throughout the world. Finally we synthesize the lessons learned about resilience-based ecosystem stewardship as a strategy for responding to and shaping change in a rapidly changing world. Change brings both challenges and opportunities for managers, resource users, and policy makers to make informed decisions that enhance sustainability of our planet.

We owe a huge debt of gratitude to Buzz Holling who originated many of the central concepts that link resilience to ecosystem stewardship, as well as to several national and international programs that have developed these ideas and applied them to education and to the real-world issues faced by a rapidly changing planet. These include the Resilience Network, the Resilience Alliance, the International Geosphere-Biosphere Programme, the Millennium Ecosystem Assessment, Assessment, Association for the Study of the Commons, the Stockholm Resilience Centre, the Beijer Institute, and the Resilience and Adaptation Program of the University of Alaska Fairbanks. Primary funding for the book came from the US National Science Foundation and the Swedish Research Council FORMAS program. In addition, many individuals contributed to the development of this book. We particularly thank our families, whose patience made the book possible, and our students, from whom we learned many of the concepts and applications presented in this book. In addition, we thank the following people for their constructively critical review of chapters in this book: Marty Anderies, Erik Anderson, Archana Bali, David Battisti, Harry Biggs, Oonsie Biggs, Shauna BurnSilver, Steve Carpenter, Melissa Chapin, Johann Colding, Graeme Cumming, Bill Dietrich, Logan Egan, Thomas Elmqvist, Walter Falcon, Victor Galaz, Ted Gragson, Nancy Grimm, Lance Gunderson, Susan Herman, Buzz Holling, Jordan Lewis, Chanda Meek, Joanna Nelson, Evelyn Pinkerton, Ciara Raudsepp-Hearne, Marten Scheffer, Emily Springer, Samantha Staley, Will Steffen, Fred Swanson, Brian Walker, Karen Wang, and Oran Young. We particularly thank Steve Carpenter for his thoughtful comments on most of the chapters in this book.

F. Stuart Chapin, III   Fairbanks, AK, USA
Gary P. Kofinas   Fairbanks, AK, USA
Carl Folke   Stockholm, Sweden

Contents

Contributors

Nick Abel CSIRO Sustainable Ecosystems, GPO Box 284, Canberra ACT 2602, Australia, Nick.Abel@csiro.au

Robert Ahrens Fisheries Centre, University of British Columbia, Vancouver, BC, Canada V6T 1Z4, ahrens@zoology.ubc.ca

Fikret Berkes Natural Resources Institute, University of Manitoba, Winnipeg, Manitoba, Canada R3T 2N2, berkes@cc.umanitoba.ca

Reinette Biggs Stockholm Resilience Centre, Stockholm University, SE 106-91 Stockholm, Sweden, oonsie.biggs@stockholmresilience.su.se

Stephen R. Carpenter Center for Limnology, University of Wisconsin, Madison, WI 53706, USA, srcarpen@wisc.edu

F. Stuart Chapin, III Institute of Arctic Biology, University of Alaska Fairbanks, Fairbanks, AK 99775, USA, terry.chapin@uaf.edu

Carl Folke Stockholm Resilience Centre, Stockholm University, SE-106 91 Stockholm, Sweden; Beijer Institute of Ecological Economics, Royal Swedish, Academy of Sciences, PO Box 50005, SE 104 05 Stockholm, Sweden, carl.folke@beijer.kva.se

J. Morgan Grove Northern Research Station, USDA Forest Service, Burlington, VT 05403, USA, mgrove@fs.fed.us

Gary P. Kofinas School of Natural Resources and Agricultural Sciences and Institute of Arctic Biology, University of Alaska, Fairbanks, Fairbanks, AK 99775, USA, gary.kofinas@uaf.edu

Rosamond L. Naylor Woods Institute for the Environment, Freeman-Spogli Institute for International Studies, Stanford University, Stanford, CA 94305, USA, roz@leland.stanford.edu

Per Olsson Stockholm Resilience Centre, Stockholm University, SE-106 91 Stockholm, Sweden, per.olsson@stockholmresilience.su.se

Evelyn Pinkerton School of Resource and Environmental
Management, Simon Fraser University, Burnaby, BC V5A 1S6,
Canada, epinkert@sfu.ca

D. Mark Stafford Smith CSIRO Sustainable Ecosystems, PO Box
284, Canberra ACT 2602, Australia, mark.staffordsmith@csiro.au

Will Steffen Climate Change Institute, Australian National
University, Canberra, ACT 0200, Australia,
will.steffen@anu.edu.au

Frederick J. Swanson Pacific Northwest Research Station, USDA
Forest Service, Corvallis, OR 97331, USA, fswanson@fs.fed.us

Brian Walker CSIRO Sustainable Ecosystems, GPO Box 284,
Canberra ACT 2602, Australia, brian.walker@csiro.au

Carl Walters Fisheries Centre, University of British Columbia,
Vancouver, BC, Canada V6T 1Z4, c.walters@fisheries.ubc.ca

Oran R. Young Bren School of Environmental Science and
Management, University of California, Santa Barbara, CA
93106-5131, USA, young@bren.ucsb.edu

Part I
Conceptual Framework

1
A Framework for Understanding Change

F. Stuart Chapin, III, Carl Folke, and Gary P. Kofinas

Introduction

The world is undergoing unprecedented changes in many of the factors that determine both its fundamental properties and their influence on society. Throughout human history, people have interacted with and shaped ecosystems for social and economic development (Turner et al. 1990, Redman 1999, Jackson 2001, Diamond 2005). During the last 50 years, however, human activities have changed ecosystems more rapidly and extensively than at any comparable period of human history (Steffen et al. 2004, Foley et al. 2005, MEA 2005d; Plate 1). Earth's climate, for example, is now warmer than at any time in the last 500 (and probably the last 1,300) years (IPCC 2007a), in part because of atmospheric accumulation of carbon dioxide (CO_2) released by the burning of fossil fuels (Fig. 1.1). Agricultural development largely accounts for the accumulation of other trace gases that contribute to climate warming (see Chapter 12). As human population increases, in part due to improved disease prevention, the increased demand for food and natural resources has led to an expansion of agriculture, forestry, and other human activities, causing large-scale land-cover change and loss of habitats and biological diversity. About half the world's population now lives in cities and depends on connections with rural areas worldwide for food, water, and waste processing (see Chapter 13; Plate 2). In addition, increased human mobility is spreading plants, animals, diseases, industrial products, and cultural perspectives more rapidly than ever before. This increase in global mobility, coupled with increased connectivity through global markets and new forms of communication, links the world's economies and cultures, so decisions in one place often have international consequences.

This **globalization** of economy, culture, and ecology is important because it modifies the **life-support system** of the planet (Odum 1989), i.e., the capacity of the planet to meet the needs of all organisms, including people. The dramatic increase in the extinction rate of species (100- to 1,000-fold in the last two centuries) indicates that global changes have been catastrophic for many species, although some species,

F.S. Chapin, III (✉)
Institute of Arctic Biology, University of Alaska
Fairbanks, Fairbanks, AK 99775, USA
e-mail: terry.chapin@uaf.edu

F.S. Chapin et al. (eds.), *Principles of Ecosystem Stewardship*,
DOI 10.1007/978-0-387-73033-2_1, © Springer Science+Business Media, LLC 2009

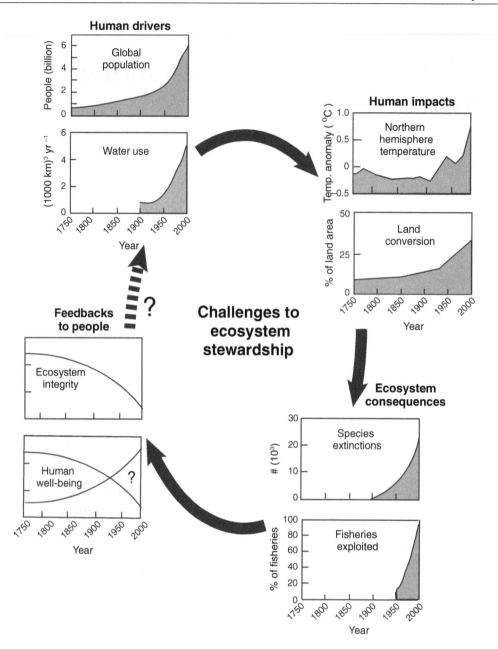

FIGURE 1.1. Challenges to ecosystem stewardship. Changes in human population and resource consumption alter climate and land cover, which have important ecosystem consequences such as species extinctions and overexploitation of fisheries. These changes reduce ecosystem integrity and have regionally variable effects on human well-being, which feeds back to further changes in human drivers. Panel inserts redrawn from Steffen et al. (2004).

especially invasive species and some disease organisms, have benefited and expanded their ranges. Human society has both benefited and suffered from global changes, with increased food production, increased income and living standards (in parts of the world), improved treatment of many diseases, and longer life expectancy being offset by deterioration in

ecosystem services, the benefits that society receives from ecosystems. More than half of the ecosystem services on which society depends for survival and a good life have been degraded—not deliberately, but inadvertently as people seek to meet their material desires and needs (MEA 2005d). Change creates both challenges and opportunities. People have amply demonstrated their capacity to alter the life-support system of the planet. In this book we argue that, with appropriate stewardship, this human capacity can be mobilized to not only repair but also enhance the capacity of Earth's life-support system to support societal development.

The unique feature of the changes described above is that they are **directional**. In other words, they show a persistent trend over time (Fig. 1.1). Many of these trends have become more pronounced since the mid-twentieth century and will probably continue or accelerate in the coming decades, even if society takes concerted actions to reduce some rates of change. This situation creates a dilemma in planning for the future because we cannot assume that the future world will behave as we have known it in the past or that our past experience provides an adequate basis to plan for the future. This issue is especially acute for **sustainable management** of natural resources. It is no longer possible to manage systems so they will remain the same as in the recent past, which has traditionally been the reference point for resource managers and conservationists. We must adopt a more flexible approach to managing resources—management to sustain the *functional* properties of systems that are important to society under conditions where the system itself is constantly changing. Managing resources to foster **resilience**—to respond to and shape change in ways that both sustain and develop the same fundamental function, structure, identity, and feedbacks—seems crucial to the future of humanity and the Earth System. **Resilience-based ecosystem stewardship** is a fundamental shift from steady-state resource management, which attempted to reduce variability and prevent change, rather than to respond to and shape change in ways that benefit society (Table 1.1). We emphasize *resilience*, a concept that embraces change as a basic feature of the way the world works and develops, and therefore is especially appropriate at times when changes are a prominent feature of the system. We address *ecosystems* that provide a suite of ecosystem services rather than a

TABLE 1.1. Contrasts between steady-state resource management, ecosystem management, and resilience-based ecosystem stewardship.

Steady-state resource management	Ecosystem management	Resilience-based ecosystem stewardship
Reference state: historic condition	Historic condition	Trajectory of change
Manage for a single resource or species	Manage for multiple ecosystem services	Manage for fundamental social–ecological properties
Single equilibrium state whose properties can be sustained	Multiple potential states	Multiple potential states
Reduce variability	Accept historical range of variability	Foster variability and diversity
Prevent natural disturbances	Accept natural disturbances	Foster disturbances that sustain social–ecological properties
People use ecosystems	People are part of the social–ecological system	People have responsibility to sustain future options
Managers define the primary use of the managed system	Multiple stakeholders work with managers to define goals	Multiple stakeholders work with managers to define goals
Maximize sustained yield and economic efficiency	Manage for multiple uses despite reduced efficiency	Maximize flexibility of future options
Management structure protects current management goals	Management goals respond to changing human values	Management responds to and shapes human values

single resource such as fish or trees. We focus on *stewardship*, which recognizes managers as an integral component of the system that they manage. Stewardship also implies a sense of responsibility for the state of the system of which we are a part (Leopold 1949). The challenge is to anticipate change and shape it for sustainability in a manner that does not lead to loss of future options (Folke et al. 2003). Ecosystem stewardship recognizes that society's use of resources must be compatible with the capacity of ecosystems to provide services, which, in turn, is constrained by the life-support system of the planet (Fig. 1.2).

This chapter introduces a framework for understanding and managing resources in a world where persistent directional changes are becoming more pronounced. We first present a framework for studying change—one that integrates the physical, ecological, and social dimensions of change and their interactions. We then describe the general properties of systems that magnify or resist change. Finally we discuss general approaches to sustaining desirable system properties in a directionally changing world and present a road map to the remaining chapters, which address these issues in greater depth.

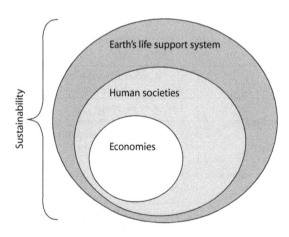

FIGURE 1.2. Social–ecological sustainability requires that society's economy and other human activities *not* exceed the capacity of ecosystems to provide services, which, in turn, is constrained by the planet's life-support system. Redrawn from Fischer et al. (2007).

An Integrated Social–Ecological Framework

Linking Physical, Ecological, and Social Processes

Changes in the Earth System are highly interconnected. None of the changes mentioned above is purely physical, ecological, or social. Therefore understanding current and future change requires a broad interdisciplinary framework that draws on the concepts and approaches of many natural and social sciences. We must understand the world, region, or community as a **social–ecological system** (also termed a coupled human–environment system) in which people depend on resources and services provided by ecosystems, and ecosystem dynamics are influenced, to varying degrees, by human activities (Berkes et al. 2003, Turner et al. 2003, Steffen et al. 2004). Although the relative importance of social and ecological processes may vary from forests to farms to cities, the functioning of each of these systems, and of the larger regional system in which they are embedded, is strongly influenced by physical, ecological, economic, and cultural factors. They are, therefore, best viewed, not as ecological *or* social systems, but as social–ecological systems that reflect the interactions of physical, ecological, and social processes (Westley et al. 2002).

Forests, for example, are sometimes managed as ecological systems in which the nitrogen inputs from acid rain or the economic influences on timber demand are considered **exogenous factors** (i.e., factors external to the system being managed) and therefore are not incorporated into management planning. Production of lumber or paper, on the other hand, is often managed as an economic system that must balance the supply and costs of timber inputs against the demand for and profits from products without considering ecological influences on forest production. Finally, local planners make decisions about school budgets and the zoning for development and recreation, based on assumptions about regional water supply, which depends on forest cover, and economic projections, which are influenced by the economic activity of forest industries. The system and its components are more vulnerable to unexpected changes (**surprises**) when each subsystem is managed in isolation. These surprises might include harvest restrictions to protect an endangered species, development of inexpensive lumber supplies on another continent, or expansion of recreational demand for forest use by nearby urban residents. More informed decisions are likely to emerge from integrated approaches that recognize the interdependencies of regional components and account for uncertainty in future conditions (Ludwig et al. 2001). Resource stewardship policies must therefore be ecologically, economically, and culturally viable, if they are to provide sustainable solutions.

In studying the response of social–ecological systems to directional change, we pay particular attention to the processes that link ecological and social components (Fig. 1.3). The environment affects people through both

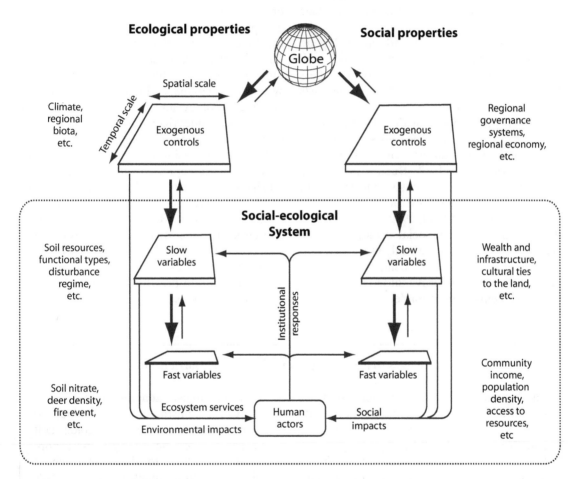

FIGURE 1.3. Diagram of a social–ecological system (the rectangle) that is affected by ecological (left-hand side) and social properties (right-hand side). In both subsystems there is a spectrum of controls that operate across a range of temporal and spatial scales. At the regional scale exogenous controls respond to global trends and affect slow variables at the scale of management, which, in turn, influence fast variables that change more quickly. When changes in fast variables persist over long time periods and large areas, these effects cumulatively propagate upward to affect slow variables, regional controls, and eventually the entire globe. Changes in both slow and fast variables influence environmental impacts, ecosystem services, and social impacts, which, together, are the factors that directly affect the well-being of human actors, who modify both ecological and social systems through a variety of institutions. Modified from Chapin et al. (2006a).

direct environmental events such as floods and droughts and ecosystem services such as food and water quality (see Chapter 2). Many economic, political, and cultural processes also shape human responses to the physical and biological environment (see Chapter 3). Human actors (both individuals and groups) in turn affect their ecological environment through a complex web of social processes (see Chapter 4). Together these linkages between social and ecological processes structure the dynamics of social–ecological systems (see Chapter 5).

The concept that society and nature depend on one another is not new. It was well recognized by ancient Greek philosophers (Boudouris and Kalimtzis 1999); economists concerned with the environmental constraints on human population growth (Malthus 1798); geographers and anthropologists seeking to understand global patterns of land use and culture (Rappaport 1967, Butzer 1980); and ecologists and conservationists concerned with human impacts on the environment (Leopold 1949, Carson 1962, Odum 1989). The complexity and importance of social–ecological interactions has led many natural and social science disciplines to address components of the interaction to both improve understanding and solve problems. For example, resource management considers the actions that agencies or individuals take to sustain natural resources, but typically pays less attention to the interactions among interest groups that influence how management policies develop or how the public will respond to management. Similarly, environmental policy analysis addresses the potential interactions of environmental policies developed by different organizations, but typically pays less attention to potential social or ecological **thresholds** (critical levels of drivers or state variables that, when crossed, trigger abrupt changes or **regime shifts**) that determine the long-term effectiveness of these policies. The breadth of approaches provides a wealth of tools for studying integrated social–ecological systems. Disciplinary differences in vocabulary, methodology, and standards of what constitutes academic rigor can, however, create barriers to communication (Box 1.1; Wilson 1998). The increasing recognition that human actions are threatening Earth's life-support system has recently generated a sense of urgency in addressing social–ecological systems in a more integrated fashion (Berkes et al. 2003, Clark and Dickson 2003, MEA 2005d). This requires a system perspective that integrates social and ecological processes and is flexible enough to accommodate the breadth of potential human actions and responses.

Box 1.1. Challenges to Navigating Social–Ecological Barriers and Bridges.

The heading of this box combines the titles of two seminal books on integrated social–ecological systems ("Barriers and Bridges" and "Navigating Social Ecological Systems"; Gunderson et al. 1995, Berkes et al. 2003). These titles capture the essence of the challenges in integrating natural and social sciences. In this book we adopt the following conventions in addressing two important challenges in this **transdisciplinary** integration (i.e., integration that transcends traditional disciplines to formulate problems in new ways).

The same word often means different things.

1. To a sociologist, **adaptation** means the behavioral adjustment by individuals to their environment. To an ecologist it means the genetic changes in a population to adjust to their environment (in contrast to acclimation, which entails physiological or behavioral adjustment by individuals). To an anthropologist adaptation means the cultural adjustment to environment, without specifying its genetic or behavioral basis. *In this book we use **adaptation** in its most general sense (adjustment to change in environment).*

2. To an engineer or ecologist describing systems with a single equilibrium, **resilience** is the time required for a system to return to equilibrium after a perturbation. To someone describing systems with multiple stable states, resilience is capacity of the system to absorb a spectrum of shocks or perturbations and still retain and further develop the same fundamental structure, functioning, and feedbacks. *We use **resilience** in the latter sense.*

3. Natural scientists describe feedbacks as being **positive** or **negative** to denote whether they are amplifying or stabilizing, respectively. These words are often used in the social sciences (and in common usage) to mean good or bad. The terminology is especially confusing for social–ecological systems, because negative feedbacks are often socially desirable (= "good") and positive feedbacks socially undesirable (= "bad"). *We therefore avoid these terms and talk about **amplifying** or **stabilizing** feedbacks.*

4. Words that represent important concepts in one discipline may be meaningless or viewed as jargon in another (e.g., **postmodern**, **state factor**). *We define each technical word the first time it is used and use* *only those technical terms that are essential to convey ideas effectively.*

Approaches that are viewed as "good science" in one discipline may be viewed with skepticism in another.

1. Some natural scientists use **systems models** to describe (either quantitatively or qualitatively) the interactions among components of a system (such as a social–ecological system). Some social scientists view this as an inappropriate tool to study systems with a strong human element because it seems too deterministic to describe human actions. *We use **complex adaptive systems** as a framework to study social–ecological systems because it enables us to study the integrated nature of the system but recognizes legacies of past events and the path dependence of human agency as fundamental properties of the model.*

2. Some natural scientists rely largely on quantitative data as evidence to test a hypothesis, whereas some social scientists make extensive use of qualitative descriptions of patterns that are less amenable to quantification. *We consider both approaches essential to understanding the complex dynamics of social–ecological systems.*

A Systems Perspective

Systems theory provides a conceptual framework to understand the dynamics of integrated systems. A social–ecological system consists of physical components, including soil, water, and rocks; organisms (plants, microbes, and animals—including people); and the products of human activities, such as food, money, credit, computers, buildings, and pollution. A social–ecological system is like a box or a board game, with explicit boundaries and rules, enabling us to quantify the amount of materials (for example, carbon, people, or money) in the system and the factors that influence their flows into, through, and out of the system.

Social–ecological systems can be defined at many scales, ranging from a single household or community garden to the entire planet. Systems are defined to include those components and interactions that a person most wants to understand. The size, shape, and boundaries of a social–ecological system therefore depend entirely on the problem addressed and the objectives of study. A watershed that includes all the land draining into a lake, for example, is an appropriate system for studying the controls over pollution of the lake. A farm, city, water-management district, state, or country might be a logical unit for studying the effects of government policies. A community, nation, or the globe might be an appropriate unit for studying barter and commerce. A neighborhood, community, or multinational region might be a logical unit for studying cultural change. Defining the most appropriate unit of analysis is challenging because key ecological and social processes often differ in scale and logical boundaries (for example, watersheds and water-management districts; Ostrom 1990, Young 1994). Most social–ecological systems are **open systems**, in the sense that there are flows of materials, organisms, and information into and out of the system. We therefore cannot ignore processes occurring outside our defined system of analysis, for example, the movement of food and wastes across city boundaries.

Social–ecological processes are the interconnections among components of a system. These may be primarily ecological (for example, plant production, decomposition, wildlife migration), socioeconomic (manufacturing, education, fostering of trust among social groups), or a mix of ecological and social processes (plowing, hunting, polluting). The interactions among multiple processes govern the dynamics of social–ecological systems. Two types of interactions among components (amplifying and stabilizing feedbacks) are especially important in defining the internal dynamics of the system because they lead to predictable outcomes (DeAngelis and Post 1991, Chapin et al. 1996). **Amplifying feedbacks** (termed positive feedbacks in the systems literature) augment changes in process rates and tend to destabilize the system (Box 1.2). They occur when two interacting components cause one another to change in the same direction (both components increase or both decrease; Fig. 1.4). A disease epidemic occurs, for example, when a disease infects susceptible hosts, which produce more disease organisms, which infect more hosts, etc., until some other set of interactions constrains this spiral of disease increase. Overfishing can also lead to an amplifying feedback, when the decline in fish stocks gives rise to price supports that enable fishermen to maintain or increase fishing pressure despite smaller catches, leading to a downward spiral of fish abundance. Other examples of amplifying feedbacks include

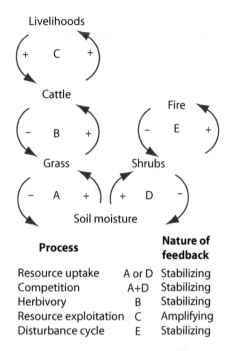

Process		Nature of feedback
Resource uptake	A or D	Stabilizing
Competition	A+D	Stabilizing
Herbivory	B	Stabilizing
Resource exploitation	C	Amplifying
Disturbance cycle	E	Stabilizing

FIGURE 1.4. Examples of linked amplifying and stabilizing feedbacks in social–ecological systems. Arrows show whether one species, resource, or condition has a positive or a negative effect on another. The feedback between two species is *stabilizing* when the arrows have opposite sign (for example, species 1 has a positive effect on species 2, but species 2 has a negative effect on species 1). The feedback is amplifying, when both species affect one another in the same direction (for example, more cattle providing more profit, which motivates people to raise more cattle; feedback loop C in the diagram).

population growth, erosion of cultural integrity in developing nations, and proliferation of nuclear weapons.

Stabilizing feedbacks (termed negative feedbacks in the systems literature) tend to reduce fluctuations in process rates, although, if extreme, they can induce chaotic fluctuations. Stabilizing feedbacks occur when two interacting components cause one another to change in opposite directions (Fig. 1.4). For example, grazing by cattle reduces the biomass of forage grasses, whereas the grass has a positive effect on cattle production. Any increase in density of cattle reduces grass biomass, which then constrains the food available to cattle, thereby stabilizing the sustainable densities of both grass and cattle at intermediate levels. Other examples of stabilizing feedbacks include prices of goods in a competitive market and nutrient supply to plants in a forest. One of the keys to sustainability is to foster stabilizing feedbacks and constrain amplifying feedbacks that might otherwise push the system toward some new state. Conversely, if the current state is socially undesirable, for example, at an abandoned mine site, carefully selected amplifying feedbacks may shift the system to a preferred new state.

Box 1.2. Dynamics of Temporal Change

The stability and dynamics of a system depend on the balance of amplifying and stabilizing feedbacks and types and frequencies of perturbations. The strength and nature of feedbacks largely govern the way a system responds to change. A system without strong feedbacks shows **chaotic behavior** in response to a random perturbation. Chaotic behavior is unpredictable and depends entirely on the nature of the perturbation. The behavior of a ball on a surface provides a useful analogy (Fig. 1.5; Holling and Gunderson 2002, Folke et al. 2004). The location of the ball represents the state of a system as a function of some variable such as water availability. In a chaotic system without feedbacks, the surface is flat, and we cannot predict changes in the state (i.e., location) of the system in response to a random perturbation (Fig. 1.5a). This system structure is analogous to theories that important decisions can be described in terms of the potential solutions and actors that happen to be present at key moments (garbage-can politics; Cohen et al. 1972, Olsen 2001).

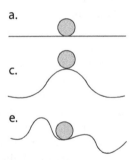
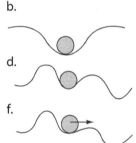

FIGURE 1.5. The location of the ball represents the state of a system in relationship to some ecological or social variable (e.g., water availability, as represented by the position along the horizontal axis). Changes in the state of the system in response to a perturbation depend on the nature of system feedbacks (illustrated as the shape of the surface). The likelihood that the system will change its state (location along the line) differs if there are (**a**) no feedbacks, (**b**) stabilizing feedbacks, (**c**) amplifying feedbacks, (**d**) alternative stable states, (**d–e**) changes in the internal feedback structure (complex adaptive system), and (**e–f**) response of a complex adaptive system to persistent directional changes in a control variable.

A system dominated by *stabilizing feedbacks* tends to be stable because the interactions occurring within the system minimize the changes in the system in response to perturbations. Using our analogy, stabilizing feedbacks create a bowl-like depression in the surface so the ball tends to return to the same location after a random perturbation (Fig. 1.5b). The resilience of the system, in this cartoon, is the likelihood that it will remain in the same state despite perturbations. This analogy characterizes the perspective of a balanced view of nature, in which there is a **carrying capacity** (maximum quantity) of fish, game, or trees that the environment can support, allowing managers to regulate harvest to achieve a **maximum sustained yield**. This view is often based on considerable depth of biological understanding but is incomplete (Holling and Gunderson 2002).

A system dominated by *amplifying feedbacks* tends to be unstable because the initial change is amplified by interactions occurring within the system. Amplifying feedbacks tend to push the system toward some new state by making the depressions less deep or creating elevated areas on the surface (Fig. 1.5c). This analogy characterizes the view that small is beautiful and that any technology is bad because it causes change. There are certainly many examples where technology has led to unfavorable outcomes, but this worldview, like the others, is incomplete (Holling and Gunderson 2002).

Many systems can be characterized by **alternative stable states**, each of which is plausible in a given environment. Neighborhoods in US cities, for example, are likely to be either residential or industrial but unlikely to be an even mix of the two. In the surface analogy, alternative stable states represent multiple depressions in the surface (Fig. 1.5d). A system is likely to return to its original state (=depression) after a small perturbation, but a larger disturbance might increase the likelihood that it will shift to some alternative state. In other words, the system exhibits a nonlinear response to the perturbation and shifts to a new state if some **threshold** is exceeded. There may also be pathways of system development, such as the stages of forest succession, in which the internal dynamics of the system cause it to move readily from one state to another. Some of

these depressions may be deep and represent irreversible traps. Others may be shallow, so the system readily shifts from one state to another through time. This worldview incorporates components of all the previous perspectives but is still incomplete.

The previous cartoons of nature imply that the stability landscape is static. However, each transition influences the internal dynamics of a complex adaptive system and therefore the probability of subsequent transitions, so the shape of the surface is constantly changing (Fig. 1.5e). Reductions in Atlantic cod populations due to overfishing, for example, increased pressures for establishment of aquaculture and charter fishing businesses, which then made it less likely that industrial-scale cod fishing would return to the North Atlantic. This analogy of a stability landscape that is constantly evolving suggests that precise predictions of the future state of the system are impossible and focuses attention on understanding the dynamics of change as a basis for stewardship (Gunderson and Holling 2002).

Now imagine that rather than having a random perturbation in some important state variable like water availability, this parameter changes directionally. This element of directionality increases the likelihood that the system will change in a specific direction after perturbation (Fig. 1.5f). The stronger and more persistent the directional changes in exogenous control variables, the more likely it is that new states will differ from those that we have known in the past. This represents our concept of system response to a directionally changing environment.

Issues of Scale: Exogenous, Slow, and Fast Variables

Changes in the state of a system depend on variables that change slowly but strongly influence internal dynamics. Social–ecological systems respond to a spectrum of controls that operate across a range of temporal and spatial

scales. These can be roughly grouped as exogenous controls, slow variables, and fast variables (Fig. 1.3). We describe these first for ecological subsystems, then consider their social counterparts.

Exogenous controls are factors such as regional climate or biota that strongly shape the properties of continents and nations. They remain relatively constant over long time periods (e.g., a century) and across broad regions and are not strongly influenced by short-term, small-scale dynamics of a single forest stand or lake. At the scale of an ecosystem or watershed, there are a few **critical slow variables**, i.e., variables that strongly influence social–ecological systems but remain relatively constant over years to decades despite interannual variation in weather, grazing, and other factors, because they are buffered by stabilizing feedbacks that prevent rapid change (Chapin et al. 1996, Carpenter and Turner 2000). Soil organic matter, for example, retains pulses of nutrients from autumn leaf fall, crop residues, or windstorms; retains water and nutrients; and releases these resources which are then absorbed by plants. The quantity of soil organic matter is buffered by feedbacks related to plant growth and litter production. Critical slow variables include presence of particular **functional types** of plants and animals (e.g., evergreen trees or herbivorous mammals); **disturbance regime** (properties such as frequency, severity, and size that characterize typical disturbances); and the capacity of soils or sediments to supply water and nutrients. Slow variables in ecosystems, in turn, govern **fast variables** at the same spatial scale (e.g., deer or aphid density, individual fire events) that respond sensitively to daily, seasonal, and interannual variation in weather and other factors. When aggregated to regional or global scales, changes that occur in ecosystems, for example, those mediated by human activities, can modify the environment to such an extent that even regional controls such as climate and regional biota that were once considered constant parameters are now directionally changing at decade-to-century time scales (Foley et al. 2005). Regardless of the causes, persistent directional changes in broad regional controls, such as climate and biodiversity, inevitably cause directional changes in critical slow variables and therefore the structure and dynamics of ecosystems, including the fast variables. The exogenous and slow variables are critical to long-term sustainability, although most management and public attention focus on fast variables, whose dynamics are more visible.

Analogous to the ecological subsystem, the social subsystem can be viewed as composed of exogenous controls, critical slow variables, and fast variables (Straussfogel 1997). These consist of vertically nested relationships, ranging from global to local, and linked by cross-scale interactions (Ostrom 1999a, Young 2002b, Adger et al. 2005). At the sub-global scale a predominant history, culture, economy, and governance system often characterize broad regions or nation states such as Europe or sub-Saharan Africa (Chase-Dunn 2000). These exogenous social controls tend to be less sensitive to interannual variation in stock-market prices and technological change than are the internal dynamics of local social–ecological systems; the exogenous controls constrain local options. This asymmetry between regional and local controls occurs in part because of asymmetric power relationships between national and local entities and in part because changes in a small locality must be very strong to substantially modify the dynamics of large regions. Regional controls sometimes persist for a long time and change primarily in response to changes that are global in extent (e.g., globalization of markets and finance institutions), but at other times change can occur quickly, as with the collapse of the Soviet Union in the 1990s or the globalization of markets and information (Young et al. 2006). As in the biophysical system, a few slow variables (e.g., wealth and infrastructure; property-and-use rights; and cultural ties to the land) are constrained by regional controls and interact with one another to shape fast variables like community income or population density. Both slow and fast social variables can have major effects on ecological processes (Costanza and Folke 1996, Holling and Sanderson 1996).

Systems differ in their sensitivity to different types of changes or the range of conditions over which the change occurs. The !Kung San of the Kalahari Desert will be much more sensitive than people of a rainforest to a 10-cm increase in annual rainfall because it represents a doubling of rainfall rather than a 5% increase. Regions also differ in their sensitivity to introduction of new biota (spruce bark beetle, zebra mussel, or West Nile virus), new economic pressures (development of aqua-culture, shifting of car manufacture to Asia, collapse of the stock market), or new cultural values. There are typically relatively few (often only three to five) slow variables that are crit-ical in understanding the current dynamics of a specific system (Carpenter et al. 2002), so management designed to reduce sensitivity to directional changes in slow variables is not an impossible task. The identity of critical con-trol variables may change over time, however, requiring continual reassessment of our under-standing of the social–ecological system. The key challenge, requiring collaborative research by managers and natural and social scientists, is to identify the critical slow variables and their likely changes over time.

Incorporating Scale, Human Agency, and Uncertainty into Dynamic Systems

Cross-scale linkages are processes that con-nect the dynamics of a system to events occur-ring at other times or places (see Chapter 5). Changes in the human population of a region, for example, may be influenced by the wealth and labor needs of individual families (fine scale), by national policies related to birth con-trol (focal scale), and by global inequalities in living standards that influence immigration (large scale). Events that occur at each scale typically influence events at other scales. The universal importance of cross-scale linkages in social–ecological systems makes it important to study them at multiple temporal and spatial scales, because different insights and answers emerge at each scale (Berkes et al. 2003).

Legacies are past events that have large effects on subsequent dynamics of social–ecological systems. This generates a **path dependence** that links current dynamics to past events and lays the foundation for future changes (North 1990). Legacies include the impact of plowing on soils of a regenerating forest, the impact of the Depression in the 1930s on economic decisions made by house-holds 40 years later, and the continuation of subsistence activities by indigenous people who move from villages to cities. Because of path

dependence, the current dynamics of a system always depend on both current conditions and the history of prior events. Consequently, dif-ferent trajectories can occur at different times or places, even if the initial conditions were the same. Path dependence is absolutely crit-ical to management, because it implies that human actions taken today, whether construc-tive or destructive, can influence the future state of the system. Good management can make a difference!

Human agency (the capacity of humans to make choices that affect the system) is one of the most important sources of path depen-dence. Human decisions depend on both past events (legacy effects) and the plans that people make for the future (**reflexive behavior**). The strong path dependence of social–ecological systems is typical of a general class of systems known as **complex adaptive systems**. These are systems whose components interact in ways that cause the system to adjust (i.e., "adapt") in response to changes in conditions. This is not black magic, but a consequence of inter-actions and feedbacks. Some of the most fre-quent failures in resource management occur because managers and resource users fail to understand the principles by which complex adaptive systems function. It is therefore impor-tant to understand their dynamics. Understand-ing these dynamics also provides insights into ways that managers can achieve desirable out-comes in a system that is responding simultane-ously to management actions and to persistent directional changes in exogenous controls.

Whenever system components with differ-ent properties interact spontaneously with one another, some components persist and oth-ers disappear (i.e., the system adapts; Levin 1999; Box 1.2). In social–ecological systems, for example, organisms compete or eat one another, causing some species to become more common and others to disappear. Similarly, purchasing or competitive relationships among businesses cause some firms to persist and oth-ers to fail. Those components that interact through stabilizing feedbacks are most likely to persist. This **self-organization** of compo-nents linked by stabilizing feedbacks occurs spontaneously without any grand design. It causes complex adaptive systems to be relatively **stable** (tend to maintain their properties over time; DeAngelis and Post 1991, Levin 1999). This **self-regulation** simplifies management challenges in many respects. A complex adaptive system like a forest, for example, tends to "take care of itself." This differs from a designed structure like a car, whose components do not interact spontaneously and where maintenance must be continually applied just to keep the car in the same condition (Levin 1999).

If conditions change enough to alter the interactions among system components, the system adapts to the new conditions, hence the term complex *adaptive* system (Levin 1998). The new balance of system components, in turn, alters the way in which the system responds to perturbations (path dependence), creating **alternative stable states**, each of which could exist in a given environment (see Chapter 5). Given that exogenous variables are always changing on all time scales, social–ecological systems are constantly adjusting and changing. Consequently, it is virtually impossible to manage a complex adaptive system to attain constant performance, such as the constant production of a given timber species. System properties are most likely to change if there are directional changes in exogenous controls. The stronger and more persistent the directional changes in control variables, the more likely it is that a threshold will be exceeded, leading to a new state.

If a threshold is exceeded, and the system changes radically, new interactions and feedbacks assume greater importance, and some components of the previous system may disappear. If a region shifts from a mining to a tourist economy, for example, the community may become more concerned about funding for education and regulations that assure clean water. The regime shifts that occur as the system changes state also depend on the past state of the system (path dependence). The presence of a charismatic leader or nongovernmental organization (NGO), for example, can be critical in determining whether large cattle ranches are converted to conservation easements or subdivisions when rising land values and taxes make ranching unprofitable.

These simple generalizations about complex adaptive systems have profound implications for resource stewardship: (1) Social and ecological components of a social–ecological system always interact and cannot be managed in isolation from one another. (2) Changes in social or ecological controls inevitably alter social–ecological systems regardless of management efforts to prevent change. (3) Historical events and human actions, including management, can strongly influence the pathway of change. (4) The thresholds and nonlinear dynamics associated with path dependence, compounded by lack of information and human volition, constrain our capacity to predict future change. Resource management and policy decisions must, therefore, always be made in an environment of **uncertainty** (Ludwig et al. 1993, Carpenter et al. 2006a).

Adaptive Cycles

The long-term stability of systems depends on changes that occur during critical phases of cycles of long-term change. All systems experience disturbances such as fire, war, recession, change in leadership philosophy, or closure of manufacturing plants that cause large rapid changes in key system properties. Such disturbances have qualitatively different effects on social–ecological systems than do short-term variability and gradual change. **Adaptive cycles** provide a framework for describing the role of disturbance in social–ecological systems (Holling 1986). They are cycles of system disruption, reorganization, and renewal. In an adaptive cycle, a system can be disrupted by disturbance and either regenerate to a similar state or be transformed to some new state (Fig. 1.6a; Holling 1986, Walker et al. 2004). Adaptive cycles exhibit several recognizable phases. The cycle may be initiated by a disturbance such as a stand-replacing wildfire that causes a rapid change in most properties of the system. Trees die, productivity decreases, runoff to streams increases, and public faith in fire management is shattered. This **release phase** occurs in hours to days and radically reduces the structural complexity of the system. Other factors that might trigger release include

a. Adaptive cycle

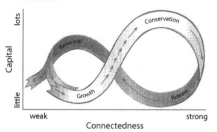

b. Panarchy

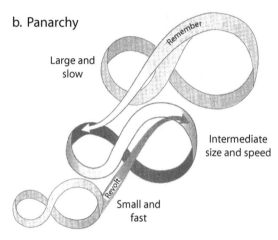

FIGURE 1.6. (**a**) Adaptive cycle and (**b**) cross-scale linkages among adaptive cycles (panarchy) in a social–ecological system. At any given scale, a system often goes through adaptive cycles of release (collapse), renewal (reorganization), growth, and conservation (steady state). These adaptive cycles of change can occur at multiple levels of organizations, such as individuals, communities, watersheds, and regions. These adaptive cycles interact forming a panarchy. For example, dynamics at larger scales (e.g., migration dynamics or wealth) provide legacies, context, and constraints that shape patterns of renewal (system memory). Dynamics at finer scales (e.g., insect population dynamics, household structure) may trigger release (revolt; e.g., insect outbreak). Redrawn from Holling and Gunderson (2002) and Holling et al. (2002b).

threshold response to phosphorus loading of a lake, collapse of the local or regional economy, or a transition from traditional to intensive agriculture. Following release, there is a relatively brief (months to years) **renewal phase**. For example, after forest disturbance, seedlings establish and new policies for managing the forest may be adopted. Many things can hap-

pen during renewal: The species and policies that establish might be similar to those present before the fire. It is also a time, however, when there is relatively little resistance to the establishment of a new suite of species or policies that emerge from the surrounding landscape (see Fig. 2.4). These innovations may lead to a system that is quite different from the prefire system, i.e., a **regime shift**. After this brief window of opportunity for change, the forest goes through a **growth phase** over several decades, when environmental resources are incorporated into living organisms, and policies become regularized. The nature of the regenerating forest system is largely determined by the species and regulations that established during renewal. During the growth phase, the forest is relatively insensitive to potential agents of disturbance. The high moisture content and low biomass of early successional trees, for example, make regenerating forests relatively nonflammable. Constant changes in the nature of the forest cause both managers and the public to accept changing conditions and regulations as a reasonable pattern. As the forest develops into the steady-state **conservation phase**, the interactions among components of the system become more specialized and complex. Light and nutrients decline in availability, for example, leading to specialization among plants to use different light environments and different fungal associations (**mycorrhizae**) to acquire nutrients. Similarly, in the policy realm, the relatively constant state of the forest leads to management rules that are aimed at maintaining this constancy to provide predictable patterns of recreation, hunting, and forest harvest. Due to the increased interconnectedness among these social and ecological variables, the forest becomes more vulnerable to any factor that might disrupt this balance, including fire, drought, changes in management goals, or a shift in the local economy. Large changes in any of these factors could trigger a new release in the adaptive cycle.

Many human organizations also exhibit cyclic patterns of change. A business or NGO, for example, may be founded in response to a perceived opportunity for profit or social reform. If successful, it grows amidst constant adjustment to changes in personnel and activities. Eventually it reaches a relatively stable size, at which time the internal structure and operating procedures are regularized, making it less flexible to respond to changes in the economic or social climate. When conditions change, the business or NGO may either enter a new period of adjustment (growth) or decline (release), followed by potential renewal or collapse.

Perhaps the most surprising thing about adaptive cycles is that the sequence of phases (release, renewal, growth, and conservation) can be used as a way of thinking about many types of social-ecological systems, including lakes, businesses, governments, national economies, and cultures, although the sequence of phases is not always the same (Gunderson et al. 1995). Clearly the specific mechanisms underlying cycles in these different systems must be quite different. One of the unsolved challenges in understanding social–ecological systems is to determine the general system properties and mechanisms that underlie the apparent similarities in cyclic patterns of different types of systems and to clarify the differences. The specific mechanisms of adaptive cycles in different types of systems are described in many of the following chapters.

One of the most important management lessons to emerge from studies of adaptive cycles is that social–ecological systems are typically most **vulnerable** (likely to change to a new state in response to a stress or disturbance) and create their own vulnerabilities in the conservation phase, where they typically spend most of their time. In this stage, managers frequently seek to reduce fluctuations in ecological processes and prevent small disturbances in order to increase the efficiency of achieving management goals (e.g., the amount of timber to be harvested; number of houses that can be built; the budget to pay salaries of personnel), increasing the likelihood that even larger disturbances will occur (Holling and Meffe 1996, Walker and Salt 2006). Flood control, for example, reduces flood frequency, which encourages infrastructure development in floodplains where it is vulnerable to the large flood that will eventu-

ally occur. Prevention of small insect outbreaks increases the likelihood of larger outbreaks. Management that encourages small-scale disturbances and innovation during the conservation phase reduces the vulnerability to larger disruptions (Holling et al. 1998, Carpenter and Gunderson 2001, Holling et al. 2002a). The specific mechanisms that link stability in the conservation phase to triggers for disruption are described in later chapters.

Release and crisis provide important opportunities for change (Gunderson and Holling 2002, Berkes et al. 2003; Fig. 1.5b). Some of these changes may be undesirable (invasion of an exotic species, dramatic shift in political regimes that decrease social equity), whereas others may be desirable (implementation of innovative policies that are more responsive to change). Recognition of these changing properties of a system through the lens of an adaptive cycle suggests that effective long-term management and policy-making must be highly flexible and adaptive, looking for windows of opportunity for constructive policy shifts.

Most social–ecological systems are spatially heterogeneous and consist of mosaics of subsystems that are at different stages of their adaptive cycles. Interactions and feedbacks among these adaptive cycles operating at different temporal and spatial scales account for the overall dynamics of the system (termed **panarchy**; Fig. 1.6b; Holling et al. 2002b). A forest, for example, may consist of different-aged stands at different stages of regeneration from logging or wildfire. In this case, the system as a whole may be at steady state (a **steady-state mosaic**) even though individual stands are at different stages in their cycles (Turner et al. 2001). In general, there are different benefits to be gained at different phases of the cycle, so policies that permit or foster certain disturbances may be appropriate. Many families contain individuals at various stages of birth, maturation, and death and benefit from the resulting diversity of skills, perspectives, and opportunities. Similarly, in a healthy economy new firms may establish at the same time that other less-efficient firms go out of business. Maintenance of natural cycles of fire or insect outbreak produces wildlife habitat in the early growth phase and prevents excessive fuel accumulation that might otherwise trigger more catastrophic fires. Perhaps the most dangerous management strategy would be to prevent disturbance uniformly throughout a region until all subunits reach a similar state of maturity, making it more likely that the entire system will change synchronously.

Sustainability in a Directionally Changing World

Conceptual Framework for Sustainability Science

A systems perspective provides a logical framework for managing changes in social–ecological systems. To summarize briefly the previous sections, the dynamic interactions of ecological and social processes that characterize most of today's urgent problems necessitate a social–ecological framework for planning and stewardship. Any sustainable solution to a resource issue must be compatible with current social and ecological conditions and their likely future changes. A resource policy that is not ecologically, economically, and culturally sustainable is unlikely to be successful. Sustainable resource stewardship must therefore be multifaceted, recognizing the interactions among ecological, economic, and cultural variables and the important roles that past history and future events play in determining outcomes in specific situations. In addition, systems undergo cyclic changes in their sensitivity to external perturbations, so management solutions that may have been successful at one time and place may or may not work under other circumstances.

The complexity of these dynamics helps frame the types of stewardship approaches that are most likely to be successful. It is unlikely that a rigid set of rules will lead to successful stewardship because key decisions must frequently be made under conditions of novelty and uncertainty. Moreover, under current rapid rates of global environmental and social changes, the current environment for decision-making is increasingly different from past conditions that may be familiar to managers or the future conditions that must be accommodated. The more rapidly the world changes, the less likely that rigid management approaches will be successful. By considering the system properties presented above, however, we can develop resilience-based approaches that substantially reduce the risk of undesirable social–ecological outcomes and increase the likelihood of making good use of unforeseen opportunities. This requires managing for general system properties rather than for narrowly defined production goals. In this section, we present a framework for this approach that is described in detail in subsequent chapters.

Sustaining the desirable features of our current world for future generations is an important societal goal. The challenge of doing so in the face of persistent directional trends in underlying controls has led to an emerging science of sustainability (Clark and Dickson 2003). Sustainability has been adopted as a central goal of many local, national, and international planning efforts, but it is often unclear exactly what it is or how to achieve it. In this book we use the United Nations Environment Programme (UNEP) definition of **sustainability:** the use of the environment and resources to meet the needs of the present without compromising the ability of future generations to meet their own needs (WCED 1987). According to this definition, sustainability requires that people be able to meet their own needs, i.e., to sustain human **well-being** (that is, the basic material needs for a good life, freedom and choice, good social relations, and personal security) now and in the future (Dasgupta 2001; see Chapter 3). Since sustainability and well-being are value-based concepts, there are often conflicting visions about what should be sustained and how sustainability should be achieved. Thus the assessment of sustainability is as much a political as a scientific process and requires careful attention to whose visions of sustainability are being addressed (Shindler and Cramer 1999). Nonetheless, any vision of sustainability ultimately depends on the life-support capacity of the environment and the generation of ecosystem services (see Chapter 2).

Types and Substitutability of Capital

Sustainability requires that the productive base required to support well-being be maintained or increased over time. **Well-being** can be defined in economic terms as the present value of future **utility**, i.e., the capacity of individuals or society to meet their own needs (Dasgupta and Mäler 2000, Dasgupta 2001). Well-being also has important social and cultural dimensions (see Chapter 3), but the economic definition enables us to frame sustainability in a systems context. Sustainability requires that the total **capital**, or productive base (assets) of the system, be sustained. This capital has natural, built (manufactured), human, and social components (Arrow et al. 2004). **Natural capital** consists of both nonrenewable resources (e.g., oil reserves) and renewable ecosystem resources (e.g., plants, animals, and water) that support the production of goods and services on which society depends. **Built capital** consists of the physical means of production beyond that which occurs in nature (e.g., tools, clothing, shelter, dams, and factories). **Human capital** is the capacity of people to accomplish their goals; it can be increased through various forms of learning. Together, these forms of capital constitute the **inclusive wealth** of the system, i.e., the productive base (assets) available to society. Although not included in the formal definition of inclusive wealth, **social capital** is another key societal asset. It is the capacity of groups of people to act collectively to solve problems (Coleman 1990). Components of each of these forms of capital change over time. Natural capital, for example, can increase through improved management of ecosystems, including restoration or renewal of degraded ecosystems or establishment of networks of marine-protected areas; built capital through investment in bridges or schools; human capital through education and training; and social capital through development of new partnerships to solve problems. Increases in this productive base constitute **genuine investment. Investment** is the increase in the quantity of an asset times its value. Sustainability requires that genuine investment be positive, i.e., that the productive base (genuine wealth) not decline over time

(Arrow et al. 2004). This provides an objective criterion for assessing whether management is sustainable.

To some extent, different forms of capital can **substitute** for one another, for example, natural wetlands can serve water purification functions that might otherwise require the construction of expensive water treatment facilities. Well-informed leadership may be able to implement more cost-effective solutions to a given problem (a substitution of human for economic capital). However, there are limits to the extent to which different forms of capital can be substituted (Folke et al. 1994). Water and food, for example, are **essential** for survival, and no other forms of capital can completely substitute for them (see Chapter 12). They therefore have extremely high value to society when they become scarce. Declines in the trust that society has in its leadership; sense of cultural identity; the capacity of agricultural soils to retain sufficient water to support production; or the presence of species that pollinate critical crops, for example, cannot be readily compensated by substituting other forms of capital. Losses of many forms of human, social, and natural capital are especially problematic because of the impossibility or extremely high costs of providing appropriate substitutes (Folke et al. 1994, Daily 1997). We therefore focus particular attention on ways to sustain these components of capital, without which future generations cannot meet their needs (Arrow et al. 2004).

Well-informed managers often have guidelines for sustainably managing the components of inclusive wealth. For example, harvesting rates of renewable natural resources should not exceed regeneration rates; waste emissions should not exceed the assimilative capacity of the environment; nonrenewable resources should not be exploited at a rate that exceeds the creation of renewable substitutes; education and training should provide opportunities for disadvantaged segments of society (Barbier 1987, Costanza and Daly 1992, Folke et al. 1994).

The concept of maintaining positive genuine investment as a basis for sustainability is important because it recognizes that the capital assets of social–ecological systems inevitably change over time and that people differ through time and across space in the value that they place on different forms of capital. If the productive base of a system is sustained, future generations can make their own choices about how best to meet their needs. This defines criteria for deciding whether certain practices are sustainable in a changing world. There are substantial challenges in measuring changes in various forms of capital, in terms of both their quantity and their value to society (see Chapter 3). Nonetheless, the best current estimates suggest that manufactured and human capital have increased in the last 50 years in most countries but that natural capital has declined as a result of depletion of renewable and nonrenewable resources and through pollution and loss of the functional benefits of biodiversity (Arrow et al. 2004). In some countries, especially some of the poorer developing nations, the loss of natural capital is larger than increases in manufactured and human capital, indicating a clearly unsustainable pathway of development (MEA 2005d). Some argue that there have also been substantial decreases in social capital as a result of modernization and urban life (Putnam 2000).

Managing Change in Ways that Foster Sustainability

Managing for sustainability requires attention to changes typical of complex adaptive systems. In the previous section we defined criteria to assess sustainability. These criteria are of little use if the system to which they are applied changes radically. Now we must place sustainability in the context of the

directional changes in factors that govern the properties of most social–ecological systems. Three broad categories of outcome are possible: (1) persistence of the fundamental properties of the current system through adaptation, (2) **transformation** of the system to a fundamentally different, potentially more desirable state, or (3) passive changes (often degradation to a less-favorable state) of the system as a result of failure of the system to adapt or transform. Intermediate outcomes are also possible, if some components (e.g., ecological subsystems, institutions, or social units) of the system persist, others transform, and others degrade (Turner et al. 2003). Sustainability implies the persistence of the fundamental properties of the system or of active transformation through deliberate substitution of different forms of capital to meet society's needs in new ways. In contrast, degradation implies the loss of inclusive wealth and therefore the potential to achieve sustainability.

How can we manage the dynamics of change to improve the chances for persistence or transformation? Four general approaches have been identified as ways to foster sustainability under conditions of directional change: (1) reduced vulnerability, (2) enhanced adaptive capacity, (3) increased resilience, and (4) enhanced transformability. Each of these approaches emphasizes a different set of processes by which sustainability is fostered (Table 1.2, Fig. 1.7). Vulnerability addresses the nature of stresses that cause change, the sensitivity of the system to these changes, and the adaptive capacity to adjust to change. Adaptive capacity addresses the capacity of actors or groups of actors to adjust so as to minimize the negative impacts of changes. Resilience

TABLE 1.2. Assumptions of frameworks addressing long-term human well-being. Modified from Chapin et al. (2006a).

Framework	Assumed change in exogenous controls	Nature of mechanisms emphasized	Other approaches often incorporated
Vulnerability	Known	System exposure and sensitivity to drivers; equity	Adaptive capacity, resilience
Adaptive capacity	Known or unknown	Learning and innovation	None
Resilience	Known or unknown	Within-system feedbacks and adaptive governance	Adaptive capacity, transformability
Transformability	Directional	Learn from crisis	Adaptive capacity, resilience

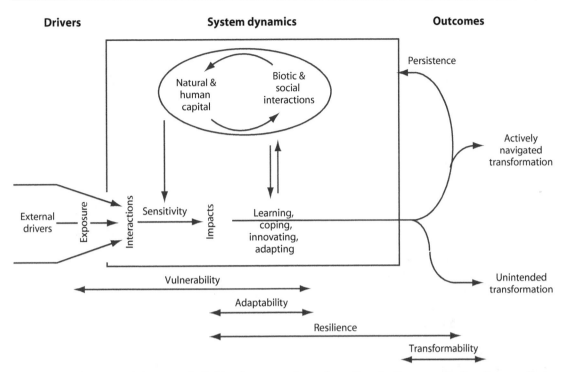

Drivers   **System dynamics**   **Outcomes**

FIGURE 1.7. Conceptual framework linking human adaptive capacity, vulnerability, resilience, and transformability. See text for definition of terms. The system (e.g., household, community, nation, etc.) responds to a suite of interacting drivers (stresses, events, shocks) to produce one of three potential outcomes: (1) persistence of the existing system through resilience; (2) actively navigated transformation to a new, potentially more beneficial state through transformability; or (3) unintended transformation to a new state (often degraded) due to vulnerability and the failure to adapt or transform. These three outcomes are not mutually exclusive, because some components (e.g., ecological subsystems, institutions, or social units) of the system may persist, others transform, and others degrade. The sensitivity of the system to perturbations depends on its exposure (intensity, frequency, and duration) to each perturbation, the interactions among distinct perturbations, and critical properties of the system. The system response to the resulting impacts depends on its adaptive capacity (i.e., its capacity to learn, cope, innovate, and adapt). Adaptive capacity, in turn, depends on the amount and diversity of social, economic, physical, and natural capital and on the social networks, institutions, and entitlements that influence how this capital is distributed and used. System response also depends on effectiveness of cross-scale linkages to changes occurring at other temporal and spatial scales. Those components of the system characterized by strong stabilizing feedbacks and adaptive capacity are likely to be resilient and persist. Alternatively, if the existing conditions are viewed as untenable, a high adaptive capacity can contribute to actively navigated transformation, the capacity to change to a new, potentially more beneficial state of the system or subsystem. If adaptive capacity of some components is insufficient to cope with the impacts of stresses, they are vulnerable to unintended transformation to a new state that often reflects degradation in conditions.

incorporates adaptive capacity but also entails additional system-level attributes of social–ecological systems that provide flexibility to adjust to change. Transformability addresses active steps that might be taken to change the system to a different, potentially more desirable state. Although anthropologists, ecologists, and geographers developed these approaches somewhat independently (Janssen et al. 2006), they are becoming increasingly integrated (Berkes et al. 2003, Turner et al. 2003, Young et al. 2006). This integration of ideas provides policy makers and managers with an increasingly sophisticated and flexible tool kit to address the challenges of sustainability in a directionally changing world. We apply the term **resilience-based ecosystem stewardship** to this entire suite of approaches to sustainability, because of its emphasis on sustaining functional properties of social–ecological systems over the long term despite perturbation and change. These issues represent the core challenges of managing social–ecological systems sustainably. We now briefly outline this suite of approaches.

Vulnerability

Vulnerability is the degree to which a system is likely to experience harm due to exposure to a specified hazard or stress (Turner et al. 2003, Adger 2006). Vulnerability theory is rooted in socioeconomic studies of impacts of events (e.g., floods or wars) or stresses (e.g., chronic food insecurity) on social systems but has been broadened to address responses of entire social–ecological systems. Vulnerability analysis deliberately addresses human values such as equity and well-being. Vulnerability to a given stress can be reduced by (1) reducing exposure to the stress (**mitigation**); (2) reducing sensitivity of the system to stress by sustaining natural capital and the components of well-being, especially for the disadvantaged; and/or (3) increasing adaptive capacity and resilience (see below) to cope with stress (Table 1.3; Turner et al. 2003). The incorporation of adaptive capacity and resilience as integral components of the vulnerability framework (Turner et al. 2003, Ford and Smit 2004) illustrates the integration of different approaches to sustainability science.

Exposure to a stress can be reduced by minimizing its intensity, frequency, duration, or extent. Prevention of pollution or banning of toxic pesticides, for example, reduces the vulnerability of people who would otherwise be exposed to these hazards. Mitigation (reduced exposure) is especially challenging when the stress is the cumulative effect of processes occurring at scales that are larger than the system being managed. Anthropogenic contributions to climate warming through the burning of fossil fuels, for example, is globally

TABLE 1.3. Principal sustainability approaches and mechanisms. Adapted from Levin (1999), Folke et al. (2003), Turner et al. (2003), Chapin et al. (2006a), Walker et al. (2006).

Vulnerability
  Reduce exposure to hazards or stresses
  Reduce sensitivity to stresses
    Sustain natural capital
    Maintain components of well-being
    Pay particular attention to vulnerability of the
      disadvantaged
  Enhance adaptive capacity and resilience (see below)
Adaptive capacity
  Foster biological, economic, and cultural diversity
  Foster social learning
  Experiment and innovate to test understanding
  Select, communicate, and implement appropriate
    solutions.
Resilience
  Enhance adaptive capacity (see above)
  Sustain legacies that provide seeds for renewal
  Foster a balance between stabilizing feedbacks and
    creative renewal
  Adapt governance to changing conditions
Transformability
  Enhance diversity, adaptation, and resilience
  Identify potential future options and pathways to
    get there
  Enhance capacity to learn from crisis
  Create and navigate thresholds for transformation

dispersed, so it cannot be reversed by actions taken solely by those regions that experience greatest impacts of climatic change (McCarthy et al. 2005). Other globally or regionally dispersed stresses include inadequate supplies of clean water and uncertain availability of nutritious food (Steffen et al. 2004, Kasperson et al. 2005).

Sensitivity to a stress can be reduced in at least three ways: (1) sustaining the slow ecological variables that determine natural capital; (2) maintaining key components of well-being; and (3) paying particular attention to the needs of the disadvantaged segments of society, who are generally most vulnerable. The poor or disadvantaged, for example, are especially vulnerable to food shortages or economic downturns, and people living in flood-plains or the wildland–urban interface are especially vulnerable to flooding or wild-fire, respectively. An understanding of the causes of differential vulnerability can lead to strategies for targeted interventions to reduce overall vulnerability of the social–ecological system.

The causes of differential vulnerability are often deeply rooted in the slow variables that govern the internal dynamics of society, such as power relationships or distribution of land-use rights among segments of society (see Chapter 3). Conventional vulnerability analysis assumes that the stresses are known or predictable (i.e., either steady state or changing in a predictable fashion). However, long-term reductions in vulnerability often require attention to adaptive capacity and resilience at multiple scales in addition to targeted efforts to reduce exposure and sensitivity to known stresses.

Adaptive Capacity

Adaptive capacity (or adaptability) is the capacity of actors, both individuals and groups, to respond to, create, and shape variability and change in the state of the system (Folke et al. 2003, Walker et al. 2004, Adger et al. 2005). Although the actors in social–ecological systems include all organisms, we focus particularly on people in addressing the role of adaptive capacity in social–ecological change, because human actors base their actions not only on their past experience but also on their capacity to *plan for the future* (**reflexive action**). This contrasts with **evolution**, which shapes the properties of organisms based entirely on their genetic responses to *past* events. Evolution has no forward-looking component. Adaptive capacity depends on (1) biological, economic, and cultural diversity that provides the building blocks for adjusting to change; (2) the capacity of individuals and groups to learn how their system works and how and why it is changing; (3) experimentation and innovation to test that understanding; and (4) capacity to govern effectively by selecting, communicating, and implementing appropriate solutions (Table 1.3) We discuss the social and cultural bases of adaptive capacity in Chapters 3 and

4 and here focus on its relationship to system properties.

Sources of biological, economic, and cultural diversity provide the raw material on which adaptation can act (Elmqvist et al. 2003, Norberg et al. 2008). In this way it defines the options available for adaptation. People can augment this range of options through learning, experimentation, and innovation. This capacity to create new options is strongly influenced by people's access to built, natural, human, and social capital. Societies with little access to capital are constrained in their capacity to adapt. People threatened with starvation, for example, may degrade natural capital by over-grazing to meet their immediate food needs, thereby reducing their potential to cope with drought or future food shortage. Rich countries, on the other hand, have greater capacity to engineer solutions to cope with floods, droughts, and disease outbreaks. Natural capital also contributes in important ways to adaptive capacity, although its role is often unrecognized until it has been degraded. Systems that have experienced severe soil erosion, for example, have fewer options with which to experiment and innovate during times of drought, and highly engineered systems that have lost their capacity to store floodwaters have fewer options to adapt in response to floods. The role of human capital in adaptive capacity is especially important. It is much more than formal education. It depends on an understanding of how the system responds to change, which often comes from experience and local knowledge of past responses to extreme events or stresses. As the world changes, and new hazards and stresses emerge, this understanding may be insufficient. Willingness to innovate and experiment to test what has been learned and to explore new approaches is crucial to adaptive capacity.

Social capital through networking to select, communicate, and implement potential solutions is another key component of adaptive capacity. Leadership, for example, is often critical in building trust, making sense of complex situations, managing conflict, linking actors, initiating partnerships among groups, compiling and generating knowledge, mobilizing broad support for change, and developing and communicating visions for change (Folke et al. 2005; see Chapter 5). It takes more than leaders, however, for society to adapt to change. Social networks are critical in effectively mobilizing resources at times of crisis (e.g., war or floods) and in providing a safety net for vulnerable segments of society (see Chapters 4 and 5).

In the context of sustainability, adaptive capacity represents the capacity of a social–ecological system to make appropriate substitutions among forms of capital to maintain or enhance inclusive wealth. In this way the system retains the potential for future generations to meet their needs.

Resilience

Resilience is the capacity of a social–ecological system to absorb a spectrum of shocks or perturbations and to sustain and develop its fundamental function, structure, identity, and feedbacks through either recovery or reorganization in a new context (Holling 1973, Gunderson and Holling 2002, Walker et al. 2004, Folke 2006). The unique contribution of resilience theory is the recognition and identification of several possible system properties that foster renewal and reorganization after perturbations (Holling 1973). Resilience depends on (1) adaptive capacity (see above); (2) biophysical and social legacies that contribute to diversity and provide proven pathways for rebuilding; (3) the capacity of people to plan for the long term within the context of uncertainty and change; (4) a balance between stabilizing feedbacks that buffer the system against stresses and disturbance and innovation that creates opportunities for change; and (5) the capacity to adjust governance structures to meet changing needs (Holling and Gunderson 2002, Folke et al. 2003, Walker et al. 2006; Table 1.3). Loss of resilience pushes a system closer to its limits. When resilience has been eroded, a disturbance, like a disease, storm, or stock market fluctuation, that previously shook and revitalized the resilient system, might now push the fragile system over a threshold into an alternative state (a **regime shift**) with a new trajectory of change. Such system changes radically alter the flow of ecosystem services (Chapter 2) and associated livelihoods and well-being of people and societies. Clearly, resilience is an essential feature of resource stewardship under conditions of uncertainty and change, so this approach to resource management is even more important today than it has been in the past.

We have already discussed the role of stabilizing feedbacks in buffering systems from change and the role of adaptive capacity in coping with the impacts of those changes that occur. Sources of diversity, which is essential for adaptation, are especially important in the focal system and surrounding landscape at times of crisis, i.e., during the renewal phase of adaptive cycles, when there is less resistance to establishment of new entities. Fostering small-scale variability and change logically contributes to resilience because it maintains within the system those components that are well adapted to each phase of the adaptive cycle—ranging from the renewal to the conservation phase. This reduces the likelihood that the inevitable disturbances will have catastrophic effects. Conversely, preventing small-scale disturbances such as insect outbreaks or fires tends to eliminate disturbance-adapted components, thereby reducing the capacity of the system to cope with disturbance.

Biophysical and social legacies contribute to resilience through their contribution to diversity. Legacies provide species, conditions, and perspectives that may not be widely represented in the current system. A buried seed pool or stems that resprout after fire, for example, give rise to a suite of early successional species that are well adapted to postdisturbance conditions but may be uncommon in the mature forest. Similarly, the stories and memories of elders and the written history of past events often provide insight into ways in which people coped with past crises as well as ideas for future options that might not otherwise be considered. This often occurs by drawing on **social memory**, the social legacies of knowing how to do things under different circumstances. A key challenge is how to foster and maintain social memory at times of gradual change, so it is available when a crisis occurs.

One of the key contributions of resilience theory to resource stewardship is the recognition that complex adaptive systems are constantly changing in ways that cannot be fully predicted or controlled, so decisions must always be made in an environment of uncertainty. Research and awareness of processes occurring at a wide range of scales (e.g., the dynamics of potential pest populations or behavior of global markets) can reduce uncertainty (Adger et al. 2005, Berkes et al. 2005), but managing for flexibility to respond to unanticipated changes is essential. This contrasts with steady-state management approaches that seek to reduce variability and change as a way to facilitate efficient harvest of a given resource such as fish or trees (Table 1.1).

Transformability and Regime Shifts

Transformability is the capacity to reconceptualize and create a fundamentally new system with different characteristics (Walker et al. 2004; see Chapter 5). There will always be a creative tension between resilience (fixing the current system) and transformation (seeking a new, potentially more desirable state) because actors in the system usually disagree about when to fix things and when to cut losses and move to a new alternative structure (Walker et al. 2004). Actively navigated transformations require a paradigm shift that reconceptualizes the nature of the system. During transformation, people recognize (or hypothesize) a fundamentally different set of critical slow variables, internal feedbacks, and societal goals. Unintended transformations can also occur in situations where management efforts have prevented adjustment of the system to changing conditions, resulting in a fundamentally different system (often degraded) characterized by different critical slow variables and feedbacks. The dividing line between persistence of a given system and transformation to a new state is sometimes fuzzy. Total system collapse seldom occurs (Turner and McCandless 2004, Diamond 2005). Nonetheless, actively navigated transformations of important components of a system are frequent (e.g., from an extractive to

a tourism-based economy). In general, diversity, adaptive capacity, and other components of resilience enhance transformability because they provide the seeds for a new beginning and the adaptive capacity to take advantage of these seeds.

Transformations are often triggered by crisis, so the capacity to plan for and recognize opportunities associated with crisis contributes to transformability (Gunderson and Holling 2002, Berkes et al. 2003). **Crisis** is a time when society, by definition, agrees that some components of the present system are dysfunctional. During crisis, society is more likely to consider novel alternatives. It is also a time when, if novel solutions are not seized, the system can become entrenched in the very policies that led to crisis, increasing the likelihood of unintended transformations. Climate-induced increases in wildfires in the western USA, for example, threaten homes that have been built in the wildland–urban interface. One potential transformation would be policies that cease assuming public responsibility for private homes built in remote fire-prone areas and instead encourage more dense development of areas that could be protected from fire and served by public transportation. This would reduce the need and cost of wildfire suppression, increase the economic efficiency of public transportation, and reduce the use of fossil fuels. Alternatively, current policies of fire suppression and dispersed residential development in forested lands might persist and magnify the risk of catastrophic loss of life and property as climate warming increases wildfire risk and fire suppression leads to further fuel accumulation.

Sometimes systems exhibit abrupt transitions (regime shifts) to alternate states because of threshold responses to persistent changes in one or more slow variables. Continued phosphorus inputs to clearwater lakes, for example, may lead to abrupt transitions to a turbid-water algal-dominated regime (Carpenter 2003). Similarly, persistent overgrazing can cause shrub encroachment and transition from grassland to shrubland (Walker et al. 2004). Regime shifts are large changes in ecosystems that include both changes in stability domains of a given system (e.g., clearwater–turbid-water

transitions; Fig. 1.7d) and system transformations (Carpenter 2003, Groffman et al. 2006).

Challenges to Sustainability

**The major challenges to sustainability vary
temporally and regionally**. Issues of sustainability are often prominent in developing nations,
especially where substantial poverty, inadequate educational opportunities, and insufficient health care limit well-being (Kasperson
et al. 2005). These situations sometimes coin-
cide with a high potential for environmental
degradation, for example, soil erosion and contamination of water supplies, as people try to
meet their immediate survival needs under circumstances of inadequate social and economic
infrastructure. **Sustainable development** seeks
to improve well-being, while at the same time
protecting the natural resources on which society depends (WCED 1987). In other words,
it seeks directional changes in some underlying controls, but not others. Questions are
often raised about whether sustainable development can indeed be achieved, given its twin
goals of actively promoting economic development while sustaining natural capital. The
feasibility of sustainable development depends
on the multiple effects of development on system properties and the extent to which these
new system properties can be sustained over
the long term. In other words, how does development influence the slow variables that govern the properties of social–ecological systems
and how can they be redirected or transformed
for improving the options of well-being without
degrading inclusive wealth? Finding sustainable solutions usually requires active engagement of **stakeholders** (groups of people affected
by policy decisions) who must live with, and
participate in, the implementation of potential
solutions.

Enhancing the sustainability of nations with
greater wealth is equally challenging. Countries such as the USA, for example, consume
fossil fuels at per-capita rates that are fivefold
greater than the world average and frequently
use renewable resources more rapidly than they
can be replenished. Here the challenge is to
avoid degradation of the ecological and cultural

bases of well-being over the long term so that
people in other places and in future generations
can meet their own needs (Plate 3).

In summary, virtually all social–ecological
systems are undergoing persistent directional
changes, as a result of both unplanned changes
in climate, economic systems, and culture and
deliberate planning to improve well-being.
Efforts to promote sustainability must therefore recognize that many of the attributes of
social–ecological systems will inevitably change
over the long term and seek ways to guide these
changes along sustainable pathways.

Roadmap to Subsequent Chapters

The first section of the book presents the general principles needed for sustainable stewardship in a changing world (Table 1.4).
Chapter 1 provides a framework for understanding change and the factors that influence sustainability under conditions of change.
A clear message from this chapter is that
social–ecological systems are complex and
require an understanding of the interactions
among ecological, economic, political, and cultural processes. Consequently, key resource-
management issues cannot be solved by disciplinary experts but require an integrated
understanding of many disciplines. Chapter 2
describes the principles of ecosystem management to sustain the delivery of ecosystem services to society. Chapter 3 describes the range
of economic, cultural, and political factors that
shape well-being and use of ecosystem services. Chapter 4 then describes the institutional
dimensions of human interactions with ecosystems. Chapter 5 explores the processes by which
social–ecological systems transform to a fundamentally different system with different controls and feedbacks.

The second section of the book applies the
general principles developed in the first section to specific types of social–ecological systems and their prominent resource–stewardship
challenges (Table 1.4), including conservation
(see Chapter 6), forests (see Chapter 7),
drylands (see Chapter 8), lakes and rivers

TABLE 1.4. Resource–stewardship challenges and the chapters in which each is emphasized.

Issue	Chapter where emphasized
Social-ecological interactions	All chapters (2–15)
Global change	Concepts (2–5), Global (14), Systems (6–13)
Ecological sustainability	Ecosystems (2), System chapters (6–14)
Ecosystem restoration	Ecosystems (2), Drylands (8)
Biodiversity conservation	Ecosystems (2), Conservation (6), Forests (7)
Invasive species	Ecosystems (2), Freshwaters (9)
Landscape management	Ecosystems (2), Drylands (8), Freshwaters (9)
Range management	Ecosystems (2), Drylands (8)
Wildlife management	Ecosystems (2), Conservation (6), Drylands (8)
Fisheries management	Freshwaters (9), Oceans (10), Coastal (11)
Water management	Ecosystems (2), Drylands (8), Freshwaters (9)
Disturbance management	Ecosystems (2), Forests (7), Freshwaters (9)
Pollution	Ecosystems (2), Agriculture (12), Cities (13)
Urban development	Livelihoods (3), Forests (7), Cities (13)
Sustaining human livelihoods	Livelihoods (3), Conservation (6), Coastal (11)
Social and environmental justice	Livelihoods (3), Coastal (11), Cities (13), Global (14)
Sustainable development	Livelihoods (3), Agriculture (12)
Local and traditional knowledge	Institutions (4), Conservation (6), Drylands (8)
Property rights and the commons	Institutions (4), Oceans (10), Coastal (11)
Natural resource policy	Institutions (4), System chapters (6–14)
Subsistence harvest	Institutions (4), Conservation (6)
Resource co-management	Institutions (4), Conservation (6), Coastal (11)
Adaptive management	Institutions (4), Drylands (8), Oceans (10)
Long-term planning	Transformation (5), Forests (7), Global (14)
Managing thresholds	Transformation (5), Drylands (8), Oceans (10)
Adaptive governance	Transformation (5), Forests (7), Global (14)
Thresholds and regime shifts	Transformation (5), Drylands (8), Freshwaters (9)

(see Chapter 9), oceans and estuaries (see Chapters 10 and 11), food production systems (see Chapter 12), cities and suburbs (see Chapter 13), and the entire Earth (see Chapter 14). Each of these chapters describes the system properties and dynamics that are especially important in that system, key management issues, and potential social–ecological thresholds. Each chapter then describes a few case studies that illustrate resilient or non-resilient management and outcomes and how the unique properties of each system shape human–environment interactions and sustainability constraints and opportunities. Each system chapter emphasizes selected general principles that were described in the first section of the book.

The final chapter (see Chapter 15) summarizes some of the major strategies that have proven valuable for managing social–ecological systems and the lessons learned from previous chapters about the role of resilience and adaptation in sustainable stewardship.

Review Questions

1. What is resilience-based resource stewardship? How does it differ from steady-state resource management, and why are these differences important in the current world?
2. How do different types of feedbacks influence the stability and resilience of a system?
3. What are the *mechanisms* by which complex adaptive systems respond to changes? Do they always respond in the same way to a given perturbation? Why or why not? In social–ecological systems, why does a given policy sometimes have different effects when implemented at different times or places?

4. Why does the sensitivity of social–ecological systems to perturbations depend on the time since the previous perturbation? What are the advantages and disadvantages of managing systems to prevent disturbances from occurring?

5. What are the processes by which vulnerability, adaptive capacity, resilience, and transformability influence sustainability?

Additional Readings

Berkes, F., J. Colding, and C. Folke, editors. 2003. *Navigating Social-Ecological Systems: Building Resilience for Complexity and Change*. Cambridge University Press, Cambridge.

Carpenter, S.R., and M.G. Turner. 2000. Hares and tortoises: Interactions of fast and slow variables in ecosystems. *Ecosystems* 3:495–497.

Chapin, F.S., III, A.L. Lovecraft, E.S. Zavaleta, J. Nelson, M.D. Robards, et al. 2006. Policy strategies to address sustainability of Alaskan boreal forests in response to a directionally changing climate. *Proceedings of the National Academy of Sciences* 103:16637–16643.

Folke, C. 2006. Resilience: The emergence of a perspective for social-ecological systems analysis. *Global Environmental Change* 16:253–267.

Gunderson, L.H., and C.S. Holling, editors. 2002. *Panarchy: Understanding Transformations in Human and Natural Systems*. Island Press, Washington.

Levin, S.A. 1999. *Fragile Dominion: Complexity and the Commons*. Perseus Books, Reading, MA.

MEA (Millennium Ecosystem Assessment). 2005d. *Ecosystems and Human Well-being: Synthesis*. Island Press, Washington.

Steffen, W.L., A. Sanderson, P.D. Tyson, J. Jäger, and P.A. Matson, editors. 2004. *Global Change and the Earth System: A Planet Under Pressure*. Springer-Verlag, New York.

Turner, B.L., II, R.E. Kasperson, P.A. Matson, J.J. McCarthy, R.W. Corell, et al. 2003. A framework for vulnerability analysis in sustainability science. *Proceedings of the National Academy of Sciences* 100:8074–8079.

Walker, B.H., C.S. Holling, S.R. Carpenter, and A.P. Kinzig. 2004. Resilience, adaptability and transformability in social–ecological systems. *Ecology and Society* 9(2):5 [online] URL: http://www.ecologyandsociety.org/vol9/iss2/art5/

Walker, B.H., and D. Salt. 2006. *Resilience Thinking: Sustaining Ecosystems and People in a Changing World*. Island Press, Washington.

2
Managing Ecosystems Sustainably: The Key Role of Resilience

F. Stuart Chapin, III

Introduction

The goal of ecosystem management is to provide a sustainable flow of multiple ecosystem services to society today and in the future. As an integral component of natural resource stewardship, ecosystem management recognizes the integrated nature of social–ecological systems, their inherent complexity and dynamics at multiple temporal and spatial scales, and the importance of managing to maintain future options in the face of uncertainty (Christensen et al. 1996; Table 2.1)—i.e., many of the factors governing the resilience and vulnerability of social-ecological systems. As a society, we have a poor track record of managing ecosystems sustainably in part because the short-term use of natural resources often receives higher priority than their long-term sustainability. Environmental degradation contributed to the collapse of many advanced human societies, including Babylon, the Roman Empire, and the Mayan Civilization (Turner et al. 1990,

Diamond 2005). More than half of the services provided by ecosystems have declined globally in the last half-century (MEA 2005a, d), raising questions about the capacity of human societies to manage ecosystems sustainably. Rapid rates of social and environmental change have magnified the challenges of sustainable management. We advocate broadening the concept of ecosystem management to **resilience-based ecosystem stewardship**. Its goals are to respond to and shape change in social–ecological systems in order to sustain the supply and opportunities for use of ecosystem services by society. Resilience-based ecosystem stewardship builds on ecosystem management by emphasizing (1) the key role of resilience in fostering adaptation and renewal in a rapidly changing world; (2) the dynamics of social change in altering human interactions with ecosystems; and (3) the social–ecological role of resource managers as stewards who respond to and shape social–ecological change. In this chapter we address key components of ecosystem stewardship, emphasizing the ecological consequences of those human actions that can tip the balance between sustainable and nonsustainable flow of ecosystem services to society. In Chapter 3, we broaden this perspective to integrate social processes that motivate human actions.

F.S. Chapin (✉)
Institute of Arctic Biology, University of Alaska
Fairbanks, Fairbanks, AK 99775, USA
e-mail: terry.chapin@uaf.edu

F.S. Chapin et al. (eds.), *Principles of Ecosystem Stewardship*,
DOI 10.1007/978-0-387-73033-2_2, © Springer Science+Business Media, LLC 2009

TABLE 2.1. Attributes of ecosystem management. Information from Christensen et al. (1996).

Sustainability	Intergenerational sustainability is the primary objective
Goals	Measurable goals are defined that assess sustainability of outcomes
Ecological understanding	Ecological research at all levels of organization informs management
Ecological complexity	Ecological diversity and connectedness reduces risks of unforeseen change
Dynamic change	Evolution and change are inherent in ecological sustainability
Context and scale	Key ecological processes occur at many scales, linking ecosystems to their matrix
Humans as ecosystem components	People actively participate in determining sustainable management goals
Adaptability	Management approaches will change in response to changes in scientific knowledge and human values

An **ecosystem** consists of organisms (plants, microbes, and animals—including people) and the physical components (atmosphere, soil, water, etc.) with which they interact. All ecosystems are influenced, to a greater or lesser degree, by social processes (i.e., are social–ecological systems), although ecosystem studies tend to focus on biological interactions. Using the ecosystem-service framework developed by the Millennium Ecosystem Assessment

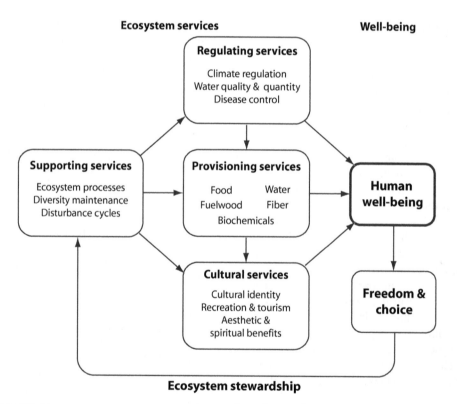

FIGURE 2.1. Linkages among ecosystem services, well-being of society, and ecosystem stewardship, a framework developed by the Millennium Ecosystem Assessment (MEA, 2005c). Supporting services are the foundation for the other categories of ecosystem services that are directly used by society. In addition, the goods harvested by people are influenced by landscape processes, which include regulatory services, and, in turn, influence people's connection to the land and sea (cultural services). Adapted from MEA (2005d).

TABLE 2.2. General categories of ecosystem services and examples of the societal benefits that are most directly affected.

Ecosystem services	Direct benefits to society
Supporting services	
Maintenance of soil resources	Nutrition, shelter
Water cycling	Health, waste management
Carbon and nutrient cycling	Nutrition, shelter
Maintenance of disturbance regime	Safety, nutrition, health
Maintenance of biological diversity	Nutrition, health, cultural integrity
Provisioning services	
Fresh water	Health, waste management
Food and fiber	Nutrition, shelter
Fuelwood	Warmth, health
Biochemicals	Health
Genetic resources	Nutrition, health, cultural integrity
Regulating services	
Climate regulation	Safety, nutrition, health
Erosion, water quantity/quality, pollution	Health, waste management
Disturbance propagation	Safety
Control of pests, invasions, and diseases	Health
Pollination	Nutrition
Cultural services	
Cultural identity and cultural heritage	Cultural integrity, values
Spiritual, inspirational, and aesthetic benefits	Values
Recreation and ecotourism	Health, values

(MEA 2005d), we first provide an overview of **supporting services**, which are the fundamental ecological processes that sustain ecosystem functioning (Fig. 2.1). We show how the degradation of certain key supporting services erodes resilience, leading to loss of other services that are used more directly by society. These services include (1) **provisioning services** (or **ecosystem goods**), which are products of ecosystems that are directly harvested by society; (2) **regulating services** that influence society through interactions among ecosystems in a landscape; and (3) **cultural services**, which are nonmaterial benefits that are important to society's well-being (Table 2.2). There is broad overlap among these categories of ecosystem services, and different authors have therefore classified them in different ways. Traditional foods, for example, function as both provisioning services that provide nutritional benefits and cultural services that sustain cultural relationships to the land or sea. The important point, however, is that the functioning of ecosystems benefits society in so many ways that human well-being cannot

be sustained without the effective functioning of the ecosystems of which people are a part.

Supporting Services: Sustaining Ecosystem Functioning

Supporting services are the fundamental ecological processes that control the structure and functioning of ecosystems. Managers and the public often overlook these services because they are not the products directly valued by society. Moreover, they are frequently controlled by variables that change relatively slowly (i.e., **slow variables**) and are therefore taken for granted by agencies tasked with managing a particular ecosystem good such as trees or fish. However, because of the fundamental dependence of all ecosystem services on supporting services, integrity of these services generally sustains many services that are valued more directly by society. In this section we focus on the slow variables that most frequently control ecosystem processes and therefore a broad suite of ecosystem services.

Slow variables
Maintenance of Soil Resources

Soils and sediments are key slow variables that regulate ecosystem processes by providing resources required by organisms. The controls over the formation, degradation, and resource-supplying potential of soils and sediments are therefore central to sound ecosystem management and to sustaining the natural capital on which society depends (Birkeland 1999, Chapin et al. 2002). The quantity of soil in an ecosystem depends largely on the balance between inputs from **weathering** (the breakdown of rocks to form soil) or deposition and losses from erosion. In addition, organisms, especially plants, add organic matter to soils through death of tissues and individuals, which is offset by losses through decomposition. If an ecosystem were at **steady state**, i.e., when inputs approximately equal outputs, the quantity of soil would remain relatively constant, providing a stable capacity to supply vegetation with water and nutrients. Natural imbalances between inputs and outputs lead to deeper soils in floodplains and at the base of hills than on hilltops. When averaged over large regions, however, changes in soil capital due to imbalances between inputs and outputs usually occur slowly—about 0.1–10 mm per century (Selby 1993). In general, the presence of a plant canopy and **litter** layer (the layer of dead leaves on the soil surface) reduces erosion. Human activities that reduce vegetation cover can increase erosion rates by several orders of magnitude, just as occurs naturally when glaciers, volcanoes, or landslides reduce vegetation cover. Under these circumstances, ecosystems can lose soils in years to decades that may have required thousands of years to accumulate, causing an essentially permanent loss of the productive capacity of ecosystems. Similarly, human modification of river channels can alter sediment inputs. In the southern USA, for example, loss of sediment inputs and subsequent soil subsidence led to the disappearance of barrier islands that had previously protected New Orleans from hurricanes. The loss

of soil resources substantially reduces resilience by reducing the natural capital by which social–ecological systems can respond to change; this increases the likelihood of a regime shift to a more degraded state.

The physical and chemical properties of soils are just as important as total quantity of soil in determining the productive potential of terrestrial ecosystems. Fine particles of mineral soils (**clay**) and organic matter are particularly important in retaining water and nutrients (Brady and Weil 2001). Clay and organic matter are typically concentrated near the soil surface, where they are vulnerable to loss by erosion. Wind and **overland flow** (the movement of water across the soil surface) transport small particles more readily than large ones, tending to remove those soil components that are particularly important in water and nutrient retention. Human activities that foster wind and water erosion, such as deforestation, overgrazing, plowing, or fallowing of agricultural fields, therefore erode the water- and nutrient-retaining capacity of soils much faster than the total loss of soil volume might suggest. Preventing even modest rates of erosion is therefore critical to sustaining the productive capacity of terrestrial ecosystems.

Accelerated soil erosion is one of the most serious causes of global declines in ecosystem services and resilience. The erosional loss of fine soil particles is the direct cause of **desertification**, soil degradation that occurs in drylands (see Chapter 8). Desertification can be triggered by drought, reduced vegetation cover, overgrazing, or their interactions (Reynolds and Stafford Smith 2002, Foley et al. 2003a). When drought reduces vegetation cover, for example, goats and other livestock graze more intensively on the remaining vegetation. Extreme poverty and lack of a secure food supply often prevent people from reducing grazing pressure at times of drought, because short-term food needs take precedence over practices that might prevent erosion. Wetter regions can also experience severe erosional loss of soil, especially where vegetation loss exposes soils to overland flow. The Yellow River in China, for example, transports 1.6 billion tons of sediment annually from agricultural areas in the loess plateau at its headwaters. Similar erosional losses occurred when grasslands were plowed for agriculture in the USA during droughts of the 1930s, creating the **dustbowl**. Changes in social processes contribute substantially to regime shifts involving severe soil erosion.

Soil erosion from land represents a sediment input to lakes and estuaries. At a global scale the increased sediment input to oceans from accelerated erosion is partially offset by the increased sediment capture by reservoirs. Therefore lakes, including reservoirs, and estuaries are the aquatic ecosystems most strongly affected by terrestrial erosion. Especially in agricultural areas, these sediments represent a large influx of nutrients (**eutrophication**) to aquatic ecosystems that can be just as problematic as the loss of productive potential on land (see Chapter 9).

Water Cycling

Water is the soil resource that is used in largest quantities by plants and which most frequently limits the productivity of terrestrial ecosystems. Water enters ecosystems as precipitation. Two of the major pathways of water loss are **transpiration**, the "green water" that supports terrestrial production, and **runoff**, which replenishes groundwater and aquatic ecosystems, the "blue water" sources that are often tapped by people for domestic and industrial uses, irrigation, and hydropower (see Chapter 9). Consequently, there are inherent tradeoffs among the ecosystem services provided by water cycling, and some of the biggest challenges in water management result from these tradeoffs.

Climatic controls over water inputs in precipitation place an ultimate constraint on quantities of water cycled through ecosystems. Within this constraint the partitioning of water between transpiration and runoff depends on (1) the degree of compaction of the soil surface, which influences water infiltration into the soil, (2) soil water-holding capacity, which depends on the quantity of soil and its particle-size distribution (see Maintenance of Soil Resources),

and (3) the capacity of vegetation to transpire water (Rockström et al. 1999).

Vegetation fosters infiltration and storage through several mechanisms. The plant canopy and litter reduce compaction by raindrops that otherwise tend to seal soil pores. In addition, roots and soil animals associated with vegetation create channels for water movement through the soil profile. By facilitating water infiltration (due to reduced compaction), water-holding capacity (due to production of soil organic matter), and reduced soil erosion (due to reduced overland flow), vegetation generates stabilizing feedbacks that sustain the productive potential of soils and reduce ecosystem vulnerability to drought. Human activities often disrupt these stabilizing feedbacks, thereby reducing resilience. For example, high densities of livestock or movement of heavy farm machinery at times (spring) or places (riparian corridors) where soils are wet can compact soils and reduce infiltration. Plowing reduces soil organic content substantially—often by 50% within a few years—thereby reducing soil water-holding capacity and the capacity of soils to support crop growth with natural rainfall (Matson et al. 1997). Alternative agricultural practices that conserve or rebuild soil organic content (e.g., no-till agriculture) or reduce compaction by livestock or equipment under wet conditions therefore increase the capacity of soils to supply water to crops or other vegetation.

Transpiration is tightly linked to the capacity of plants to fix carbon and therefore to their productive potential. This explains why productive agricultural systems are such prodigious consumers of water and why streamflow increases after logging. At a more subtle level, any factor that increases the productive potential of vegetation (e.g., nutrient additions from fertilizers, introduction of exotic nitrogen-fixing species; atmospheric deposition of nitrogen; replacement of shrublands by forests) will increase transpiration (green water flows) and reduce water movement to groundwater and runoff (blue water flows). The tradeoffs between transpiration and runoff have implications for the role of ecosystems in regulating water flow, as discussed later.

Carbon and Nutrient Cycling

Within a climate zone, the availabilities of belowground resources (water and nutrients) are the main factors that constrain terrestrial carbon cycling and ecosystem productivity. The carbon, nutrient, and water cycles of terrestrial ecosystems are tightly linked (Fig. 2.2; Chapin et al. 2002, 2008). The major controls are (1) climate, which governs water inputs, rates of soil development, and cycling rates of carbon, nutrients, and water; (2) the water- and nutrient-holding capacity of soils (see Maintenance of

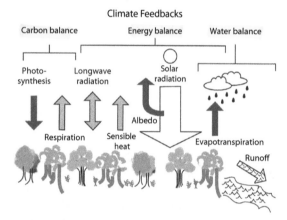

FIGURE 2.2. Three major categories of climate feedbacks (each shown by the arrows beneath the bracket) between ecosystems and the climate system. Carbon balance is the difference between CO_2 uptake by ecosystems (photosynthesis) and CO_2 loss to the atmosphere by respiration and disturbance. Energy balance is the balance between incoming solar radiation, the proportion of this incoming solar radiation that is reflected (albedo), and the transfer of the absorbed radiation to the atmosphere as sensible heat (warming the surface) or evapotranspiration (cooling the surface). Longwave radiation from the ecosystem or clouds depends on the temperature of these surfaces. Water balance between the ecosystem and atmosphere is the difference between precipitation inputs and water return in evapotranspiration; the remaining water leaves the ecosystem as runoff. Each of these ecosystem–atmosphere exchanges influences climate. Cooling effects on climate are shown by black arrows; warming effects by gray arrows. Arrows show the direction of the transfer; the magnitude of each transfer differs among ecosystems. Redrawn from Chapin et al. (2008).

Soil Resources); and (3) the productive capacity of vegetation. In addition, carbon and nutrient cycles are physically linked because carbon forms the skeleton of organic compounds that carry nitrogen and phosphorus among plants, animals, and soils.

Because both water and nutrient availability depend on the fine particles in soils, the factors that sustain water cycling (i.e., maintenance of vegetation and prevention of erosion) also sustain nutrient cycles. This is typical of many of the **synergies** among ecosystem services: Management practices that protect the basic integrity of ecosystem structure foster sustainability of multiple ecosystem services. This simplifies the task of ecosystem management, because most services "take care of themselves" if ecosystem structure and functioning are not seriously disrupted.

Plants are an important control point in the cycling of carbon through ecosystems, because they are the entry point for carbon and determine the chemistry of dead organic matter that eventually becomes food for decomposers. However, plant production in most intact ecosystems is limited by water and/or nutrient availability, so the productive capacity of vegetation typically adjusts to the availability of water and nutrients that a given climate and soil type provide. Consequently, across a broad range of ecosystem types, vegetation absorbs most of the nutrients that are released by the **decomposition** (chemical breakdown of dead organic matter by soil organisms). Consequently, groundwater and runoff leaving these systems have relatively low concentrations of nutrients. If, however, plant production is reduced below levels that the climate and soils can support, as for example in a fallow field or overgrazed pasture, or if nutrients are added to the system at rates that exceed the absorptive capacity of the vegetation, the excess nutrients leave the system in groundwater and runoff or as trace gases to the atmosphere (e.g., N_2O, a potent greenhouse gas that contributes to climate warming). Certain **nitrogen-fixing** plant species form mutualistic relationships with soil microorganisms that convert atmospheric nitrogen to plant-available nitrogen. The expansion of soybean and other nitrogen-fixing species has substantially increased nitrogen inputs to many agricultural regions (see Chapter 12). Industrial fixation of nitrogen, primarily to produce fertilizers, is an even larger source of nitrogen inputs to managed ecosystems. Ammonia volatilizes from fertilizers and cattle urine and enters downwind ecosystems in precipitation. In addition, fossil-fuel combustion produces nitrogen oxides (NO_x) that are a major component of acid rain. Together these anthropogenic sources of nitrogen have doubled the naturally occurring rates of nitrogen inputs to terrestrial ecosystems (Fig. 2.3; Schlesinger 1997, Vitousek et al. 1997). This massive change in global biogeochemistry weakens the internal stabilizing feedbacks that confer resilience to ecosystem processes (see Chapter 1). Reducing nitrogen inputs to levels consistent with the productive capacity of vegetation, for example, by reducing fertilizer applications, reducing air pollution, or preventing the spread of nitrogen-fixing exotic species, reduces the leakage of nitrogen from terrestrial to aquatic ecosystems.

Carbon and nutrient cycles in aquatic ecosystems run on the leftovers of terrestrial nutrient cycles. In intact landscapes with tight terrestrial nutrient cycles, the small quantity of nutrients delivered to streams spiral slowly downstream, moving through decomposers, stream invertebrates, algae, and fish (Vannote et al. 1980). Lakes are typically more nutrient-impoverished than streams, because they receive relatively little leaf litter and groundwater per unit of water surface, and sediments chemically fix much of the phosphorus that enters the lake (see Chapter 9). Aquatic organisms are well adapted to these low-nutrient conditions. Algae efficiently absorb nutrients from the water column, are eaten by invertebrate grazers that in turn are eaten by fish. When phosphorus inputs to lakes exceed the chemical fixation capacity of sediments, algae grow and reproduce more rapidly than grazers can consume them, reducing water clarity and increasing the rain of dead organic matter to depth. Here bottom-dwelling decomposers break down the dead organic matter, depleting oxygen below levels required by fish, which causes fish to die. The high phosphorus-fixation capacity of lake sediments makes most lakes quite resilient to individual eutrophication events. Once this phosphorus-sequestration mechanism is saturated, however, the sediments become a source rather than a sink of phosphorus, causing the lake to shift to a eutrophic state (Carpenter et al. 1999, Carpenter 2003; see Chapter 9).

Estuaries and the coastal zone of oceans, which are the final dumping ground of terrestrially derived nutrients, are typically quite productive. The rapid decomposition and nutrient release from sediments supports a productive bottom fishery and a rich nutrient source for the overlying water column. Chesapeake Bay, for example, was historically an extremely rich fishery that supported dense populations of Native Americans and later of European settlers. Just as in lakes, however, excessive nutrient and organic matter inputs to estuaries can be too much of a good thing, depleting oxygen and killing the organisms that would otherwise decompose and recycle the accumulating organic matter. The Mississippi River Delta, for example, has undergone a regime shift from a productive shrimp fishery to a dead zone with insufficient oxygen to support much biological activity (Rabalais et al. 2002). The biological mechanisms that limit the resilience of lakes and estuaries are well understood, but the failure of social–ecological systems to prevent regime shifts demonstrates the need to incorporate social processes into management planning (see Chapters 3 and 4).

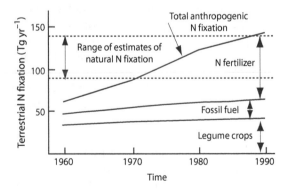

FIGURE 2.3. Anthropogenic fixation of nitrogen in terrestrial ecosystems over time compared with the range of estimates of natural biological nitrogen fixation on land. Redrawn from Vitousek et al. (1997).

Carbon and nutrient cycling in the water column of deep ocean basins is similar to that described for lakes, except that it is often even more nutrient-impoverished, because of the large vertical separation of ocean sediments from the surface where algal production occurs (Valiela 1995). Productive fisheries are often concentrated in zones of **upwelling** where deep nutrient-rich water moves to the surface near the edges of continental shelves or at high latitudes, where wind-driven mixing is most pronounced.

Maintenance of Biological Diversity

The known effects of biodiversity on ecosystem functioning relate most strongly to traits that govern the *effect* of species on ecosystem processes and traits that govern the *response* of species to environmental variation. These known effects of biodiversity change (both loss of key species or invasion of species with large impacts) can be fostered by maintaining species that span the spectrum of **effect diversity** and **response diversity** present in the system (Elmqvist et al. 2003, Suding et al. 2008).

Keystone species are species that have disproportionately large *effects* on ecosystems, typically because they alter critical slow variables. Ecosystems are most likely to sustain their current properties if current keystone species (or their functional equivalents) are maintained, and new ones are not introduced. Species that influence the supply of growth-limiting resources generally have large effects on the functioning of ecosystems. *Myrica faya*, for example, is a nitrogen-fixing tree introduced to nitrogen-poor ecosystems of Hawaii by the Polynesians. The resulting increases in nitrogen inputs, productivity, and canopy shading eliminated many plant species from the formerly diverse understory of this forest (Vitousek 2004). Highly mobile animals, such as salmon and sheep, act as keystone species governing nutrient supply by feeding in one place and dying or defecating somewhere else. Similarly, species that modify disturbance regime exert strong effects on ecosys-

tem functioning through their effects on the adaptive cycle of release, renewal, and growth. The introduction of flammable grasses into a tropical forest, for example, can increase fire frequency and trigger a regime shift from forest to savanna (D'Antonio and Vitousek 1992). One of the main ways that animals affect ecosystem processes is through physical disturbance (Jones et al. 1994). Gophers, pigs, and ants, for example, disturb the soil, creating sites for seedling establishment and favoring early successional species. Elephants trample vegetation and remove portions of tree canopies, altering the competitive balance between trees and grasses in tropical savannas. Many keystone species exert their effect by modifying species interactions, for example, by eating other species (e.g., forest pests) or competing or facilitating the growth of other species (Chapin et al. 2000). Conserving key **functional types** (i.e., a group of species that have similar effects on ecosystem processes) and preventing the invasion of novel functional types reduces the likelihood of large changes in ecosystem services.

Diversity in the environmental *response* of species can stabilize ecosystem processes. Many species in a community appear functionally similar, for example, algal species in a lake or canopy trees in a tropical forest (Scheffer and van Nes 2004). What are the ecosystem consequences of changes in diversity *within* a functional type (i.e., **functional redundancy**)? Differences in environmental responses among functionally similar species provide resilience by stabilizing rates of ecosystem processes (McNaughton 1977, Chapin and Shaver 1985). In midlatitude grasslands, for example, cool-season grasses are particularly productive under cool, moist conditions, and warm-season grasses under hot, dry conditions. As environmental conditions fluctuate within and among years, different species attain a competitive advantage over other functionally similar species (i.e., other grasses), thus stabilizing rates of ecosystem processes by the entire community (Ives et al. 1999).

The functional redundancy associated with species diversity also provides insurance against more drastic changes in environment, such as those that may occur in the event of human mismanagement of ecosystems or change in climate. Radical changes in environment are unlikely to eliminate all species of a given functional type in a diverse ecosystem, allowing the surviving species to increase in abundance and maintain functions that might otherwise be seriously compromised. Overgrazing in Australian grasslands, for example, eliminated the dominant species of palatable grass, severely reducing the quantity of cattle that the ecosystem

TABLE 2.3. Examples of biodiversity effects on ecosystem services. We separate the diversity effects into those due to functional composition, numbers of species, genetic diversity within species, and landscape structure and diversity. Modified from Díaz et al. (2006).

Ecosystem service	Diversity component and mechanism
1. Production by societally important plants	*Functional composition*: (a) fast-growing species produce more biomass; (b) species differ in timing and spatial pattern of resource use (complementarity allows more resources to be used)
	Species number: large species pool is more likely to contain productive species
2. Stability of crop production	*Genetic diversity*: buffers production against losses to pests and environmental variability
	Species number: Cultivation of multiple species in the same plot maintains high production over a broader range of conditions
	Functional composition: species differ in their response to environment and disturbance, stabilizing production
3. Maintenance of soil resources	*Functional composition*: (a) fast-growing species enhance soil fertility; (b) dense root systems prevent soil erosion
4. Regulation of water quantity and quality	*Landscape diversity*: Intact riparian corridors reduce erosion.
	Functional composition: Fast-growing plants have high transpiration rates, reducing stream flow
5. Pollination for food production and species survival	*Functional composition*: Loss of specialized pollinators reduces fruit set and diversity of plants that reproduce successfully
	Species number: Loss of pollinator species reduces the diversity of plants that successfully reproduce (genetic impoverishment)
	Landscape diversity: Large, well-connected landscape units enable pollinators to facilitate gene flow among habitat patches
6. Resistance to invasive species with negative ecological/cultural effects	*Functional composition*: Some competitive species resist the invasion of exotic species
	Landscape structure: Roads can serve as corridors for spread of invasive species; natural habitat patches can resist spread
	Species number: Species-rich communities are likely to have less unused resources and more competitive species to resist invaders
7. Pest and disease control	*Genetic diversity or species number*: Reduces density of suitable hosts for specialized pests and diseases
	Landscape diversity: Provides habitat for natural enemies of pests
8. Biophysical climate regulation	*Functional composition*: Determines water and energy exchange, thus influencing local air temperature and circulation patterns
	Landscape structure: Influences convective movement of air masses and therefore local temperature and precipitation
9. Climate regulation by carbon sequestrations	*Landscape structure*: Fragmented landscapes have greater edge-to-area ratio; edges have greater carbon loss
	Functional composition: Small, short-lived plants store less carbon
	Species number: High species number reduces pest outbreaks that cause carbon loss
10. Protection against natural hazards (e.g., floods, hurricanes, fires)	*Landscape structure*: Influences disturbance spread and/or protection against natural hazards
	Functional composition: (a) extensive root systems prevent erosion and uprooting; (b) deciduous species are less flammable than evergreens

could support and exposing the soil to wind erosion. Fortunately, a previously rare grass species that was less palatable, survived the overgrazing and increased in abundance, when the dominant grass declined, thus maintaining grass cover and reducing potential degradation from erosion (Walker et al. 1999). These examples of diversity effects on resilience provide hints of the general importance of diversity in stabilizing ecosystem processes and associated ecosystem services (Table 2.3) and suggest that management that sustains diversity is critical to long-term sustainability.

Maintenance of Disturbance Regime

Disturbance shapes the long-term fluctuations in the structure and functioning of ecosystems and therefore their resilience and vulnerability to change. Disturbances are relatively discrete events that alter ecosystem structure and cause changes in resource availability or physical environment (Pickett and White 1985). Disturbance is not something that "happens" to ecosystems but is an integral part of their functioning. Species are typically adapted to the **disturbance regime** (i.e., the characteristic severity, frequency, type, size, timing, and intensity of disturbance) that shaped their evolutionary histories. Grassland and boreal species, for example, resprout rapidly after fire or have reproductive strategies that enable them to colonize recent burns (Johnson 1992). In contrast, many tropical tree species produce a pool of young seedlings that grow slowly in the understory, "waiting" until a hurricane or other event creates gaps in the canopy. The tropical tree strategy is poorly adapted to fire, which would kill the understory seedlings, and the boreal trees are poorly adapted to wind, which would leave an organic seedbed unfavorable for post-disturbance germination. Naturally occurring disturbances such as fires and hurricanes are therefore not "bad"; they are normal properties of ecosystems and indeed are essential for the long-term resilience of species and community dynamics that characterize a particular ecosystem.

The adaptive cycle that is triggered by disturbance both generates and depends upon landscape patterns of biodiversity. After disturbance (release phase) and colonization (renewal phase), ecosystems undergo **succession** (growth phase), a directional change in ecosystem properties resulting from biologically driven changes in resource supply. Succession is accompanied by changes in the sizes and types of plants, leading to a diversity of food and habitat for animals and soil microbes. These changes in species composition and diversity both cause and respond to the changing availability of light, water, and nutrients as succession proceeds, leading to characteristic changes in the cycling of carbon, water, and nutrients and the associated supply of ecosystem services. The scale of this successional dynamic ranges from individual plants (e.g., gap-phase succession in moist temperate and tropical forests) to extensive stands (e.g., flood plains or conifer forests characterized by large stand-replacing crown fires) and from years (e.g., grasslands) to centuries (e.g., many forests). Subsequent renewal of a disturbed patch draws on both on-site legacies (e.g., buried seeds and surviving individuals) and colonization from the surrounding matrix (Fig. 2.4; Nyström and Folke 2001, Folke et al. 2004). The resilience of the integrated disturbance-renewal system depends on both a diversity of functional types capable of sustaining the characteristic spectrum of ecosystem functions (effect diversity) and functional redundancy *within* functional types (response diversity). In coral reefs, for example, storms cause local extinctions that are repopulated by dispersal from the surrounding matrix. If overfishing or eutrophication eliminates some grazer species, such as parrot fish that remove invading algae to produce space for recolonizing coral larvae, the grazer-functional group is less likely to provide the conditions for successful coral recruitment. In the absence of parrot fish, algae overgrow the corals, leading to a regime shift that supports substantially less biodiversity and ecosystem services (Bellwood et al. 2004).

As climate-driven stresses become more pronounced, and local extinctions occur more frequently, the functional redundancy and

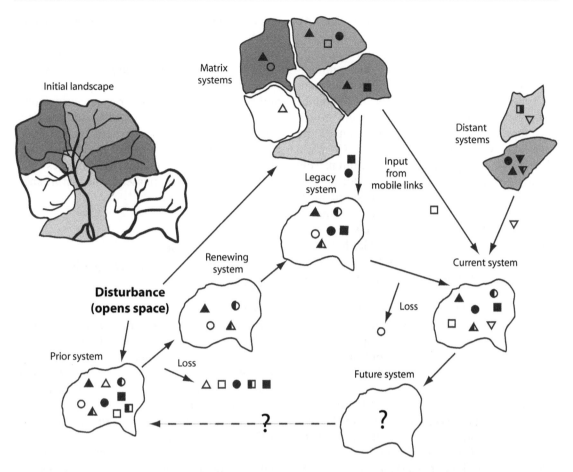

FIGURE 2.4. Roles of biodiversity in ecosystem renewal after disturbance (Folke et al. 2004). A disturbance such as a fire, hurricane, volcanic eruption, or war opens space in a social–ecological system. In this diagram, each shape represents a different functional group such as algal-grazing herbivores in a coral reef, and the different patterns of shading represent species within a functional group. After disturbance, some species are lost, but an on-site legacy of surviving species serves as the starting point for ecosystem renewal. For example, after boreal fire, about half of the vascular plant species are lost. The larger the *species diversity* of the pre-disturbance ecosystem, the more species and functional groups are likely to survive the disturbance. (In this figure all functional groups except "squares" survived the disturbance.) *Landscape diversity* of the matrix surrounding the patch is also important to

ecosystem renewal because it provides a reservoir of diversity that can recolonize the disturbed patch through the actions of **mobile links** (biological or physical processes that link patches on a landscape). In this figure the "square" function was renewed by colonization from the matrix surrounding the ecosystem. Through time some additional species may be gained or lost. In addition, new functional groups (inverted triangles in this diagram) may invade from a distance. The greater the species diversity of the patch, the less likely is this invasion (and associated functional change). In summary, diversity in both the patch and the surrounding matrix are essential to maintaining ecosystem functioning over the long term. Although we have described the importance of diversity from an ecological perspective, the same logic holds true for economic and institutional diversity (see Chapters 3 and 4).

biodiversity of the matrix become increasingly important to landscape resilience (Elmqvist et al. 2003). Postage-stamp reserves in a matrix

of agricultural monoculture, for example, are less likely to sustain their functional diversity than in a diverse landscape, especially during times of rapid environmental change (Fig. 2.4). A patchwork of hedge rows, forest patches, and riparian corridors in an agricultural landscape or of urban gardens, cemeteries, and seminatural landscaping of lawns in cities can substantially increase landscape diversity and resilience of human-dominated landscapes, even though they may occupy only a tiny fraction of the land area (Colding et al. 2006; see Chapters 12 and 13). In a rapidly changing, intensively managed world, **assisted migration** of species with low migration potential can supplement landscape biodiversity as a source of renewal (McLachlan et al. 2007). Intensively managed Fennoscandian forests, for example, have lost 70% of the insect and bird biodiversity in their wood-decomposer food webs. Climate warming now creates conditions that are conducive to migration of more southerly wood decomposers. This wood-dependent biodiversity could be regenerated if the niche becomes available (through protection of older stands and retention of green and dead trees and coarse woody debris after logging) and if the species can arrive [management of the matrix to foster migration of late successional species supplemented by assisted migration (introduction) of southerly taxa; Chapin et al. 2007].

Society typically derives benefits from most phases of a disturbance cycle. Disturbance itself reduces population densities of certain pests and diseases. Early successional stages are characterized by rapidly growing species that have tissues that are nutritious to herbivores, such as deer, and fleshy fruits that are dispersed by birds and harvested by people. Later successional stages are dominated by species that provide other types of goods, such as timber or medicines. **Swidden** (slash-and-burn) **agriculture** is a cultural system that can be an integral sustainable component of some tropical forests, as long as the traditional disturbance regime is maintained (see Chapter 7). Human activities that alter disturbance regimes, however, modify the suite of goods and services provided by the landscape. Prevention of small insect outbreaks, for example, increases the continuity of susceptible individuals and therefore the probability of larger outbreaks (Holling 1986). A **command-and-control** approach to resource

management that prevents disturbances characteristic of an ecosystem often reduces resilience at regional scales by producing new conditions at all phases of the successional cycle to which local organisms are less well adapted (Holling and Meffe 1996; see Chapter 4). Decisions that alter disturbance regimes tend to address only the short-term benefits to society, such as flood control, pest control, and fire prevention, and ignore the broader context of changes that propagate through all phases of the disturbance cycle.

Ecosystem "Restoration": The Reconstruction of Degraded Ecosystems

In degraded ecosystems such as abandoned mines and degraded wetlands, active transformation to a more desirable state may be a central management goal. In this case, introduction of new functional types can foster transformation for ecosystem reconstruction or renewal (Bradshaw 1983). Introduction of nitrogen-fixing trees to abandoned mine sites in England, for example, greatly enhanced the accumulation of soil nitrogen and soil organic matter, providing the soil resources necessary to support forest succession. Planting of metal-tolerant grasses on metal-contaminated sites can play a similar role. Planting of beach grasses in coastal developments can stabilize sand dunes that might otherwise be eroded by winds and storms. Similarly, planting of salt marsh plants with different salinity tolerances along a salinity gradient can renew coastal salt marshes that were eliminated by human disturbance or by sediment deposition from upstream land-use change.

At larger scales, the introduction of a Pleistocene-like megafauna has been suggested as a strategy to convert moss-dominated unproductive Siberian tundra into a more productive steppe-like ecosystem. This regime shift could support greater animal production and partially compensate for the loss of economic subsidies to indigenous communities in the post-Soviet Russian North (Zimov et al. 1995). Parks in cities are savanna-like environments that provide important cultural services for all residents, and city gardens provide valuable nutritional and cultural benefits to people who have moved to cities from rural agricultural areas (Colding et al. 2006). In a world dominated by rapid human population growth and directional environmental change, the deliberate introduction of new functional types creates path dependence for ecosystem transformation to a new, potentially more desirable state (Choi 2007). However, deliberate species introductions have a history of creating unintended undesirable side effects. Exotic grasses used to stabilize roadsides may expand into adjacent ecosystems, or biological control agents may expand their diet to nontarget species. Consequently, the introduction of novel functional types to trigger ecosystem transformation is a tool that requires caution and can often be avoided through use of locally adapted species and genotypes.

Provisioning Services: Providing the Goods Used by Society

Provisioning services are the goods produced by ecosystems that are consumed by society. They are the most direct link between ecosystems and social systems and are therefore the ecosystem properties that receive most direct attention from managers and the public. They are fast variables that depend on supporting services in ecosystems and often exhibit rapid nonlinear responses to fluctuations in environment. Large changes may be difficult to reverse if thresholds in supporting services are exceeded. In this section we identify the major provisioning services and discuss ways to sustain their supply.

Fresh Water

Water is the ecosystem service that is most likely to directly limit well-being in the twenty-first century. Although water is the most abundant compound on Earth, only a small fraction of it (0.1%) is available to people, primarily in lakes, rivers, and shallow groundwater

(see Chapter 9). People currently use 40–50% of available freshwater, with use projected to increase to 70% by 2050 (Postel et al. 1996). The shortage of clean water is particularly severe in developing nations, where future population growth and water requirements are likely to be greatest. The projected increases in human demands for fresh water will strongly impact aquatic ecosystems through eutrophication and pollution, diversion of fresh water for irrigation, and modification of flow regimes by dams and reservoirs (see Chapter 9).

Maintaining the essential role of intact ecosystems in the hydrological cycle is the single most effective way to sustain the supply and quality of fresh water for use by society. Intact ecosystems that surround reservoirs minimize sediment input, serve as a chemical and biological filter that removes pollutants and pathogenic bacteria, and buffer seasonal fluctuations in river flows, as described later. There are tradeoffs between the quantity and the quality of water provided by ecosystems. Forest clearing is sometimes suggested as a way to increase runoff and therefore the blue-water flows that can be withdrawn for human use. The clear-cutting of an experimental watershed at Hubbard Brook in the northeastern USA did indeed increase runoff fourfold (Likens et al. 1977). However, it also increased stream nitrate fluxes 16-fold because of a fourfold increase in nitrate concentration—to levels exceeding health standards for drinking water and led to loss of a spectrum of ecosystem services provided by the intact forest. Understanding the tradeoffs among water-related ecosystem services derived from green-water and blue-water flows is critical to ecosystem stewardship (Rockström et al. 1999, Gordon et al. 2008).

Expansion of human populations into arid regions is often subsidized by tapping groundwater supplies that would otherwise be unavailable to surface organisms. In dry regions 80–90% of this water is used to support irrigated agriculture, which can be highly productive once the natural constraints of water limitation are removed. The substantial cost of irrigated agriculture in turn creates incentives for intensive management with fertilizers and pesticides, leading to a cascade of associated social and economic consequences. The sustainability of irrigated agriculture depends on the rate of water use relative to resupply to the groundwater and the downstream consequences of irrigation (see Chapters 9 and 12). Many irrigated areas are supported by **fossil groundwater** that accumulated in a different climatic regime and is being removed much more rapidly than it is replenished, a practice that clearly cannot be sustained.

More than half of the water diverted for human consumption, industrial use, or agriculture is wasted. Most irrigation water, for example, evaporates rather than being absorbed by plants to support production. Management actions that increase the efficiency of water use and/or reuse water for multiple purposes can increase the effective water supply for human use without additional fresh-water diversion from ecosystems (see Chapter 9).

Food, Fiber, and Fuelwood

Management of ecosystems for the production of food, fiber, and fuelwood cause greater changes in ecosystem services and the global environment than any other human activity. Humans have transformed 40–50% of the ice-free terrestrial surface to produce food, fiber, and fuelwood (Vitousek et al. 1986, Imhoff et al. 2004). We dominate (directly or indirectly) about a third of primary productivity on land and harvest fish that use 8% of ocean production (Myers and Worm 2003). Most of the nitrogen that people add to the environment is to support agriculture, either as fertilizer or as nitrogen-fixing crops. The global human population increased fourfold during the twentieth century to 6.1 billion people, with corresponding increases in the harvest of ecosystem goods to feed, clothe, and house these people.

Two general categories of ecosystem change have enabled food and fiber production to keep pace with the growing human population. There has been **intensification** in use of existing agricultural areas through inputs of fertilizers, pesticides, irrigation, and energy-intensive technology and **extensification** through land-use conversion or modification of existing ecosys-

tems to provide goods for human use. Meeting the needs for food and fiber of the projected 60–70% increase in human population by 2050 will further increase the demands for agricultural and forestry production. Recent increases in food production have come primarily from intensification of agriculture, with much of the expected future increase expected to come from extensification in marginal environments where largest population increases may occur. There are critical tradeoffs between intensification, which often creates pollution problems, and extensification, which eliminates many of the services associated with natural ecosystems. Appropriate management can reduce the impacts of intensive agriculture (Matson et al. 1997). For example, no-till agriculture reduces soil disturbance and therefore the decomposition of soil organic matter, enhancing the water- and nutrient-retaining capacity of soils (see Chapter 12). Careful addition of water and nutrients to match the amounts and timing of crop growth can substantially reduce losses to the environment. There are also ways in which the extensification of agriculture can minimize impacts on ecosystem services by considering the landscape framework in which it occurs. Swidden (slash-and-burn) agriculture in tropical forests, for example, can provide food in newly cleared lands and forest products in regenerating forests. With appropriate rotation length, this practice has been sustained for thousands of years (Ramakrishnan 1992). However, in areas where rotation length declined from the traditional 30-year periodicity to 10 years or less in response to recent human population growth, soils had insufficient time to regain fertility, forest species with long life cycles disappeared, and the system underwent a regime shift to intensive agriculture of cash crops with radically different social and ecological properties (Ramakrishnan 1992). Just as with water, some of the greatest opportunities to minimize tradeoffs associated with enhancing agricultural production are to explore practices that maximize the effectiveness of lands and resources to support food production in ways that are consistent with ecological sustainability and local cultural norms and values.

About 70% of marine fisheries are overexploited (see Chapter 10). Much of this fishing pressure results from the globalization of markets for fish and from perverse subsidies that enable fishermen to continue fishing even for stocks that would otherwise no longer be profitable to harvest. This illustrates the importance of social and economic factors in driving increased harvest of many provisioning services and a need for improved resource stewardship of marine ecosystems to sustain provisioning services.

Other Products

Ecosystems provide a diverse array of other products that are specific to individual ecosystems and societies. These include aesthetically and culturally valuable items such as flowers, animal skins, and shells. In addition, ecosystems constitute a vast storehouse of genetic potential to deal with current and future conditions. This includes genes from traditional cultivars or wild relatives of crops and other wild species that produce products that benefit society (see Chapter 6). For example, about 25% of currently prescribed medicines originate from plant compounds that evolved as defenses against herbivores (Dirzo and Raven 2003) and have substantial potential for bioprospecting in regions of high biodiversity (Kursar et al. 2006).

Regulating Services: Sustaining the Social–Ecological System

Regulating services influence processes beyond borders of ecosystems where they originate. They constitute some of the key cross-scale linkages that connect ecosystems on a landscape and integrate processes across temporal scales. They are, however, largely invisible to society and generally ignored by managers, so failures to sustain regulating services often have devastating consequences.

Climate Regulation

The cycling of water, carbon, and nutrients has important climatic consequences. About half of the precipitation in the Amazon basin, for example, comes from water that is recycled by evapotranspiration from terrestrial ecosystems (Costa and Foley 1999). If tropical forests were extensively cut and replaced by pastures with lower transpiration rates, this could lead to a warmer, drier climate more typical of savanna, making forest regeneration more difficult (Foley et al. 2003b, Bala et al. 2007). At high latitudes, tree-covered landscapes absorb more radiation and transfer it to the atmosphere than does adjacent snow-covered tundra. The northward movement of treeline 6,000 years ago is estimated to have contributed half of the climate warming that occurred at that time (Foley et al. 1994). Extensive human impacts on ecosystems can have similar large effects. In Western Australia the replacement of native heath vegetation by wheatlands increased regional **albedo** (reflectance). As a result, the dark heathlands absorbed more radiation than the cropland, causing air to warm and rise over the heathland and drawing moist air from the adjacent wheatlands. The net effect was a 10% increase in precipitation over heathlands and a 30% decrease in precipitation over croplands (Chambers 1998). Many vegetation changes, if they are extensive, generate a climate that favors the new vegetation, making it difficult to return vegetation to its original state (Chapin et al. 2008). This suggests that ecosystem integrity is critical to resilience of the climate system at regional scales.

Ecosystems are also important sources and sinks of greenhouse gases that determine the heat-trapping capacity of the atmosphere and therefore the temperature of our planet. Approximately half of the CO_2 released by burning fossil fuels is captured and stored by ecosystems—half on land and half in the ocean. The capacity of ecosystems to remove and store this carbon therefore exerts a strong influence on patterns and rates of climate change. Forests and peatlands are particularly effective in storing large quantities of carbon, in trees and soils, respectively. Maintaining the integrity of these ecosystem types or restoring them on degraded lands enhances the capacity of the terrestrial biosphere to store carbon. Increased recognition of the value of this climate regulatory service has led to a market in carbon credits for activities that enhance the capacity of ecosystems to store carbon (see Chapter 7).

Regulation of Erosion, Water Quantity, Water Quality, and Pollution

As discussed earlier (see Water), intact ecosystems regulate many water-related services by buffering stream flows to prevent floods and soil erosion and by filtering ground water to reduce pollutant concentrations (Rockström et al. 1999). Many of the compounds in agricultural and urban runoff are identical or similar to compounds that naturally cycle through ecosystems and are therefore used by organisms to support their growth and reproduction. Ecosystems therefore have a natural capacity to absorb these pollutants, cleansing the air or water in the process (see Chapter 9). Ecosystems also process some novel chemicals, with potentially positive and negative environmental consequences. Oil spills, for example, select for oil-degrading bacteria that use oil as an energy source, although their capacity to do so is generally limited by nutrient availability. Polychlorinated-hydrocarbon pesticides (PCBs) are a potential energy source for those organisms that evolve resistance to their toxicity. The rapid evolution that typifies microbial populations in soils and sediments sometimes selects for populations capable of degrading or converting these compounds to other products. Their activity reduces pollutant concentration. Sometimes, however, the breakdown products are even more toxic to other organisms than was the original compound, as in the conversion of insecticide DDT to DDE, or are environmentally stable and accumulate in ecosystems, as in the fat-soluble PCBs that accumulate in food chains and have caused reproductive failure in many marine birds (Carson 1962). Society therefore cannot count on ecosystems to provide a "quick fix" that solves pollution problems.

Those pollutants that are processed by ecosystems frequently alter their structure, diversity, and functioning. The processing of nitrogen derived from acid rain or agricultural pollution, for example, increases productivity, altering the competitive balance and relative abundance of species. The dominant plants often increase in size and abundance and outcompete smaller organisms, leading to a loss of species diversity. Moreover, as ecosystems cycle more nitrogen, soil nitrate concentrations increase, leading to emissions of more nitrous oxide (a potent greenhouse gas) to the atmosphere and leaching of nitrate to groundwater. As nitrate (an anion) leaches, it carries with it a cation to maintain charge balance, reducing the availability of cations such as calcium and magnesium, which can be replaced only by slow soil weathering. This is representative of many ecological responses to change, in which apparently beneficial effects of ecosystems (e.g., removal of nitrogen-based pollutants or PCBs) can initiate a cascade of unanticipated consequences. Ecological research that recognizes the complex adaptive nature of ecosystems (see Chapter 1) can increase the likelihood of anticipating some of these effects as a basis for informed policy decisions.

Natural-Hazard Reduction

Adaptations of organisms to the characteristic disturbance regime of their environment often reduce the societal impacts of disturbance. As discussed earlier, every ecosystem has a particular disturbance regime to which organisms are adapted. However, these same disturbances, such as hurricanes, wildfires, and floods, often have negative societal impacts on the built environment that people create. Incorporation of organisms adapted to a particular disturbance regime into the built environment sometimes reduces the impact of disturbances when they occur. Kelp forests in the intertidal zone, for example, dissipate energy from storm surges and protect beaches from coastal erosion, just as beach grasses protect sands from wind erosion. Floodplain trees and shrubs reduce the speed of flood waters, leading to deposition of sediments and reducing the water energy that would otherwise cause erosional changes in channel morphology. Some of the greatest challenges in managing disturbance are to identify and separate those locations where disturbances have large negative effects on human-dominated environments (e.g., towns and cities) from areas where disturbances have greater societal benefits and/or are most likely to occur. For example, concentrating suburban development in areas that are unlikely to experience fire or flooding reduces risks to the built environment. Similarly, allowing regular small disturbances (e.g., floods, prescribed fires, and insect outbreaks) to occur periodically reduces the likelihood of larger disturbances that are more difficult to control (Holling and Meffe 1996).

Regulation of Pests, Invasions, and Diseases

Biodiversity often enhances pest resistance in agricultural systems through both ecological and evolutionary processes (Díaz et al. 2006). An increasing tendency in intensive agriculture is to reduce weeds, pests, and pathogens using agrochemical pesticides. Due to their high population densities and short life cycles, however, insects and weeds typically evolve resistance to synthetic biocides within 10–20 years, necessitating continuing costly investments to develop and synthesize new biocides as current products become less effective. An alternative more resilient approach is to make greater use of natural processes that regulate pests. Increased genetic diversity of crops nearly always decreases pathogen-related yield losses (Table 2.3). Recently, a major and costly fungal pathogen of rice, rice blast, was controlled in a large region of China by planting alternating rows of two rice varieties (Zhu et al. 2000). Similarly, a high diversity of crop species reduces the incidence or severity of impact of herbivores, pathogens, and weeds (Andow 1991, Liebman and Staver 2001, Díaz et al. 2006). Sometimes these diversity effects have multiple benefits. Crop diversity treat-

ments that reduce the abundance of insect herbivores also suppress the spread of viral infection, because plant-feeding insects transmit most plant viruses. This leads to lower viral densities in polycultures than monocultures (Power and Flecker 1996).

The species richness of **natural enemies** (pathogens, predators, and parasitoids) of pests tends to be higher in species-rich agroecosystems than in monocultures and higher in natural vegetation buffers than in fields, leading to higher ratios of natural enemies to herbivores and therefore lower pest densities. The spraying of biocides can increase vulnerability because it reduces the abundance of natural enemies more than the pests that are targeted. This allows pest populations to rebound rapidly, sometimes causing more damage than if no pesticides had been used (Naylor and Ehrlich 1997; see Chapter 12).

Invasive exotic species cost tens of millions of dollars annually in the USA, primarily in crop losses and pesticide applications (Pimentel et al. 2000). Ecosystems have mechanisms, as yet poorly understood, that reduce invasibility (Díaz et al. 2006). These include disturbance adaptations that enable certain native species to colonize and grow rapidly after disturbance, a time when invasive species might otherwise encounter little competition from other plants. In a given environment, more diverse ecosystems are less readily invaded, perhaps because local species already fill most of the potential biological roles and utilize most of the resources that might be available (Díaz et al. 2006). Invasive species most frequently colonize environments that naturally support high levels of diversity. Together these observations suggest that (1) hot spots for diversity are particularly at risk of invasion by introduced species, and (2) the loss of native species may increase invasibility.

Natural enemies also reduce ecosystem invasibility. One reason that exotic species are often so successful is that they escape their specialized natural enemies that constrain success in their region of origin. Natural enemies in the new environment often prevent these exotic species from becoming noxious pests (Mitchell and Power 2003).

Maintaining the integrity of natural ecosystems may reduce disease risk to people. Lyme disease, for example, is caused by a pathogen that is transmitted by ticks from mice and deer to people (Ostfeld and Keesing 2000). Although forest mice are the largest reservoir of the disease, deer move the disease from one forest patch to another. Exploding deer densities have increased the incidence of lyme disease in the northeastern USA in response to agricultural abandonment, expansion of suburban habitat, and elimination of the natural predators of deer. In Sweden, climate warming has increased overwinter survival of ticks, further increasing disease incidence (Lindgren et al. 2000, Lindgren and Gustafson 2001).

Currently 75% of emerging human diseases are naturally transmitted from animals to humans (Taylor et al. 2001). Clearly, efforts to control these diseases require improved understanding of social–ecological dynamics (Patz et al. 2005).

Pollination Services

Much of the world's food production depends on animal pollination, particularly for fruits and vegetables that provide a considerable portion of the vitamins and minerals in the human diet. The value of these pollination services is likely to be billions of dollars annually (Costanza et al. 1997). Pollination by animals is obviously essential for the success of plants that are not wind- or self-pollinated. These pollination services often extend well beyond a given stand and are often important in pollinating adjacent crops. Temperate orchards and tropical coffee plantations adjacent to uncultivated lands or riparian corridors often have more pollinators and are more productive than are larger orchards (Ricketts et al. 2004). Large monocultures reduce pollination services by reducing local floral diversity and nesting sites and by using insecticides that reduce pollinator abundances. Pollination webs often connect the success of a wide range of species in multiple ecosystem types (Memmott 1997).

Cultural Services: Sustaining Society's Connections to Land and Sea

Cultural Identity and Cultural Heritage

Cultural connections to the environment are powerful social forces that can foster stewardship and social–ecological sustainability (see Chapter 6). Because people have evolved as integral components of social–ecological systems, this human–nature relationship is often an important component of **cultural identity**, i.e., the current cultural connection between people and their environment (Plate 4). Cultural identity in turn links to the past through **cultural heritage**, i.e., the stories, legends, and memories of past cultural ties to the environment (de Groot et al. 2005). Cultural identity and heritage are ecosystem services that strongly influence people's sense of stewardship of social–ecological systems (Ramakrishnan 1992, Berkes 2008) and therefore offer an excellent opportunity for natural resource managers to both learn from and contribute to stakeholder efforts to sustain their livelihoods and environment (see Chapter 6). Many indigenous peoples, for example, have **traditional ecological knowledge** based on oral transmission of their cultural heritage in ways that inform current interactions with the environment (i.e., cultural identity). This cultural heritage provides information about how people coped with past environmental and social–ecological challenges and about important values that influence likely future responses to changes in both the environment and the resource management policies (Berkes 1998; see Chapter 4). Integration of traditional knowledge with the **formal knowledge** (i.e., "scientific knowledge") that often informs resource management decisions is not easy, because the "facts" (e.g., the nature of the human–environment relationship) sometimes differ between the two knowledge systems (Berkes 2008). Both knowledge systems are important if they influence the ways in which stakeholders perceive and interact with their environment. The linkage between knowledge systems (as informed by cultural heritage), perceptions, and actions is at least as important to understanding and predicting human actions as are the biophysical mechanisms that are believed to underlie scientific knowledge. Social learning that builds new frameworks to sustain social–ecological systems is most likely to occur if both traditional and formal knowledge are treated with respect rather than subjugating traditional knowledge to western science.

Local knowledge held by farmers, ranchers, fishermen, resource managers, engineers, and city dwellers is also valuable. As in the case of indigenous knowledge, local knowledge consists of "facts" based on cultural heritage and observations that determine how people respond to and affect their environment. Many local residents, whether indigenous or not, spend more time interacting with their environment than do policy makers and therefore have a different suite of observations and perceptions. **Co-management** of natural resources by resource managers and local stakeholders provides one mechanism to integrate these knowledge systems and perspectives in ways that increase the likelihood of effective policy implementation (see Chapter 4).

Traditional knowledge systems are being eroded by social and technological changes. Many indigenous traditional knowledge systems are maintained orally and are therefore tightly linked to language. There are about 5,000 indigenous languages, half of them in tropical and subtropical forests (de Groot et al. 2005). Many of these languages are threatened by national efforts to assimilate people into one or a few national languages. Language loss and cultural assimilation generally erode traditional knowledge, so efforts to sustain local languages and cultures can be critical to sustaining the knowledge and practices by which people traditionally interacted with ecosystems. Similarly, sustaining opportunities for locally adapted approaches to farming, ranching, and fishing preserves practices that may sustain local use of natural resources (Olsson et al. 2004b). Obviously, many local practices, whether indigenous or otherwise, do not contribute to sustainable management in a modern world, but they nonetheless pro-

vide information about perceptions, tradeoffs, and institutions that are a source of resilience for developing new frameworks to sustain social–ecological systems in a rapidly changing world.

Spiritual, Inspirational, and Aesthetic Services

The spiritual, inspirational, and aesthetic services provided by ecosystems are important motivations for conservation and long-term sustainability. "The most common element of all religions throughout history has been the inspiration they have drawn from nature, leading to a belief in non-physical (usually supernatural) beings" (de Groot et al. 2005). These spiritual services provided by ecosystems for both personal reflection and more organized experiences have proven to be a powerful force for conservation. Sacred groves in northeast India and Madagascar, for example, are major reservoirs of biodiversity and places where people maintain their sense of connection with the land in landscapes that are increasingly converted to agriculture to meet the food needs of local people (Ramakrishnan 1992, Elmqvist et al. 2007). Often these sacred groves are maintained and protected by local institutions (see Chapters 6 and 7). Well-meaning national and international efforts to protect these few remaining sites of high conservation value by placing them under national control sometimes undermine local institutions and lead to degradation rather than protection. This underscores the importance of understanding the social context of cultural services, when addressing conservation and sustainability goals (Elmqvist et al. 2007).

The inspirational qualities of landscapes that motivated ancient Greek philosophers, Thoreau's writings, French Impressionist paintings, and Beethoven's symphonies continue to inspire people everywhere through both direct experience and increasingly television, films, and the Internet. This provides natural resource managers with a diverse set of media that can supplement personal experience in reinforcing the human–environment connection (Swanson et al. 2008).

The aesthetic and inspirational properties of landscapes are closely linked. Research suggests that aesthetic preferences are surprisingly similar among people from very different cultural and ecological backgrounds (Ulrich 1983, Kaplan and Kaplan 1989, de Groot et al. 2005). For example, when asked which is more aesthetically pleasing, people generally prefer natural over built environments and park/savanna-like environments over arid or forest environments, regardless of their background. People do differ in aesthetic preferences, of course. Farmers and low-income groups generally prefer human-modified over natural landscapes, whereas city dwellers and high-income groups have an aesthetic preference for natural landscapes. In addition, the view of western Euro-Americans toward wilderness has changed through history from a perception of wilderness as a hostile land until the late seventeenth century toward a more romantic view of wilderness in the eighteenth and nineteenth centuries. Even today, wilderness views are changing as people move to cities and change their patterns of use of remote lands (de Groot et al. 2005). At a time of rapid global change, it seems important to explore potential changes in the spiritual, inspirational, and aesthetic benefits that people derive from ecosystems and the resulting human decisions and actions that influence their environment.

Iconic species are species that symbolize important nature-based societal values. Protection of these species can mobilize public support for protection of values that might otherwise be difficult to quantify and defend as management goals. Polar bears, wolves, Siberian tigers, eagles, and whales, for example, are top predators whose population dynamics are sensitive to habitat fragmentation and to factors that might affect their prey species. Panda bears and spotted owls require ecosystems with structural properties typical of old-growth ecosystems. Ecosystem management that protects the habitat of these species often sustains a multitude of other services.

Recreation and Ecotourism

Recreation and tourism have always been important benefits that people gain from ecosystems, sometimes as rituals and pilgrimages, sometimes just for pleasure and enjoyment. For example, about a billion people (15% of the global population) visit the Ganges River annually. Nature tourism accounts for about 20% of international travel and is increasing 20–30% annually. At a more local scale, people use parks and other ecosystems as important components of daily life. Cultural and nature-based tourism constitutes 3–10% of GDP (gross domestic product) in advanced economies and up to 40% in developing economies. It is *the* main source of foreign currency for at least 38% of countries (de Groot et al. 2005). Clearly there are both personal and economic incentives to manage the recreational opportunities provided by ecosystems in ways that do not degrade over time.

Synergies and Tradeoffs among Ecosystem Services

Ecosystems that maintain their characteristic supporting services provide a broad spectrum of ecosystem services with minimal management effort. At a finer level of resolution, **bundles of services** can be identified that have particularly tight linkages. This creates **synergies** in which management practices that sustain a few key services also sustain other synergistic services. For example, management of fire and grazing in drylands to maintain grass cover minimizes soil erosion (sustaining most supporting services), sustains the capacity to support grazers, and reduces vulnerability to invasion by exotic shrubs (see Chapter 8). Management of fisheries to sustain bottom habitat through restrictions on trawling or to maintain populations of top predators reduces the likelihood of fishery collapse. In general, management that sustains slow variables (soil resource supply, disturbance regime, and functional types of species) sustains a broad suite of ecosystem services. Managers are often tasked with managing one or a few fast variables such as the supply of corn, deer, timber, or water, each of which might be augmented in the short term by policies that reduce the flow of other ecosystem services (**tradeoffs**). However, even these fast variables that are the immediate responsibility of managers are best sustained over the long term through attention to slow variables that govern the flow of these and a broader suite of services.

Tradeoffs most frequently emerge when people seek to enhance the flow of one or a few services (Table 2.4). For example, agricultural production of food typically requires the replacement of some naturally occurring ecosystem by a crop with the corresponding loss of some regulatory and cultural services. Management of forests to produce timber as a crop (short rotations of a single species) involves similar tradeoffs between the efficient production of a single species and the cultural and regulatory services provided by more diverse forests (see Chapter 7). Many management choices involve **temporal tradeoffs** between short-term benefits and long-term capacity of ecosystems to provide services to future generations. Management of lands to provide multiple services (i.e., multiple-use management) requires identification of tradeoffs among services and decisions that reflect societal choices among the costs and benefits associated with particular options.

Given the large number of essential services provided by ecosystems (e.g., about 40 identified by the MEA), which services should

receive highest management priority? One approach is to sustain supporting services that underpin most other ecosystem services, as described earlier. During times of rapid social or environmental change, however, inevitable tradeoffs arise that make objective decisions difficult. Under these circumstances, it may prove valuable to identify **critical ecosystem services**, i.e., those services that (1) society depends on or values; (2) are undergoing (or are vulnerable to) rapid change; and/or (3) have no technological or off-site substitutes (Ann Kinzig, pers.com.). Stakeholders often disagree about which ecosystem services are most critical, so identification of these services benefits from broad stakeholder participation (Fischer 1993, Shindler and Cramer 1999).

Facing the Realities of Ecosystem Management

Sustainability: Balancing Short-Term and Long-Term Needs

Sustainability requires the use of the environment and resources to meet the needs of the present without compromising the ability of future generations to meet their own needs (WCED 1987; see Chapter 1). Balancing the temporal tradeoff between short-term desires and long-term opportunities is a fundamental challenge for ecosystem stewards. The demands by current stakeholders are always more certain

TABLE 2.4. Examples of synergies and tradeoffs among ecosystem services.

Synergies

Supporting services: maintenance of soil resources, biodiversity, carbon, water, and nutrient cycling

Water resources: water provisioning, maintenance of soil resources, regulation of water quantity and quality by maintaining intact ecosystems, flood prevention

Food/timber production capacity: food/timber provisioning, maintenance of soil resources, genetic diversity of crops/forest

Climate regulation: maintenance of soil resources, regulation of water quantity by maintaining ecosystem structure

Cultural services: maintenance of supporting services (including biodiversity), suite of cultural services

Tradeoffs

Efficiency vs sustainability:

Short-term vs long-term supply of services

Food production vs services provided by intact natural ecosystems

Intensive vs extensive management to provide food or fiber

Recreation vs traditional cultural services

and outspoken than those of future generations, creating pressures to manage resources
for short-term benefits. Ecosystem stewardship
implies, however, a responsibility to sustain
ecosystems so that future generations can meet
their needs. How do we do this, if we do not
know what future generations will want and
need? The simplest approach is to sustain the
inclusive wealth of the system, i.e., the total capital (natural, manufactured, human, and social)
that constitutes the productive base available to
society (see Chapter 1). Since natural and social
capital are the most difficult components of capital to renew, once they are degraded, these are
the most critical components of inclusive wealth
to sustain. Social capital is discussed in Chapter
3; here we focus on natural capital.

Future generations depend most critically on
those components of natural capital that cannot be regenerated or created over time scales
of years to decades. These always include (1)
soil resources that govern the productive potential of the land; (2) biodiversity that constitutes the biological reservoir of future options;
(3) regulation of the climate system that governs future environment; and (4) cultural identity and inspirational services that provide a
connection between people and the land or
sea. Earlier we described strategies for sustaining each of these classes of ecosystem services.
Other more specific needs of future generations, such as specific types of food or recreation, are less certain and therefore less critical
to sustain in precisely their current form.

At intermediate time scales (e.g., years to
decades), it is more plausible to assume that the
needs of people in the future will be similar to
those of today, leading to a more constrained
(and therefore more precisely defined) set of
tradeoff decisions. Depletion of fossil groundwater to meet irrigation needs today reduces
the water available in the future—for example, for domestic water consumption. Forests or
fish stocks that are harvested more rapidly than
they can regenerate reduce the services in coming years to decades. **Maximum sustained yield**
was a policy that sought to maximize the harvest of forests, fish, and wildlife to meet current needs, while sustaining the potential to
continue these yields in the future. Although

intended to prevent overharvest, these policies
often proved unsustainable because of overly
optimistic assumptions about the current status
and recovery potential of managed populations
(see Chapters 7 and 10).

Economists often **discount** (i.e., reduce) estimates of the future value of manufactured
goods and services because of opportunity costs
(i.e., the opportunities foregone to spend the
money on current goods). Discounting the
future is *not* appropriate, however, for temporal tradeoffs involving those ecosystem services
that cannot be restored once they are lost, for
example, fossil ground water, biodiversity, or
sacred groves in a highly modified landscape
(Heal 2000). These services will be at least as
valuable, and perhaps much more valuable, in
the future than they are today (Heal 2000).

Multiple Use: Negotiating Conflicting Desires of Current Stakeholders

**Tradeoffs among ecosystem services valued by
society generate frequent conflicts about management of ecosystem services.** Clearing forests
to provide new agricultural land, for example, represents a tradeoff between the benefits of harvested timber and increased potential for food production and the loss of other
services previously provided by the forest such
as regulation of water quality and climate.
Recreational use of an area by snow machines
and wilderness skiers creates tradeoffs between
the services sought by each group. There are
also tradeoffs in allocation of fresh water
among natural desert landscapes, agricultural-
ists, ranchers, and urban residents. These are
just a few of the many tradeoffs faced by ecosystem stewards. How does a manager balance
these tradeoffs or choose among them? There
is no simple answer to this question, but the
following questions raise issues that warrant
consideration. What are the gains and losses in
bundles of ecosystem services associated with
alternative management options? How much
weight do we give to gains and losses that
are uncertain but potentially large? Are there
new options that might provide many of the
same benefits but reduce potential losses in ser-

vices? Who wins and who loses? Can conflicts be reduced by landscape approaches that separate in time or space the services that different stakeholders seek to sustain? The social dimensions of tradeoffs among ecosystem services are at least as important as the ecological ones, so we address these questions in a social–ecological context in Chapter 4. We then provide numerous examples of challenges and strategies for managing synergies and tradeoffs in Chapters 6–14 and summarize the lessons learned in Chapter 15.

Adjusting to Change

Managing ecosystems in the context of dynamic and uncertain change is an integral component of resilience-based ecosystem stewardship. Many changes, such as those associated with interannual variability in weather, cycles of disturbance and succession, and economic fluctuations are widely accepted as normal components of the internal variability of social–ecological systems. Facilitating change is sometimes an explicit management objective, for example, in sustainable development projects, where the goal is to enhance well-being, while sustaining the capacity of ecosystems to provide services (WCED 1987). In reclamation and restoration projects, the goal is to enhance ecological integrity, while sustaining the social and economic benefits accrued from development. In still other cases, regional change may be driven by exogenous changes in climate, introduction of exotic species, human migration, and/or globalization of culture and economy. Many of these latter changes are persistent trends that will inevitably cause changes in social–ecological systems (see Chapters 1 and 5). The critical implications of managing change include (1) identifying the persistent changes that are occurring or might occur and assessing their plausible trajectories; (2) understanding that attempts to prevent change or to maintain indefinitely the current flow of ecosystem goods (e.g., production of cattle or pulpwood) may be unrealistic; and (3) continuously reassessing the social–ecological long- and short-term goals of ecosystem stewardship in light of the challenges and opportunities associated with change.

Identifying persistent (directional) trends in important controls such as climate and human demography and projecting plausible trends into the future is often a more realistic basis for planning than assuming that the future will be like the past. Although the future can never be predicted with certainty, planning in the context of expected changes can be helpful. Some of these changes may be difficult for individual managers to alter (e.g., climate change), but the trajectory of other changes, for example, in sustainable development or ecological restoration projects, can be strongly influenced by management decisions because these decisions create legacies and path dependencies that influence future outcomes. That, after all, is the purpose of management. In general, those changes that sustain or enhance natural and social capital will likely benefit society [see Sustainability (this chapter), Chapters 4 and 5].

One limitation of maximum sustained yield as a management paradigm is that it usually fails to account for variability and change (particularly persistent directional changes) in those yield-influencing processes that managers cannot control (see Chapter 7). Recognition of this problem was one of the primary motivations for a move toward ecosystem management as a more comprehensive approach to natural resource management.

Ecosystems are not the only parts of social–ecological systems to undergo persistent directional changes. Changes in societal goals are essential to consider. This requires active communication and engagement with a variety of stakeholders to enable them to participate meaningfully in the management process (see Chapters 4 and 5).

Regulations and Politics: Practicing the Art of the Possible

Many of greatest challenges faced by resource stewards reflect the social and institutional environment in which they work. Differences among stakeholders in goals and norms, power relationships, regulatory and financial constraints, personalities, and other social and institutional factors often dominate the day-to-day challenges faced by resource managers. Social processes are therefore an integral component of ecosystem stewardship (see Chapters 3 and 4). At times of rapid social or environmental change, frameworks for managing natural resources may become dysfunctional, requiring communication with a broader set of stakeholders and openness to new ways of doing things (see Chapter 5).

Summary

Resilience-based ecosystem stewardship involves responding to and shaping change in social–ecological systems to sustain the supply and opportunities for use of ecosystem services by society. The capacity of ecosystems to supply these services depends on underlying supporting services, such as the supply of soil resources; cycling of water, carbon, and nutrients; and the maintenance of biological diversity at stand and landscape scales. The resilience of these supporting services depends on maintaining a disturbance regime to which local organisms are adapted. Directional changes in these supporting services inevitably alter the capacity of ecosystems to provide services to society. An understanding of these linkages provides a basis for not only sustaining the services provided by intact ecosystems but also enhancing the capacity of degraded ecosystems to provide these services by manipulating pathways for ecosystem renewal.

Both managers and the public generally recognize the value of provisioning services such as water, food, fiber, and fuelwood. It is therefore not surprising that the links between supporting and provisioning services are generally well understood by ecosystem managers and local resource users. This provides a strong local and scientific basis to manage ecosystems sustainably for these goods. In contrast, regulatory services (e.g., the regulation of climate and air quality, water quantity and quality, pollination services, and risks of disease and of natural hazards), although generally recognized as important by society, are often overlooked

when discussions focus on the short-term supply of provisioning services. Some of the greatest opportunities and challenges in ecosystem management involve the stewardship of ecosystems to provide bundles of services that both meet the short-term needs of society and sustain regulatory services that are essential for their secure supply at larger temporal and spatial scales.

The long-term well-being of society depends substantially on the cultural services provided by ecosystems, including aesthetic, spiritual, and recreational values. These values often motivate support for sustainable stewardship practices, if the basic material needs of society are met. Understanding this hierarchy of societal needs, which provides the social context for ecosystem stewardship is the central topic of Chapter 3.

Review Questions

1. What slow variables are usually most important in sustaining ecosystem services? How are these likely to be altered by climate change in the ecosystem where you live? What are the likely societal consequences of these changes?
2. Describe a strategy for enhancing the supply of ecosystem services for a degraded ecosystem such as an abandoned mine or an overgrazed pasture.
3. What are the advantages and disadvantages of managing an ecosystem to maximize the production of a specific resource such as trees or fish? How would your management strategy differ if your goal were to maximize harvest over the short term (e.g., 5 years) vs the long term (several centuries). How might directional social or environmental change influence your long-term strategy?
4. If the future is uncertain and you do not know what future generations will want or need, how can you manage ecosystems sustainably to meet these needs?
5. Explain the mechanisms by which biodiversity influences supporting services, provisioning services, and cultural services.

Additional Readings

Carpenter, S.R. 2003. *Regime Shifts in Lake Ecosystems: Pattern and Variation. International Ecology Institute*, Lodendorf/Luhe, Germany.

Chapin, F.S., III, P.A. Matson, and H.A. Mooney. 2002. *Principles of Terrestrial Ecosystem Ecology*. Springer-Verlag, New York.

Christensen, N.L., A.M. Bartuska, J.H. Brown, S.R. Carpenter, C. D'Antonio, et al. 1996. The report of the Ecological Society of America committee on the scientific basis for ecosystem management. *Ecological Applications* 6:665–691.

Daily, G.C. 1997. *Nature's Services: Societal Dependence on Natural Ecosystems*. Island Press, Washington.

Díaz, S., J. Fargione, F.S. Chapin, III, and D. Tilman. 2006. Biodiversity loss threatens human well-being. *Plant Library of Science (PLoS)* 4:1300–1305.

Heal G. 2000. *Nature and the Marketplace: Capturing the Value of Ecosystem Services*. Washington: Island Press.

Holling, C.S., and G.K. Meffe. 1996. Command and control and the pathology of natural resource management. *Conservation Biology* 10:328–337.

Olsson P., C. Folke, and T. Hahn. 2004. Social-ecological transformation for ecosystem management: The development of adaptive co-management of a wetland landscape in southern Sweden. Ecology and Society 9(4):2. [online] URL: www.ecologyandsociety.org/vol9/iss4/art2/.

Schlesinger, W.H. 1997. *Biogeochemistry: An Analysis of Global Change*, 2nd edition. Academic Press, San Diego.

Szaro, R.C., N.C. Johnson, W.T. Sexton, and A.J. Malk, editors. 1999. *Ecological Stewardship: A Common Reference for Ecosystem Management*. Elsevier Science Ltd, Oxford.

3
Sustaining Livelihoods and Human Well-Being during Social–Ecological Change

Gary P. Kofinas and F. Stuart Chapin, III

Introduction

Social processes strongly influence the dynamics of social–ecological responses to change. In Chapter 2, we described the ecological processes that govern the flow of ecosystem goods and services to society. Sustainability of these flows depends not only on ecosystems, but also on human actions that are motivated, in part, by desires and needs for these services. Many of the social and ecological slow variables that determine the long-term dynamics of social–ecological systems act primarily through their effects on human well-being (see Chapter 1). Today humans are *the* dominant force driving changes in the Earth System, with the biophysical changes during the last 250 years so fundamental that they define a new geologic epoch—the **Anthropocene** (Crutzen 2002; see Chapter 14). Social process play a key role in driving these changes, so it is essential to understand them as critical determinants of

sustainability from local to planetary scales. This chapter focuses on those social processes that affect well-being, society's vulnerability and resilience to recent and projected changes, and strategies for sustainable development that seek to enhance well-being, while sustaining the capacity of ecosystems to meet human needs. Institutional dimensions of well-being, which are also critical to sustainability, are addressed in Chapter 4.

Well-being, or quality of life, is more than human health and wealth. In the context of ecosystem stewardship and sustainability, well-being also includes happiness, a sense of fate control, and community capacity. **Livelihoods** of individuals and households include their capabilities, tangible assets, and means of living (Chambers and Conway 1991). Well-being and livelihoods are therefore key elements that set the stage for sustainability, resilience, and adaptability of people to change.

Livelihoods are both complex and dynamic (Allison and Ellis 2001). People both respond to and cause many ecological changes as a result of resource consumption by a growing population and efforts to meet their desires and needs in new ways. Perceptions of well-being are shaped by material conditions, history, and culture. For these reasons, the relationship

G.P. Kofinas (✉)
School of Natural Resources and Agricultural Sciences and Institute of Arctic Biology, University of Alaska Fairbanks, Fairbanks, AK 99775, USA
e-mail: gary.kofinas@uaf.edu

F.S. Chapin et al. (eds.), *Principles of Ecosystem Stewardship*,
DOI 10.1007/978-0-387-73033-2_3, © Springer Science+Business Media, LLC 2009

between well-being, livelihoods, and natural and social capital can define the prospects for long-term sustainability (Janssen and Scheffer 2004). **Vulnerability,** the degree to which a system is likely to experience harm due to exposure to a specified hazard or stress (Turner et al. 2003, Adger 2006), and **resilience**, the capacity of a system to respond to and shape change and continue to develop, are important concepts in understanding these relationships (see Chapter 1). **Vulnerability analysis** has typically focused on the potential threats (e.g., climate change) and endowments (e.g., assets, social capital) that affect livelihoods and well-being (Turner et al. 2003), whereas resilience thinking has emphasized the capacity of groups to cope with ecological change and avoid dramatic and undesirable social–ecological shifts (Walker and Salt 2006). We integrate the vulnerability and resilience approaches in this chapter to provide a framework for understanding livelihoods and well-being in the context of ecosystem stewardship and argue that well-being based on adequate livelihoods is essential to reduce vulnerabilities to the point that people can engage in long-term planning for ecosystem stewardship. We also note that, because stakeholders differ in their values, perceptions of needs, and capacity to fulfill them, ecosystem stewardship requires a participatory place-based approach to improve well-being, plan for change, and cope with inevitable surprises (Maskrey 1989, Smit and Wandel 2006).

From Human Ecology to Sustainable Development to Social–Ecological Resilience

The context for understanding social–ecological interactions is shifting from studies of ecological effects on society (and vice versa) to development planning that presumes predictable effects on society to resilience-based stewardship that seeks to respond to and shape a rapidly changing world. The anthropological subdisciplines of human ecology, cultural ecology, and political ecology have studied the coevolution of human–environment relations

and their effects on the well-being of human groups. These examinations have helped us to understand how ecosystems shape social organization, economies, and the exploitation of ecosystems and how human exploitation affects ecological conditions (Rappaport 1967, Turner et al. 1990). Anthropologists point out that, although each case is unique, there are common patterns that characterize types of livelihood strategies, such as sharing among hunting and gathering societies, specialized division of labor among family members in pastoralism, and specialization of technical skills in urban-based capitalism. The study of changes in these livelihood strategies has highlighted the current rapid changes in social complexity and emerging issues of disparity both within and across groups, particularly between those of developing and developed societies.

In the 1960s and 1970s, developed countries experienced a transformation in their thinking about resources use and environmental quality (Repetto 2006), in part because of the awareness raised by the books *Silent Spring* (Carson 1962) and the Club of Rome's *Limits to Growth* (Meadows et al. 1972). These books and several large environmental catastrophes, such as oil spills, captured the public's attention through mass media (Bernstein 2002). The global scope and consequences of environmental degradation from pollution and the central role of technology and population growth led to the UN Conference on Human Development in Stockholm in 1972, and later the UN Commission on the Environment and the Economy, which produced the historic document, *Our Common Future* (WCED 1987). Calling for the need to balance economic development with environmental quality, the commission popularized the term **sustainable development**, defining it as the means by which society improves conditions without sacrificing opportunities for future generations (see Chapter 1). The dilemmas of these potentially conflicting objectives have served both as a catalyst for discussions on sustainability, and the basis for fierce debate about the extent to which economic growth can provide for quality of life, while ensuring that ecosystem services are sustained to meet current and future needs. These issues underscore the societal dilemmas represented by rising material consumption and current positive trends of population growth.

A dual focus on social–ecological resilience and well-being puts the debates on sustainable development into a dynamic context, raising questions about the sources of both social and ecological resilience available to groups seeking to shape change and navigate critical thresholds that may affect well-being (see Chapter 5). They also highlight the behavioral traps that may emerge and ultimately impede human development. The focus on vulnerability and resilience adds important insight to these discussions by directing attention to exposure to risks, potentials for shocks and pulses of change, and the capacity of the system to absorb and shape those forces. Although much of the sustainable development debate has focused on issues of well-being and livelihoods in the context of poverty, these issues have equally important implications for all societies and their life support systems. In the section below, we present information on current global trends in quality of life and describe some elements of well-being with relevance to ecosystem stewardship in a changing world. We also present an analytical framework for assessing vulnerability and resilience in the context of sustainable livelihoods, well-being, and social–ecological change.

Well-Being in a Changing World

Well-being integrates many dimensions of the quality of life. Disciplinary scholars, such as economists, anthropologists, and political scientists, typically focus on specific indices of well-being such as levels of consumption and expenditures, availability of employment for cash, sense of cultural continuity, and choice, respectively. Although each is important, so is a broader array, including one's sense of security, freedom, fate control, and good social relations (Levy et al. 2005, MEA 2005a). People differ in the ways that they define a sense of well-being, depending on culture, age, gender, household structure, and other factors. How-

ever, there is surprising agreement throughout the world that this broader array of elements is essential for a good life (Narayan et al. 2000, Dasgupta 2001). The social psychologist Abraham Maslow (Maslow 1943) framed well-being as a hierarchy of motivations, stating that basic physiological needs such as food and water are the most fundamental, followed by perceptions of safety and security, then love and belonging through social connections with family and community, then the need for self-esteem and the respect of others, and finally, self-actualization through pursuits such as creative action and reflective morality. Although Maslow's focus was mostly on individual motivation and well-being, there is an extensive literature linking individual and community resilience (e.g., Luthar and Cicchette 2000) that has logical consequences for social–ecological sustainability. Opportunities for sustainable stewardship increase as more of Maslow's components of well-being are met. In the following sections, we summarize the connections between these components of well-being and ecosystem stewardship.

Material Basis of Well-Being

The material basis of well-being includes people's essential needs, such as clean air and water and adequate food supply. Ecosystem services, generated at regional and local levels, contribute substantially to the material bases of well-being. Failure to meet the basic material needs of life generally has negative implications for ecosystems because people tend to prioritize the fulfillment of basic needs above long-term stewardship goals. In the absence of basic material needs, people are more likely to be trapped in perpetuating conditions of poverty, poor health, and social instability.

In the global aggregate, the *material basis* of well-being has increased substantially in terms of income and food supplies. **Gross domestic product** (GDP—a standard economic measure of income), for example, has doubled in less than 50 years and continues to increase in many regions. Much of the increase in income and associated consumption, however, has occurred in developed nations and affluent segments of society, where consumption reflects desires and preferences rather than the fulfillment of essential material needs. However, in some areas there has been a genuine reduction in poverty. In East Asia income poverty declined by 50% between 1990 and 2005 (Levy et al. 2005). Yet, in about 25% of countries, particularly in sub-Saharan Africa, vulnerability as estimated by the Human Development Index is increasing (Fig. 3.1; UNDP 2003).

Economic disparity between rich and poor nations has increased since the beginning of the Industrial Revolution, a trend that continues today (Levy et al. 2005). In other words, poor countries are becoming poorer, relative to

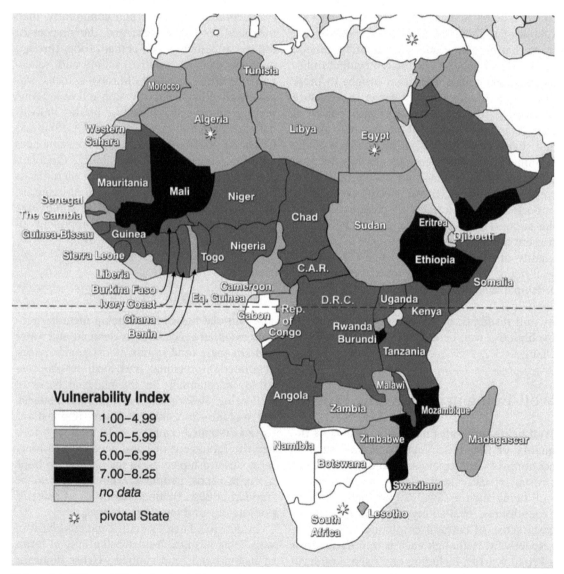

FIGURE 3.1. Vulnerability index for African countries, composed from 12 indicators (food import dependency ratio, water scarcity, energy imports as percentage of consumption, access to safe water, expenditures on defense vs. health and education, human freedoms, urban population growth, child mortality, maternal mortality, income per capita, degree of democratization, and fertility rate. Reprinted from Kasperson et al. (2005).

more developed nations. Within poor countries and some rich countries, economic disparity is also increasing. For example, as population increases, income has declined in low-productivity systems such as drylands, relative to more productive coastal areas (see Chapters 8 and 11). In extremely impoverished areas, people lack the resources to move to cities, which has been the predominant demographic trend in most of the world. Consequently, the greatest human pressure on the environment is occurring in those regions that are particularly prone to degradation of ecosystem services. Of the 20% of the global population that lives in poverty (here defined as living on less than $1 per day), 70% live in rural areas, where they depend directly on ecosystem services for daily survival (MEA 2005a). Large disparities between rich and poor can also create political instability and low resilience to shocks. These same regions of poverty and economic disparity are commonly subject to violent conflict, severe hardship, and chronic rates of high mortality.

Poverty and its implications for the degradation of ecosystems services can also have cascading and long-term effects on most other components of well-being. As people spend more time meeting their basic survival needs for food, for example, they have less energy to pursue education and other opportunities. Inadequate nutrition and education, in turn, can impact health and social relations. In about 25% of countries, the decline in natural capital has been larger than increases in manufactured plus human capital, indicating a clearly unsustainable trajectory associated with declines in assets for production (see Chapter 1). Many critical ecosystem services (e.g., pollination services, pest regulation, maintenance of soil resources) appear to be declining (Foley et al. 2005), although data gaps make it difficult to assess overall trends. This widespread decline in ecosystem services worldwide clearly threatens the future well-being of society, particularly the disadvantaged, and widens the gap between haves and have-nots.

The material basis for well-being is important but to some extent subjective. Below a per capita income of about $12,000, there is a strong correlation between the wealth of nations and the average happiness of their citizens (Diener and Seligman 2004). Similar correlations are observed within countries. There is, however, no significant increase in happiness once an individual's basic material needs are satisfied (Easterlin 2001). As people acquire greater wealth, they aspire to achieve even greater wealth, which, in turn, seems to reduce their happiness and overall satisfaction. In addition to the pursuit of consumption for its own sake, other factors become important once basic needs are met, such as the quality of social relations or one's sense of fate control. These findings suggest two basic approaches to achieving a better match between the flow of ecosystem services and the material needs of society: (1) assure that the basic material needs of poor people are met and (2) reduce the upward spiral of consumption by people whose material needs are already met.

Safety and Security

Perceptions of insecurity and risk vis-a-vis ecosystem services affect decisions that govern well-being. High risks of starvation, disease, armed conflict, economic crises, or natural disasters, for example, have psycho-social implications for all aspects of life and inevitably undermine society's commitment to ecosystem stewardship. Some hazards and risks have declined, and others have increased.

Stable livelihoods are one of the strongest determinants of safety and security. Lack of food security, in particular, is a critical component of well-being (IFPRI 2002; see Chapter 12). Families with insufficient income to buy food and no suitable land to grow crops are more vulnerable to environmental hazards that influence food supply than are people with greater access to these assets. Although the early literature on risk and vulnerability focused primarily on exposure to environmental hazards, there is increasing recognition that livelihood security is generally a stronger determinant of vulnerabilities and risks such as famine than are climatic events such as floods and droughts (Sen 1981, Adger 2006).

Clearly, basic material resources, security, and well-being are tightly linked.

Disease risk interacts with other components of well-being. In aggregate, human health has improved substantially. Reductions in infant and child mortality have contributed to a doubling of average life expectancy during the twentieth century. The associated increases in human population, however, augment demands for ecosystem services and nonrenewable resources and therefore rates of ecological change (Fig. 3.2). Changes in land use, irrigation, climate, and human settlement patterns in turn contribute to spread of infectious diseases that now constitute 25% of the global burden of disease (Patz et al. 2005). Most of these diseases are treatable, but ecological changes frequently increase the abundance or proximity of disease vectors, leading to high infection rates that overpower the medical capacity for treatment. Deforestation, for example, often brings forest animals in contact with livestock, increasing the risk of transfer of wildlife diseases to livestock and then to people; about 75% of human diseases have links to wildlife or domestic animals (Patz et al.

2004). Malaria, schistosomiasis, and tuberculosis are widespread despite active control programs. Other diseases such as trypanosomiasis and dengue in the tropics and Lyme disease and West Nile Virus in temperate countries are emerging health risks that appear related to climatic and ecological change. Current trends of climate warming create significant uncertainty about future trajectories of disease.

Several current global trends place increasing pressure on natural resources, adding risk to livelihoods and well-being: (1) an increasing human population, (2) a decline in basic environmental resources such as clean water, and (3) increasing demand for energy. The resulting increase in competition for renewable and nonrenewable resources substantially increases the likelihood of local and global conflicts. Oil-producing states, for example, now host a third of the world's civil wars (up from 20% 15 years ago), and half of the oil-exporting countries (OPEC) are poorer than they were 30 years ago (Ross 2008). Although there are international organizations (e.g., the UN), national and non-governmental organizations (NGOs), and treaties that could reduce the likelihood of conflicts, these are not always effective in the absence of basic social order. In summary, the risks of environmentally induced conflict are increasing, necessitating the adaptation and transformation of governance processes to alleviate these risks (see Chapters 4 and 5).

Many factors contribute to overall risk and insecurity, leading to social–ecological vulnerability. The impacts of climate change, for example, depend strongly on poverty and well-being. Sub-Saharan Africa will likely be much more vulnerable to climate change than will the drylands of Australia or the USA, even where the climatic changes and local ecology might be relatively similar. These differences arise out of the vastly different endowments that are available to people to deal with these challenges, for example, government services, better infrastructure, and more personal assets. The effects of drought on food supply may also be dramatically increased by civil strife. Wars accounted for about half of the major famines in the twentieth century (Hewitt 1997). Because vulnerability is always a multifactor social–ecological

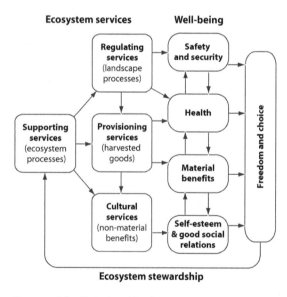

FIGURE 3.2. Relationship between ecosystem services and well-being. Adapted from the framework developed by the Millennium Ecosystem Assessment (MEA 2005d).

problem, potential solutions should be place-based and pay careful attention to local context and slow variables such as climate, property rights, and livelihoods (Turner et al. 2003, Adger 2006).

Despite the complex nature of vulnerability, repeatable syndromes emerge (Petschel-Held et al. 1999). The Sahel and many other arid zones, for example, are characterized by a set of processes that result in the overuse of agriculturally marginal land and make people vulnerable to various hazards, including drought, war, and epidemics of AIDS/HIV. When patterns of correlation among 80 social and environmental symptoms of global change were examined, about 16 syndromes emerged that represented repeatable sets of interactions among these symptoms. These syndromes were categorized based on how humans were using nature: as a source of resources, a medium for economic development, and environmental sinks for human pollutants (Table 3.1). They provide a basis for identifying intervention strategies that might apply to broad categories of social–ecological situations. A combination of soil-conserving agricultural techniques and poverty reduction actions, for example, were thought to

be most promising in reducing vulnerability in the Sahel of Africa and other similar social–ecological situations (Petschel-Held et al. 1999).

Good Social Relations

Good social relations involve social cohesion, mutual respect, equitable gender relations, strong family associations, the ability to help others and provide for children, and the ability to express and experience aesthetic, spiritual, and cultural values. These higher-order motivations (good social relations, self-esteem, need for mutual respect, and actualization) in Maslow's hierarchy are critical to ecosystem stewardship. They are the basis for **social capital** — the capacity of a group to work collectively to address and solve problems (Coleman 1990). One tangible measure of social capital is the extent to which social relations provide access directly to resources or to those with skills not held by individuals or members of a household. Social relationships and social capital are typically slow to build, but can deteriorate quickly, especially in times of rapid social or environmental change. Individuals and organizations interact through **social networks,** the

TABLE 3.1. Major patterns of social–ecological interactions grouped by the nature of human use of nature: as a source for production, as a medium for socio-economic development, or as a sink for outputs of human activities. Names are derived from either prototypical regions or catchwords for characteristic features. Adapted from Petschel-Held et al. (1999).

Nature as a source for production	
Sahel syndrome	Overuse of marginal land
Overexploitation syndrome	Overexploitation of natural ecosystems
Rural exodus syndrome	Degradation through abandonment of traditional agricultural practices
Dust bowl syndrome	Non-sustainable agro-industrial use of soils and bodies of water
Katanga syndrome	Degradation through depletion of nonrenewable resources
Mass tourism syndrome	Development and destruction of nature for recreational ends
Scorched earth syndrome	Environmental destruction through war and military action
Nature as a medium for socio-economic development	
Aral Sea syndrome	Damage of landscapes as a result of large-scale projects
Green revolution syndrome	Degradation through the transfer and introduction of inappropriate farming methods
Asian tiger syndrome	Disregard for environmental standards in the course of rapid economic growth
Favela syndrome	Socio-ecological degradation through uncontrolled urban growth
Urban sprawl syndrome	Destruction of landscapes through planned expansion of infrastructure
Disaster syndrome	Singular anthropogenic environmental disasters with long-term impacts
Nature as a sink for outputs of human activities	
Smokestack syndrome	Environmental degradation through large-scale diffusion of long-lived substances
Waste-dumping syndrome	Environmental degradation through controlled and uncontrolled disposal of wastes
Contaminated land syndrome	Local contamination of environmental assets at industrial locations

linkages that establish relations among individuals, households and organizations across time and space (see Chapter 4). Networks can also serve important functions at the landscape scale, as for example in the regulation of water use for rice production in the Balinese water temple networks (Lansing 2006). It is therefore important to avoid evaluations of well-being and livelihoods that focus solely on individual conditions. The social processes that form the basis of social relations and their underlying social networks (See Chapters 4 and 12) are some of the critical endowments that define the sensitivity of a household, community, and society to risk and vulnerability.

Self-Esteem and Actualization

Fostering self-esteem and providing outlets for actualization of creative abilities and actions motivated by a person's sense of place provide some of the greatest opportunities to enhance ecosystem stewardship. The highest-order elements in Maslow's hierarchy of human motivations, self-esteem and self-actualization, are most fully expressed when people have met their immediate material needs, feel secure in their capacity to continue meeting these needs, and have strong social relations. Wealth, however, is not a prerequisite for these higher-order motivations. Many preindustrial cultures found time and ways to express their relationship with the environment (e.g., Coastal Indians of the Northwest Coast of North America) through art and the celebration of stories that are part of oral traditions. When basic needs are met, there is a sense of fate control, and people feel a sense of group and space to practice and appreciate music, literature, theater, and others arts that embody the importance of people's relationship with ecosystems. These activities help to embed stewardship as a part of culture by providing the social space for reflection on the value of ecosystems and people's relationship with their environment. People's appreciation for their relationship to nature (i.e., a land ethic; Leopold 1949) is often a strong motivation for ecosystem stewardship and sustainability and

can be a powerful stabilizing feedback for managing rates of change.

As noted above, well-being and ecosystem stewardship are defined primarily around issues of social conditions, yet Maslow's work was to a great extent focused on the individual. Indeed, ecological stewardship depends substantially upon individual resilience—the capacity of individuals to cope with change while remaining healthy. Although the connections of individual resilience and ecosystem stewardship may seem distant for professional resource managers, they are nonetheless worth considering when implementing and evaluating specific programs and policies. Where communities face epidemic rates of suicide or substance abuse, addressing issues of individual resilience under conditions of rapid socioeconomic change may prove to be the key to community-scale resource stewardship.

Assessing Vulnerabilities to Well-Being

Vulnerability is the degree to which a system is likely to experience harm due to exposure to a specified hazard or stress (Turner et al. 2003, Adger 2006; see Chapter 1). Social dimensions of vulnerability are intimately tied to elements of well-being, particularly to livelihoods; safety and security; and strong social relations. In this section, we link the concepts of livelihoods and well-being to social–ecological vulnerability in a rapidly changing world (Fig. 3.3). Vulnerability analysis includes three evaluative steps: identifying (1) the type and level of exposure of a system to hazards or stresses, (2) the system's sensitivity to these hazards based on the social and economic endowments available to them, and (3) the adaptive capacity of the system to make adjustments to minimize the impacts of hazards (Turner et al. 2003, Adger 2006).

An important strength of vulnerability analysis is its focus on specific hazards and their interactions and on the differential impacts of these interacting hazards on specific social groups at a specific place. Vulnerability analysis of poor coastal communities of Bangladesh to climate

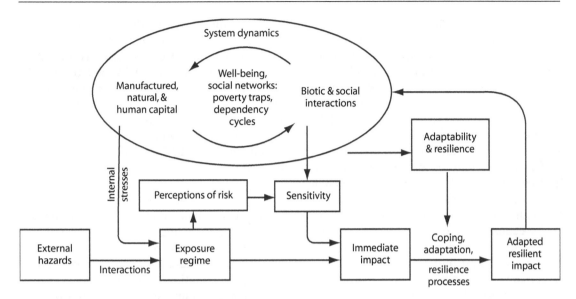

FIGURE 3.3. Components of vulnerability. Vulnerability depends on exposure to external hazards or internal stresses; sensitivity of the system to these interacting stresses; and the adaptive adjustment in response to the resulting impacts.

change, for example, would consider the intensity and frequency of storm surges (exposure); the capability of local infrastructure to withstand storms and flooding (infrastructure sensitivity); and the capacity of the community to mobilize resources in anticipation of the storm and capacity of aid workers to arrive at the scene after the incident has resulted in damage (social sensitivities). The goal of vulnerability analysis is to provide the type of information necessary for targeted intervention strategies that foster adaptation to changing conditions.

Exposure to Hazards and Stresses

Social–ecological systems experience hazards that depend on both external events and internal dynamics. Exposure is the nature and degree to which the system experiences environmental or socio-political hazards (Adger 2006). **Mitigation** (reduction in exposure to a hazard) is therefore an important strategy for reducing vulnerability. Hazards can be characterized in terms of their magnitude, frequency, duration, and spatial extent (Burton et al. 1993). Exposure can reflect **exoge-**

nous hazards that a system experiences (e.g., heat waves, earthquakes, tsunamis, colonization, wars, novel market forces) or **endogenous stresses** (e.g., food or water shortage, lack of financial resources, high levels of internal conflict; Turner et al. 2003, Kasperson et al. 2005), or both. Internal stresses often reflect scarcity of one or more forms of capital or other components of well-being (Fig. 3.3). Hazards, whether they be exogenous or endogenous, can range from events or pulses that shock the system, such as an earthquake or collapse of a financial market, to chronic stresses that undermine well-being and social–ecological integrity of the system, such as rising sea level from climate change or increased energy cost.

Exogenous Hazards

The increased frequency of extreme events adds to the vulnerability of social–ecological systems. Model simulations suggest that human-induced climate warming contributes strongly to many environmental components of this trend (IPCC 2007a, b; see Chapter 2). Extreme heat killed 35,000 people in Europe in 2003, for example, and extreme droughts are occurring more frequently. Flooding is also occurring more frequently in low-lying countries like Bangladesh. Increased sea-surface temperatures may intensify tropical storms, perhaps contributing to their fourfold increase in frequency. Melting of sea ice makes hunting of marine mammals more hazardous for arctic subsistence hunters (Krupnik and Jolly 2002). Each of these changes exposes people to more severe environmental hazards.

Climatically induced disturbances are integral components of social–ecological systems, so exposure to these disturbances is shaped by both exogenous climate and endogenous social–ecological interactions (see Chapter 2). The frequency and extent of wildland fire, flooding, and pest outbreaks, for example, are increasing in response to climatic warming, but the extent to which people are exposed to these disturbances also depends on management decisions. Frequently, there is public pressure to prevent these changes by increasing effort to suppress wildland fires, building larger flood control structures, and using pesticides to reduce populations of pests and diseases (see Chapter 1). In the short term, these policies may reduce exposure, but in the long term, they increase the risk of larger, more severe disturbances (see Chapter 2; Holling and Meffe 1996). An important arena of policy debate is therefore how best to minimize vulnerability that results from climatically induced increases in exposure to disturbances. In some cases there are potentially simple solutions such as policies that minimize development in floodplains or fire-prone wildlands. In other cases, policies designed to reduce exposure simply shift the vulnerability to certain segments of society (typically the disadvantaged) that occupy the lands with greatest exposure to hazards. Active public discourse on these issues can provide the basis for evaluating vulnerabilities and informing those potentially affected.

Air and water pollution also increase the hazards to which people are exposed. As with climatically sensitive disturbances, disadvantaged segments of society tend to live and work in places where they experience greater exposure to these environmental hazards. In some cases regulatory and market approaches can reduce the release of pollutants into the environment and therefore the exposure of society and ecosystems to their effects. For example, international treaties such as the 1987 Montreal Protocol, which banned the use of ozone-destroying chlorofluorocarbons (CFCs), led to a decline in the production and use of these compounds. As with all long-lived pollutants, there has been a long lag-time in the decline in atmospheric concentrations of CFCs in response to these "cleanup" policies. The rate of ozone destruction by these compounds is only now beginning to decline, 20 years after policy implementation. This indicates the importance of acting early to reduce pollution, as soon as environmental consequences are identified. There is currently active debate and experimentation with market mechanisms and regulatory limits to the quantities of CO_2 that a country or firm can emit. In summary, there are a variety of regulatory and market mechanisms, as well as the necessary technology, to reduce vulnerability to pollution. Unfortunately, there are often strong economic incentives at the levels of nations, regions, and firms not to pay the cost of pollution mitigation or to continue polluting illegally (see Chapter 4). Reducing vulnerability to pollution is thus more of a political and social issue than a scientific challenge.

Economic globalization increasingly links economic stresses in vulnerable nations to global economic fluctuations. Although these links may improve living conditions in the short run, they may also create vulnerabilities as communities become dependent on services supplied by central agencies. Food aid, for example, can eliminate markets for local farmers, leading to a decline in the in-country capacity to produce food (see Chapter 12).

Endogenous Stresses

Endogenous stresses are affected by both ecological and social processes (Kasperson et al. 2005). Declines in ecosystem services (e.g., the productive capacity of soils, the filtration of pollutants by wetlands, and pollination of food crops) exposes people to shortages of food and clean water. Such shortages interact with socioeconomic factors to contribute to the vulnerability of some segments of society, as described in general in Chapter 2, and in greater detail in Chapters 6–14. Similarly, chronic poverty and or the lack of freedom to make lifestyle choices are stresses that contribute substantially to vulnerability of disadvantaged segments of society. *Exposure* to endogenous stresses is difficult to separate from *sensitivity* to these stresses, so we treat both exposure to endogenous stresses and sensitivity as sources of vulnerability in the next section.

Sensitivity to Hazards and Stresses

Sensitivity to hazards and stresses depends on ecological (see Chapter 2) and social sensitivity (previous section on livelihoods and well-being), as well as the degree of coupling between social and ecological components of the system. A group with a high dependence of locally harvested crops and limited interaction with cash markets, for example, may have a high sensitively to drought. Sensitivities can also depend on economic and social conditions (e.g., wealth stratification, strength of social capital, infrastructure availability), or power relationships (e.g., caste systems).

Group members with greater authority and influence generally command more resources and are therefore less sensitive to hazards and stresses. Within a given location, people of different race, gender, and age generally differ in their power to access resources and/or to influence decision making. Women, children, and disempowered ethnic groups, for example, usually have less access to resources, and this difference in livelihood makes them more vulnerable to hazards and stresses. For example, ties to children and social norms associated with travel may limit the mobility of women in some regions of Asia, whereas men have greater mobility to pursue jobs in urban areas. Livelihoods and power relationships are deeply intertwined, and together determine **entitlements**, which are the sets of alternative resources that people can access by legal and customary means, depending on their rights and opportunities (Sen 1981, Adger 2006). Entitlements

depend in part on the availability of resources in terms of built, human, and natural capital and in part on institutions and power relationships that allow people to access resources (see Chapter 4).

Power relationships can vary geographically. People living in remote areas generally have less influence to affect change than those in urban or government centers. Similarly, those in inner city ghettos are marginal to those in more wealthy suburbs (see Chapters 8 and 13). These power relationships can translate into higher vulnerabilities to environmental shocks of people at periphery (e.g., high dependence on central government support for crisis relief) than those at core areas (Cutter 1996; see Chapter 8). Nonetheless, those at the margin may also have options for subsistence harvesting in conditions of scarcity that are unavailable to the rich and poor of the inner cities. The distinctions between exogenous and endogenous stresses are difficult to tease apart in an increasingly connected world, where local groups with sufficient resources can garner regional-to-international support for their needs and interest through use of the media, policy communities, and strong leadership (Young et al. 2006). While a greater connectivity and stronger voice can improve conditions for those at the margin, the power imbalances remain.

Cycles of Dependency

Endogenous stresses can be exacerbated when repeated patterns of dysfunctional social behavior become reinforced through time. Cycles of dependency are often precipitated by welfare or land-use policies that can lure individuals, households, and entire communities away from traditional livelihood practices into systems with disincentives for self-sufficiency, instead of the safety net or development that is intended. In many regions of the underdeveloped world, cycles of dependency have been linked to the philanthropic activities of NGOs, which result in unanticipated consequences. The resulting conditions can entrap individuals and in some cases entire social groups, increasing their vulnerabilities to change. Breaking patterns of dependency is not easy and needs to consider the broader social–ecological context. For example, in the 1880's loss of traditional livelihoods by several Native Americans tribes of the US West and the resulting cycles of dependency resulted from both the tribes' exposure to foreign pathogens (e.g., small pox) and the Manifest Destiny policies of the US government. Because it is impossible to restore fully any social–ecological system, finding a way to escape from these patterns for Native Americans of the US West requires an integration of both traditional and novel solutions. Development of credible and responsive institutions that support self-determination of those affected is an important ingredient in escaping dependency (see Chapter 4).

Social and Environmental Justice

The impacts of social and environmental changes are unevenly distributed both locally and globally, making some regions and segments of society more vulnerable than others. Environmental injustice—the uneven burden of environmental hazards among different social groups (Pellow 2000)—results from differences in both exposure and sensitivity. For example, the arctic, mountains, and drylands, where climatic change is occurring most rapidly and climatic extremes shape local culture and adaptation, are more exposed to the risks of climate change than are some other regions (McCarthy et al. 2005). Similarly, there is a net transport of persistent organic pollutants to high latitudes, where they are further concentrated through food chains in animals harvested for food by indigenous residents (AMAP 2003). Local variations in exposure also occur, especially for point sources of contaminants and pollution and for risks such as flooding that are locally heterogeneous. Disadvantaged segments of society are often most exposed to environmental hazards because they lack the resources to avoid exposure to existing hazards or lack the political influence to prevent hazards from increasing.

Those groups that are most exposed to environmental hazards are also disproportionately sensitive to these hazards, because they lack the material resources and safety from risks to cope and adapt effectively. These same individuals are also frequently disengaged from and in many cases disillusioned with the political process, or lack the power to make their voices heard. Consequently, addressing the issues of social and environmental injustices generally requires a concerted and personalized effort to engage stakeholders in the decision-making process to solve problems.

Sources and Strategies of Resilience and Adaptive Capacity

Sources of resilience in social systems provide people with the means to buffer against change. Sources of resilience can be material, social, cultural, ecological, and intellectual. Resilience can also follow from a group's capacity to innovate in the face of new or rapidly changing social–ecological conditions. The central issue here is to avoid innovations and adaptations socially and economically that provide short-term benefits at the cost of the longer-term capacity of ecosystems to sustain societal development.

Spectrum of Adaptive Responses

Social–ecological systems exhibit a broad spectrum of responses to hazards and stresses. These responses range from the immediate impacts of hazards and stresses without adaptation, to short-term coping mechanisms, to adaptation or transformation, in which adaptive responses foster favorable long-term social–ecological adjustments. As described in the previous section, the **immediate impacts** of hazards and stresses are a product of interactions between the exposure and sensitivity of the social–ecological system. Although, by definition, immediate impacts involve no adaptation, they differ regionally and among segments of society largely due to differences in poverty and well-being.

Coping is the short-term adjustment by individuals or groups to minimize the impacts of hazards or stresses. Coping mechanisms enable society to deal with fluctuations that fall within the normal range of experience (Fig. 3.4; Smit and Pilifosova 2003). Coping sometimes occurs by drawing on savings or social relationships, for example, by harvesting animals at times of drought or famine, borrowing from neighbors or family after a flood, or relying on extended social networks during a regional crisis. However, stresses that occur over a sustained time period can lead to a decline in the availability of these resources. Alternatively, an increase in the population that is dependent on these resources constrains a system's coping capacity and the range of conditions with which it can cope (Smit and Wandel 2006). Other factors that reduce the coping range include external events and political factors (e.g., wars or loss of a leader) and increased frequency of stresses (e.g., drought).

Adaptation is a change in a social–ecological system that reduces adverse impacts of hazards or stresses or takes advantage of new opportunities (Adger et al. 2005). In this way, adaptation acts as a stabilizing feedback that confers resilience and reduces vulnerability. Adaptation occurs through social learning, experimenting, innovating, and networking to communicate and implement potential solutions (Fig. 3.5). These processes depend critically on patterns of social interactions and group behavior, which are the main subjects of Chapter 4. **Adaptive capacity** refers to preconditions that

enable adaptation. Adaptations can differ in many ways. For example, they can be reactive, based on past experience, or anticipatory, based on expected future conditions. Adaptation can also be local or widespread; technological, economic, behavioral, or institutional (Smit and Wandel 2006). There is a continuum in the degree of adjustment and complexity of adaptation ranging from simple coping strategies by individual actors to complex system-level changes. Some authors prefer to treat the entire spectrum of adjustment as "adaptation" (Gallopin 2006). Substantive adaptation often requires a substitution among forms of capital or a reorganization of the institutional frameworks that influence the patterns of use of manufactured, human, and natural capital. Hence, adaptation is a continuous stream of activities, actions, decisions, and attitudes that inform decisions about all aspects of life and that reflect existing social norms and processes. Adaptation for resilience requires directing a social–ecological system in a way that provides flexibility and responsiveness in dealing with disturbance and that allows a way to take advantage of the latent diversity within the system and the range of opportunities available (Nelson et al. 2007).

Transformation is the conversion to a new, potentially more beneficial, state with new feedbacks and controls when existing ecological, economic, or social structures become untenable (Fig. 3.5; see Chapter 5). Transformation in the social context may include radical changes in governance, such as the American Revolution or the fall of the Soviet Union or shifts from an extractive to a tourism-based economy. There is often a tension between efforts to fix the current system and a decision to cut losses and move to a qualitatively different structure (see Chapter 1; Walker et al. 2004). This is discussed in Chapter 5 in the context of social–ecological transformation.

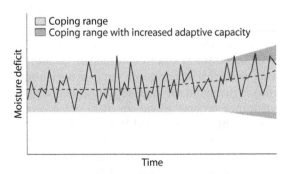

FIGURE 3.4. Coping range and extreme events. Redrawn from Smit and Wandel (2006).

Fostering Diversity: Seeds for Adaptation and Renewal

Diversity provides the raw material or building blocks on which adaptation can act (Fig. 3.5).

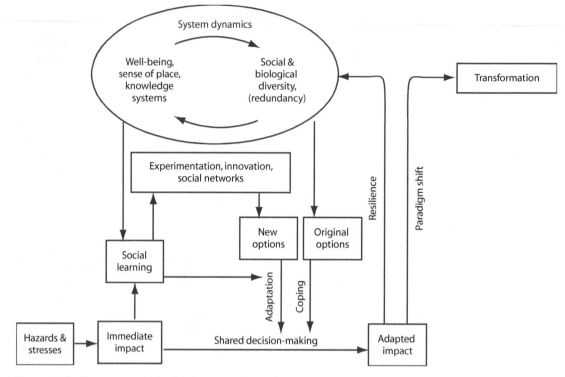

FIGURE 3.5. Components of adaptive capacity and resilience. Adaptive capacity depends on capacity to cope with the normal range of variation in hazards and stresses and the capacity to adapt through social learning, experimentation, and innovation.

It increases the range of options, at least some of which are likely to be successful under whatever new conditions arise, thereby reducing the likelihood of radical degradation of the system. It also provides **redundancy**, that is, multiple means of accomplishing the same ends, which augments the likelihood that valuable functions will be retained during times of rapid change. Chapter 2 describes the role of biological diversity in fostering resilience and adaptation. Economic and social components of diversity can also be important. The role of diversity in social–ecological resilience is well-illustrated in economics. Economic diversity reduces vulnerability and facilitates adaptation to economic or environmental change (Chapin et al. 2006a). For example, a nation whose economy is overly dependent on a single product is vulnerable to changes in the demand for, or value of, that product. Economic diversity can be enhanced by incentives that encourage entrepreneurship and initiation of new types of businesses. This role of subsidies contrasts strikingly with their more typical role in supporting production in sectors that have been adversely affected by climatic, ecological, or economic change (see Chapter 10). Agricultural subsidies in Europe and the USA, for example, typically support the continued production of crops that would otherwise not be able to compete on the global market with foods produced by countries with lower labor and production costs. In some cases, subsidies for non-competitive sectors of the economy are maintained simply because of the political power of groups that benefit from them. In other cases, there may be good reasons to subsidize traditional sectors. Scandinavian countries, for example, support agriculture in part to maintain an agricultural capacity as insurance against global or regional conflicts that might interrupt food imports. Agriculture or other traditional activities may also be subsidized for their cultural value to society, as in the case of agriculture in northern Norway.

Once decisions have been made to provide incentives to foster economic innovation and diversity, there are many potential mechanisms to accomplish this, including "carrot and stick" combinations, voluntary agreements, and tax incentive schemes (Baliga and Maskin 2003). Investment in education and infrastructure can increase the capacity of local residents to diversify their economy and explore options. Economic diversification potentially enhances resilience because it is unlikely that all economic activities would be equally sensitive to any given change in the natural environment or in economic and political conditions.

Diversification of livelihoods is a common strategy among groups in anticipation of stressors and shocks to the system (Ellis 1998). Most rural peoples have considerable experience managing exposure to risk that occurs within "normal" levels (Fig. 3.4). The development of diverse portfolios of livelihood strategies among members of a household or kinship group is common. Diversification may include, for example, planting a variety of crops because of uncertainty in crop yields or markets. Another form of diversification is the cash employment by one or more members of a household in an off-site enterprise while remaining members work family-farm operations. A similar role diversification is found among some contemporary hunting cultures where some family members work full-time jobs while others are full-time subsistence harvesters, sharing their take with the extended family.

Diversification can include strategies other than increasing sources of cash. Households may build their resilience by intentionally extending their social networks to new groups that can provide additional resources or potential support in times of hardship. It is noteworthy that diversification can be used by both the wealthy and the poor and can contribute to equities and inequities. The elimination of constraints that allow for choice in diversification can enhance adaptive capacity and reduce vulnerabilities (Ellis 1998). Diversification of livelihoods may broaden the capacity of a group to cope with change (Fig. 3.4), but its effectiveness may be limited if the shocks to the system are extreme. The capabilities of rich versus poor households to diversify are also categorically different (BurnSilver 2009). Well-off families usually have the benefits of larger asset portfolios on which to diversify their activities, and this can lead to greater wealth stratification between wealthy and poor over time.

The role of cultural diversity in social–ecological resilience is complex. As with economic and biological diversity, cultural diversity provides a range of perspectives and approaches to addressing problems (see Chapter 4). For example, industrial fishing typically uses nets with a single mesh size capable of catching all large fish, while allowing small fish to escape. In contrast, Cree Indians in Canada use mixed-mesh nets that allow escapement of a range of size classes. The single-mesh nets are much more efficient at catching valuable fish. However, large fish are also disproportionately important in reproduction so, as fishing pressure increases, the loss of large fish can decimate the fishery (see Chapter 10). The diversity of fishing approaches provided by two cultures proved important in conserving the fish on which the Cree depended (Berkes 1995). Sometimes, however, cultural perspectives are so radically different or are so strongly rooted in historical conflict that cultural diversity creates more challenges than opportunities, especially where institutions for shared decision making are lacking. The major point here is that cultural diversity broadens the range of perspectives and experiences with which to address change.

Institutional diversity provides a range of mechanisms for fostering innovation and implementing favorable outcomes and will be discussed in Chapter 4.

Legacies and Social Memory: Latent Sources of Diversity

Social legacies and social memory contribute to resilience by providing insight into the alternatives for responding to disturbance and change. **Social legacies** are the lasting effects of past events affecting current social conditions. **Social memory** is the collective memory of past experiences that is retained by groups (Folke et al. 2003; see Chapter 4). Because it endures longer than the memory of individual people, it contributes strongly to resilience and sustainability. Social memory includes both the written record (e.g., books, articles, reports, and regulations) and oral traditions and stories of events, strategies for coping with those events, and their lasting lessons. Thus, social memory provides a wealth of ideas on how social–ecological systems and their components have responded and adapted to changes in the past. This memory extends the range of potential future options beyond those that dominate the current system (Fig. 3.5). For example, the stories and memories of elders of many cultures provide historical accounts of coping strategies as well as insights that inspire innovation in problem solving. Social memory can be particularly important in times of crisis, when current options, by definition, are insufficient, and a broader range of possibilities is essential to discover pathways to favorable outcomes (see Chapter 4). Social memory constantly evolves in response to system change. Social memory is important because it provides a context in which to frame novelty and change (Taylor 2000).

People who are negotiating social–ecological systems that are subject to disturbance and change may draw on their internal social memory as sources of resilience for renewal and reorganization. Others may use diverse memories and recombine them in ways that lead to social–ecological transformations through, for example, new inventions or leadership with new visions or worldviews.

Individual actors, communities, and whole societies may successfully adapt to changing social and economic circumstances in the shorter term, but such an adaptation may be at the expense of the essential capacity of ecosystems to provide support in the longer term. For example, the unprecedented expansion of human activities since the Second World War made possible through the exploitation of fossil energy has in many respects been an impressive human adaptation, with numerous local and regional transformations (see Chapter 14). However, the cumulative effects of these changes have contributed to global climate change in the present.

To what extent can social memory contribute to such adaptations? People's knowledge of resource uses and ecosystem dynamics is often tacit or implicit, linked to norms and rules, embedded in rituals and social capital, and framed by worldviews and belief systems. This is true for both **traditional knowledge** and the **local knowledge** of contemporary societies. It is also true for the more formal science-based approaches to knowledge production. Having the capacity to tap into a group's local and traditional knowledge, while cultivating its development, is fundamental for integrating livelihoods and well-being into resilience-based ecosystem stewardship. In many local communities there are management practices with tacit ecological knowledge that respond to ecosystem feedbacks in ways that can enhance social–ecological resilience (Berkes and Folke 1998a, Berkes et al. 2000). Yet as sources of resilience, legacy and memory are subject to interpretation. Indentifying key individuals in a group who hold a rich understanding of the events of history and facilitating group processes that lead to reflection on past experiences can provide access to these sources. Through social learning, actors, networks, and institutions can integrate traditional knowledge with current practices to find new ways to collaborate (Schultz et al. 2007).

Innovation and Social Learning: Creating and Sharing New Options

Learning to cope with uncertainty and surprise is a critical component of adaptive capacity. Individuals and groups are generally familiar with short-term variability in the hazards and stresses to which they are exposed and have coping mechanisms and adaptations that do not require learning new things. However, when stresses or conditions change directionally (or variability increases) and move beyond past experience, people may lack the social memories and legacies to formulate an effective response. Learning is therefore essential to deal with **surprises**—unanticipated outcomes or conditions that are outside the normal range of variability. There are at least three potential responses to surprise (Gunderson 2003): (1) **No effective response** occurs when inertia or ineffectiveness in resource management prevents a response or when vested interests block changes that are attempted. This preserves the *status quo* and maintains or augments vulnerability. (2) **Response without experience** occurs if actions are taken in response to novel circumstances or if institutions fail to incorporate understanding gained from previous experience. The outcomes from inexperienced responses can be good, bad, or neutral. (3) **Response with experience** occurs when **social learning** has retained insights gained from previous experiences (Fig. 3.5; Lee 1993, Berkes and Folke 2002). It occurs through the retention and sharing by groups and organizations of knowledge gained from individual learning (see Chapter 4). Social learning is therefore important in increasing the likelihood that the system will respond based on experience and understanding of system dynamics. Social learning can occur in many ways. Considerable learning occurs through the cumulative experience gained as local knowledge by observing, managing, and coping with uncertainty and surprise. It can also occur through scientific research or by reading or hearing about the experiences of others. Local knowledge is critical for effective implementation of new solutions because it provides information about local conditions or "context" that determine how to implement new actions effectively. Local knowledge also incorporates people's **sense of place** reflecting how people relate to the social–ecological system in which they are embedded and therefore how they are likely to respond to, or participate in, potential new approaches to ecosystem stewardship.

Social learning that contributes effectively to social–ecological adaptive capacity almost always requires the integration of multiple perspectives and knowledge systems that represent different views about how a system functions (Fig. 3.5). Examples of complementary perspectives and knowledge systems include the natural and social sciences; academics, practitioners, and local residents; and Western, Eastern, and indigenous knowledge systems.

Resilience learning is a form of social learning that fosters society's capacity to be pre- **pared for the long term by enhancing its capacity to adapt to change while maintaining sustainability**. Like other forms of social learning, resilience learning benefits from cumulative experience, learning from others, and integration of diverse perspectives and knowledge systems. However, resilience learning is particularly challenging because it requires that individuals collectively consider the interactions of different components of the entire social–ecological system, rather than a single aspect such as forest productivity or poverty, which are complex suites of issues in their own right. The core of resilience learning is developing the social mechanisms to make decisions that enhance long-term sustainability and resilience under conditions of uncertainty and surprise. This is radically different than a paradigm of postponing decisions until the group believes that it knows enough not to make mistakes. The latter approach is never sufficient for addressing long-term complex issues and is often used as an excuse for avoiding short-term risks and maintaining the *status quo*. The decision *not to take action* is just as explicit a decision as acting decisively.

There are at least two important components of resilience learning: The first is to reduce uncertainty and improve understanding of the dynamics of social–ecological systems under conditions of variability and change so as to explore new options. This can occur through observation, experimentation, and modeling (see Chapter 4). The second is the process of social learning that requires **framing** the issues of sustainability and resilience in a context that conveys its value to the public and to decision makers (Taylor 2000).

Research on complex adaptive systems is challenging because everything changes simultaneously, and you can only guess what future hazards and stresses may be. There are, however, logical approaches to improved understanding of resilience by (1) identifying critical slow variables, (2) determining their responses to, and effects on, other variables (often nonlinear and abrupt); and (3) identifying processes occurring at other scales that might alter the identity or dynamics of critical slow variables.

There are typically only a few (about 3–5) slow variables that are critical in understanding long-term changes in ecosystems (Gunderson and Pritchard 2002). Critical slow variables are frequently ignored, assumed to be constant, or taken for granted by policy-makers, who tend to focus on the fast variables that fluctuate more visibly and are often of more immediate concern to the public. The dynamics of ecological slow variables were discussed in Chapter 2. The identity of critical **social slow variables** is currently less clear but probably includes variables that constrain well-being and effective governance. Critical social slow variables are often stable or change slowly over long periods of time, but may change abruptly if thresholds are exceeded. For example, property rights of many pastoral societies in Africa changed abruptly as a result of European colonization and policy debates over how to promote pastoral development, causing changes in land use, social disruption and declines in well-being (Mwangi 2006). Oil development or building fences to create game parks for ecotourism can today cause similar disruption to pastoral livelihoods. Many of the factors that disrupt well-being were discussed earlier in this chapter, and factors influencing governance are addressed in Chapter 4.

Interactions among key variables operating at different scales are also important causes of abrupt changes in social–ecological systems (Gunderson 2003). For example, interactions among changes in global climate and local population density may cause soil loss in drylands to exceed the changes that would occur based on either factor acting alone. Research and awareness of processes occurring at a wide range of scales (e.g., the dynamics of potential pest populations to behavior of global markets) can reduce uncertainty and enhance resilience (Adger et al. 2005, Berkes et al. 2005). Often these cross-scale effects are difficult to anticipate, so managing to maintain or enhance resilience of desired social–ecological states may be the most pragmatic approach to minimizing undesirable effects of cross-scale interactions.

Planning for the long term requires an understanding of the tradeoffs between short-term and long-term costs and benefits. Actions that contribute to long-term sustainability often incur short-term costs. For example, managing forests or economies for diversity usually incurs short-term costs in terms of reduced efficiency and profits. Monospecific single-aged forest plantations, for example, can be harvested efficiently, and the harvested material can be transported and processed to produce one or a few products, with substantial economies of scale. These plantation forests are, however, more prone to losses from windthrow, disease, and changes in global markets than are more diverse multi-species stands and therefore lack resilience to long-term changes in these hazards and stresses (Chapin et al. 2007; see Chapter 7). Substantial social learning must occur before people or groups are willing to look beyond the short-term costs and benefits to address long-term resilience and sustainability. In economic terms people tend to discount future value to such a degree that many decisions are based primarily on short-term rather than long-term benefits. Politicians may be interested in maximizing benefits over their term of office (a few years), and community planners seldom think beyond 5–20 years. Resilience learning requires a reassessment of these discount rates and exploration of new options that extend the time horizon over which benefits are valued by society. This is most likely to occur if strategies are developed that maximize **synergies** (win-win situations) and reduce the magnitude of tradeoffs between short-term and long-term benefits (see Chapter 1). Now may be the best time in history to convince society of the importance of long-term thinking because the magnitudes of directional changes that previously occurred over many generations (centuries and millennia) are now evident within a single human lifetime due to the accelerated rates of change and increased human longevity (see Fig. 1.1).

Incorporating Novelty in Social–Ecological Systems

Experimentation and innovation allow people to test what they have learned from observations and social learning. Managers are often reluctant to engage in social–ecological experiments under conditions of uncertainty because of the risk that outcomes will be unfavorable to some stakeholders (see Chapter 4). An alternative approach is to learn from the diversity of approaches that inevitably occur as a result of temporal and spatial variation in land ownership, culture, access to power and resources, and individual ingenuity. Multiple suburbs of a city or school districts within a region, for example, represent different efforts to address a common set of problems. In general, this diversity in approaches leads to improved levels of performance, despite the inefficiency associated with redundancy (Low et al. 2003). Similarly, social–ecological differences between production forests, multiple-use forests, and wilderness conservation areas provide opportunities to learn about resilience to climatic, ecological, and economic shocks and openness to innovation and renewal (see Chapter 7). As expected in complex adaptive systems, those experiments that are successful persist, and the failures disappear. Social learning further increases the likelihood that successful solutions will be adopted by others. Providing opportunities for a diversity of social–ecological approaches spreads the risk of failure and the motivations for success.

Networking increases the efficiency of social learning. Two essential features of adaptation in any physical, biological, or social–ecological system are the presence of a diversity of components and the selection and persistence of those components that function effectively. In social–ecological systems, diversity is created through experimentation and innovation, and selection occurs through the process of **governance**—the pattern of interaction among actors that steer social and environmental processes within a particular policy arena. System-level adaptation requires that successful innovation to cope with surprise be communicated broadly, so other people and groups can avoid the mistakes of unsuccessful experiments and build on the successes. These decisions and innovations take place at the level of livelihood choices of groups and individuals. Identifying successful options requires leadership and appropriate participation to select adaptation strategies and an appropriate governance structure to share information and resources and implement solutions. These topics are central features of Chapters 4 and 5.

Participatory Vulnerability Analysis and Resilience Building

A key challenge of applying vulnerability and resilience frameworks to ecosystem stewardship is developing qualitative and quantitative measures of the framework elements. It is also important to move beyond an examination of the vulnerability of only one aspect of the system (e.g., a fishing community's fishing activities) to a more holistic assessment (Allison and Ellis 2001). The scale of vulnerability analysis must also be appropriate to inform those who are vulnerable. National or global assessments, for example, might show that Bangladesh is more vulnerable than Nepal, which is more vulnerable than Greenland. These top-down approaches to vulnerability analysis provide little direct help to those at the local level. Given the importance of informal social networks for coping with shocks to a system and the tendency of bureaucracies to capture the lion's share of financial resources associated with disaster relief, engaging local communities in the assessment links theory and research with action.

Participatory vulnerability analysis is a systematic process of involving potentially affected communities and other stakeholders in an in-depth examination of conditions. Participatory vulnerability analysis provides a basis for formulating actions that support resilience-building. It can also be the basis for mitigation. The steps for the analysis include assessments of (1) exposure and sensitivities to determine past and current levels of risk, (2) historic strategies and constraints for coping with risks, (3) identification of possible future risks, and (4) development of plans for mitigating and or adapting to future risks (Smit and Wandel 2006). To be effective, the analysis should use multiple sources of knowledge (local and traditional knowledge, modeling techniques, science-based analysis, remote sensing imagery) and address the multiple levels at which local, regional, and national decision makers might be engaged to reduce vulnerability. The objective is not to arrive at a single vulnerability score, but to build the resilience of the system to cope with anticipated and potentially surprising change. Addressing issues that reduce vulnerability and build resilience is usually incremental, so the process must track emerging conditions and make the necessary adjustments. The participatory approach can be helpful in energizing a community into action, but it must be approached with a grounded sense of realism that recognizes the day-to-day demands of local residents (see Chapter 4).

Summary

Well-being based on adequate livelihoods is essential to reduce vulnerabilities to the point that people can engage in long-term planning for ecosystem stewardship. Well-being depends on the acquisition of basic material needs including ecosystems services, as well as other social factors such as freedom of choice, equity, strong social relations, and pursuit of livelihoods. Once these basic human needs are met, people have greater flexibility to think creatively about options for ecosystem stewardship to meet the needs of future generations in a rapidly changing world. There has been a tendency however, for people to seek greater levels of consumption, once their basic needs are met, even though this leads to no measurable increase in happiness. Two important strategies to enhance sustainability are therefore to meet basic human needs and to shift focus from further increases in consumption to social–ecological stewardship, once the basic needs are met.

Social dimensions of vulnerability are intimately tied to elements of well-being, particularly to livelihoods; safety and security; and good social relations. Vulnerability can be assessed by identifying (1) the type and level of exposure of a system to hazards or stresses, (2) the system's sensitivity to these hazards, and (3) adaptive capacity and resilience of the system to make adjustments to minimize the impacts of hazards. Although many systems are experiencing increased stress from exogenous drivers like climate change and globalization of markets, the impacts of these stresses are strongly shaped by levels of well-being and associated endogenous stresses, such as poverty and conflict. Both exposure and sensitivity to stresses vary tremendously with time and place and among social groups. Reducing vulnerability therefore requires an understanding of spatial context and engagement of local stakeholders, particularly the disadvantaged, who might otherwise lack the power and resources to cope with change.

Resilience thinking adds to resource stewardship by addressing the capacity of social groups to cope with change, learn, adapt, and possibly to transform challenging circumstances into improved social–ecological situations.

Review Questions

1. What factors determine quality of life for an individual, household, or social group? Why is it important to understand people's quality of life and livelihood in a region if your goal is to foster ecosystem stewardship?
2. Describe how well-being influences exposure and sensitivity of a group to external hazards and stresses. Why is it often insufficient to simply provide food and shelter to people who have experienced an event such as a hurricane or drought if the goal of the aid is to reduce vulnerability?
3. What economic policies would be most likely to increase regional resilience to uncertain future changes in climate and a globalized economy? How might this be fostered?
4. How do conditions of rapid change enhance and or confound a household and a community's capacity to adapt?
5. What are some of the potential problems and practical applications of vulnerability and resilience frameworks?

Additional Readings

Adger, W.N. 2006. Vulnerability. *Global Environmental Change* 16:268–281.

Berkes, F., and C. Folke. 1998. Linking social and ecological systems for resilience and sustainability. Pages 1–25 *in* F. Berkes and C. Folke, editors. *Linking Social and Ecological Systems: Management Practices and Social Mechanisms for Building Resilience*. Cambridge University Press, Cambridge.

Kasperson, R.E., K. Dow, E.R.M. Archer, D. Caceres, T.E. Downing, et al. 2005. Vulnerable peoples and places. Pages 143–164 *in* R. Hassan, R. Scholes, and N. Ash, editors. *Ecosystems and Human Well-Being: Current State and Trends. Millennium Ecosystem Assessment*. Island Press, Washington.

Nelson, D.R., W.N. Adger, and K. Brown. 2007. Adaptation to environmental change: Contributions of a resilience framework. *Annual Review of Environment and Resources* 32:doi 10.1146/annurev.energy.1132.051807.090348.

Petschel-Held, G., A. Block, M. Cassel-Gintz, J. Kropp, M.K.B. Ludeke, et al. 1999. Syndromes of global change: A qualitative modeling approach to assist global environmental management. *Environmental Modeling and Assessment* 4:295–314.

Smit, B. and J. Wandel. 2006. Adaptation, adaptive capacity and vulnerability. *Global Environmental Change* 16:282–292.

Turner, B.L., II, R.E. Kasperson, P.A. Matson, J.J. McCarthy, R.W. Corell, et al. 2003. A framework for vulnerability analysis in sustainability science. *Proceedings of the National Academy of Sciences* 100:8074–8079.

4
Adaptive Co-management in Social–Ecological Governance

Gary P. Kofinas

Introduction

Directional changes in factors that control social–ecological systems require a flexible approach to social–ecological governance that promotes collaboration among stakeholders at various scales and facilitates social learning. Previous chapters showed that environmental and social changes are rapidly degrading many ecosystem services on which human livelihoods depend. However, simply knowing that degradation is occurring seldom leads to solutions. People have tremendous capacity to modify their environment by changing the rules that shape human behavior, yet much of the conventional thinking on resource management offers limited insights into how to steward sustainability in conditions of rapid change. Consequently, there is a critical need to understand the role of people and their social institutions as mechanisms for negotiating social–ecological change. Designing and implementing appro-

priate resource management in conditions of change requires an understanding of both the processes by which groups make decisions and the mechanisms by which these decision-making processes adjust to change. It also requires moving beyond notions of resource management as control of resources and people, toward an approach of adaptive social–ecological governance.

Social–ecological governance is the collective coordination of efforts to define and achieve societal goals related to human–environment interactions (Young et al. 2008). It is a process by which self-organized citizen groups, NGOs, government agencies, businesses, local communities, and partnerships of individuals and organizations are part of a stewardship process. This process may or may not involve government. In the context of sustainability, social–ecological governance addresses problems of maintaining social and natural assets while sustaining ecosystem services (Folke et al. 2005). *Adaptive co-management* emerges as an approach for meeting this challenge through the intentional efforts at and across multiple scales to understand emergent conditions, learn from experience, and act in ways that maintain the desirable properties of social–ecological systems (Folke et al. 2002, Olsson et al. 2004a, Armitage et al. 2007). The ideas of adaptive

G.P. Kofinas (✉)
School of Natural Resources and Agricultural Sciences and Institute of Arctic Biology, University of Alaska Fairbanks, Fairbanks, AK 99775, USA
e-mail: gary.kofinas@uaf.edu

F.S. Chapin et al. (eds.), *Principles of Ecosystem Stewardship*,
DOI 10.1007/978-0-387-73033-2_4, © Springer Science+Business Media, LLC 2009

co-management build, in part, on the concept of *co-management*, which is the sharing of power and responsibility among local resource user communities and resource management agencies (Pinkerton 1989b, 2003) for more collaborative and coordinated actions, and *adaptive management*, which is an approach to resource management based on the science of learning by doing (Holling 1978, Walters 1986, Lee 1993). A successful adaptive co-management process is (1) flexible and responsive to change, (2) focused holistically on social–ecological interactions, (3) well informed by a diversity of perspectives, (4) reflective in decision making, and (5) innovative in problem solving. It is achieved through networks of decision-making arrangements that are guided by good leadership to effectively link communities of resource users with regional-, national-, and global-scale governance (Olsson et al. 2004b, Armitage et al. 2007). In practice, adaptive co-management is implemented through a range of management activities, such as social and ecological monitoring and data collection; research, model building, and data analysis; habitat protection and impacts assessment; enforcement; resource allocation; education; and policy making. To build resilience, adaptive co-managers must integrate these activities within and across organizational, spatial, and temporal scales to facilitate learning, adaptation, and, where needed, social–ecological transformation (see Chapter 5). In these ways adaptive co-management is particularly suited to addressing conditions of dynamic change.

Realizing successful adaptive social–ecological governance is not easy because it requires collaboration and social learning across multiple scales. Although there is no single governance arrangement appropriate for all resource situations (Ostrom et al. 2007), there are basic concepts, frameworks, and principles that can guide the design, implementation, and evaluation of such a process. This chapter begins by exploring some of the shortcomings of conventional approaches to resource management, then presents six key challenges of adaptive social–ecological governance. These challenges include (1) building responsive institutions that provide for collec-

tive action; (2) finding a good fit of institutions with social–ecological systems; (3) building and maintaining strong community-based resource governance; (4) linking scales of governance for communication, responsiveness, and accountability; (5) facilitating adaptive learning; and (6) generating innovation.

Why Adaptive Social–Ecological Governance?

Decades of research and practical experience provide resource managers with many of the tools necessary to address critical resource challenges, but are insufficient to manage social–ecological systems under conditions of rapid change. Conventional resource management (sometimes called "scientific management") often seeks to maintain ecosystems in a steady state by perpetuating current desired ecological conditions. This approach is based on well-accepted ecological principals and adjusts practices to fit local conditions. When applied appropriately, steady-state resource management often organizes information in ways that efficiently address specific problems, such as compaction and erosion of overgrazed rangelands, site conditions that favor establishment of preferred forest trees after logging, or the maintenance of desired stocking levels of fish or deer (see Chapter 2). These approaches to management have been somewhat effective in addressing short-term variability during the growth and conservation phases of the adaptive cycle, when conditions are either relatively constant or change in a predictable fashion (see Chapter 1). However, under conditions of rapid directional social–ecological change (the release and renewal phases of the adaptive cycle), new conditions emerge for which steady-state resource management is no longer adequate. Under these conditions, adaptive co-management provides mechanisms to adjust to change (Brunner et al. 2005, Armitage et al. 2007). Dietz et al. (2003) liken the challenge of finding appropriate management systems to a co-evolutionary race in which governance systems must keep pace with social, economic and technological changes that increase the potential for human impact on the biosphere, including the ingenious efforts by people to evade rules for conservation.

In many cases, steady-state resource management occurs through top-down control of people and resources by bureaucratically structured government agencies that operate through a set of well-defined rules and regulations (Gunderson et al. 1995, Gunderson and Holling 2002). A command-and-control approach to management tends to focus on environmental and ecological controls and views human behavior as exogenous to ecosystems (see Chapter 6). In a top-down approach, agency personnel typically assume responsibility for enforcement of rules, with controversial decisions sometimes leading to compliance problems, limited trust relations, and protracted conflict. Consequently, the top-down approach can be insensitive and unresponsive to local conditions, human livelihoods, and community concerns.

Conventional approaches to resource management can also become overly focused on "science of the parts," while ignoring more integrated approaches to knowledge production. An integration of the parts requires recognition of the range of social-ecological interactions that can occur and the importance of uncertainty in limiting our capacity to predict outcomes of these interactions (Holling 1998). It also requires a culture of respect for different disciplinary and cultural approaches to knowledge production. In the absence of these considerations, resource managers and resource users frequently face surprises, such as significant loss of resource productivity and shifts in social conditions that previously supported resource conservation (Holling 1986).

In some cases, the strategies of resource management also include a narrow set of indicators of social welfare, employ worldviews that separate people from ecosystems, and assume that it is possible to find substitutes for the loss of ecosystems and the services they generate (Costanza et al. 1997, Folke et al. 1998). Such policies have been inadequate in addressing social incentives, especially at the local level. These approaches have also assumed

a view that technological fixes would resolve future problems. Five "recurring nightmares" can emerge from such inadequate approaches (Yaffee 1997): (1) a process in which short-term interests outcompete long-term visions and concerns; (2) conditions in which competition supplants cooperation because of the conflicts that emerge in management issues; (3) the fragmentation of interest and values; (4) the fragmentation of responsibilities and authorities (sometimes called *functional silos* or *stove pipes*); and (5) the fragmentation of information and knowledge, which leads to inferior solutions.

These nightmares, including the widespread assumption in resource management that systems are approximately at equilibrium and can be maintained in their current condition, are exacerbated in conditions of directional change. Together they suggest that implementing an adaptive co-management strategy is an important way to adjust to change, but one that will be challenging to actors at all levels of the system. Implementation of adaptive co-management therefore requires more than simple public participation in decision making. It also requires collaborative processes of knowledge co-production and problem-solving at multiple scales, with a diversity of parties and incorporating knowledge of ecosystem dynamics. For local resource-based communities, directional changes require an evaluation and readjustment of long-standing rules for interacting with ecological components of the system while at the same time innovating to sustain traditional ways of life. For subnational to national government agencies, directional change requires expanding the scope of resource management to consider a diversity of knowledge and to engage a broad set of public and private actors in making policies at various levels. At the international level, these changes necessitate new forms of governance that assess global-scale phenomena while linking those assessments with regional-to-local processes in social–ecological systems (see Chapter 14). Working toward the objectives of adaptive co-management of ecosystems also requires careful consideration of the match between institutions for decision making and their specific social–ecological conditions. For these reasons, adaptive co-management must be evaluated not simply on its structural features (e.g., centralized to decentralized organization) or by a set of management outcomes, but also by the extent to which its process provides effective feedbacks that contribute to robust human responses in relation to changes in ecosystem dynamics. Achieving these ends requires effective communication among key actors, some level of intergroup cooperation, as well as political know-how and maneuvering. This cooperation, in turn, requires an understanding of the functions of institutions, human organizations, and their social networks in relation to ecosystem feedbacks. Many of these ideas have been developed through the international synthesis efforts, such as the International Association for the Study of the Commons (NRC 2002, Ostrom et al. 2007) and the Institutional Dimensions of Global Environmental Change program (Young et al. 2008).

Building Responsive Institutions that Provide for Collective Action

Institutions, organizations, and social networks affect the feedbacks of social–ecological systems and human well-being. Institutions are socially constructed constraints that shape human choice (North 1991, Young et al. 2008) and are among the most important elements of governance systems. Institutions are "the rules of the game" that assign the roles assumed by individuals and organizations in society, direct the allocation of resources to individuals and organizations, and affect human interactions (Ostrom 1990, Young 2002b). In the language of institutionalists, the rules of the game (i.e., institutions) affect the play of the game (i.e., human behavior, including human–environment interactions). Institutions give rise to repeatable patterns of human behavior that become the political, economic, and social interactions of society (NRC 2002). For example, **property rights**, one of the most important types of institutions, are

those rules that govern access to and use of resources (Bromley 1989, 1992) and thus reflect power relations that affect social–ecological dynamics and resource sustainability. In many respects, institutions define a group's human–environment relations, which in turn affect social relations among individuals and groups (Berkes 1989). Institutions also function as problem-solving devices, providing road maps of established processes by which emerging issues and conflicts are confronted, evaluated, and potentially resolved. The regularity of their use can produce a level of predictability among actors that can result in trust among groups to engage in a range of social practices. To a large extent, institutions dictate the rigidity and/or flexibility of a group's decision making and thus determine the responsiveness of a group to change. Institutions are also historical artifacts of human experience, evolving through time and in a particular place or time. The coevolution of social–ecological conditions in a particular context establishes a **path dependence** of events that shape the direction and transformation of institutional legacies (see Chapter 1; North 1990).

Institutions can be formal or informal, with the **interplay** (interactions among institutions) between the two affecting social–ecological governance. **Formal institutions,** such as constitutions, laws, contracts, and legally based conventions, are typically written rules enforced by governments, such as the US National Environmental Policy Act, the World Trade Organization Free Trade Agreement, and a local county's ordinance on pollution prevention. **Informal institutions** are the unwritten rules of social life, such as sanctions, taboos, customs, traditions, and codes of conduct that are part of the fabric of all groups. Examples of informal institutions include the customary rules for managing sacred forests and the unwritten sanctions for responding to social misconduct (e.g., shunning). Informal rules are always present and in some cases may be inconsistent with formal rules. Because formal and informal rules may emerge both as espoused rules and as rules-in-use (Argyris and Schöen 1978), understanding their respective roles and interplay may be critical in assessing the effectiveness of a social–ecological governance system. For example, formal rules or policies such as a government regulation for land management can be imposed on a group, while informal rules may or may not change as a result of those imposed formal rules. Inconsistency between formal and informal rules can lead to dysfunctional behavior in a governance process, such as conflicts over regulations and enforcement and covert activities, such as poaching. Effective stakeholder engagement, e.g., through co-management, is critical to resolving problems that arise from inconsistency between formal and informal rules.

Institutions play out in the context of **organizations**—social collectives with membership and resources, which thus differ from rules of the game (Scott 2000). Human organizations, whether they are a small community, a corporation, a club, or citizens of a state, have structural and behavioral characteristics that affect their performance in addressing social–ecological governance problems. The organizational environment is the context in which people assume roles of status and authority, amass resources and power, network to form coalitions, and exercise property rights. Openness of membership can vary in ways that affect use of and relationship to resources. In some cases there is an ebb and flow of membership, such as a rural community experiencing boom-bust economic cycles. In other cases membership can be restricted, as in a closed family clan, private club, or nation that limits access to resources. As already noted, some organizations are hierarchical, such as government agencies, others decentralized and uncoordinated such as some community groups, and others are multidimensional, highly complex, interorganizationally linked, and dynamically responsive.

Typically, large organizations with systems of centralized authority function with predetermined procedures for decision making and are less responsive to change than smaller, more decentralized organizations (Table 4.1). They also tend to have greater disparity between formally stated rules and rules-in-use. Bureaucracies are generally predictable in their behavior, and, because of their centralized authority, they can be highly effective in mobilizing abun-

TABLE 4.1. Frequently observed differences between centralized and decentralized organizations.

Centralized organizations	Decentralized organizations
Hierarchical decision-making	Democratic decision-making
Tends to become bureaucratic	Tends to remain flexible
Decision-making is often efficient	Decision-making is time-consuming
Common in large organizations	Common in small organizations
Prone to functional silos	Effective integration of information
Susceptible to agency capture	
Relatively predictable in outcomes	Relatively unpredictable in outcomes
Resistant to change	Potentially responsive to change

dant resources to address a problem, as with a military response. Bureaucracies are also subject to **goal displacement**—a condition in which the survival of the organization assumes greater priority than efforts to meet its stated mission. Where a resource management bureaucracy is focused on the needs of a single group, that agency is subject to **agency capture**, a condition in which a special interest group establishes a controlling relationship, and the agency works almost exclusively on the interest group's behalf. Goal displacement and agency capture can reduce the resilience of the system by limiting the responsiveness of governance to change. Conversely, highly democratic organizations generally require small membership, a high diffusion of knowledge among members, and considerable time to deliberate and make decisions. All organizations are potentially subject to the forces of **bureaucratization**—a process by which, over time, an organization has an increased dependence upon formal rules and procedures. This evolution frequently leads to functional silos and related communication problems and in some cases increased rigidity. The collapse of a social–ecological system can have its roots in these institutional and organizational dysfunctions, as noted in the adaptive cycle described in Chapter 1. In general, centralized organizations work most effectively under relatively constant predictable conditions, where there is broad agreement on management goals, whereas decentralized organizations are more effective in addressing uncertainty and change and therefore in conferring social–ecological resilience.

Individuals and organizations interact through **social networks**, the links that establish relations among individuals and organizations (and their institutions) across time and space. Social networks and their supporting formal and informal institutions distribute information and resources and can be important to the resilience of a community. For example, institutions for the sharing of harvested animals among the !Kung San hunters of the Kalahari ensure survival of the community when only some hunters are successful (Lee 1979). Market networks that distribute goods and services globally can function in a similar manner but can also create vulnerabilities where there is overdependence on external sources without adequate alternatives. The principle of "six degrees of separation," which states that everyone on Earth is separated from any other person by a maximum of six linkages (Watts 2003), shows mathematically how interactions among large groups and individuals can be closely linked. The development and disappearance of linkages generates constantly shifting patterns of change in social networks, as would be expected in any complex adaptive system. Linkages of social networks can vary in strength. The **theory of weak ties** (Granovetter 2004) states that critical information is most commonly received from those outside one's stronger social network, suggesting that resilience is maintained through both **bonding networks** of long-standing familiar relations and **bridging networks** that connect individuals and groups to sustain more generalized forms of reciprocity (Putnam 2000). Weak ties developed through bridging networks can therefore generate novelty and resilience and shape change.

The interplay of institutions, organizations, and networks constitutes the critical social dynamics of a social–ecological governance system. Said another way, human organizations define and modify institutions. Social networks provide critical pathways for information exchange, stimulation for innovation, and resource sharing; and institutions shape individual, organizational, and interorganizational behavior. Social–ecological governance unfolds in highly complex clusters of interactions involving institutions, individuals, organizations, networks, and ecosystems within and across multiple scales. Infrastructure is another critical element in the interactions of social–ecological governance by enabling communication among parties and affecting ecological process and the capacity of groups to respond (Anderies et al. 2004). When these interactions involve a particular area of interest or a particular resource, they are referred to as **resource regimes** (Young 1982; e.g., resource regime for governance of Atlantic cod; see Chapter 11). Although it is possible and sometimes helpful to examine the amplifying and stabilizing feedbacks of a resource regime, it is important to remember that all interactions of a particular institutional arrangement play out as open systems affected by formal and informal as well as internal and external forces.

A focus on the interplay of formal and informal institutions, organizations, infrastructure, and social networks shifts the discussion beyond social–ecological system processes as a mechanistic set of components and feedbacks and places **human agency**, the capacity of people to make choices and to impose their choices on the world, as a central driver of the system. Humans are unique in their ability to transform their social–ecological environments. Through human agency and **collective action** (i.e., the cooperation of individuals to pursue a goal), people hold considerable capacity to change the rules by which they live, mitigate the causes of change, and/or adapt to change. By understanding better the relationship among institutions, organizations, networks, and human choice, we begin to explore how people can intervene to sustain the desired features of their world and how they can learn from experience and work to shape emergent conditions.

No One Model, Discipline, or Solution

There is no one optimal institutional arrangement for sustaining resources and building

social–ecological resilience (Ostrom et al. 2007, Young et al. 2008). There has been substantial debate about the role of institutions in resource management and which institutional arrangements are best suited to achieve the sustainable use of resources. Much of this debate has centered on understanding how to avoid a **tragedy of the commons**, an outcome in which the self-interest of resource consumers and poorly defined property rights result in significant degradation of collectively shared or **common-pool resources** (termed **common-property resources** in the early literature; NRC 2002, Young et al. 2008). These include fresh water, marine fish, migratory species, the global atmosphere, biodiversity, and public lands — resources whose integrity is critical to the life-support system of the planet and therefore to human well-being. Common-pool resources constitute some of the most important natural resources but are highly vulnerable to over-exploitation because of an array of problems associated with securing their sustainability. In the extreme case these problems can lead to a tragedy of the commons (Hardin 1968) if (1) potential users cannot be excluded from resource use — a situation called **open access**, (2) users have the capacity to use the resource more rapidly than its rate of renewal, and (3) all users seek to maximize their self-interest to the detriment of society as a whole (Acheson 1989). Hardin considered the case of individual herders who raise cattle on a village commons (i.e., grazing land shared by all villagers). In his account of the commons, economic incentives and the absence of institutions supporting resource conservation motivate each herder to add additional cattle to his/her herd, even though each addition would lead to a degradation of the village commons at a cost to all. Hardin argued that because of the danger of over-exploitation, all common-pool resources should either be privatized or be managed with full government control.

Although Hardin's model for managing the commons was initially widely accepted, it ultimately generated many questions. What is the role of small-scale institutions in management of a commons? Should common-pool resources be managed by privatizing resource owner-ship or primarily by state government agencies, or should small communities with a dependence on those resources have the lead authority? Are there other alternatives? What are the prospects for public–private partnerships or hybrid management arrangements in providing for robust social–ecological governance? Underlying these questions are others on how to define the nature of human rationality — the logic by which people make decisions in resource management institutions. To what extent are people means-ends oriented when making decisions? When do they pursue individual benefits versus electing to self-sacrifice and act in the interests of their social collective? How does **bounded rationality** (deciding with incomplete information) affect decision making? To what extent are people cost-avoidant? And to what extent do culture and social conditions shape human choice situations?

Finding *the* appropriate solution for social–ecological governance is anything but simple. Extensive research shows that, although there are many models of social–ecological governance and many forms of institutional design that can be applied to a resource problem, there is no single institutional arrangement (or organizational structure) that will guarantee the sustainability of a social–ecological system in all cases (Ostrom et al. 2007). There are simply too may unique variables in each situation that are together too dynamic to allow for a generalized prescriptive solution. This "no one solution fits all" principle has important implications in the assessment of any social–ecological governance problem. First, it demands an in-depth understanding of the particular context to assess conditions (Young et al. 2008). Who are the players? What are their vulnerabilities? What are the values for the resources in question? What is the relationship between the use of the resource and the greater social order? What are the power relationships among players? To what extent do ecological conditions create bottlenecks or opportunities for alternative actions? What are the possible future states that may challenge the system? Simply transferring the "answers" and solutions of resource management from one situation to another may result in a host of unintended consequences.

TABLE 4.2. Lens for assessment of institutional performance.

Lens	Economic	Cultural	Political
Assumptions	Rational choice; utility seeking	Group values and norms guide behavior	Conflicting interests; winners and losers
Key concepts	Incentives; free riding; transaction costs	Membership; identity; worldview	Property rights; interdependence

Moreover, the complexity of social–ecological governance and the limitations of any one frame of analysis underscore the need for an interdisciplinary approach when assessing a resource-governance problem. To address these questions, there is a growing body of literature that integrates multiple approaches to understanding commons management.

The three sections below present economic, cultural, and political lenses for assessing the functionality and performance of resource regimes (Table 4.2). Considered together, they provide an integrated framework for assessing the capacity of a governance system to be responsive to change and ensure important ecosystem services and allow for the integrated understanding of resilient social–ecological governance. In practice they require that analysts see problems from a diversity of perspectives and embrace a variety of disciplinary approaches to understanding problems.

The Economic Lens: Incentives, Transaction Costs, and Social Dilemmas

Institutions set up incentives that affect human behavior, functioning as reward systems that motivate certain behaviors and/or dissuade and punish other behaviors (Bromley 1989). Incentive systems can reinforce individualistic behavior or encourage cooperation and collective action among actors. Institutional incentives can be monetary in nature, work through moral suasion, and/or subject violators of rules to significant penalties.

Institutional arrangements that place the responsibilities for sustainability on individuals without making them sufficiently accountable to the group can lead to the shirking of group responsibilities. The **free-rider problem** occurs when individuals or groups of actors do not assume their fair share of responsibilities while consuming more than their share of the resources. A free rider, for example, is an able-bodied village resident who gleans the benefits of a communal irrigation system but shirks his or her responsibilities to help maintain the system. **Rent seeking** is the seeking of profit or gain through manipulation of an economic or legal system rather than through trade or production. For example, profiteering or monopolization of natural resources as a result of land trades is a form of rent seeking by providing benefits to the individual without contributing to the productivity of society. At modest levels both free riding and rent seeking occur frequently and may have little effect on group behavior or the ecological sustainability of the system. At high levels, they can have dramatic impacts and be indicators of institutional failure. Institutional arrangements can also set up incentives that encourage the discounting of future resource benefits for the pursuit of current resource values. For example, a policy that provides incentives to encourage economic development, such as tax incentives for clear-cutting of a forest, may generate immediate financial benefits to a firm, increase the cash flow, and transfer natural capital into financial capital, but not provide rewards for maintaining the value of those resources for future generations. Such institutions do not account for the **opportunity costs** of their actions—i.e., the potential benefits that might have been gained from some alternative action. Conversely, institutions that provide strong systems of accountability are based on an adequate level of social–ecological monitoring and are oriented toward a multigenerational view of human actions. Designing and transforming institutions to address the social dilemmas between individual and/or group gain versus social benefits are among the greatest challenges in achieving a resilient form of social–ecological governance.

Decision making involves **transaction costs**, such as search costs, information costs, negotiation costs, monitoring costs, and enforcement costs. High transaction costs can be a barrier to the emergence of collective action because they can create difficulties in obtaining information or in monitoring the behavior of resource users or ecological conditions. The regularity and predictability provided by institutions make the work of society more efficient by lowering transaction costs of decision making. Attention to transaction costs reorients rational choice assumptions of human behavior away from an exclusive focus on benefits by recognizing that costliness of information gathering is a key impediment in decision making (Williamson 1993).

Conditions that commonly lower transaction costs in social–ecological governance (and thus improve the effectiveness of a resource regime) include homogeneity of the values and interests among actors and the longevity and multiplicity of relationships among interacting individuals and groups (Taylor and Singleton 1993). Collective action theorists argue that up-front investments in communication and negotiation will lower the transaction costs of enforcement, leading to higher compliance with rules (Hanna et al. 1996). In addition, groups that make up-front investments in detecting and understanding the dynamics of social–ecological change will face fewer transaction costs when confronting the challenges of responding to those changes. Over time, groups that continue to interact can improve their institutions for governance and further develop their capacity for problem solving. The capacity to solve problems and act collectively is termed *social capital*, which provides the basis for responding to unanticipated events (e.g., the emergence of social–ecological surprise). Social capital is an important but unusual slow variable of social–ecological systems. It is typically slow to develop but can erode quickly if there is a breakdown in institutions. It may develop quickly when social actors perceive a common threat. For example, public response to the pro-posed nuclear waste dump in New England led an unlikely coalition of local residents with little mutual understanding or respect to speak in one voice and successfully confront the project proponents.

The Cultural Lens: Identity, Mental Maps, and Ideology

Institutions represent cognitive, symbolic, and ideological conditions embedded within a specific social–ecological context. They are more than simply a set of rules. From a cultural perspective, institutions shape the identity of actors, generate common discourses, and draw individuals into routinized activities that do not involve the calculations (e.g., cost–benefit evaluations) of decisions on a day-to-day basis (Douglas 1986, McCay 2002). Accounting for the cultural dimensions of institutions therefore highlights a suite of considerations that would otherwise be overlooked in a purely economic analysis. For example, language and the rules of interpersonal and intergroup communication shape patterns of communication, which in turn determine *how* issues are discussed (e.g., by what terms); which aspects of those discussions are ultimately deemed legitimate in decision making; and which are marginalized. For example, while some Asian cultures have traditions of indirect communication and deferential behavior in conflict situations, western cultures often reward for directness, competition, and confrontation. The **cognitive or mental maps** of a culture provide blueprints for group thinking that are typically implicit in group transactions. Rules for group membership establish social and organizational boundaries in the transmission of ideas and information and can limit the extent of social networks. These institutional attributes are found in all groups, from the family to the professional organization (e.g., Association of Professional Foresters), to the nation state, and beyond. For example, in resource management, organizational culture in government agencies, such as a national forest service, is typically pronounced and entrenched, and has a significant effect on the interpretation of policies. These **norms** (i.e., rules for social behavior) translate into how staff members act on organizational policies, such as the appropriateness of public participation in decision making or the role of traditional knowledge in research (Mahler 1997). Embodied in the institutions of each culture group is its **worldview**, the framework by which the group interprets events and interacts with its social–ecological system. For example, indigenous hunters and gatherers commonly view animals as sentient beings whose actions can include emotion, revenge, and respect, analogous to human behavior. Resource managers who rely solely on the methods of western science may reject these perspectives as "irrational" because of the problem of validating such ideas with scientific test. Worldview embodies whether human–environment relations convey a sense of dominion over nature or one of deference and risk avoidance.

Negotiating the differences in worldview between groups sufficiently to achieve effective social–ecological governance may require the development of common vocabulary and mutually agreed upon protocols to establish shared visions of problems. A diversity of perspectives can contribute to problem solving. In other cases, these differences are irreconcilable. Problems not withstanding, institutions that allow for the insights of multiple perspectives, while also managing for conflict, potentially add to the resilience of a social–ecological system.

The Political Lens: Power and Power Sharing

Power relations are an unavoidable aspect of social–ecological governance that can either build or undermine social–ecological resilience. **Power** takes many forms such as the imposition of one's will on others, the use of direct and/or indirect influence, the threat or use of harm, the power of personal characteristics such as charisma, the power of argument, the power of information and resource hoarding, and the power of rewards (Lukes 2002). Power can be intentionally imposed or achieved through passive or disruptive behavior. Questions of power in social–ecological governance include who controls the process, who ultimately decides, and who are the winners and losers. Political process may also involve issues of contested meanings. How a group perceives power, either as a finite resource or something that is unlimited and to be shared, often shapes its use. Power generally influences resource management decisions where players depend on one another to some degree, differ in goals and objectives or in values about technologies or the decision-making process, and consider the issue to be important (Pfeffer 1981).

Institutions affect power relations either directly or indirectly. For example, institutions can impose ideological agendas, such as through the implementation of policies that foster the introduction of market economies to third-world nations for economic development. Powerful interest groups can advance policies that privatize collectively held fishing rights and make them transferable, which can indirectly result in a greater ownership of a fishery by the industrial fleet and a crisis of sustainability for local coastal communities (see Chapters 10 and 11). Although power relations are always present, power can also be shared, contributing to collective action and cooperation.

Because sustainability, resilience, and vulnerability are **normative** concepts (i.e., they assume a values orientation), they require a process for defining their meaning and thus raise the political question "Sustainability for whom?". Many of the questions in defining sustainability are social value choices that lack a single "correct" answer. For example, determining a community's critical level of approval for a particular issue (e.g., having a 51 vs 66% majority) is best determined through a collective-choice process rather than scientific analysis. That being said, **equity** in power relations in social–ecological governance generally contributes to social–ecological resilience. Equity fosters resilience primarily by maintaining diversity of opinion, cultural orientation, socio–economic class, and empowering disadvantaged people to contribute to ecosystem stewardship. **Pluralism** in social–ecological governance is an operating principle that affirms and accepts a diversity of perspectives (Norgaard 1989). Institutional diversity can add to resilience by providing an array of approaches to problem solving, which may be critical in unanticipated future conditions. Practices of conflict resolution, such as principle-based negotiation and third-party mediation or arbitration, are helpful tools in avoiding political gridlock when institutionalized as part of the resource-management process (Wondolleck 2000).

In summary, economic, cultural, and political dimensions of institutions, when applied in isolation, provide an incomplete understanding of the system and can lead to different policy outcomes (Young 2002b). Focusing exclusively on economic dimensions may result in a set of unintended consequences that do not account for cultural factors that drive group behavior. Conversely, images of social choice dictated solely by cultural considerations, such as group identity or values, can overlook economic incentives that may lead to unsustainable behaviors. Given that power relations permeate all human decision-making processes and determining goals of sustainability and resilience are at some levels questions of values, attention to the distribution of power highlights the issues of social–ecological justice and the maintenance of diversity. Striving to assemble a multidimensional or pluralist view can bridge these models, leading to more robust forms of governance.

Finding a Good Institutional Fit with the Social–Ecological System

The effectiveness of institutions in governance of natural resources depends to a great extent on how governance fits with physical, ecological, and social conditions. Is the scope of the arrangement a good match for the problems being addressed? Does the geographic scale of a resource regime match with the distribution and movement of the resource? Are institutions adequately sustaining the supporting ecosystem services for important resources? Can the institutional arrangement be responsive in a way that matches expected rates of social–ecological change?

The "problem of fit" can confound efforts to maintain essential ecosystem services,

lead to high transaction costs, rent seeking, and unpleasant surprises (Folke et al. 2007). Achieving a good fit between institutions, social-ecological systems, and their dynamic characteristics requires careful attention to the specific conditions in question about the system, its exogenous drivers, and internal interactions (Young 2002a). Because social–ecological systems are complex adaptive systems with nonlinear dynamics, even the most careful considerations may lead to institutional misfits (Galaz et al. 2008) and surprise.

Policy analysts have traditionally used three terms to evaluate the appropriateness of particular institutional arrangements to ecological conditions—subtractability, exclusion, and congestion (Fig. 4.1). **Subtractable resources** are those that, when used, become less abundant or are degraded. Fish and trees, for example, are subtractable because, when used by one person, they are no longer available for use by others. Sunsets and radio programs, on the other hand, are nonsubtractable because they are still fully available to others, no matter how many people experience them. Many resources are intermediate in their subtractability, depending on how they are used. Water, for example, can be used for hydropower generation and still be available for other functions. Institutions for management of subtractable resources

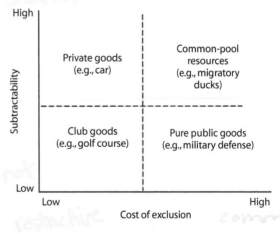

FIGURE 4.1. Institutional arrangements that differ depending on subtractability and cost of exclusion of goods.

should include design characteristics that allow decision makers to be sensitive to the nature of its subtractability, such as potential rates of renewal and nonreversible tipping points.

The degree of subtractability depends in complex ways on many social and ecological variables. The productivity of wildlife available for human consumption, for example, habitat conditions, life cycle bottlenecks in energy budgets, population cycling, and interspecific competition all influence production, so production and consumption are both highly variable and difficult to determine from their current values. To some extent, all resources are coupled to their ecological context, and in some cases they are tightly coupled with specific ecosystem properties (Berkes and Folke 1998b). For example, the cyclical predator–prey relations of hare and lynx constitute an interrelationship between two resources, requiring a holistic approach to understanding ecosystems. Multiple forces for change, such as human harvesting with climate change, add to the challenge. The complex, nonlinear, and chaotic behavior of ecosystems adds to the difficultly in calculating subtractability in natural resources (Wilson 2002). Assessments of subtractability can also be confounded by the limitations of human constructs used to understand biophysical and social conditions (Young 2007). The use of ecological concepts such as carrying capacity or maximum sustained yield, for example, may lead to policy solutions that do not accurately capture ecological dynamics (see Chapters 2, 6, and 11). The resulting misfit of institutions with ecological dynamics suggests the need of a more adaptive approach to social–ecological governance.

Excludable resources are those from which potential users can be excluded at low cost. Neighbors, for example, can be excluded from harvesting potatoes from a private garden. On the other hand, it is more difficult to exclude someone from watching a sunset or harvesting fish from the open ocean. Excludability depends on the transaction costs of the exclusion process. It is much less costly to exclude neighbors from a potato patch than to prevent an industry in another country from polluting the atmosphere, although both types of users could conceivably be excluded. Such costs depend to a great extent on a group's or individual's access to power (e.g., money, influence, authority) over others. Governance that does not account for transaction costs associated with exclusion may result in failures because of unenforceable rules. **Fugitive resources**, such as anadromous fish, migratory waterfowl, and rivers, move across a range of jurisdictions and can have many user groups. Fugitive resources entail high costs of exclusion. In general, institutional arrangements for resources with high transaction costs, such as fugitive resources, require a multilateral approach and a high level of collective action among groups.

Congestible resources decrease in quality when the number of resource users reaches a threshold, for reasons other than ecological productivity. Congestion can be a special management problem for nonconsumptive resource users. For example, one's wilderness experience may be unaffected by other visitors at low visitation levels, but increased visitation and the lack of appropriate institutions can cause congestion, which subtracts from the experience of all visitors. An unacceptable level of congestion is, of course, a societal value judgment. Catch-and-release sport fishing may be ecologically sustainable but not desirable when high numbers of fishermen (combat fishing) appear in a prized fishing location. Nonlocal hunters at low levels may not be problematic to resident harvesters, but at higher levels their presence may subtract from the traditional hunters' ability to transmit knowledge and values about the need to respect and care for land and animals. Congestion problems can be addressed with a policy for **limited entry**, such as requiring permits in zones of use. Limited entry systems can be effective but may entail high administrative costs. Alternatively, reducing (or simply not improving) physical access to use areas, such as eliminating the maintenance of roads, can produce the same outcome without the cost of permits and enforcement.

More recent analyses of institutional fit have expanded on these ideas to address a broader and more complex set of social–ecological

TABLE 4.3. Misfits in Governance Adapted from Galaz et al. (2008).

Type of misfit	Definition	Examples	Possible solutions
Spatial	Does not match the spatial scales of ecosystem processes	• Administrative boundaries mismatch with resource distribution creating collective-action problems • Local institutions unable to cope with roving bandits engaged in global market • Central managers apply one-size-fits-all solutions that are inappropriate for local users	• Extend or create broader jurisdictional authority • Establish multiple-scale restraining institutions • Develop plan for co-management that allows for power sharing and multiscale governance
Temporal	Does not match the temporal scales of ecosystem processes	• Speed of invasive species faster than responsiveness of policy-making process	• Refocus policy to include consideration of slower variables • Improve monitoring program
Threshold behavior	Does not recognize or is unable to avoid social-ecological regime shifts	• Sustained yield harvesting policies lead to collapse of keystone species • Runoff from nitrogen fertilizer results in sudden change in aquatic ecosystem	• Engage in integrated assessments with scenario analysis • Implement adaptive management to identify solutions through experimentation
Cascading effects	Unable to buffer or amplifies cascading effects between domains	• Reduction in polar ice cap opens northern sea route to shipping, creating new forms of land use change and impacts on local harvesters	• Engage in SES modeling to understand multidimensional dynamics of system

conditions (Folke et al. 1998, Young 2002b, Galaz et al. 2008). Young's (2002b) framework examined the relationships between ecosystems, institutions, and human system characteristics, as well as their interactions, pointing out how efforts to achieve a good fit are confounded by the many dynamics of the system. Galaz et al. (2008) noted that mismatches between institutions and ecosystems can be (1) spatial, where social–ecological governance does not match the spatial scales of ecosystem processes (e.g., too small to capture a full watershed); (2) temporal, where social–ecological governance does not match the temporal scales of ecosystem processes (e.g., the speed of invasive species encroachment outpaces the responsiveness of society); (3) threshold behavior, where the social–ecological governance system does not recognize or cannot avoid abrupt shifts or social–ecological transformations (e.g., over-harvesting of keystone species); or (4) cascading, in which governance has limited capacity to buffer against or amplifies cascading effects through the system (Table 4.3). In short, finding a good fit is difficult in all cases and, if social–ecological conditions change, requires the coevolution of institutional arrangements with changing governance challenges.

Building and Maintaining Strong Community-Based Social–Ecological Governance

Local communities are a key component of any social–ecological governance system. Communities represent the social space in which people interact regularly, where people have access to and make use of resources for livelihoods, and where the use of ecosystem services translates into social well-being. Strong local-scale systems of social–ecological governance increase the likelihood that governance at other scales will be successful (Dietz et al. 2003).

Major insights have emerged from recent research on the diversity of ways in which commons are managed at a local scale (e.g., McCay and Acheson 1987, Berkes 1989, Ostrom 1990). This research has shown that small-scale commons have been managed, in many cases sustainably for decades to centuries, primarily because local institutions evolved to overcome the problems that Hardin had hypothesized in the tragedy of the commons. An important element of success in community-based management of commons is that resource users found ways to exclude other individuals, preventing open access to all users, while at the same time having strong internal institutions for resource stewardship. There are cases, however, in which common-pool resources have been degraded, frequently at times of rapid social or economic change when long-standing self-regulating local institutions were undermined. These examples illustrate ways in which people can manage (or fail to manage) trade-offs between individual and collective well-being.

As mentioned, there is no universal solution for sustainable management of a commons. When a commons is successfully managed, however, local-level institutions often prove to be important, especially when local residents depend on the resources for their well-being. These local institutions often involve the prevention of open access; the diffusion of information in ways that facilitate shared decision making; the development of cooperation and social capital and mechanisms to prevent free riding; and the minimizing of transaction costs associated with communication, negotiation, monitoring, coordination, and enforcement. A high dependence on the resource, the total number of resource users, their heterogeneity, leadership, and systems of reciprocity are also important to the success of these systems. Through comparative case studies, Ostrom (1990) and others (e.g., Agrawal 2002) have generated design principles for successful management of commons (Box 4.1). Figure 4.2 relates those principles to activities associated in social–ecological governance.

Box 4.1. Design Principles for Successful Management of a Commons (Ostrom 1990, Dietz et al. 2003)

1. **Clearly Defined Boundaries.** The boundaries of the resource system (e.g., irrigation system or fishery) or households with rights to harvest resource units are clearly defined. Resource users understand the borders clearly, with respect to both geographical extent of use and boundaries of acceptable behavior.

2. **Proportional Equivalence between Benefits and Costs.** Rules specifying the amount of resource products that a user is allocated are related to local conditions and to rules requiring labor, materials, and/or money inputs. This principle addresses the need for a sufficient level of equity in allocation and uses of resources.

3. **Collective-Choice Arrangements.** Many of the individuals affected by harvesting and protection rules are included in the group that can modify these rules. Although the level of participation in rule making is situational, there is an effort to avoid decision making by arbitrary or capricious means.

4. **Monitoring.** Monitors who actively audit biophysical conditions and user behavior are at least partially accountable to the users and/or are the users themselves. This sharing of roles has implications for the effectiveness of social control mechanisms.

5. **Graduated Sanctions.** Users who violate rules-in-use are likely to receive graduated sanctions (depending on the seriousness and context of the offense) from other users, from officials accountable to these users, or from both. Graduated

sanctions provide opportunities for individuals and the social system as a whole to respond to changes and learn from experience.

6. **Conflict-Resolution Mechanisms.** Users and their officials have rapid access to low-cost, local arenas to resolve conflict among users or between users and officials.

7. **Minimal Recognition of Rights to Organize.** The rights of users to devise their own institutions are not challenged by external governmental authorities, and users have long-term tenure rights to the resource.

8. **Nested Enterprises.** Appropriation, provision, monitoring, enforcement, conflict resolution, and governance activities are organized in multiple layers of nested enterprises.

Several crosscutting themes emerge from these principles (Box 4.1): the need for resource users to comply with rules; the need for resource users to have a role the creation and development of rules; and the problem of externally imposed rules that may be perceived as illegitimate by resource users. These design principles constitute a stabilizing feedback loop that supports **social learning**—the process by which groups assess social–ecological conditions and respond in ways that support their well-being. Monitoring of resource use by users may result in detection of overexploitation or inappropriate uses, which, when appropriately punished, reduces the likelihood that these users or their neighbors will again overexploit the resource. Similarly, graduated sanctions and conflict resolution mechanisms provide opportunities for individual and group learning through experience by allowing people to change their behavior over time. Changes that frequently threaten common-pool resource management include changes in slow variables that are more rapid than the system can adapt to; failure to transmit knowledge across generations, which may limit the capacity of communities to understand the historical basis of ecosystems and governance; and the lack of large-scale and well-linked institutions that relate community actions to regional-, national- and global-scale interactions. As previously noted, cultural factors also shape the outcomes of these processes, such as the extent to which a group perceives a threat to overexploitation or resource degradation, the degree to which the group shares a common identity, and the depth and richness of local knowledge in the governance process.

Community social–ecological governance, like all forms of social–ecological governance, is not a panacea (Ostrom et al. 2007). Local leaders can be subject to corruption, resource users can discount future values of resources if presented with the wrong incentives, and local knowledge holders can be wrong about the causes of social–ecological dynamics. Nevertheless, strong local institutions for governance of common-pool resources serve as an important foundation in a world in which cross-scale interactions (down- and up-scale effects) increasingly affect social–ecological processes. During the past several decades there has been a global trend to decentralize decision making from national-level governance to regional- and local-level social–ecological governance, as well as an emergence of a more global governance process (see Chapter 14). These trends suggest new opportunities to develop local-scale social–ecological governance while raising challenges in understanding the role of community governance in a global context. Although the lessons learned from local social–ecological governance are, to some extent, unique to the local scale, there are interesting similarities to global-scale governance. These similarities include a substantial dependence on consensus forms of decision making and informal institutions (Koehane and Ostrom 1995). In contrast, national and subnational governance are generally more centralized and top-down in authority and hierarchal in structure.

While the role of communities in social–ecological governance is important, the forces

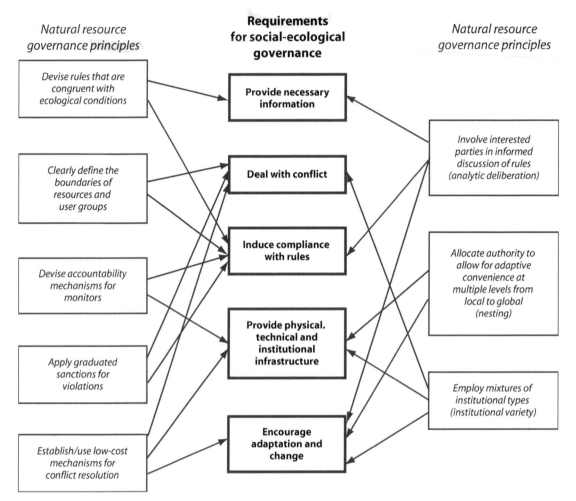

FIGURE 4.2. The general principles for robust governance of resource regimes (boxes on *right* and *left*) interact with requirements for governance of social–ecological systems. The likely connections between principles and requirements are presented with arrows, which are relevant to local to regional- and global-scale problems. Redrawn from Dietz et al. (2003).

of globalization are making the conditions that support self-governing small-scale resource systems increasingly rare (Dietz et al. 2003). For example, **roving bandits** in the form of pirate fishers that range widely to exploit resources without regard to established institutions, represent a significant challenge to local sustainability (Berkes et al. 2006; see Chapter 10). Concurrently, there is interest in forms of global governance that address global-scale issues, such as climate change (Biermann 2007; see Chapter 14). Together the potential contribution of strong local-scale systems of governance and the need for global-scale governance points

to the importance of coordination and accountability between local, regional, national, and international institutions.

Linking Scales of Governance for Communication, Responsiveness, and Accountability

The increase in geographic connectivity associated with globalization and the complexity of social–ecological governance necessitates effec-

tive cross-scale interactions among institutions and organizations. In such an environment, local communities, regional and national government agencies, research institutes, and international NGOs increasingly interact, forming social networks with ties of different strength. Linkages between institutions, organizations, and social networks (similar to what Ostrom refers to as "nested systems"), facilitate interactions, providing opportunities for those at one scale to deliberate and solve problems with others at another scale. **In a directionally changing world, neither top-down nor bottom-up interactions are the preferred direction of interaction, but instead two-way transactions are needed to account for observations, understandings, and human needs as perceived at the various levels.**

There are two general categories of cross-scale linkages (Young 1994, Berkes 2002, Young 2002a). **Horizontal linkages** occur at the same level of social organization and across spatial scales, for example, in treaties among countries or regional compacts. **Vertical linkages** occur as part of a hierarchy of interorganizational relationships at the same location, such as in the transactions between local- and national-level management systems that operate in a given region. Linkages that include rich information exchanges and shared processes for defining problems and seeking solutions to problems contribute to the effectiveness of social–ecological governance as do those that address accountability across scales. The linkages are typically more complex than local–national or lateral relationships and are oriented around specific issues, creating **polycentric governance**, a complex array of interacting institutions and organizations with overlapping and varying objectives, levels of authority, and strengths of linkages (Ostrom 1999b).

Deciphering cross-scale institutional interplay helps to identify inconsistencies in rules, asymmetries or gaps in information flow, political inequities, and their respective implications for sustainability. Once these inconsistencies are identified, there are often high transaction costs required to resolve differences in jurisdiction of authority, address power inequities, and develop trust and effective communica-

tion. Sometimes these challenges associated with effective linkages are overwhelming, especially when time is limited. For example, the institutional failures in the post-Katrina hurricane events in New Orleans in 2005 are largely attributable to inadequate coordination among scales of governance ranging from neighborhoods, enforcement agencies, and emergency relief organizations to the policies of state- and federal-elected officials. However, overlaps in authority can also provide resilience through redundancy, because when one of these institutions is ineffective, other institutions may still be able to accomplish many of the tasks required for effective social–ecological management (Berkes et al. 2003).

Strong linkages can also have negative consequences that cascade downward to local-scale implementation of governance. These may involve the shifting of management to exclude local knowledge; a colonial perspective that subjugates local authority; the nationalization of resources that were formally governed successfully through local institutions; and the expansion of commodity-based markets for development, with development defined by central (nonlocal) authorities. To counterbalance these problems, there are several strategies that can allow communities to affect actions at other scales, strengthen stewardship by keeping local systems accountable to broader principles of sustainability, revitalize cultural and political empowerment, build capacity of local organizations to engage in activities (political and otherwise) at higher levels, and help in the establishment and development of institutions (Berkes 2002, Table 4.4).

Co-management is the sharing of power and responsibility between resource users' communities and government agencies in management of natural resources (Pinkerton 2003). Power sharing through co-management arrangements can take many forms. It can be established through formal contractual agreements that outline the authority of the arrangement in various functions of management [e.g., indigenous land-claim settlements (see Chapter 6) or water management districts (see Chapter 9)]. Co-management can also be informal, based

TABLE 4.4 Strategies for cross-scale linkages (Adapted from Berkes 2002).

Organization	Goal	Organizational relationship	Typical type	Short- or long-term
Co-management	Sharing of power	Agency-community	Long-term collaboration	Short and long
Special interest pressure groups	Influencing management	User organization	Issue-based	Short
Policy communities	Influencing management	Multiorganization	Issue-based network	Short and long
Social movements	Achieve a social goal	Multiorganization	Issue-based network	Short
Multistakeholder bodies	Scope issues	Agency-initiated;	Short-term solutions sought	Generally short, depending on mandate
Public consultation processes	Hear from community	Agency-initiated	Short-term concerns; open participation	Short
Development and empowerment	Capacity building	Vertical	NGO or government aid	Short and long
Citizen science	Improve understanding	Vertical	Task-focused	Short and long

on personal relationships between agency personnel and local community members. "Formal co-management agreements" do not always achieve power-sharing, so care must be taken to clarify the terms of working relationships and develop the social capital necessary for effective joint problem-solving. Effective involvement of co-management partners in policy making, especially with respect to policies on resource allocation and in the interpretation of research findings, is critical to achieving "complete co-management" (Pinkerton 2003). Many formal co-management arrangements are implemented with **bridging** or **boundary organizations** (boards and councils of representatives) that link local communities to policy processes at a regional scale. Co-management often requires the modification or development of new pathways of communication as well as strong leaders who hold bicultural perspectives and broker cross-scale solutions. Commonly accompanying these and other types of linkages are **shadow networks**, groups that are indirectly involved in decision making, but supportive to the process, such as academic researchers whose work supplements the process (Olsson et al. 2006).

Cultural differences, communication problems, or differing perceptions of appropriate forms of decision making (e.g., consensus vs bureaucratic) can confound efforts to achieve effective representation in these processes. Even with these challenges, co-management has proven in many cases to build common understanding of problems across scales and can be important where a commons requires a high degree of coordinated action among geographically dispersed parties.

Other strategies serve different functions in linking groups with decision making. **Multistakeholder bodies**, for example, provide policy makers a way to scope issues, seek solutions to problems, achieve broader public participation, and foster public consensus. They are often short-lived bodies that lack management authority and therefore tend to function primarily as forums for discussion, planning, and conflict resolution. However, in cases where they have a mandate to negotiate and some authority to affect policy outcomes, they have the potential to resolve conflicts, set direction for future actions, and achieve "win–win" solutions (Wondolleck and Yaffee 2000). **Public consultation processes**, another strategy for linking across scales, can be used by government management agencies to inform decision making about the interests and concerns of local communities. For example, environmental impact assessments typically provide for public review and commenting, allowing a formal documentation of local users' perceptions of potential impacts. In some cases, public issues raised in these reviews must be included in the public record and require that agencies respond to them in writing as a part of the final impact assessment. Focus groups, public meetings, mail-out, and other forms of survey research can also be used for accessing public input. The strength of the linkage between the public voice in public consultation processes and the final decision depends to a great extent on the receptiveness of government administrations and the ability of public interest groups to organize and articulate their concerns and interests.

Development and empowerment organizations can work with stakeholder groups to build their capacity to address problems, both current and future, and communicate better internally and with other groups. These organizations are typically NGOs, although some federal programs of developed countries work in developing countries to achieve the same objectives (see Chapter 6). The Nature Conservancy of the USA, for example, has facilitated a range of "community partnership" programs that convene key players to address present and future needs of community development and the habitat conservation actions that can best meet those objectives. Larger-scale organizations with a development and/or empowerment mission provide important networking services to groups in ways that bridge decision making across scales. To be successful, they must work through a process that avoids dependence on those organizations that exacerbate problems.

Special interest groups, policy communities, and social movements seek management policies and actions that foster their specific interests. **Special interest groups** typically self-organize to advocate for policies that serve their interests at local to regional scales. Hunting organizations and the timber industry, for example, can organize to seek policies that facilitate harvest of specific resources; conservation NGOs may seek to reduce these harvests; and the tourist industry may seek regulations that foster infrastructure development for recreation. **Policy communities** are loosely affiliated groups that share an interest in one or more specific issues and collaborate to change policy. They are typically interorganizational networks of NGOs, international governments, agencies, and local communities. Collaborative alliances of this kind typically must go through stages of development, including problem definition, direction setting, implementation, and evaluation (Gray 1989, Gray and Wood 1991). Failure to move through these stages can limit the groups' collaboration (Kofinas and Griggs 1996). **Social movements** operate over broader geographic scales than special interest groups and can be cultivated by organizations and/or strong leadership to advance a particular ideological perspective. Those involved in social movements may have little or no history of working together and no recognized leader or organizational center. However, they can take on a life of their own to become a powerful force for realizing a common vision. For example, many distinct indigenous groups have created social movements in their efforts to achieve international indigenous empowerment. The wilderness movement has similarly created a push toward wildland preservation. Similarly, many sovereign states have coordinated to create a global free-market economy through the establishment of the World Trade Organization.

Programs in **citizen science** provide opportunities for local residents to share observations and, in some cases, local understanding of environmental process with groups operating at other scales. They have been extensive in their reach and provide information in the assessment of environmental change. The Christmas Bird Count, for example, has operated since 1900 as a citizen-based network of bird watchers in the USA that annually documents the presence of bird species in their communities. Frog Watch is another geographically extensive program of citizen science that has bridged local observations with regional- to global-scale observations of environmental change. Similar programs have been established with indigenous hunters of Africa and the Arctic. These programs can parallel advocacy for special interest (e.g., Audubon and its advocacy for avian fauna) or be a complementary part of a co-management arrangement. An example of one of the most successful areas of citizen science is watershed stewardship programs (see Chapter 9). Increased connectivity through the internet is extending the capacity of citizen groups to have influence across scales. In the context of citizen science, it is also providing an educational function through classroom programs such as "GLOBE"—an Earth-System Science education program that facilitates international interactions among students. Another side benefit of citizen science is that citizens learn about environmental issues, and their participation builds commitment to resource stewardship.

The effectiveness of each of these strategies in linking across scales depends to a great extent on the sources (money, infrastructure, human capital, organizational capacity) and balance of power among groups. Formal institutions that recognize the rights of resource users may enable local communities to actively engage regional- and national-level decision makers in problem solving, yet the organization of an active policy community or an agency culture sympathetic to local interest may provide strong linkages in the absence of formal institutions (see Chapters 6 and 11). The benefits of cross-scale linkages follow when the interaction of parties facilitates the building of trust, mutual respect among parties, and social learning.

Facilitating Adaptive Learning

Adaptive learning as a part of resource management provides the means for coping with uncertainty and change in a social–ecological environment. Adaptive learning in governance occurs when one or more groups (1) carefully and regularly observe social–ecological conditions, (2) draw on those observations to

improve understanding of the system's behavior, (3) evaluate the implications of emergent conditions and the various options for actions, and (4) respond in ways that support the resilience of the social–ecological system. Face-to-face deliberation is a vital component of this process, providing for reflection on past experience, revisiting of basic operating assumptions, and careful evaluation of policy alternatives. Adaptive learning in social–ecological governance can occur within and across a number of geographic and organizational scales, as noted in the section above. For example, adaptive learning within a single organization, such as a single agency may occur through periodic and well-structured planning processes. Bridging organizations, such as management councils, can function as central nodes of cross-scale network interactions in these processes. Broader societal-level social learning can also unfold through a more diffuse process of **communities of practice** or **learning networks**. At this large scale the process may be less structured but still part of a governance process in which a society sets its course for the future. Linking adaptive learning at the level of the bridging organization with society at large can be a considerable challenge where there is limited social capital and no effective communications strategy.

Adaptive learning in social–ecological governance follows from the idea of **adaptive management**, which is defined as resource management based on the science of learning by doing (Walters 1986, Gunderson 1999, Lee 1999). Embracing uncertainty is a central tenet of adaptive management. Uncertainty in resource management can result from a lack of, or incorrect, information about environmental factors that influence a given situation (observational error), a lack of understanding of the underlying social and/or ecological processes (process error), a lack of knowledge about the effects of incorrect decisions (model error), and/or the inability of the decision maker to assess the likelihood that a factor will affect the success or failure of the system (implementation error; Francis and Shotton 1997). From a statistical perspective, uncertainty involves **Type 1 errors**, which are the claims that something is true when in actuality it is not (a false positive), and **Type 2 errors**, which are failures to disprove a null hypothesis when another is true (a false negative). Beyond statistics there is also the error of getting the right answer to a wrong question. Uncertainty in resource management can also be cognitive (i.e., mentally constructed), such as ambiguity in terminology of concepts used by groups in the decision-making process.

While it may be compelling to reduce uncertainty, an alternative approach is to assume that uncertainty is the norm and therefore make decisions based on risk evaluation and reduction rather than predicting a precise future path. In such an approach, uncertainty is less of a central problem to resolve and more a condition to navigate (see Chapters 1 and 5). Even if assumed, uncertainty presents a number of challenges for resource managers, particularly when management is undertaken on an ecosystem basis. The challenges follow from the complex adaptive nature of interacting components of a social–ecological system and the limited ability of people to make predictions about the future (Wilson 2002). Yet, in the face of uncertainty, managers and resource users alike need to act, typically *before* scientific consensus is achieved (Ludwig et al. 1993); inaction (or no decision) is in fact a decision that can lead to surprise (see Chapter 11).

Adaptive management assumes conditions of uncertainty by viewing policy actions as hypotheses that are tested through policy implementation and later evaluated to improve the state of knowledge and practice. Key objectives of adaptive management are to anticipate major surprises where possible, limit their negative consequences when they do occur, and use change as an opportunity for adaptive learning. Over time, adaptive learning is a cyclical process by which observations inform understanding, which informs decision making, and so on. Adaptive learning ultimately requires a deliberative process by which the state of knowledge is reviewed, evaluated, and, where needed, modified based on improved understanding and the desired future. The distinguishing feature of adaptive management is **double-loop learning**, the feedback process in decision making by

which practitioners reflect on the consequence of past actions before taking further action (Argyris and Schöen 1978, Senge 1990, Argyris 1992). Double-loop learning differs from **single-loop learning**, which adjusts actions to meet identified management goals (e.g., modifies harvest rate to conform to specified catch limits) but does not evaluate basic assumptions and approaches (Fig. 4.3). It also differs from decision making by "muddling through," a process of trial-and-error decision making in which decisions are based primarily on politics without the benefits of careful reflection (Lindblom 1959, 1972).

Ideally, policy actions are viewed as experiments that are analyzed in subsequent cycles of adaptive learning. Policy changes can be operational, such as a change in a stipulation for construction of new infrastructure; or regulatory, such as a change in the harvesting period. Building on the idea of policies as experiments, the process can be undertaken

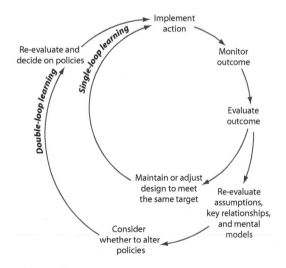

FIGURE 4.3. Single- and double-loop learning. Single-loop learning involves changing actions to meet identified management goals (e.g., modifying harvest rate to conform to specified catch limits), often through trial and error. Double-loop learning includes a reflection process of evaluating underlying assumptions and models that are the basis of defining problems (e.g., revising the indicators and simulation models used to calculate allowable catch from stock assessments).

as **active adaptive management** through the intentional manipulation of the system to test its response or **passive adaptive management** through an intensive examination of historic cause–effect relationships. Scale is an important determinant in the appropriateness of active and passive approaches to adaptive management. For example, experimentation in management of hydrological systems for an oil field may glean good insights into improved designs that maintain ecological properties, whereas efforts at regional-to-global scale experimentation are inappropriate because of the potential irreversibility of negative outcomes (see Chapter 14).

The ideas of adaptive learning can become operational by integrating the activities of monitoring, research, and policy making. Ongoing social–ecological **monitoring** provides the basis for observing current conditions and the patterns and trajectories of change. Monitoring can assess if there is compliance with policy, if management actions are having the desired effect, and if models and their assumptions are valid (Busch and Trexler 2002). Multiple sources of information (e.g., local knowledge, field-based monitoring, remotely sensed data) enhance the monitoring process by providing multiscaled understanding of phenomena as well as interdisciplinary and cross-cultural perspectives. Programs of citizen science, for example, can contribute to this process, as do natural science monitoring programs (e.g., Kofinas et al. 2002). **Indicators** in monitoring characterize the structural and compositional parts of the system and their processes. In many parts of the world "sustainability indicator programs" are being implemented by municipalities to track change by integrating social–ecological conditions. In some cases, benchmarks of performance are identified and reviewed periodically to assess change and make adjustments as needed to meet desired targets. This is a valuable application of single-loop learning.

Linking observations of change with analysis and evaluation can be improved with the development and use of models. **Models** are simplified representations of the real world that provide insight into social–ecological processes and change. Conceptual models, statistical models, simulation models, and spatial models, when effectively used, can serve as **decision-support tools** that help assess the state of knowledge, assess management actions, direct future monitoring and research, and explore the implications of alternative futures. Effective adaptive management draws on multiple models to test alternative hypotheses about why given changes occurred or are likely to occur in the future. As with monitoring, modeling can be based on multiple knowledge systems or the integration of **expert knowledge** with empirically based relationships. Models, developed in conjunction with **scenario analysis**, help to conceptualize emergent conditions and support a visioning process (see Chapter 5). The use of scenarios can be undertaken through (1) **backcasting**, which identifies societal objectives and uses models and scenarios to explore how these objectives and past events led to the current situation or some desired future condition, and (2) **forecasting**, which projects future conditions and their social–ecological consequences based on recent trends (Robinson 1996). Making models complex enough to capture the nature of the problems while simple enough to understand the drivers of change is a significant challenge in their use (Starfield et al. 1990). Participation in the development and use of models as well as model transparency are also important in making assumptions and key relationships clear for all users.

In practice, the adaptive cycle of social–ecological governance occurs imperfectly in different phases and within and across different scales. Adaptive management becomes adaptive co-management when groups at various scales of social–ecological systems, including local communities, participate in processes of social learning. This broader participation potentially enhances the learning process, as well as confounds it. An effective adaptive co-management process typically requires that subgroups have sufficient organizational capacity to engage in these activities at any one level (e.g., regional) and across scales. Bridging organizations, such as co-management boards or councils, can be critical to facilitating these processes.

Although the concept of adaptive management has been widely accepted by resource managers as a good idea, it is difficult to implement (see Chapter 10; Walters 1997, Lee 1999). Elected officials and the public at large tend to prefer stability since it conveys a sense of well-being. The idea of "policies as experiments" can be unsettling for policy makers who are trying to satisfy their myopically oriented electorate. Ultimately, the success of adaptive co-management depends to a great extent on the degree of shared management goals and objectives among managers and resource users and their support for the long-term implementation of the adaptive management program (Lee 1999). This suggests that any adaptive management process must first seek to identify goals and objectives, at least among those directly involved with decision making (e.g., members of a bridging organization). The adaptive management of the Everglades, the Grand Canyon, and the Columbia River Basin in the USA are a few cases that illustrate the potential success and challenges of successful adaptive management. Projects undertaken in Wisconsin provide good examples of participatory scenario analysis (Peterson et al. 2003). The monitoring of ecosystems and development of models for reflexive learning are less useful if undertaken only for brief periods, given the need to understand system dynamics and translate policy experiments into new knowledge. Yet funding of long-term programs such as observation systems can be problematic, and agreement on management objectives can unravel in the heat of new political controversies, changes in funding priorities, or the recognition of new ecological threats. Moving from adaptive management to adaptive co-management represents an additional challenge since it suggests the need to develop learning processes while concurrently sharing power and responsibility in governance between resource users and other parties involved in management of ecosystems. While these conditions are challenging, they are not impossible. In practice passive and active adaptive management represents endpoints on a spectrum, with passive adaptive management (intentionally learning from past experience) regarded as responsible resource management by most professionals.

Generating Innovative Outcomes

The success of social learning through adaptive co-management depends on more than having the right process and ultimately rests on arriving at solutions to resolve short- and long-term problems. Given the high rate and novelty of global-to-local social–ecological change faced today, those involved in adaptive co-management will increasingly need to scan their environment to detect change, understand it in ways that are meaningful to decision making, and ultimately achieve robust policy responses that sustain the properties of social–ecological systems deemed desirable by society. Facing novel conditions, adaptive co-managers will also need to think beyond political agendas and reflect more carefully on alternative solutions to novel problems and the processes by which novel solutions are readily discovered and tested. Thus, the capacity of social–ecological governance systems to generate innovative solutions to novel problems may be one of the most important outcomes of adaptive co-management (Kofinas et al. 2007). While research on innovation in social–ecological governance is somewhat new, initial findings suggest that innovative outcomes follow from a number of facilitating conditions, including the productive friction of cultural diversity, power sharing, and power politics. Box 4.2 presents a set of conditions that facilitate innovation through adaptive co-management. The quest for social learning and innovation through adaptive governance raises the question of whether the production of innovation through adaptive co-management can be sustained or whether it waxes and wanes in response to social need and conditions. It is probably both: Human responsiveness to address emergent problems is alerted through the waxing and waning of perceived impending crisis, while the quality of our responses toward resilience is enhanced by the development of our capacity to innovate (Kofinas et al. 2007).

Box 4.2 Conditions that Facilitate Innovation in Adaptive co-management

The novel problems facing society from rapid directional change require that managers and resource users work together toward innovation in problem solving. Several factors contribute to the production of innovative outcomes in adaptive co-management processes. Some are listed below (Kofinas et al. 2007).

1. Interdependence of actors' needs and interests and sufficient levels of social capital (i.e., trust) provide the basis for creative engagement in an adaptive co-management process.
2. Appropriate levels of social heterogeneity and productive conflict provide for the comparison of perspectives and stimulation of novel solutions.

3. A culture of openness to new ideas and the taking of risk promote an environment in which innovation can be cultivated.
4. Policy leaders and policy entrepreneurs promote and guide innovative problem solving and gain the acceptance of innovative solutions by the greater public.
5. Reflection and innovation do not just happen, but require the allocation of time and careful facilitation of process.
6. Decision-support tools, such as the use of scenario analysis with simulation models, can help in anticipating possible futures and stimulating creative thinking.
7. Prior experience with successful innovation builds confidence to experiment and learn in the future.

Summary

In spite of the significant challenges of rapid change, there are conceptual frameworks and practical tools available to resource managers, resource–user communities, NGOs, and others for building resilience and sustaining the flow of ecosystem services to society. This chapter explored the critical components with which social–ecological governance is implemented and assessed, with a focus on the social mechanisms by which institutions are linked with ecosystems and social learning.

Institutions — the rules of the game — function as incentive structures in human decision making, shaping the choices of society and affecting individual and collective responsiveness to change. Institutions also convey cultural identity, shape patterns of communication, and define power relations that affect the distribution of costs and benefits that follow from policy choices. Institutions can both constrain and/or enhance the capacity of groups to respond to change and innovate for adaptation. Because institutions are artifacts of human invention and creativity, their use and manipulation provide a significant opportunity for society to increase its sensitivities to change and be proactive in anticipation of surprise. Institutions, human organizations, and social networks are components of highly dynamic and complex adaptive systems. The role of human agency and the nonlinear nature of change in these systems add to the unpredictability that groups face in ecosystem stewardship. Because of the heterogeneity and complexity of social–ecological systems, there are no "one size fits all" panaceas for resolving the problems associated with social–ecological governance. Developing institutional arrangements that are well suited for particular situations require careful attention to context, the underlying social dilemmas that affect choice arenas, and the institutional fit and interplay of social–ecological systems across multiple scales.

The ongoing flow of ecosystem services to society depends on institutions that are sufficiently robust, that are highly adaptive, and that link local communities to other scales. Adaptive co-management has been suggested as an approach well suited to address the problems of rapid change. Adaptive co-management has the potential to shift human conflict toward increased cooperation, facilitate social learning through experience and reflection, and serve as a feedback that shapes the trajectories of social–ecological change. Adaptive co-management is not simply a set of rules or a management structure, but is a process by which groups draw on a systems perspective to think holistically and learn about social–ecological interactions and gain better insight from a diversity of perspectives. When effective, the outcomes of adaptive co-management stimulate innovation in problem solving. Realizing the ideals of adaptive co-management requires an ongoing reevaluation of conditions, assumptions, and human values associated with management decisions. It also requires a process of continuous reflection on the performance of institutions to improve and in some cases transform them to better suit emergent conditions.

Review Questions

1. How can adaptive co-management approaches to social–ecological governance address the shortcomings of steady-state resource management?
2. How do economic incentives created by institutions for management of a commons affect the sustainability of important resources and the social and political systems that are part of human communities?
3. What are some of the potential strengths and limitations of local communities having full authority over resource management?
4. How can the misfits in institutions and social–ecological systems be avoided in conditions of rapid change?

5. Given the multiple ways to achieve cross-scale linkages of institutions in social–ecological governance, what are the conditions in which each approach is likely to be most effective?
6. Provide an example of how the practice of adaptive co-management, including the use of community-based monitoring, simulation modeling, and scenario analysis, could be used as part of a broad-scale social learning process.

Additional Readings

Armitage, D., F. Berkes, and N. Doubleday, editors. 2007. *Adaptive Co-Management: Collaboration, Learning, and Multi-Level Governance.* University of British Columbia Press, Vancouver.

Berkes, F. 2008. *Sacred Ecology: Traditional Ecological Knowledge and Resource Management.* 2nd Edition. Taylor and Francis, Philadelphia.

Dietz T., E. Ostrom, and P.C. Stern. 2003. The struggle to govern the commons. *Science* 302:1907–1912.

NRC (National Research Council); E. Ostrom, T Dietz, N. Dolšak, P.C. Stern, S. Stovich, et al., editors. 2002. *The Drama of the Commons.* National Academy Press, Washington.

Olsson, P., C. Folke, and F. Berkes. 2004. Adaptive co-management for building resilience in social-ecological systems. *Environmental Management* 34:75–90.

Wilson, D.C., J.R. Nielsen, and P. Degnbol, editors. 2003. *The Fisheries Co-Management Experience.* Kluwer Academic Publishers, Dordrecht.

Young, O.R. 2002. *The Institutional Dimensions of Environmental Change: Fit, Interplay, and Scale.* MIT Press, Cambridge, MA.

5
Transformations in Ecosystem Stewardship

Carl Folke, F. Stuart Chapin, III, and Per Olsson

Introduction

Changes in governance are needed to deal with rapid directional change, adapt to it, shape it, and create opportunities for positive transformations of social–ecological systems. Throughout this book we stress that human societies and globally interconnected economies are parts of the dynamics of the biosphere, embedded in its processes, and ultimately dependent on the capacity of the environment to sustain societal development with essential ecosystem services (Odum 1989, MEA 2005d). This implies that resource management is not just about harvesting resources or conserving species but concerns stewardship of the very foundation of a prosperous social and economic development, particularly under conditions of rapid and directional social–ecological change (Table 5.1). We first discussed the integration of the ecological (see

Chapter 2) and social (see Chapters 3 and 4) aspects of ecosystem stewardship in relation to directional change and resilience in a globally interconnected world (see Chapter 1), emphasizing processes that reduce the likelihood of passive degradation that might lead to socially undesirable regime shifts. In this chapter we identify ways to enhance the likelihood of constructive transformative change toward stewardship of dynamic landscapes and seascapes and the ecosystem services that they generate. Rapid and directional changes provide major challenges but also opportunities for innovation and prosperous development. Such development requires systems of governance of social–ecological dynamics that maintain and enhance adaptive capacity for societal progress, while sustaining ecological life-support systems.

An Integrated Social–Ecological Perspective

Over decades, segregated approaches have dominated policy and the structure of governmental departments, agencies, and decision-making bodies, with little communication between sectors. This was true in science

C. Folke (✉)
Stockholm Resilience Centre, Stockholm University, SE-106 91 Stockholm, Sweden; Beijer Institute of Ecological Economics, Royal Swedish Academy of Sciences, PO Box 50005, SE 104 05 Stockholm, Sweden
e-mail: carl.folke@beijer.kva.se

F.S. Chapin et al. (eds.), *Principles of Ecosystem Stewardship*,   103
DOI 10.1007/978-0-387-73033-2_5, © Springer Science+Business Media, LLC 2009

TABLE 5.1. Features of shift in perspective from command-and-control to complex systems. Adapted from Folke (2003).

From command-and-control	To complex systems
Assume stability, control change	Accept change, manage for resilience
Predictability, optimal control	Uncertainty, risk spreading, insurance
Managing resources for sustained yield	Managing diversity for coping with change
Technological change solves resource issues	Adaptive co-management builds resilience
Society and nature separated	Social–ecological coevolution

as well, with reward systems stimulating within-discipline knowledge generation and limited collaboration across disciplines (Wilson 1998). Resource and environmental management have been subject to similar divisions. Only recently have managers appreciated the significance of a broader systems perspective to deal with rapid and directional change. The integration of the human and the environmental dimensions for ecosystem stewardship is still in its infancy, so analyses of social–ecological systems are not as well developed as those of social or ecological systems alone (Costanza 1991, Ludwig et al. 2001, Westley et al. 2002). A focus that is restricted to the social dimension of ecosystem stewardship without understanding how it is coupled to ecosystem dynamics will not be sufficient for sustainable outcomes. For example, the development of fishing cooperatives in Belize was considered a social success by managers. However, the local mobilization of coastal fishers into socially desirable and economically effective fishing cooperatives became a magnet of fishing efforts to capture **economic rent** (the income gained relative to the minimum income necessary to make fishing economically viable) and resulted in a short-term resource-exploitation boom of lobster and conch, causing large-scale resource-use problems and increased vulnerability (Huitric 2005).

Similarly, focusing only on the ecological aspects as a basis for decision making for sustainability leads to conclusions that are too narrow. Ecosystems can pass a threshold and shift from one state to another, often triggered by human actions (Folke et al. 2004). When a lake shifts from a clearwater state attractive for fishing and recreation to a state of unde-

sired algal blooms and muddy waters, it may look like an ecologically irreversible transition. However, if there is sufficient adaptive capacity in the social system to respond to the shift and foster social actions that return the lake to a clearwater state, the social–ecological system is still resilient to such change (Carpenter and Brock 2004, Bodin and Norberg 2005; see Chapter 9).

Hence, in a social–ecological system with high adaptive capacity, the actors have the ability to renew and reorganize the system within desired states in response to changing conditions and disturbance events. However, there are also situations where it would be desirable to move away from the current conditions and transform the social–ecological system into a new configuration. **Transformation** is the fundamental alteration of the nature of a system once the current ecological, social, or economic conditions become untenable or are undesirable (Walker et al. 2004, Nelson et al. 2007; see Chapter 1). The capacity to learn, adapt to and shape change, and even transform are central aspects of social–ecological resilience and require that learning about resource and ecosystem dynamics be built into management practices (Berkes and Folke 1998b) and supported by flexible governance systems (Folke et al. 2005). **Transformative learning** is learning that reconceptualizes the system through processes of reflection and engagement. It relates to **triple-loop learning** (Fig. 5.1), which directs attention to redefining the norms and protocols upon which single-loop and double-loop learning (see Chapter 4) are framed and governed. This draws together human agency and individual and collective learning with processes of change, uncertainty, and surprise (Keen et al. 2005, Armitage et al. 2007).

FIGURE 5.1. Triple-loop learning involves the same reevaluation of assumptions and models as double-loop learning (see Fig. 4.2) but considers whether to alter norms, institutions, and paradigms in ways that would require a fundamental change in governance (e.g., a shift from an agricultural systems focused on supporting farmers to a tourist-based economy requiring a broader, more inclusive form of governance).

Dealing with Uncertainty and Surprise

Recognizing and accepting the uncertainty of future conditions is the primary motivation for incorporating resilience thinking into ecosystem stewardship. We are nowhere close to a predictive understanding of the complex interactions and feedbacks that govern trajectories of change in social–ecological systems nor able to anticipate the future human actions that will modify these trajectories. Uncertainty is therefore a central unavoidable condition for ecosystem stewardship. There are several sources of uncertainty, only some of which can be readily reduced (Carpenter et al. 2006a). Both scientific research and the observations and experience of managers and other people provide data that inform our understanding. However, there are many uncertainties regarding the validity of any dataset and its representativeness of the real world (Kinzig

et al. 2003). Models, both quantitative computer models and conceptual models of how the world works, also have many uncertainties in assumptions and structure. Models are most useful when based on observations and other data, but there are always important processes for which data are unavailable and data that do not fit our current understanding. Surrounding these uncertainties in data and models are uncertainties in other factors that we know to be important but for which we have neither data nor models—the "known unknowns" (Fig. 5.2). There are also "unknown unknowns" that we cannot anticipate—the surprises that inevitably occur (Carpenter et al. 2006a).

There are several types of surprises (Gunderson 2003). **Local surprises** occur locally. They may be created by a narrow breadth of experience with a particular system, either temporally or spatially. Local surprises have a statistical distribution, and people respond to these surprises by forming subjective probabilities that are updated when new

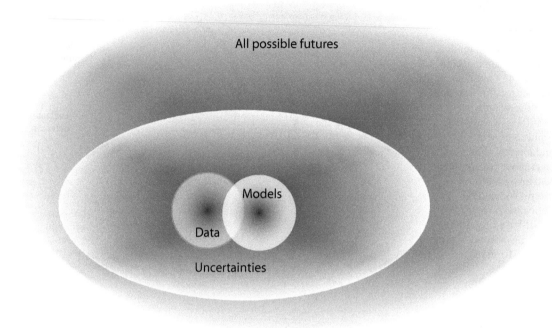

FIGURE 5.2. The full set of possible futures of social–ecological systems is only partially represented in available data and models. Together, the data and models allow us to project some uncertainties (knowable unknowns). The probability that any model projection of future conditions will actually occur depends on the full set of all possible futures, most of which are unknown. Redrawn from Carpenter et al. (2006a).

information becomes available. Based on these estimates, there is a wide range of adaptations to risk that are economically rational to individuals, including risk-reducing strategies and risk-spreading or risk-pooling among independent individuals. Local surprises are manageable by individuals and groups of individuals. Their detection requires a comprehensive systems perspective (e.g., an ecosystem rather than a single-species approach). Adaptation-to-risk strategies fail when surprises are not local.

Cross-scale surprises occur when there are cross-scale interactions, such as when local variables coalesce to generate an unanticipated regional or global pattern, or when a process exhibits contagion (as with fire, insect outbreak, and disease). Cross-scale surprises often occur as the unintended consequences of the independent actions of many individual agents who are managing at different scales.

Although individual responses are generally ineffective, individuals acting in concert can address these surprises, if appropriate cross-scale institutions are available or are readily formed (see Chapter 4).

True-novelty surprises constitute never-before-experienced phenomena for which strict preadaptation is impossible. However, systems that have developed mechanisms for reorganization, learning, and renewal following sudden change may be able to cope effectively with true-novelty surprises. These are the social–ecological features that nurture resilience to deal with unexpected change.

Directional change in the context of global and climatic change creates a situation of increased likelihood of unknowable surprises. It is within these sources of uncertainty and surprise that ecosystem stewardship must function and where a resilience approach becomes essential.

Preventing Social–Ecological Collapse of Degrading Systems

The Interplay Between Gradual and Abrupt Change

Theories, models, and policies used in resource management have historically been developed for situations of gradual or incremental change with implicit assumptions of linear dynamics. These approaches generally disregard interactions that extend beyond the temporal and spatial scales of management focus. The resilience approach to ecosystem stewardship and the adaptive-cycle framework outlined in Chapters 1 and 4 indicate that there are times when change is incremental and largely predictable and other times when change is abrupt, disorganizing, or turbulent with many surprises. It also draws attention to ways in which such social–ecological dynamics interact across temporal and spatial scales (the concept of **panarchies**, see Chapter 1; Gunderson and Holling 2002). This dynamic interaction challenges manage-

ment to learn to live with uncertainty, be adaptive, prepare for change, and build it into ecosystem stewardship strategies. Periods of large abrupt changes in social and ecological drivers, including climatic change and economic globalization, are occurring more frequently (Steffen et al. 2004; see Chapter 14), increasing the likelihood of abrupt social–ecological change. In the absence of resilience-based stewardship, these changes are quite likely to trigger shifts from one state to another that may be socially and ecologically less desirable. A focus on resilience in social–ecological systems is needed to deal with the challenging new global situation of rapid and directional social–ecological change.

Behaviors that reduce the adaptive capacity to deal with interactions between gradual and abrupt change may push systems toward a threshold that precipitates regime shifts or critical transitions. The existence of thresholds between different regimes or domains of attraction has been described for several ecological systems (Scheffer et al. 2001; Fig. 5.3).

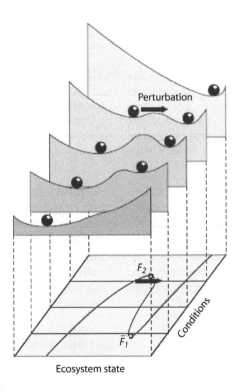

FIGURE 5.3.

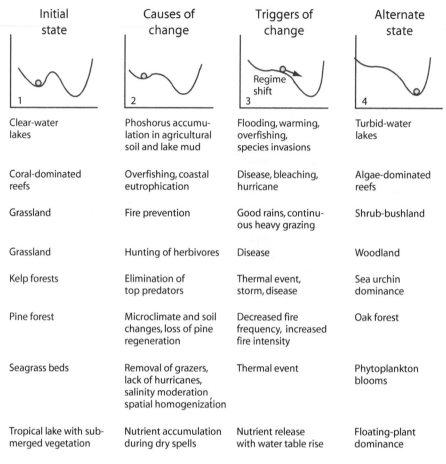

Initial state	Causes of change	Triggers of change	Alternate state
Clear-water lakes	Phoshorus accumu-lation in agricultural soil and lake mud	Flooding, warming, overfishing, species invasions	Turbid-water lakes
Coral-dominated reefs	Overfishing, coastal eutrophication	Disease, bleaching, hurricane	Algae-dominated reefs
Grassland	Fire prevention	Good rains, continu-ous heavy grazing	Shrub-bushland
Grassland	Hunting of herbivores	Disease	Woodland
Kelp forests	Elimination of top predators	Thermal event, storm, disease	Sea urchin dominance
Pine forest	Microclimate and soil changes, loss of pine regeneration	Decreased fire frequency, increased fire intensity	Oak forest
Seagrass beds	Removal of grazers, lack of hurricanes, salinity moderation spatial homogenization	Thermal event	Phytoplankton blooms
Tropical lake with sub-merged vegetation	Nutrient accumulation during dry spells	Nutrient release with water table rise	Floating-plant dominance

FIGURE 5.3. Regime shifts in ecosystems: **Previous page**: Cup-and-ball model illustrating the shift from one ecosystem regime to another. The bottom plane shows the hysteresis-curve, which highlights the nonlinear relation of moving from one regime to another. Modified from Scheffer et al. (2001).

Above: Alternate states in different ecosystems (1, 4) and some causes (2) and triggers (3) behind loss of resilience and regime shifts. For more examples, see Thresholds Database on the Web site www.resalliance.org. Modified from Folke et al. (2004).

Thresholds may also occur in resilient ecological systems, but fostering resilience reduces the likelihood of this occurring. Experience suggests that critical ecosystem transitions are occurring increasingly often as a consequence of human actions and seem to be more common in human-dominated landscapes and seascapes (Folke et al. 2004). Human actions that are most likely to cause loss of resilience of ecosystems include

- introduction or removal of functional groups of species that reduce effect and response

diversity, including loss of entire trophic levels (top-down effects),

- impact on ecosystem resources and toxins via emissions of waste and pollutants (bottom-up effects), and
- alteration of the magnitude, frequency, and duration of disturbances to which the biota is adapted.

Similarly there are features of social systems that impact social–ecological resilience like

- degradation of the components of human well-being, including public education and income levels,

- erosion of social capital and adaptive capacity, for example, through corruption, rent seeking, and loss of new opportunities for the future, and
- dysfunctional institutions, causing, for example, weak or insecure property rights and high inequality in power and wealth.

The loss of resilience through the combined and synergistic effects of such factors makes social–ecological systems increasingly vulnerable to changes that previously could be absorbed. As a consequence they are more likely to shift from desired to less desired states. In some cases, these regime shifts or critical transitions may be irreversible or too costly to reverse. Irreversibility is a reflection of changes in critical slow variables (e.g., biogeochemical, hydrological, climatic, constitutional, cultural) and loss of diversity and of social–ecological interactions that support renewal and reorganization into desired states. A challenge for resilience-based stewardship is to address such changes in a more integrated fashion than is usually done today.

In light of the risk for irreversible shifts and their implication for well-being, the self-repairing capacity of social–ecological systems in the face of directional change should not be taken for granted. It must be nurtured. Critical features have been identified (Folke et al. 2003; see Chapter 1) for fostering adaptive capacity and resilience in social–ecological systems:

- learning to live with change and uncertainty,
- cultivating diversity for reorganization and renewal,
- combining different types of knowledge systems for learning,
- creating opportunity for self-organization toward social–ecological sustainability, and
- experimenting and innovating to test understanding and implement solutions.

Adaptive management (see Chapter 4) and adaptive governance of resilience (discussed later in this chapter) will be required, for example, in the context of scenarios of plausible futures to prevent social–ecological degradation or to transform systems into more desired states.

Multiple Regimes

Shifts between social–ecological states can occur because of external perturbations, like a climatic event or a political crisis. They can also occur because of complex cross-scale dynamics within the social–ecological system, with myriad localized interactions among smaller entities serving as a source of adaptation and novelty, and larger-scale emergent constructs such as political systems or climatic conditions serving to frame the behavior and dynamics at smaller scales.

Critical transitions or regime shifts have been described primarily for either ecological systems (e.g., Folke et al. 2004) or social and economic ones (e.g., Repetto 2006). However, shifts also occur due to interactions and feedbacks between social and ecological processes that are triggered by external events or internal dynamics that cause a loss of resilience (Gunderson and Holling 2002, Kinzig et al. 2006). For example, resource-management institutions that perform in a socially and economically resilient manner, with well-developed collective action and economic incentive structures, may in ignorance degrade the capacity of ecosystems to provide ecosystem services. Such behavior may cause a transition to a degraded ecosystem state that in turn feeds back into the social and economic systems, causing unpleasant surprises and social–ecological regime shifts. As a consequence, the social–ecological system may fall into a rigidity or poverty trap (Gunderson and Holling 2002). In **rigidity traps** people and institutions try to resist change and persist with their current management and governance system despite a clear recognition that change is essential. The tendency to lock into such a pattern comes at the cost of the capacity to adjust to new situations. This behavior constrains the ability of people to respond to new problems and opportunities. A **poverty trap**, a social–ecological system with persistent poverty, also reflects a loss of options to develop or deal with change (Bowles et al. 2006). It is locked into persistent degraded conditions and would need external support to get out of it. However, simply providing money, technical expertise, infrastructure, and public education is seldom sufficient to move out of a poverty trap. Escape from rigidity and poverty traps depends on the capacity of people within the social–ecological system to create continuously new opportunities. New opportunities, in turn, are strongly linked to the existence of sources of resilience and adaptive capacity (see Chapter 3) to help people find ways to move out of traps. Hence, the risks or possibilities of sudden shifts between social–ecological states have profound implications for stewardship of essential ecosystem services in a world of rapid and directional change.

Thresholds and Cascading Changes

The movement of a social–ecological system across a threshold to a new regime alters social–ecological interactions, triggering new sets of feedbacks with cascading social and ecological consequences. Once a system has exceeded a threshold, many changes occur that can only be understood or predicted in the local context. However, certain repeatable patterns emerge that provide a basis for designing appropriate management strategies (see Chapters 3 and 4). New interactions frequently become important, and social–ecological processes become sensitive to different slow variables. These changes require a reassessment of management goals and priorities and flexibility to seek new solutions through innovation in institutions and approaches (double-loop learning; see Fig. 4.3). In the absence of these resilience-based strategies, the system may continue to degrade. In Western Australia, for example, extensive areas of native heath vegetation were converted to wheatlands (Kinzig et al. 2006; see Chapter 8), leading to a radically different system, both socially and ecologically, than the shrub savanna occupied by Aborigines. Replacement of deep-rooted heath by shallow-rooted cereals altered the hydrologic cycle by reducing transpiration rate, causing the saline water table to rise close to the surface, creating saline soils that reduced productivity. Declines in production, in turn, caused people to leave the region. The combination of declining population and productivity (33% of the land too saline to farm), coupled with changes in national farm policy, led to amalgamation of farms into larger units that could remain

profitable due to economies of scale (a social transformation). As population declined, many towns could no longer support a service sector, causing still more people to leave and towns to be abandoned. This made it difficult for people who stopped farming to find other jobs, compounding the levels of social stress. This example of cascading consequences of exceeding a social–ecological threshold illustrates several points (Kinzig et al. 2006).

- New sets of interactions come into play when a social–ecological system crosses a threshold, leading to a cascading set of social and ecological consequences.
- The reorganization of the system after crossing the threshold increases the vulnerability to further degradation and threshold changes.
- Each successive transformation is more resilient, in the sense that it would be more difficult to return it to its original state or to some other more desirable state.

The cascading changes that occur when a threshold is exceeded are typical of the behavior of complex adaptive systems (Levin 1999; see Chapter 1), in which any change in the system triggers additional changes in its fundamental properties and feedback structure. New feedbacks then develop that stabilize the system in a new state, making it progressively more difficult to return to the original state.

Directional changes in external drivers such as climate can trigger similar regime shifts and passive degradation. Changing sea surface temperatures in the Atlantic Ocean, for example, reduced rainfall in the Sahel region of sub-Saharan Africa. This reduced precipitation, causing declines in vegetation, which increased regional albedo, weakening the monsoon and stabilizing the drought conditions (Foley et al. 2003a; see Chapter 2). The declines in vegetation caused people to concentrate their herds on the remaining vegetation, causing further increases in albedo and strengthening the drought. These drought-induced feedbacks probably contributed to the long (30-year) duration of drought. Fortunately, large-scale circulation changes eventually ended the drought. Other potential interventions might have included strategies to reduce albedo by increasing vegetation cover (e.g., through planting of forests that tap deep groundwater or introduction of drought-resistant crops that might provide enough food that grazing pressure by cattle could be reduced). It is quite likely that current rates of climate change, if they continue, will trigger regime shifts and cascading social–ecological changes in many parts of the planet. If these are extensive, they could exceed a global **tipping point**—i.e., in this context a threshold for transformational change to a new system, leading to novel global changes in climate, economy, and politics (Plate 3).

Endogenous changes in social–ecological systems can also trigger regime shifts with cascading effects. Several ancient societies such as the Roman Empire and Mayan Civilization appear to have collapsed at least in part because of unsustainable practices that caused environmental degradation and loss of the productive potential of the ecosystems on which they depended (Janssen et al. 2003, Diamond 2005). Similarly, collapse of the Soviet Union in 1990 was a regime shift with many social, economic, and political consequences.

Institutional Misfits with Ecosystems: A Frequent Cause of Regime Shifts

It is no longer rational to manage systems so they will remain the same as in the recent past, which has traditionally been the reference point for managers and conservationists (see Chapter 1). We must instead adopt a more flexible approach to managing resources—management to sustain and enhance the *functional* properties of integrated social–ecological systems that are important to society under conditions where the system itself is constantly changing. Sustaining and enhancing such properties and recombining them in new ways are the essence of sustaining social–ecological development and the very core of the resilience approach to ecosystem stewardship (Folke 2006).

The problem of fit between institutions and ecosystem dynamics in social–ecological systems (see Chapter 4) is one of the most frequent causes of undesirable regime shifts. This interplay takes place across temporal and spatial scales and institutional and organizational levels in a dynamic manner (see Table 4.4).

Temporal Misfits in Social–Ecological Systems

The implementation of conventional resource management tends to lead to governance systems that invest in controlling a few selected ecosystem processes, often successfully in the short term, in order to fulfill immediate economic or social goals, such as the production of wood by forests. But this success tends to turn into a longer term failure through the erosion of social–ecological resilience and key functions (Holling and Meffe 1996; see Chapter 2). "Science-of-the-parts" perspectives (see Chapter 4) have contributed to resource-management systems that focus on producing a narrow set of resources, often in vast monocultures like tree plantations or resource-intensive systems like chicken farms, or salmon aquaculture operations. The widespread approach of "optimal production of single resources" underlying these production systems (Table 5.1) may be successful during periods of stable environmental and economic conditions (Holling et al. 1998). In situations of uncertainty and surprise, however, they become vulnerable because they lack backup systems and sources of reorganization and renewal. These systems are therefore seriously challenged by rapid directional change.

This challenge may also hold true for more diverse management systems with seemingly flexible institutional and organizational arrangements. The Maine lobster fishery, for example, is a sophisticated collective action and multilevel governance system that has sustained and regulated the economically valuable lobster fisheries. It has been considered one of the classic cases of successful people-oriented local management of common-pool resources. However, when the linkage of the social domain to the production of lobsters is taken into account, the Maine fishery seems to have followed the historical pattern of fishing-down food webs (Jackson et al. 2001). Depletion of the cod fishery opened up space for the expansion of species lower down in food webs, like lobsters. Currently the coastline is massively dominated by lobsters, like a coastal monoculture, with the bulk of the lobster population artificially fed with herring supplied as bait in lobster pots. The lobster has a high market price and sustains the social organization and the fishery. However, such simplification of marine systems through removal of functional diversity (see Chapter 2) has created a highly vulnerable social–ecological system waiting for an accident, like a lobster disease, to happen. If such a "surprise" occurs, the lobster population might be decimated over huge areas, perhaps triggering a shift into a very different social–ecological system in which coastal waters no longer provide a viable livelihood for local fishers (R. Steneck and T. Hughes, pers. comm.). Because lobster fishing is central to regional identity, the potential loss of lobster fishing could have severe social as well as economic impacts.

Similar mismatches between short-term success (a governance system that delivers short-term economic and social benefits) and long-term failure of resource management (lack of an ecosystem approach and ecosystem stewardship leading to erosion of resilience) have occurred in forests and lake fisheries (Regier and Baskerville 1986), other coastal and regional fisheries (Finlayson and McCay 1998), crop production (Allison and Hobbs 2004), and a range of other situations (Gunderson et al. 1995; see Chapters 6–14). The question remains to what extent such patterns of resource and ecosystem exploitation can foster adaptive capacity to either prevent passive degradation or actively transform landscapes and seascapes to more beneficial states through sustainable ecosystem stewardship.

Spatial Interdependence of Social–Ecological Systems

Human societies are now globally interconnected, through technology, financial markets, and systems of governance with decisions in one place influencing people elsewhere. However, the interplay between globally interconnected societies and the planet's ecological life-support systems is not yet fully appreciated.

The seriousness and challenges of the climate issue have begun to mentally reconnect people to their dependence on the functioning of the biosphere. The common policy response to climate change has been to focus on mitigation of greenhouse gases through technical means or on social and economic adaptation to climatic change. The urgent policy response that is beginning to emerge as a critical complement is the stewardship of the social–ecological capacity to sustain society with ecosystem services and its links to adaptation, resilience, and vulnerability in the face of unprecedented directional changes (see Chapters 2, 3, and 14). It requires systems of governance that are adaptive and that allow for ecosystem-based management of landscapes and seascapes (see Chapter 4).

Governance to address global issues must be aware of, account for, and relate to the dynamic interactions of people and ecosystems across local-to-global scales. For example, the efforts by large chains of food stores in developed regions and urban centers to reduce temporal fluctuations in the supply of fish, fruits, and other commodities has increased both the extraction and the exploitation of resources in remote areas, creating spatial dependence on other nations' ecosystems (Folke et al. 1997, Deutsch et al. 2007). People in the cities of Sweden, for example, depend on ecosystem services over an estimated area about 1000 times larger than the actual area of the cities, corresponding to about 2–2.5 ha of ecosystem per person (Jansson et al. 1999; see Chapter 13). In this broader context it becomes clear that patterns of production, consumption, and well-being depend not only on locally sustainable practices but also on managing and enhancing the capacity of ecosystems *throughout the world* to support societal development. For example, salmon and shrimp produced in aquaculture operations in temperate and tropical regions, respectively, are traded on global markets and consumed in developed regions and urban centers (Lebel et al. 2002). The feed input to produce these aquaculture commodities comes from coastal ecosystems all across the planet. Shrimp produced in ponds in Thailand, for example, use meal from fish caught in the North Sea. Similar globally interconnected patterns, made possible by fossil-fuel-based technology and supported by information technology, exist in agricultural food and energy production. Demand in one corner of the world shapes landscapes and seascapes in other parts of the planet (see Chapters 12 and 14).

Stewardship of ecosystems is continuously subject to global drivers (Lambin et al. 2001). In this context it becomes important to address the underlying social causes challenging ecosystem capacity to generate services. They include the structure of property rights; macro-economic, trade, and other governmental policies; economic and legal incentives; the behavior of financial markets; causes behind population pressure; transfer of knowledge and technology; misguided development aid; patterns of production and consumption; power relations in society; level of democracy; and worldview, lifestyle, religion, ethics, and values.

In the UK in the 1980s, for example, tax concessions on afforestation were increased but not for the purchase of land. Investors therefore minimized land purchases and located forest plantations on economically low-valued land, such as wetlands, heath, and moorland, thereby depleting "unpriced" wildlife values (Wibe and Jones 1992). In the Brazilian Amazon, one could only acquire a title to land by living on and "using" the land, with logging as a proof of the land being occupied and used. A farm containing "unused" forests was taxed at higher rates than one containing pastures or cropland. The real interest rate on loans for agriculture was lower than for other land uses, and agricultural income was almost exempt from taxation (Binswanger 1990). Hence, policies and activities that, at first glance, seem to be unrelated to the capacity of ecosystems to generate services may indirectly counteract ecosystem stewardships. Such policies serve as subsidies from society to use living resources and ecosystem capacity in unsustainable manners. They need to be redirected into incentives for more sustainable resource use.

Global market drivers sometimes operate so quickly that local governance responses do not have time to respond or adapt, as illustrated by the "roving bandits" phenomena in coastal fisheries (Berkes et al. 2006), where exploiters linked to global markets rapidly move from one fish stock to another over wider and wider spatial scales. This implies that sustainable adaptive governance systems for ecosystem stewardship need to be prepared to deal in a constructive manner with sudden external shocks like the rapid development of a new market demand or sudden shifts in governmental policies.

Fostering Desirable Social–Ecological Transformations

Identifying Dysfunctional States

Dysfunctional states occur when society cannot meet the basic needs of human well-being or when environmental, ecological, social, or political determinants of well-being are degraded to the point that loss of well-being is highly likely to occur. Some social–ecological systems persist in dysfunctional states, such as dictatorships, persistent civil strife, and extreme poverty, for extended periods of time. Other dysfunctional states result from natural disasters such as floods, hurricanes, and tsunamis or from social disasters such as wars. When such systems experience shocks and surprises they may lack the adaptive capacity to reorganize, or they may reorganize in ways that increase the likelihood of future shocks. Getting out of dysfunctional states often requires external institutional, financial, and/or political support, and many bodies from local nongovernmental organizations (NGOs) to international aid organizations like the Red Cross or economic ones like the World Bank work actively to support such transformations. However, external aid is insufficient. Escape from dysfunctional states also requires local development of adaptive capacity for innovation.

Systems can also degrade due to gradual loss of ecosystem services and resilience; increased demand for ecosystem services because of population growth or excessive consumption of services; and various social or political trends. As this degradation proceeds, there is often a spectrum of opinion among stakeholders about whether to fix the current system by incrementally addressing specific problems or enhancing resilience to deliberately explore transformation to a new social–ecological state. In the Goulburn-Broken Catchment in Australia, the agricultural development trajectory was strongly embedded socially and culturally and economically supported, making it difficult to explore new ways to manage the land. Local resources and institutions were initially focused on maintaining a system that fostered a continual downward spiral into a dysfunctional and nonresilient state. More recently, crisis awareness at the system level triggered shifts in perception and action and transformed whole management and governance systems toward ecosystem stewardship of the social–ecological system (Walker and Salt 2006). Similar shifts toward ecosystem stewardship at regional scales are evident in landscapes of southern Sweden and the vast Great Barrier Reef seascape of Australia (Olsson et al. 2006, 2008).

Recognizing Impending Thresholds for Degradation

Scenarios of plausible future changes provide a starting point for exploring policy options that reduce the likelihood of undesirable regime shifts. Global, national, and local assessments often provide clear evidence of trends in environmental, ecological, and social conditions that are leading in unsustainable directions for social–ecological systems at local-to-planetary scales. The causes of many of these trends are increasingly well understood, providing a basis for quantitative or conceptual models that project some of these trends into the future, assuming that people continue their current patterns of behavior ("business-as-usual" scenarios). Continuation of current trends in fossil-fuel use, for example, will likely cause "dangerous" climatic change within the next few decades by altering Earth's climate system beyond a tipping point that would have serious ecological and societal consequences and be difficult to reverse (Stern 2007, IPCC 2007a, b). Similarly, current declines in biodiversity and ecosystem services are degrading livelihoods and well-being of social–ecological systems globally, particularly for underprivileged segments of society (MEA 2005a; see Chapters 2, 3 and 4).

Scenarios represent plausible futures that are based on our understanding of past and current trends. They are not predictions because of the considerable uncertainties that surround future trajectories. Scenarios are most useful for assessing gradual changes that are controlled by processes for which we have both data and models. Trends in global climate, for example, reflect predictable biophysical interactions that can be projected with reasonable confidence over decadal time scales. These trends include resource consumption, fossil-fuel emissions, land-use change, and local-vs-global resource dependence. Scenarios can also be defined that assume a suite of policies and human actions with predictable biophysical, ecological, and social outcomes. These "what-if games" allow comparisons of alternative potential future states, depending on policies that society chooses to implement. Two scenarios commonly accepted by policy makers and the public are that (1) there will be no directional change in controls over social–ecological processes (today's world will remain unchanged) or (2) people will continue their current behavior (business as usual). Scenarios can explore the logical social–ecological consequences of these assumptions or alternative policies that might lead to more desirable outcomes.

The greatest shortcomings of scenarios are that they do not capture (1) the uncertainty associated with processes that are well-understood; (2) the effects of processes that are missing from assumptions and models; (3) many of the complexities of social–ecological interactions and feedbacks; and (4) the unknowable surprises that are an increasingly common property of social–ecological dynamics. Given these severe shortcomings in the capacity of scenarios to predict the future, why would anyone want to use them? Clearly, scenarios should be treated as plausible futures rather than predictions. Scenarios are most useful when used comparatively to explore the logical consequence of *differences* in assumptions about how the world works or policy options that might differ in their social–ecological consequences.

World leaders in industry, government, and the environment disagree about how best to achieve social–ecological sustainability. The Millennium Ecosystem Assessment (MEA) sought to explore the consequences of this spectrum of world opinion by describing four general scenarios of policy strategies intended to enhance social-ecological sustainability (Bennett et al. 2003, MEA 2005c). These scenarios differed in the extent to which policies were global or regional in their design/implementation and whether ecosystem management was reactive (responding to ecosystem degradation after it occurred) or proactive (deliberately seeking to manage ecosystem services in sustainable ways (Cork et al. 2006; Table 5.2). In Global Orchestration, there is global economic liberalization with strong policies to reduce poverty and inequality and substantial investment in public goods such as education. In Order-from-Strength, economies become more regionalized, and nations emphasize their individual security.

Adapting Mosaic also has more regionalized economies, but there is emphasis on multi-scale, cross–sectoral efforts to sustain ecosystem services. In TechnoGarden, the economy is globalized, with substantial investments in sound environmental technology, engineered ecosystems, and market-based solutions to environmental problems (MEA 2005c, Carpenter et al. 2006a). A combination of quantitative and qualitative modeling suggested that these alternative policy options would lead to quite different ecological and social outcomes. In each of them there are tradeoffs among ecosystem services and among social benefits. The most encouraging result was that all scenarios except the Order from Strength would reduce the current net degradation of ecosystem services and improve human well-being, relative to today's uncoordinated spectrum of global policies. Nonetheless, these net improvements result from quite different patterns of tradeoffs and social equity. Some of the key lessons from these scenarios were (Cork et al. 2006):

TABLE 5.2. Defining characteristics of four scenarios[1].

	Global orchestration	Order from strength	Adapting mosaic	Technogarden
Dominant approach for sustainability	Sustainable development, economic growth, public goods	Reserves, parks, national-level policies, conservation	Local-regional co-management, common-property institutions	Green-technology, ecoefficiency, tradable ecological property rights
Economic approach	Fair trade (reduction of tariff boundaries), with enhancement of global public goods	Regional trade blocs, mercantilism	Integration of local rules regulates trade; local nonmarket rights	Global reduction of tariff boundaries, fairly free movement of goods, capital, and people, global markets in ecological property
Social policy foci	Improve world; global public health; global education	Security and protection	Local communities linked to global communities; local equity important	Technical expertise valued; follow opportunity; competition; openness
Dominant social organizations	Transnational companies (Companies that spread seamlessly across many countries): global NGO and multilateral organizations	Multinational companies (Companies that consist of loose alliances of largely separate franchises in different countries)	Local cooperatives, global partnerships, and collaborations established as local groupings recognize the need to share experiences and solutions	Transnational, professional associations, NGOs

Reprinted from Cork et al. (2006).

- No utopian solution is likely to emerge because of tradeoffs among ecosystem services and among social benefits.
- Global cooperation to deal with social and environmental challenges would lead to better outcomes than lack of cooperation.
- Proactive environmental policies would lead to lower risks of major environmental problems and loss of well-being than would reactive policies.
- Participation by a breadth of stakeholders in designing the scenarios clarified the variation in visions about how to achieve sustainability and acceptance of the conclusions of the study.
- Comparison of a small number (2–4) scenarios allowed a diverse but manageable set of options to be considered.

Creating Thresholds for Transformation from Undesirable States

How can ecosystem stewardship help people, communities, and societies escape rigidity and poverty traps? In a rigidity trap there is a tendency to lock management and governance into their existing attitudes or worldviews, making it difficult to respond to changing conditions. Even if there is a general feeling that something needs to be done, it can be surprisingly difficult to get a group out of such gridlock, and the investment in a certain perspective or behavior may be so strong that it is hard to create incentives that are strong enough to change it. Rigidity is deeply rooted because it develops as a way to ensure consistency. In such situations, the "exceptional few" individuals play an important role in catalyzing tipping points and shifting management and governance over a threshold into a new direction. Some individuals appear to be able to mobilize groups to remove the inertia and change management behavior and world views. They may, for example, be particularly well connected, have high social capital, be innovators or early adopters by nature, or have the charisma to cause emotional contagion. The absence of such leaders makes a social group

as a whole rigid and weak when adaptation to change is required (Scheffer and Westley 2007).

It is more difficult to create tipping points to escape from a poverty trap, because these traps are generally characterized by very low levels of social capital and adaptive capacity; initial poverty is often self-reinforcing; and concentrated poverty tends to undermine processes of community organization (Bowles et al. 2006). Even if the group or community can mobilize internally and build adaptive capacity to get out of the trap, it may be overwhelmed by broader-scale factors, such as changes in regional and global markets, governmental corruption, or low level of education in a country as whole. There is often a long historical path dependence of political and economic goals and institutional structures that push a social-ecological system into a poverty trap (Engerman and Sokoloff 2006). The challenges of moving out of poverty traps are huge. Although economic capital and technology support are important, they are often insufficient to help social-ecological systems escape such traps (Bowles et al. 2006).

Moving out of traps requires not just a shift in the social (including economic) dimensions but also active stewardship of ecosystem processes. A major challenge is to secure, restore, and develop the capacity of ecosystems to generate ecosystem services because this capacity constitutes the very foundation for the social and economic development needed to escape from poverty traps (Enfors and Gordon 2007). Ecological restoration and ecological engineering are subdisciplines of ecology that focus on enhancing the capacity of ecosystems to provide services. These fields tend to emphasize the growth and conservation phases of the adaptive cycle. More recently, research on biological diversity as sources of resilience is gaining momentum (Folke et al. 2004). Biological diversity provides the ingredients for regenerating an ecosystem within its current state after disturbance or the seeds for alternative states that might be more viable under new conditions (see Chapter 2). Hence, biodiversity plays a central role in the release and renewal phases of the adaptive cycle. Diverse landscapes and seascapes with resilience have higher capacity to regenerate in the face of disturbance and thereby sustain the supply of ecosystem services. Management that focuses on using protected areas and reserves as reservoirs of biodiveristy to strengthen resilience is gaining ground (e.g., Bengtsson et al. 2003). For example, to enhance the resilience to climate change and secure ecosystem services of the Great Barrier Reef in Australia, the seascape has been rezoned into 70 habitats, each of which has fully protected areas as insurance for ecosystem regeneration after disturbance. A major task of ecosystem stewardship is to identify and manage the role of functional groups of organisms, their redundancy, and their response diversity in relation to ecosystem services at the landscape and seascape scale (Walker 1995, Naeem 1998, Elmqvist et al. 2003, Nyström 2006; see Chapter 2).

Resource management for poverty reduction has tended to focus on water use, food production, or management of other crucial resources, but, in our view, these resources need to be managed in the broader social-ecological context as part of ecosystems and landscape dynamics. Managing for ecosystem resilience is a necessary but insufficient condition for social-ecological transformations from poverty traps into improved states.

Recent work on social interactions also reveals the significance of diversity in human interactions, in institution building, and for collective action (Ostrom 2005, Page 2007). Diversity is a crucial element of resilience for coping with extreme events in a world characterized by accelerating directional change. **Redundancy** (backups), functional diversity of roles (of species in ecosystems, or of people and institutions in social systems), and response diversity (different responses of species, landscape elements, individuals or institutions to suites of disturbances) provide options for flexible outcomes and help social-ecological systems reorganize and develop (see Chapter 2).

These insights illustrate that the search for blueprint solutions, i.e., uniform solutions to a wide variety of problems that are clustered under a single name based on one or more successful exemplars, lead to resource-management failures (Ostrom et al. 2007; see Chapter 4). Yet, diversity seems to be erod-

ing in many dimensions. It is declining systematically in agriculture and most land- and seascapes (MEA 2005a; see Chapter 12). At the same time, in human societies, the benefits of efficiency, rationality, and standardization have resulted in an emphasis on "best practice", efficiency, and a tendency toward monoculture and a dominance of the few. All this challenges resilience, because it leaves us with an impoverished set of sources of novelty for renewal.

Such erosion of resilience can, for example, be counteracted by increasing the diversity of problem solvers in a team, community or society, thereby stimulating a wide range of mental models and also allowing for transparency regarding conflicting viewpoints (e.g., disciplinary background, methodology, conflict and learning styles, age, gender, and cultural background). **Complex problems** (problems with many potential solutions that are quite different in execution and rankable in quality of outcome) may be solved more effectively by a diverse team of competent individuals than by a team composed of the best individual problem solvers (Page 2007). In this sense, social diversity contributes to the sources of resilience that strengthen social-ecological systems. Combining social, ecological, and economic sources of resilience in times of directional and often unexpected change provides the seeds not only for adaptive capacity but also for transforming social-ecological systems into new and potentially more desirable states.

New global institutional structures emerge during rapid globalization from financial markets, multinational companies, trade agreements, IT-developments, and intergovernmental treaties. Currently, however, we lack or have primarily weak international institutions to deal with ecosystem stewardship for sustainability (see Chapter 14). Important advances like UN declarations and the IPCC are still largely disconnected from powerful economic and political institutions. We envision that, in pace with climatic change and associated disturbances, new regional and global governance structures will emerge that will truly merge the ecological and social dimensions for improved stewardship of ecological life-support systems and ecosystem services. For example, structures such as the European Water Directive and the MEA are already emerging. Institutional scholars talk about these structures as multilevel governance systems, and some propose a **polycentric governance structure**, in which citizens are able to organize in multiple democratic governing bodies at differing scales in a specified geographical area to deal with common pool resources and stewardship of ecosystems. Selforganized resource governance systems within a polycentric system may be organized as special districts, nongovernmental organizations, or parts of local governments. These are nested in several levels of general-purpose governments that provide civil equity as well as criminal courts. The smallest units can be viewed as parallel adaptive systems that are nested within ever-larger units that are themselves parallel adaptive systems. The strength of polycentric governance systems in coping with complex, dynamic biophysical systems is that each of the subunits has considerable autonomy to experiment with diverse rules for using a particular type of resource system and with different response capabilities to external shock. In experimenting with rule combinations within the smaller-scale units of a polycentric system, citizens and officials have access to local knowledge, obtain rapid feedback from their own policy changes, and can learn from the experience of other parallel units. Redundancy builds in considerable capabilities and small-scale disasters that may be compensated by the successful reaction of other units in the system (Ostrom 2005).

ecological system into a new regime. **Adaptive governance** has emerged as a framework for understanding transformations by expanding the focus from adaptive management of ecosystems to address the broader social contexts that enable shifts in governance systems toward ecosystem-based management (Folke et al. 2005).

By **governance systems** we mean the interaction patterns of actors, their sometimes conflicting objectives, and the instruments chosen to steer social and environmental processes within a particular policy area. Institutions are a central component in this context, as are the interactions between actors and the multilevel institutional setting, creating complex relationships between people and ecosystem dynamics (Galaz et al. 2008, see Chapter 4).

A transformation of governance may include both shifts in perceptions or mental models and changes in institutions and other essential social features. Adaptive governance research addresses transformations of entire governance systems from one state to another. Transformations that increase the capacity to learn from, respond to, and manage ecosystem feedbacks generally require shifts in social features such as perception and meaning; social network configurations and patterns of interactions among actors; and associated institutional arrangements and organizational structures. In this book we are concerned with transformations that redirect governance into restoring, sustaining, and developing the capacity of ecosystems to generate essential services.

Navigating Transformation Through Adaptive Governance

The capacity to adapt to and shape change is a central component of resilience of social–ecological systems. When there is high adaptability, actors have the capacity to reorganize the system within desired states in response to changing conditions and surprises. But high adaptability may also be recombined with innovation and novelty to transform a social–

Path Dependence and Windows of Opportunity

We still know relatively little about how social–ecological transformations can be orchestrated and the enabling social processes that make it possible for actors to actively push systems from one trajectory to another. Why do certain strategies succeed and suddenly take off, while others utterly fail? There is a need to understand transformative capacity, the capacity to shift from trajectories of unsustainable resource use to sustainable ones in the face of increased resource depletion and global change (Chapin et al. 2006b). Path-dependence characterizes most institutional development and public policy-making (Duit and Galaz 2008). These paths often show a **punctuated equilibrium**, in which long periods of stability and incremental change are separated by abrupt, nonincremental, large-scale changes (Repetto 2006).

Windows of opportunity often trigger these large-scale changes. Sometimes windows open due to exogenous shocks and crises, including shifts in underlying economic fundamentals like a rapid rise in energy price, a change in the macro-political environment, new scientific findings, regime shifts in ecosystems, or rapid loss of ecosystem services. For example, a window for changing direction opened up in water management for agricultural and urban areas in California. Water management had been locked for decades into a highly engineered infrastructure that reinforced one policy and excluded others and pushed the social–ecological system into a crisis. As a response, a new awareness emerged among multiple stakeholders in Californian water management that business-as-usual was no longer a viable option. The window of opportunity opened through the awareness of the crisis. As a necessity, policy and management shifted and broadened to incorporate a wider array of state and federal agencies as well as private and public organizations to address the crisis (Repetto 2006). The social–ecological system seems to be going through a transformation.

Leadership, Actor Groups, Social Networks, and Bridging Organizations

Leadership in Transformations

The interplay between individual actors, organizations, and institutions at multiple levels is central in social-ecological transformations. A literature on the role of leadership strategies in transformations to ecosystem-based management is emerging (Westley 2002, Olsson et al. 2004b, Fabricius et al. 2007; see Chapter 15), with a focus on the relationship between social structures and **human agency** (see Chapters 1 and 4). In the governance systems of the

Everglades in Florida in the USA and Kristianstad Vattenrike in southern Sweden, successful transformations occurred because of the ability of leaders to

- reconceptualize key issues,
- generate and integrate a diversity of ideas, viewpoints, and solutions,
- communicate and engage with key individuals in different sectors,
- move across levels of governance and politics, i.e., span scales,
- promote and steward experimentation at smaller scales, and
- recognize or create windows of opportunity and promote novelty by combining different networks, experiences, and social memories.

Leaders who navigate transformations are able to understand and communicate a wide set of technical, social, and political perspectives regarding the particular ecosystem stewardship issues at hand. Visionary leaders fabricate new and vital meanings, overcome contradictions, create new syntheses, and forge new alliances between knowledge and action.

Diversity of Actor Groups and Social Networks

People with different social functions operating in teams or actor groups play significant roles in mobilizing social networks to deal with change and unexpected events and to reorganize accordingly. Social roles in networks also interact to create tipping points and transformations. Gladwell (2000) identified tipping point roles and labeled them **mavens** (altruistic individuals with social skills who serve as information brokers, sharing and trading what they know) and **connectors** (individuals who know many people (both numbers and especially types of people). They enhance the information base of their social network. Mavens are data banks and provide the message. Connectors are social glue and spread the message. There are also **salesmen** who have the social skills to persuade people unconvinced of what they are hearing. Other social roles of key individuals operating in actor groups include knowledge carriers, knowledge genera-tors, stewards, leaders, people who make sense of available information, knowledge retainers, interpreters, facilitators, visionaries, inspirers, innovators, experimenters, followers, and rein-forcers (Folke et al. 2005).

Social capital (see Chapter 4) focuses on relationships among groups, i.e., the bridging and bonding links between people in **social net-works** (Wasserman and Faust 1994). Applied to adaptive governance, these relationships must be fed with relevant knowledge about ecosys-tem dynamics. This is related to the capacity of teams to acquire and process information, to make sense of scientific data and connect it to a social context, to mobilize the social memory of experiences from past changes and responses, and to facilitate adaptive and inno-vative responses. Social roles of actor groups are all important components of social net-works and essential for creating the conditions that we argue are necessary for adaptive gover-nance of ecosystem dynamics during periods of rapid change and reorganization. Linking dif-ferent societal levels and knowledge systems requires an active role of individuals as coordi-nators and facilitators in co-management pro-cesses. Intermediaries, or middlemen can, for example, play a role in linking local commu-nities to outside markets (Crona 2006). Bring-ing together different actor groups in networks and creating opportunities for new interactions are important for dealing with uncertainty and change and critical factors for learning and nur-turing integrated adaptive responses to change.

Bridging Organizations Connect Different Levels of Governance

Bridging organizations coordinate collabora-tions among local stakeholders and actors at multiple organizational levels (Westley 1995, Hahn et al. 2006). Bridging organizations pro-vide arenas for trust-building, vertical and hor-izontal collaboration, learning, sense-making, identification of common interests, and con-flict resolution. As an integral part of adap-tive governance of social–ecological systems, bridging organizations reduce transaction costs

of collaboration and provide social incentives to participate in ecosystem stewardship. The initiative behind a bridging organization may come from bottom-up, top-down, or from, for example, NGOs or companies that bridge local actors with other levels of governance to gener-ate legal, political, and financial support. Such bridging organizations may also filter external threats and redirect them into opportunities and help transform social–ecological systems toward resilience-based stewardship (Olsson et al. 2004b, 2008). Their role in resilience and sustainability needs further investigation.

Interplay Between the Micro and the Macro

How do new multilevel governance systems emerge? What are the enabling conditions for the emergence of innovative initiatives to deal with ecosystem change, uncertainty and cri-sis, and the social mechanisms that diffuse innovations across scales? The micro level involves encounters and patterned interaction among individuals (which include communica-tion, exchange, cooperation, and conflict), and the macro level refers to structures in soci-ety (groups, organizations, institutions, and cul-tural productions) that are sustained by mech-anisms of social control and that constitute both opportunities and constraints on indi-vidual behavior and interactions (Münch and Smelser 1987). Could social innovations gen-erated at local/regional scales influence and transform governance at a global scale? Can multi-actor experiments be designed that gen-erate new knowledge, network across scales, pressure governance regimes, and ultimately lead to tipping points and transformations to more ecosystem-benign management and gov-ernance? These issues are beginning to be addressed in the context of adaptive gover-nance (Dietz et al. 2003, Folke et al. 2005).

Learning Platforms as Part of Adaptive Governance

The adaptive governance framework suggests a learning approach that includes fostering a

diversity of approaches and creating "learning platforms" to experiment with social responses to uncertainty and change. Such a learning approach has great potential to enhance the resilience of interconnected social–ecological systems and enhance the capacity of ecosystems to produce services for human well-being. For example, initiatives like UNESCO's Man and the Biosphere Programme identifies potential learning sites and policy laboratories. The creation of transition arenas in the Netherlands is another example of experimenting with new approaches for managing and governing water resources (van der Brugge and van Raak 2007).

Successful large and long-lived companies that depend on continuous innovation, such as Phillips or IBM, have addressed the tension between moving forward in a conventional fashion and exploring by "encapsulating" creative or explorative units (Epstein 2008). They often physically separate the research and development departments or teams from the production teams and train special managers who can champion and shepherd the innovation process while buffering it from the demands of production. This allows the company to build up a bank of new ideas and products to draw upon in future launches, while simultaneously producing and marketing successful initiatives (Kidder 1981, Kanter 1983, Quinn 1985). Others implement the ideas when they are successful enough, thereby avoiding the diversion of energy from creativity to production (Mintzberg and Westley 1992).

Three Phases of Social–Ecological Transformation

Transformation can be triggered by perceived threats to an area's cultural and ecological values. In the wetland landscape of Kristianstad in southern Sweden, for example, people of various local steward associations and local government responded to perceived threats by mobilizing and moving into a new configuration of ecosystem management within about a decade (Olsson et al. 2006). This self-organizing process was led by a key individual who provided visionary leadership in directing change

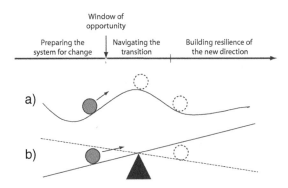

FIGURE 5.4. Three phases of social-ecological transformation, linked by window of opportunity–preparing the system for change, navigating the transition, and building resilience of the new direction. The transformation is illustrated in two ways: (**a**) as a regime shift between multiple stable states, passing a threshold or (**b**) as a tipping point.

and transforming governance. The transformation involved three phases, where phases (a) and (b) are linked by a window-of-opportunity (Fig. 5.4):

(a) preparing the system for change,
(b) navigating the transition, and
(c) building resilience of the new governance regime.

Trust-building dialogues, mobilization of social networks with actors and teams across scales, coordination of ongoing activities, sense making, collaborative learning, and creating public awareness were part of the process.

A comprehensive framework with a shared vision and goals that presented conservation as development and turned problems into opportunities was developed and contributed to a shift in values and meaning of the wetland landscape among key actors. The shift was facilitated through broader-scale crises, such as seal deaths and toxic algal blooms in the North Sea, which raised environmental issues to a top national political priority at the time that the municipality was searching for a new identity. This coincidence of events opened a window of opportunity at the political level, making it possible to tip and transform the governance system into a trajectory of adaptive co-management of the landscape with extensive social networks of practitioners engaged in multilevel governance (Fig. 5.4).

A broader analysis of five case studies confirms this pattern of social–ecological transformation (Olsson et al. 2006). The strategies of preparing for change are shown in Table 5.3. Key leaders and **shadow networks** (informal networks that are politically independent from formal organizations) play a key role in preparing a system for change by exploring alternative system configurations and developing strategies for choosing among possible futures. Key leaders can recognize and use or create windows of opportunity and navigate transitions toward adaptive governance. Leadership functions include the ability to span scales of governance, orchestrate networks, integrate and communicate understanding, and reconcile different problem domains. Successful transformations are often preceded by the emergence of informal networks that help to facilitate information flows, identify knowledge gaps, and create nodes of expertise in ecosystem management that can be drawn upon at critical times. In the Kristianstads Vattenrike Biosphere Reserve and in the Everglades, these networks were politically independent from the fray of regulation and implementation in places where formal networks and many planning processes fail. These shadow networks serve as incubators for new approaches to governing social–ecological systems (Gunderson 1999). These informal, outside-the-fray shadow groups are places where new ideas often arise and flourish. These groups often explore flexible opportunities for resolving resource issues, devise alternative designs and tests of policy, and create ways to foster social learning. Because the members of these networks are not always under scrutiny or the obligations of their agencies or constituencies, they are freer to develop alternative policies, dare to learn from each other, and think creatively about how to resolve resource problems.

In Australia, a flexible organization, the Great Barrier Reef Marine Park Authority, was crucial in initiating a tipping point of governance toward ecosystem-based management (Olsson et al. 2008). This agency was also instrumental in the subsequent transformation of governance systems, from the level of local fisher organizations to national political processes, and provided leadership throughout the process. The Great Barrier Reef study identifies social features and strategies that made it possible to shift the direction of an already existing multilevel governance regime toward large-scale ecosystem-based management. Strategies involved active internal reorganization and management innovation, leading to

- an ability to coordinate the scientific community,
- increased public awareness of environmental issues and problems,
- involvement of a broader set of stakeholders, and
- maneuvering the political system for support at critical times.

The transformation process was driven by increased pressure on the Great Barrier Reef (from terrestrial runoff, overharvesting, and global warming) that triggered a new sense of urgency to address these challenges. It shifted the focus of governance from protection of selected individual reefs to ecosystem stewardship of the larger-scale seascape.

The study illustrated the significance of stewardship that can change patterns of interactions among key actors and allow new forms of management and governance to emerge in response to environmental change. The study also showed that enabling legislation or other forms of social bounds were essential, but not sufficient for shifting governance toward adaptive co-management of complex marine ecosystems.

In contrast to the Great Barrier Reef case, marine zoning in the USA has been severely constrained due to inflexible institutions, lack of public support, difficulties developing acceptable legislation, and failures to achieve desired results even after zoning is established. Understanding successes and failures of governance systems is a first step in improving their adaptive capacity to secure ecosystem services in the face of uncertainty and rapid change.

The case studies discussed here show that transformation is more complex than simply changing legislation, providing economic

TABLE 5.3 Key factors for preparing five different social–ecological systems for tansformation[1].

Social–ecological system	Key factors building knowledge	Networking	Leadership
Kristianstads Vattenrike (creation of the KV)	A new perspective on ecosystem management of integrated landscape-level solutions guided the development of knowledge. It included identifying knowledge gaps for managing the KV and initiating studies to fill them.	The emergence of the network in the mid-1980s connected actors with different interests. This included vertical links and horizontal links between government agencies, NGOs, the municipality, and landowners.	Leadership emerged that was important for connecting people, developing, and communicating a vision of ecosystem management, and building trust and broad support for change.
The Everglades (ecosystem restoration)	A few key scientists were frustrated by continuing ecosystem degradation, which they tried to address in workshops. The ecosystem restoration (resilience) perspective guided modeling workshops in which information was synthesized and used to develop composite policies.	A network of scientists emerged in the late 1980s and formed the adaptive management group. In 1992, networking was extended into the management and political areas to spread the ideas of the adaptive management group, link actor groups operating at different organizational levels, and represent different interests.	Leadership emerged that brought in a novel perspective of ecosystem resilience, built trust, and connected people. The leaders were weary of ongoing legal actions and wished to pursue alternative ways of management. They focused on ensuring the engagement of all groups, not just a few special interest groups.
The Northern Highlands Lake District (sustainable futures)	The polarization among different actor groups hinders the sharing of new ideas and innovations. However, a few bridging efforts are developing, and these could nucleate to provide the necessary institutions for building and sharing knowledge.	Networking at a regional scale that connects different groups of actors is poorly developed.	Leadership for collective action and ecosystem management at the regional level has not emerged. Instead, leadership has emerged for pursuing specific interests.
Mae Nam Ping Basin (sustainable water management)	Knowledge based on the ecosystem approach has been assimilated from a wide range of sources, and innovative ecosystems approaches exist but do not guide networking at the regional level.	Networking at the basin level is lacking. Instead, networks that serve and protect specific interests are developing.	Leadership for collective action and ecosystem management at the basin level has not emerged. Instead, leadership has emerged for pursuing specific interests.
Goulburn-Broken Catchment (sustainable agriculture)	There was a lack of innovation that made it impossible to explore new configurations of the system, in particular, ways to address ecosystem processes. Building knowledge to support the status quo approach to ecosystem management was emphasized.	Networks emerged that connected people and interests at different levels. These networks were later formalized into decision-making and implementing organizations.	Leadership did emerge for collective action at the catchment level, but not to provide a novel approach to ecosystem management.

1. Reprinted from Olsson et al. (2006).

incentives or introducing new restrictions on resource use. As observed by McCay (1994), the change of perceptions and mental models of the significance of ecosystems for human well-being is an important part of ecosystem stewardship and can change human behavior on a fairly large scale without involving the political processes of making and changing institutions.

Actions that foster successful transformations of social–ecological systems toward adaptive governance of landscapes and seascapes often include:

- Change attitudes among groups to a new, shared vision; differences are good, polarization is bad.
- Check for and develop persistent, embedded leadership across scales; one person can do it for a time, but several are better locally, regionally, and politically.
- Design resilient processes, e.g., discourse and collaborations, not fixed structures.
- Evaluate and monitor outcomes of past interventions and encourage reflection followed by changes in practices.
- Change is both bottom-up and top-down. Otherwise, scale conflicts ultimately compromise the outcome; globalization is good but can destroy adaptive capacity both regionally and locally.
- Develop and maintain a portfolio of projects, waiting for opportunities to open.
- Always check larger scales in different sectors for opportunities; this is not science, but politics.
- Know which phase of an adaptive cycle the system has reached and identify thresholds; talk about it with others.
- Plan actions for surprise and renewal differently than growth and conservation; efficiency is on the last part and resilience on the first.
- The time horizon for effect and assessment is at least 30–50 years; restructuring resilience requires attention to slow dynamics.
- Create cooperation and transform conflict, but some level of conflict ensures that channels for expressing dissent and disagreement remain open.

- Create novel communication face-to-face, individual-to-individual, group-to-group, and sector-to-sector.
- Encourage small-scale revolts, renewals and reorganizations, not large-scale collapses.
- Try to facilitate adaptive governance by allowing just enough flexibility in institutions and politics.

These generalizations can help managers navigate more effectively the periods of uncertainty and turbulence that are unavoidable components of any social–ecological transformation.

Summary

This book emphasizes the need for ecosystem stewardship to generate a deeper understanding of integrated social–ecological systems undergoing change. It requires an expansion of focus from managing natural resources to stewardship of dynamic and evolving landscapes and seascapes in order to sustain ecosystem services. Such stewardship requires governance systems that actively support ecosystem-based management and allow for learning about resource and ecosystem dynamics. A challenge is to develop governance systems that are flexible, adaptive, and have the capacity to transform. It requires dealing with change—not just incremental and predictable change, but uncertain, abrupt, and surprising change. Chapter 5 has identified and discussed features of social–ecological systems that create barriers and bridges for transformations to more desired states and presents strategies for building and enhancing their resilience in times of directional change.

The first five chapters of the book presented existing and emerging theory and concepts in relation to resilience-based ecosystem stewardship and serve as the foundation for the remaining chapters. The following chapters, structured into major types of social–ecological resource systems covering local-to-global scales, illustrate this foundation and provide insights, challenges, and implementation strategies for improved stewardship of terrestrial and marine ecosystems and the services and fundamental support that they provide to humanity.

Review Questions

1. How does transformation differ from adaptation?
2. What types of surprises occur in social–ecological systems? How do management strategies differ in preparing for and responding to each type of surprise?
3. What human actions can change resilience? In what ways might they interact and lead to regime shifts of social–ecological systems? What intervention strategies might address these interactions and reduce the likelihood of undesirable regime shifts?
4. How might appropriate intervention strategies differ between poverty traps, rigidity traps, and cascading effects of regime shifts?
5. In what ways can mismatches of institutions and ecosystems lead to surprises and regime shifts?
6. What tools are available to deal with true uncertainty?
7. What is adaptive governance and how does it differ from adaptive co-management?
8. Which are the major phases of transformations and their social features? In what ways can management actions foster social–ecological resilience to facilitate actively navigated transformations?

Additional Readings

Carpenter, S.R., E.M. Bennett, and G.D. Peterson. 2006. Scenarios for ecosystem services: An overview. *Ecology and Society* 11(1):29 [online] URL: http://www.ecologyandsociety.org/vol11/iss1/art29/

Folke, C., T. Hahn, P. Olsson, and J. Norberg. 2005. Adaptive governance of social–ecological systems. *Annual Review of Environment and Resources* 30:441–473.

Gunderson, L.H., and C.S. Holling, editors. 2002. *Panarchy: Understanding Transformations in Human and Natural Systems*. Island Press, Washington.

Olsson, P., L.H. Gunderson, S.R. Carpenter, P. Ryan, L. Lebel, et al. 2006. Shooting the rapids: Navigating transitions to adaptive governance of social-ecological systems. *Ecology and Society* 11(1):18. [online] URL: http://www.ecologyandsociety.org/vol11/iss1/art18/

Repetto, R. 2006. *Punctuated Equilibrium and the Dynamics of U.S. Environmental Policy*. Yale University Press, New Haven.

Scheffer, M., S.R. Carpenter, J.A. Foley, C. Folke, and B.H. Walker. 2001. Catastrophic shifts in ecosystems. *Nature* 413:591–596.

Walker, B.H., C.S. Holling, S.R. Carpenter, and A.P. Kinzig. 2004. Resilience, adaptability and transformability in social–ecological systems. *Ecology and Society* 9(2):5 [online] URL: http://www.ecologyandsociety.org/vol9/iss2/art5/

Part II
Stewarding Ecosystems
for Society

6
Conservation, Community, and Livelihoods: Sustaining, Renewing, and Adapting Cultural Connections to the Land

Fikret Berkes, Gary P. Kofinas, and F. Stuart Chapin, III

Introduction

Since most of the world's biodiversity is not in protected areas but on lands used by people, conserving species and ecosystems depends on our understanding of social systems and their interactions with ecological systems. Involving people in conservation requires paying attention to livelihoods and creating a local stake for conservation. It also requires maintaining cultural connections to the land and at times restoring and cultivating new connections. This chapter addresses human–wildlife–land interactions across a range of hinterland ecosystems, from relatively undisturbed "wildlands" to more intensively manipulated rural agricultural areas where local communities are an integral component of the landscape. These regions commonly comprise unique ecosystems that are in many cases important hotspots of global biodiversity. Here we examine three case studies – Ojibwa and Cree use of boreal forest biodi-

versity, the community-based programs for elephant conservation in sub-Saharan Africa, and Gwich'in engagement in international management of the Porcupine Caribou Herd in Arctic North America. The three cases highlight the relationship between conservation and community livelihoods to illustrate strategies that communities have used and the challenges they face to sustain land, resources, and their own well-being.

Historically, hinterland regions of low human density have been the homelands of people who are highly dependent for their livelihoods on ecosystem services through small-scale agriculture, forest resource use, small-scale fisheries, and subsistence economies of hunting and gathering. By some estimates, indigenous peoples of these hinterland regions inhabit 20–30 percent of the earth's surface, about four to six times more area than is included in all formally designated protected areas of the world (Stevens 1997). These areas also provide cultural and sometimes spiritual connections to land, plants, and animals. But in the globalized world of today, no place is isolated from external drivers such as global market forces. Thus, challenges to sustainability are ever-present, even when the social–ecological system is distant from densely inhabited areas. In many cases, these areas

F. Berkes (✉)
Natural Resources Institute, University of Manitoba, Winnipeg, Manitoba, Canada R3T 2N2
e-mail: berkes@cc.umanitoba.ca

F.S. Chapin et al. (eds.), *Principles of Ecosystem Stewardship*,
DOI 10.1007/978-0-387-73033-2_6, © Springer Science+Business Media, LLC 2009

are increasingly encroached upon by nonlocal interests. In other cases, problems are due to local population increase, land-use intensification, introduction of new markets, persistent social inequities at community and regional levels, and competition over scarce resources. And in still others, the challenges are related to the effects of global-scale processes, such as climate change. A serious challenge in conservation for people in all these regions is how to sustain the connections between people and the land. Approaches to these challenges illustrate several principles presented in Chapters 1–5:

1. People and nature are integral components of all social–ecological systems, from wilderness to cities.
2. Local and traditional knowledge, cultural legacies, social institutions, and social networks play a critical role in sustaining the use of ecosystem services.
3. Resource uses compatible with conservation objectives can be important sources of resilience by fostering ecological renewal cycles.
4. Participatory approaches that include resource users in monitoring, research, and policy-making improve understanding, responsiveness to change, and compliance with rules.

5. Changes in wildlife habitat often reflect regional and global processes, so sustainability requires attention to cross-scale interactions.

Strategies for Conservation and Biodiversity

Different strategies have been implemented across the globe to achieve conservation and maintain biodiversity (Box 6.1). One strategy is to establish protected areas that strictly exclude human use and provide for the preservation of wilderness. In other cases, such as Biosphere Reserves, human use areas are established adjacent to core protected areas, with restricted-use areas as an intervening buffer zone. In yet other cases, intensive human use across a region occurs with strategies that seek the related goals of human livelihoods and biodiversity protection. In the face of rapid change in land use and the extinction of species, there is considerable debate about which strategies are best for achieving conservation objectives. Climate change complicates this debate, because areas considered to be sensitive habitat today may shift geographically in the future.

Box 6.1. Categories of Protected Areas

Protected areas are defined by World Conservation Union as "*An area of land and/or sea especially dedicated to the protection and maintenance of biological diversity, and of natural and associated cultural resources, and managed through legal or other effective means.*" The World Conservation on Protected areas and the World Conservation Union (IUCN) have provided leadership in classifying protected areas into six categories (see below), ranging from those dedicated exclusively to preserving ecologically unique areas to those in which people are part of important landscapes. This classification system has been modified since first established in 1978, reflecting shifts in thinking about ecological change, the recognized importance of unique ecosystems providing for hinterland people's livelihoods, and the involvement of local communities in conservation (IUCN 1991, Stevens 1997).

CATEGORY Ia: Strict Nature Reserve: protected area managed mainly for science.

Definition: Area of land and/or sea possessing some outstanding or representative ecosystems, geological or physiological features, and/or species, available primarily for scientific research and/or environmental monitoring.

CATEGORY Ib: Wilderness Area: protected area managed mainly for wilderness protection.

Definition: Large area of unmodified or slightly modified land, and/or sea, retaining its natural character and influence, without permanent or significant habitation, which is protected and managed so as to preserve its natural condition.

CATEGORY II: National Park: protected area managed mainly for ecosystem protection and recreation.

Definition: Natural area of land and/or sea, designated to (a) protect the ecological integrity of one or more ecosystems for present and future generations, (b) exclude exploitation or occupation inimical to the purposes of designation of the area, and (c) provide a foundation for spiritual, scientific, educational, recreational, and visitor opportunities, all of which must be environmentally and culturally compatible.

CATEGORY III: Natural Monument: protected area managed mainly for conservation of specific natural features.

Definition: Area containing one, or more, specific natural or natural/cultural feature which is of outstanding or unique value because of its inherent rarity, representative or aesthetic qualities or cultural significance.

CATEGORY IV: Habitat/Species Management Area: protected area managed mainly for conservation through management intervention.

Definition: Area of land and/or sea subject to active intervention for management purposes so as to ensure the maintenance of habitats and/or to meet the requirements of specific species.

CATEGORY V: Protected Landscape/ Seascape: protected area managed mainly for landscape/seascape conservation and recreation.

Definition: Area of land, with coast and sea as appropriate, where the interaction of people and nature over time has produced an area of distinct character with significant aesthetic, ecological and/or cultural value, and often with high biological diversity. Safeguarding the integrity of this traditional interaction is vital to the protection, maintenance, and evolution of such an area.

CATEGORY VI: Managed Resource-Protected Area: protected area managed mainly for the sustainable use of natural ecosystems.

Definition: Area containing predominantly unmodified natural systems, managed to ensure long-term protection and maintenance of biological diversity, while providing a sustainable flow of natural products and services to meet community needs.

The American national parks model of protected areas is considered by many as synonymous with biodiversity and preservation. Variants of this approach have been applied throughout the globe, and in some cases have significantly enhanced biodiversity in the face of global change. Yellowstone National Park, among the first national parks in the world, was established out of concern for the western expansion of settlers and the impacts of that expansion on Native American societies, wildlife, and wilderness. In spite of many park-management problems, such as loss of species and the suppression of most fires, the protection of these wildlands has supported the persistence of unique species in a wilderness landscape that is of high cultural value to society.

The national park approach, however, comes with limitations. First, the vast numbers of visitors may impact a region even where no extractive use is allowed (2.8 million visitors to Yellowstone in 2006). Second, exclusion of people and their associated disturbances may not be sufficient to maintain a particular ecological state. Ecosystems are naturally dynamic, and sustaining ecosystem resilience requires the maintenance of ecological variability and cycles (see Chapter 2). Third, effective protection in some cases may require that professional resource managers work closely with local residents, rather than excluding them or their livelihood activities from the protected area.

The idea of protected areas and notions of wilderness as free from people have in some regions created serious conflicts with local land and resource use. People-free parks are referred to by critics as the "fences and fines" approach to conservation.

The American national park model is not the only way to achieve conservation. National and international approaches to conservation have increasingly used a mix of strategies to maintain biodiversity and conservation, including those in which community-based management serves as a foundation (Borgerhoff Mulder and Coppolillo 2005). The appropriateness of these strategies depends to a great extent on scale. In the absence of adequate community institutions for conservation, national-level strategies may be critical, especially where internal conflict and or national military activities threaten endangered species. But where local communities have a close relationship with land and resources, and where institutions provide for sufficient local incentives to conserve, community-based approaches can support biodiversity while addressing issues of human livelihoods and community sustainability. In those hinterland regions where there is poverty or marginal economic opportunities, stewardship of nature may be an effective means of addressing the problems of livelihoods (WRI 2005).

Critical Social–Ecological Properties that Support Both Conservation and Livelihoods

Many small-scale indigenous and other rural societies throughout the world live in close relationship with lands and resources. These hinterlands support the rural economy of much of the developing world, as well as the parts of the developed world where people make a livelihood using a variety of local resources – forests, wildlife, fisheries, and rangelands, along with small-scale agriculture. In many regions today, these activities represent mixed economies of subsistence harvesting and small-scale cash markets.

Human–environment relationships among many hinterland peoples are reflected in their social organization, modes of production, and cultural worldview. These lands can include abandoned agricultural and restored environments, such as rehabilitated wetlands, as well as the more "pristine." They can support agroforestry systems that are low-intensity mixes of agriculture and tree crops, such as those found in many parts of Asia, Latin America, and Africa. The ecosystems discussed in this chapter usually (but not always) have low human population densities. These lands often retain a relatively high proportion of their native biodiversity and a relatively productive set of ecosystem services. They are the rural areas of the world that support a diversity of economic activities that enable the inhabitants to make a livelihood largely from their local ecosystems.

Although many systems are losing biological diversity and associated ecosystem services at an accelerating pace (MEA 2005d), the outlook is not entirely negative. There is an enormous reservoir of cultural understanding of ecosystems (Posey 1999, Berkes 2000) and an associated diversity of environmental ethics (Taylor 2005). Through many generations, many societies have developed sensitivities to ecological change and strategies for responding to them in ways that foster sustainability. On the one hand, we need to study and build on this experience. On the other, we need to seek significant changes in existing policies, institutions, and practices to meet the challenge of reversing the degradation of ecosystems in the face of increasing demands for ecological services. As important as they are, setting aside protected areas alone will not be sufficient to meet this challenge.

There is an increasing interest in conservation on lands used by rural people. Major international organizations have been exploring various options (WRI 2005). One UN agency, the UN Development Programme (UNDP), has encouraged grassroots projects that combine biodiversity conservation and poverty alleviation goals in tropical countries. This project, called the UNDP Equator Initiative (Timmer and Juma 2005, Berkes 2007, UNDP 2008), is one of many ways in which the international community has seriously begun to explore ways to reconcile conservation and livelihoods (Brown 2002). There are active debates on the appropriate role of local communities in this kind of integrated conservation development. While the track record of community-based conservation as practiced in the last 20 or so years is mixed, there is an emerging consensus that rural peoples of the world and their communities need to be involved as major partners in any conservation effort (Berkes 2004).

Conservation biologists Bawa and colleagues (2004) have been exploring the implications of different approaches to conservation. They observed that a number of conservation models have been developed by large international conservation organizations. These models seek unifying conservation principles and have been partly driven by a need for general application. However, these generalized, monolithic models tend to view the world through relatively coarse filters and may be at odds with the emergence of fine-grained models adapted to local conditions (Bawa et al. 2004). The implementation of such monolithic models can lead to the opposite effect, ultimately eroding biological diversity and ecosystem health.

Conservation efforts at all levels are confronted by complexity and heterogeneity of ecosystems. In lands that serve both conservation and livelihood needs, the situation is further complicated by a myriad of social, economic, and institutional factors that influence the prospects for conservation. Monolithic conservation models that are broadly applied are not equipped to deal with human issues. However, many protected areas include human uses, and the management of most kinds of protected areas, even national parks, involves dealing with people who live in or around them. Conservation science has proceeded to develop as a technical field with narrowly trained technical experts, whereas practical conservation requires interdisciplinary skills, including the

ability to understand social, economic, and institutional processes (Berkes 2004).

Before the advent of conservation science, many local groups successfully conserved biodiversity and protected ecological processes by following a diversity of approaches, ranging from the strict preservation of sacred areas (Ramakrishnan et al. 1998) to sustainable resource use. Obviously, in many areas, human activities have caused the degradation of ecosystems and loss of biodiversity. But this is not always the case, and it is important to document and understand how some of the more successful traditional systems work.

The Western Ghats in southern India is one of the "biodiversity hotspots" of the world. These low-lying beautiful green mountains support high levels of biodiversity as well as a relatively high population density (Plate 5). Much of the landscape is in diverse agroforestry systems (as opposed to monoculture plantations). Annual, perennial, and tree crops are grown together by landholders in species combinations that have evolved over hundreds of years. The patchwork of these agroforestry operations are interspersed with government-designated protected areas and traditional sacred groves where people refrain by social custom from exploitation. Bhagwat et al. (2005) compared the biodiversity of several groups of organisms in three kinds of areas: forest reserves, sacred groves, and coffee agroforestry plantations. They found surprisingly high levels of biodiversity, comparable to that found in forest reserves, in sacred groves and in coffee plantations (Fig. 6.1). Although endemic tree species were more abundant in the forest reserve than in sacred groves, threatened trees were more abundant in sacred groves than in the forest reserve. They concluded that sacred groves maintained by tradition and the multifunctional landscapes produced by centuries-old systems of agroforestry should be considered an integral part of conservation strategies in this region (Bhagwat et al. 2005).

Based on findings such as these, Bawa et al. (2004) argue that the best hope for conservation in a complex and rapidly changing world is to focus on locally driven approaches that (1) draw

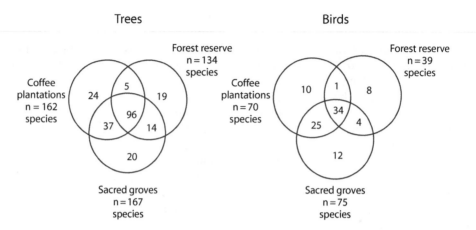

FIGURE 6.1. Numbers of shared and restricted tree species and bird species found in forest reserves, sacred groves, and coffee plantations in Kodagu, Western Ghats of India. Adapted from Bhagwat et al. (2004).

on local and traditional practices as well as science; (2) are locally adaptive; and (3) seek to increase human and social capital in addition to natural capital. They argue that initiatives that incorporate participatory approaches, conflict resolution, partnerships, and mutual learning are likely to garner broad support. Such approaches are relevant to many parts of the world. Empowerment of local communities and strengthening local institutions are preconditions to the success of long-term conservation goals. These goals may be best achieved through a focus on building resilience in social–ecological systems to be able to respond to shocks and surprises, and maintaining a reciprocal relationship between humans and nature.

The sections below present three dimensions of linking conservation, communities, and livelihoods. Each is supported by a case study to illustrate (1) how community-based approaches can support conservation and livelihoods, (2) how community institutions for conservation of threatened species can evolve when attention is paid to the incentive structures that provide for livelihoods, and (3) how co-management arrangements can serve to link communities to the national and international arenas to give local interests a voice in conservation. A fourth section illustrates how conventional single-species approaches to conservation of wild resources can be counter-productive if social–ecological interactions are neglected.

Addressing Conservation Issues

Traditional Management of Biodiversity

Many traditional societies have used ecosystem services without destroying the ecosystems of which they have been part for thousands of years. Otherwise, there would have been no resources left for us to conserve today. A landscape produces a variety of goods and services for livelihood needs, and some human actions maintain ecological processes and biodiversity. In practice, people actively disturb and manage their landscape through the triggering of successional processes. Disturbance helps maintain spatial and temporal diversity at both landscape and site levels. The specific mechanisms by which biodiversity is conserved in these multifunctional landscapes vary with the ecological and cultural settings. The Cree and Ojibwa (Anishnaabe) people of the boreal forest of central Canada have developed practices that sustain biodiversity. For example, the use of fire helps maintain ecological functions and renewal cycles of the boreal forest ecosystem while enabling the Cree and Ojibwa to meet their livelihood needs (Box 6.2). The case study illustrates the potential of using local cultural practices and values to protect biodiversity.

Box 6.2. Boreal Case Study of Indigenous People Keeping the Land Healthy

Sources: Berkes and Davidson-Hunt (2006); Davidson-Hunt et al. (2005); O'Flaherty et al. (2008); Lewis and Ferguson (1988).

The central Canadian boreal zone is populated mainly by the Cree and the Ojibwa (Anishnaabe), two large and related groups of indigenous people. There is little agriculture because of the short growing season and poor soils in this glaciated landscape. Traditionally, up until 1960s or so, people were mobile, following an annual cycle of wildlife harvesting activities. The subsistence system was based on terrestrial and aquatic animals (moose, caribou, and beaver), waterfowl (geese and ducks), fish, medicinal plants, berries, and other nontimber forest products (NTFPs). The cash economy depended on the fur trade. Senior hunters or stewards provided leadership, oversaw proper land use practices, and rules of respect and reciprocity. Social organization was by family groups, loosely organized into local bands, and since the 1960s, into permanent villages consisting of several local bands.

In recent decades, traditional family-based land use practices and institutions have become weaker but persisted, with the hunters relying on rapid transport such as snowmobiles and motor canoes to access resources seasonally. The land-based economy is still important for food, cash (forestry, commercial fishing, hunting and fishing outfitting, ecotourism), and a sense of cultural identity. Much of the cash economy is now based on service industries in permanent towns and villages. Store-bought food is gaining in importance, even though it provides a poorer diet, high in fat and carbohydrates, than the one based on fresh meat.

The use of many NTFPs is linked to the ecological processes of disturbance and succession. Various species of trees and shrubs are distributed in space and time relative to disturbance. For example, in the lands of the Shoal Lake Ojibwa people in northwestern Ontario, fireweed (*Epilobium angustifolium*

L.) occurs in the early years following a disturbance, ginseng (*Panax quinquefolius* L.) is found under mature forest canopies, while highbush cranberry (*Viburnum trilobum* L.) often occurs along riverbanks disturbed periodically by spring flooding. Fireweed and ginseng are utilized as medicinal plants while highbush cranberry is an edible berry.

Many NTFPs typically depend on succession management, and a huge diversity of succession-management systems have been documented from all parts of the world. Ecologically speaking, these succession-management systems all involve renewal cycles initiated by a disturbance event. The disturbance could be a natural fire, a pest infestation, a blowdown, or it could be a human-made fire or a forest cut. A typical renewal cycle starts with an early successional phase of rapidly growing herbaceous plants (see Chapter 2). Gradually, shrubby plants take over, shading out the grasses and other pioneer species. Larger trees gradually take over, approaching a climax phase before the cycle repeats.

In the boreal zone, the climax phase is not biologically productive. To provide fresh browse for wildlife and NTFPs for people, a disturbance is needed to release nutrients and start the renewal cycle over again. In the central Canadian boreal zone, the typical fire frequency is about 300 years; insect infestations may be more frequent. Traditionally, some groups of boreal indigenous people used small-scale fires to clear the land, attract grazing animals and waterfowl, but this practice was banned in the 1950s by resource-management authorities. As rediscovered in recent years, and now used by parks authorities, frequent small disturbance events actually help ecosystem renewal. Conversely, the prevention of small disturbances makes a forest ecosystem increasingly vulnerable to large, potentially disastrous disturbances. A classical example is Sequoia National Park in the USA, where a century of fire prevention allowed white fir trees to form a ladder of fuels from the understory to the canopy of giant sequoia trees that had resisted understory fires for thousands of years (Stephens 1998).

In resilient ecological systems, small disturbances precipitate the release phase that helps system renewal by leading to a reorganization phase in which the memory of the system enables ecological cycles to start over (see Chapter 2). The memory can consist, for example, of pine cones, anything that helps the forest ecosystem to perpetuate itself. It can also consist of the "social memory" of traditional practices such as those of the Ojibwa that help renew forest ecosystems.

Senior hunters and land experts among the Cree and Ojibwa have an understanding of renewal cycles that is similar to that of contemporary ecologists. Although they rarely practice fire management (it is illegal), they use naturally occurring disturbances and forestry clear-cuts to get what they need from the forest. Their ecological understanding is mixed with indigenous spirituality and land ethics. For example, in the perspective of the Iskatewizaagegan (Shoal Lake) Ojibwa, the Creator provided everything that the people would need for their survival. In return, the Ojibwa hold the responsibility to maintain these gifts that were given to them. Practices that harm these gifts can lead to certain undesirable consequences for an individual or the individual's family. There is a basic duty upon the Ojibwa not to try to "control" nature. The people believe that the forest changes all the time but follows a natural pattern. The same principle is followed in Pikangikum, a different Ojibwa group, and is concisely translated into English as: "as was, as is".

The creation of blueberry patches through repeated burning is not seen as a contradiction of this principle. Burning or other disturbance simply reveals the different combinations of plants that naturally occur. "As was, as is" means that all that was on the land before should be still be there today and also tomorrow. While a fire may destroy a forest, it also reveals the natural pattern that blueberries follow fire on sandy soils. Then other plants follow the blueberries, and then the forest returns. The cycle can be modified by humans but should not be disrupted.

Ojibwa elders hold that all plant species are equally important. It cannot be said that some are more important than others. What is important is the protection of the full suite of plant species. This is because every habitat and species has a reason to be there, known or unknown to humans, and for that reason the full set of plant species should be maintained into the future. There is a second major argument for maintaining the full suite of biodiversity. Many species have known uses, including survival value in difficult times. But the value of a plant, such as its medicinal value, cannot be known ahead of time. This knowledge is accessed during the healing process, as a healer may receive a vision during a dream in which a plant, or other being, offers itself. Hence one must hold an attitude of humility for the mysteries of nature, as one never knows the real value of a plant given by the Creator.

The Ojibwa practice of site-specific burning, in combination with landscape-scale natural fires, would increase the temporal diversity of the boreal forest. The combined outcome is a landscape that is diverse spatially. Within that spatial diversity, there are habitats at different temporal stages that increase the overall landscape biodiversity. Small disturbances that create habitat diversity may also help fireproof the landscape and increase resilience by reducing the fuel load on the forest floor, thus averting large disturbances.

The first mechanism of conserving and enhancing biodiversity is to maintain all successional stages. Since each stage in forest succession represents a unique community, having forest patches at different successional stages maintains overall biodiversity. At the same time, it contributes to the continued renewal of ecosystems by conserving the building blocks for renewal and reorganization. The second is the creation of what landscape ecologists might call patches, gaps, and mosaics. It is well known in landscape ecology that low and intermediate levels of disturbance often increase biodiversity as compared to nondisturbed areas. Boreal hunters know this mechanism intuitively and use it effectively, for example, through the creation of moose yards by the use of fire (Lewis and Ferguson. 1988). The third is the creation of edges (**ecotones**). Edges exist in nature but new edges can also be created by disturbance. Boundaries between ecological zones are commonly characterized by high diversity, both ecologically and culturally (Turner et al. 2003).

Detailed studies with one boreal group, the Shoal Lake Ojibwa, show that their principle of maintaining the full suite of biodiversity is similar to the scientific practice of multifunctional landscape management. But the approach is from a different angle. Ojibwa elders do not seek to connect biodiversity to known functional properties of a habitat or species, as scientists do. Rather, the elders believe that, since the Creator has provided everything that the Ojibwa need to survive, there must be a reason for the existence of every plant, animal, and other beings (Davidson-Hunt et al. 2005). The maintenance of these species and the health of the land on which they grow are part of their ethical responsibility.

Introduction of new values and approaches and global economic drivers have been impacting the boreal forest and creating new challenges. Logging is the major management issue facing the boreal forest of central Canada. Large-scale timber production makes little allowance for the ecological processes that produce berries and medicinal plants. Logging operations have been gradually moving northward on Ojibwa traditional lands. What determines a "commercial forest" is the international price for pulp and paper products. Paper from the boreal forest is not used locally but is exported worldwide. Logging is only one use of the forest ecosystem but an increasingly dominant one. The economic impact of forestry dwarfs any monetary benefits from wildlife, ecotourism, berries, and medicinal plants.

Can these forests be managed sustainably – in a way that conserves ecosystem processes while at the same time providing for timber production and other ecological services? Learning from traditional management systems, such as those of the Ojibwa, is important for broadening conservation objectives and approaches. The use of local and traditional ecological knowledge is an effective mechanism for the empowerment of indigenous communities so they can participate in making the decisions that impact their resources and livelihoods.

Threatened Species and Community-Based Conservation

Threats to some of the world's endangered species are found in social–ecological systems of the developing world. Elephants of east Africa, turtles of Costa Rica, blue sheep of the Himalayas, pandas in China, and other threatened species are commonly found where human populations face severe poverty; there are limited options for economic development, and in some cases ecosystem services have been dramatically reduced in quality (WRI 2005). Historically, nongovernmental organizations (NGOs) have taken a lead role in research and international advocacy for the conservation of threatened species. Species such as elephants and whales have become symbolic in fund raising for many of these organizations, with some advancing the ideals of animal rights and the prohibition of harvestings (Freeman and Kreuter 1994). To meet the conservation challenges of threatened species, protected areas, often in the form of national parks, have been established throughout the developing world, some successful and others failing to achieve their conservation objectives (Robinson and Bennett 2000). In some of these cases, imposed protected areas have exacerbated problems by relocating residents from their homelands, creating heightened conflicts among those seeking to conserve resources and potential community resource stewards (Songorwa et al. 2000). Conservation issues can be especially problematic adjacent to national parks where wildlife populations are increasing and where local residents view these species as pests that transform culti-vated lands (Getz et al. 1999).

The paradigm of **community-based wildlife management** assumes that high levels of community involvement in resource management and explicit incentives for conservation encourage stewardship behavior and protect threatened species from possible extinction. During the past 20 years the World Bank has shifted away from supporting top-down arrangements for economic development and adopted this new approach. The related shift in funding for community-based programs is striking. In 1996, US$125 million was loaned by the World Bank for participatory environmental projects; by 2003, the total of loans had grown to $7 billion (Mansuri and Rao 2004).

Community-based wildlife management is intended to be a bottom-up approach in which communities benefit from the use of wildlife and sustainable management. In some cases these programs are implemented adjacent to protected areas without provisions for contributing to livelihoods, but instead are focused on reducing opposition to the protected areas (Songorwa et al. 2000). In other cases, community livelihoods have been developed hand-in-hand with conservation initiatives and in partnership with government agencies, safari hunting operators, and environmentally orientated NGOs. Ideally, these approaches should be adaptive through ongoing cycles of experimentation, regular review, and modification.

The CAMPFIRE program of community-based management in Zimbabwe involving elephant conservation (Box 6.3) demonstrates both the extraordinary opportunities of these programs and the significant difficulties in their successful long-term implementation. The case of Zimbabwe (and with most other countries of Africa) illustrates how traditional resource-use systems with long-standing biodiversity can be undermined through colonization and unregulated markets of animal products. The case of CAMPFIRE shows how an innovative a program that is attentive to both conservation and livelihoods can be initiated first as an experiment and later implemented to successfully reduce conflicts and provide community benefits, and how it can be taken up as a model of resource management by other countries in the region. Yet the case also illustrates how a project can face difficulties because of corruption of regional leaders, and ultimately flounder because of the collapse of national-level institutions that support it.

Box 6.3. CAMPFIRE: Elephant Conservation and Community Livelihoods in Southern Africa

Sources: Freeman and Kreuter (1994); Berger (2001); Getz et al. (1999) Peterson (1994); Thomas (1994); Hulme and Murphree (2001); WRI (2005)

Because of their huge size, high food intake, potential damage to farmlands and human infrastructure, and occasional aggressive behavior, elephants have significant impacts on their social–ecological systems. Yet these large animals have coexisted for thousands of years with Africans without significant damage to one another (Freeman and Kreuter 1994). In the precolonial period, human populations were small, elephant harvest levels were low, traditional institutions for managing harvests were intact, and the human harvesting of elephants was presumably sustainable. The export of ivory for European markets, starting in Napoleonic times and intensifying with the rise of the global economy, changed the traditional relationship between humans and elephants and began to threaten the future of the animals in many parts of Africa.

At the time of Kenya's independence in 1963, the elephant population was estimated at 170,000. However, by 1989 elephant numbers in Kenya had been reduced to 16,000, with the decline attributed almost entirely to poaching for ivory. During that same time the value of ivory went from $3 a pound to $300. Similar trends were seen in other countries as well. For example in Zaire, the elephant population plummeted from about 150,000 in 1989 to near extinction during the same period. In 1977, the Kenyan Government, with pressure from developed countries and nongovernmental conservation organizations, put a ban on hunting. Several other African governments did the same. Colonial governments dismissed the knowledge of indigenous people as related to conservation and neglected the potential benefits from non-preservationists' forms of management.

In 1989, international concern for elephants increased and parties to the Convention on International Trade in Endangered Species (CITES) voted to ban all trade in African elephant products. Elephants took on a high symbolic role in developed countries like the USA and the UK, embodying environmental organizations' rationale for preservation. Animal rights advocates noted the large brain size and social behavior of elephants as evidence of their high intelligence. Attempting to address the problem of elephant conservation during the 1970s to 1980s, many African countries established national parks and other forms of protected area that restricted all local uses of elephants. In some cases villages were relocated outside parks to aid in enforcement of illegal hunting. However, in spite of the CITES designation and protected areas with enforcement officers, an organized illegal trade of elephant products grew, driven by black market buyers. These buyers sold their contraband internationally at high rates, little of the profit going to illegal hunters themselves. Although governments spoke of their efforts to implement elephant conservation programs, such efforts were accompanied with high degrees of government corruption. Since government officials administered access rights to elephant habitat, poaching and the ivory trade continued and in many cases thrived.

Zimbabwe was the first country in Africa to recognize **conservation by utilization**, which acknowledged that landowners should benefit from wildlife. The transformation of policies toward utilization began in the 1960s and was enacted in 1975 with legislation that gave private commercial ranchers an economic rationale to conserve resources. The population of elephants in Zimbabwe was not near extinction, and hunting had historically taken place on private reserves, with operations owned by Africans who were descendents of colonists and catered to foreign big game hunters for considerable profit. In the late 1970s, government officials of various departments and wildlife researchers, with local communities began experimenting with the idea of approaching elephant conservation by creating incentives that would benefit local villagers and encourage village participation. In its initial phase it was called "Operation Windfall" in which elephant culling was linked to participation by communities. The rationale for the new process followed from the acknowledgement that elephants on communal use areas caused hardship for communities, destroying crops, creating fear among residents, and in some cases killing people by trampling. By the late 1970s the Zimbabwe Department of National Parks and Wildlife Division initiated a 2-year experiment on elephant conservation under the Communal Areas Management Program for Indigenous Resources (CAMPFIRE).

At the time it started, CAMPFIRE was unique in Africa in empowering local communities by distributing the benefits of trophy hunts to villages near game reserves, thus encouraging conservation. One of the immediate effects of the program was a dramatic increase in cited violations for breaking hunting regulations because community residents were beginning to help enforce regulations to safeguard *their* resource.

In one community, the local management committee decided that benefits were to be used for community development projects, including construction of a new school and a grinding mill. As the program became established and the benefits became clear, it was expanded to other regions of the country. As a part of the program, villagers contributed ecological monitoring data for research supported by The World Wildlife Fund. In one region residents decided to relocate to create enhanced elephant-hunting opportunities that would bring the financial benefits directly to residents.

Although CAMPFIRE was criticized by some international NGOs for compromising pure conservation objectives, it was touted by many as a huge success of community-based conservation. The program was not limited to elephants but extended to multiple wildlife species and safari hunting in general, distributing back to the communities revenues as well as meat from the hunt. The program did encounter problems associated with bureaucratic delays in delivering benefits to villages, the interception of funding by regional organizations, and inequities stemming from corrupt local leaders. It was also becoming clear that CAMPFIRE was run as a top-down program that did not effectively devolve the authority to manage wildlife below the district level of government. It also did not go far in devolving resource tenure on communal lands (Hulme and Murphree 2001). Nevertheless, from the 1970s to about 2000, CAMPFIRE was a model that inspired community-based wildlife-management programs in many southern and eastern African countries.

Since about 2000, the repressive political regime in Zimbabwe has led to a loss of safari tourism and ecotourism and international NGO support. These political and economic changes resulted in the collapse of many institutions of Zimbabwe, and with it, most of the tourism and safari industry that provided the economic benefits of the CAMPFIRE program. In other parts of southern and eastern Africa, including Namibia, Zambia, and Mozambique, community-based systems originally inspired by CAMPFIRE were established and continue to support conservation while providing for livelihoods. For example, Namibia's conservancies, established through the *Nature Conservation Act* of 1996, are considered to be among the most successful African efforts at community-based wildlife management, with significant devolution of management power to local communities, and effective in poverty alleviation (WRI 2005). Elephant conservation is still a global problem. But in several of the countries of the sub-Saharan Africa, elephant numbers have actually increased, and the current problem has ironically shifted from a concern about too few elephants to the management problem of too many.

Community-based wildlife management (like co-management described in the next section) is neither simple nor automatic. In their evaluation of seven community-based wildlife management programs in Africa, Songorwa et al. (2000) identified insights about the potential for success and failure of these programs. These included problems associated with implementers not adopting a bottom-up approach, lack of true participation, and the inability of the program to address basic community needs and distribute benefits equitably. In other cases, the authors found a mismatch between the goals and the objectives of NGOs that funded the programs and the livelihood goals of the community. An associated problem was the interruption of funding from NGOs because of changes in the organization's budget and/or mission. At the community level, issues can arise around the difficulties of villagers having to enforce their own rules on one another but continue to live as neighbors. Others arose when the job obligations of part-time wildlife enforcement officers competed with the responsibilities of tending family farms and caring for the welfare of kin. Economic incentives are also important. For example, Getz et al. (1999) note that programs like CAMPFIRE are most likely to succeed where there are marginal opportunities for agricultural development and thus higher incentives to conserve wildlife that might otherwise be a hardship.

Ecologically and socially, there have been a number of successes with community conservation. In some countries, community-based wildlife management has contributed to sustainable use and biodiversity conservation. In many cases, it has provided a mechanism for decentralizing resource management, so that the people dependent on wildlife participate in management decisions that affect their livelihoods. It has also led to the development of community-based monitoring systems, involving the participation of local harvesters as monitors of wildlife, community empowerment, and an increased appreciation of the local and traditional knowledge of people who live in the hinterlands and interact daily with wildlife (Kofinas et al. 2003).

Although the track record of community-based resource management globally is not very strong, we know that it can work when communities have partners who help them build their management capabilities and have co-management arrangement with government agencies (Berkes 2004) Where these conditions are fulfilled, community-based management has led to new governance arrangements that improve the long-term prospects for both people and biodiversity (WRI 2005).

Resource Co-management

The social systems of small, resource-based communities are embedded within a broader social–ecological context. In today's world, all local communities have some level of interaction with regional and national governments, NGOs, and neighboring resource users. The global commons has no remote locations. Effective linkages across scales of management that foster the evolution of local-level conservation strategies are therefore critical to sustainability (Dietz et al. 2003). In most cases, existing indigenous-state relationships constitute the legacies of past and current colonialism, including imposed harvesting policies implemented with little awareness of community livelihoods. These impositions can make the prospects of community-based conservation efforts difficult

to realize since state agencies typically resist devolving power to local communities. Top-down imposed policies can also undermine traditional institutions of community management that previously supported resource conservation. In several cases around the world, indigenous peoples of hinterland regions have negotiated self-government agreements with national governments that specify rights for use and self-management of important resources. But even in cases of negotiated self-government and comprehensive land claims, communities are embedded within broader legal institutions, creating a need for community rights that secures access to resources and systems that provide for cross-scale communication and accountability.

In cases where communities share use of commons with other communities and there are potential issues of resource scarcity, there is a need to establish vertical (across levels of organization) and horizontal (across the same level) linkages of resource governance among resource user communities, state agencies, and others. These arrangements fall within the rubric of **co-management**, defined as the sharing of power and responsibility in decision-making between state governments and communities in the functions of resource management. Co-management functions may include monitoring, habitat protection, impact assessment, research, enforcement, and policy making. When implemented with a focus on learning-by-doing, these arrangements are referred to as **adaptive co-management** (see Chapter 4). Formal and informal co-management has proven critical in the development of strategies that support livelihoods and conservation initiatives, contributing to social learning and social–ecological resilience at local and regional scales. We give two examples.

In the American Southwest, the US National Park Service and Navaho Nation co-manage Canyon de Chelly National Monument, a protected area that is situated entirely on Navajo Tribal Trust Land and that celebrates the spiritual significance and traditional uses of the area for the Navajo. The Park's co-managers increasingly recognize that climate changes will alter the unique ecosystem of the canyon, including orange groves dating back to the time of Cortez, and therefore the ecosystem services to Navaho residents.

In the American North, the US–Canada Polar Bear Agreement established a linkage between the two countries and among Alaskan Iñupiat and Canadian Inuvialuit indigenous peoples for conservation, helping to coordinate the setting of quotas and to facilitate discussions about habitat issues. With climate change now threatening the future of the Southwest and certain Arctic species, the terms of these agreements and discussions among comanagers become increasingly critical in reducing the risk of species loss. With the partner-

ships in place, all parties should be better situated to address these challenges.

Co-management systems are difficult to implement (Pinkerton 1989a, Wilson et al. 2003), even without the stresses of social–ecological change. Because state governments are reluctant to give up authority in resource management to communities and national interests typically take precedence over local interests, communities involved in co-management may have to be creative and fully engaged to be heard in highly charged political processes. Such a process is illustrated in Box 6.4 on caribou co-management involving the Gwich'in people of Alaska, Yukon, and the Northwest Territories.

Box 6.4. The Case of Porcupine Caribou Herd Co-management Across an International Border

Sources: Kofinas (1998, 2005); Kofinas et al. (2007); Griffith et al. (2002).

The Gwich'in, indigenous people of northwestern Canada and northeastern Alaska, have depended on caribou of the US–Canada Borderlands (Alaska, Yukon, and the Northwest Territories) for millennia. The Gwich'in's close relationship with the Porcupine caribou, an internationally migratory herd of that region, is central to Gwich'in livelihoods, identity, worldview, and sense of cultural survival. Like many of the barren ground caribou herds of the Circumpolar North, the Porcupine herd is a commons of tens of thousands of animals that overwinter in boreal forests and migrate north to coastal plain tundra for calving and summer insect relief. Scientific research shows that calving grounds are important habitat for caribou, with a strong relationship between tundra forage quality at calving grounds and caribou reproductive success.

The traditional hunting practices of the Gwich'in have changed dramatically since 1900, although aspects of the relationship between Gwich'in and caribou have per-

sisted. In the precontact period, caribou were harvested in the fall by family groups, often using an extensive system of caribou fences or corrals into which migrating caribou were directed and snared. Caribou were then butchered, dried, and stored for the winter by family groups. Strict rules dictated the management of waste to ensure that capture areas were clean and without odors that would disturb subsequent caribou drives. Decisions about hunting were directed by leaders and informed by traditional knowledge, which provides an intergenerational understanding of annual herd movements, decadal patterns of shifting winter habitat, the effects of severe climatic events, indicators of body condition and animal health, and the implications of human disturbance to animals and hunters. From a Gwich'in perspective, people's livelihoods and the well-being of caribou are interrelated and inseparable.

With the introduction of firearms, the social organization of harvest was transformed, with the eventual abandonment of caribou fences. Group hunting with leadership remained important until the introduction of the snowmobile, which made hunting a more individualized activity. In spite of changes in hunting organization and technology, the sharing of harvested meat remains a central element of the Gwich'in contemporary subsistence-cash economy. The "lucky hunter" (i.e., successful) is the one who shares with fellow community members, is not boastful, and is respectful of land and animals.

Beginning in the 1930s, and more intensively from the 1960s to present, national and international interest in mineral and oil and gas resources of the region has been a concern for the Gwich'in because of its potential impacts on caribou. In the 1950s the Canadian Government constructed a highway through the winter grounds of the Porcupine herd, with no formal consultation with local communities of the region. US federal and state government efforts to open the "northern frontier" to industrial development in the herd's calving grounds of Alaska in 1980 led to reestablishing kinship ties between Canada and the US Gwich'in communities to address threats to caribou. The early issues of industrial development also initiated a 20-year negotiation process with governments that led to the signing of a Canadian Porcupine caribou co-management agreement in 1985 and a US–Canada international agreement for the conservation of Porcupine Caribou in 1987. These agreements and their management boards were intended to provide local-to-international linkages for joint decision-making about caribou and important caribou habitat, to ensure the ongoing availability of caribou for the Gwich'in and other indigenous peoples of the region.

US–Canada bilateral activities of these co-management arrangements have been largely dominated by the issue of proposed oil and gas development in concentrated calving areas of the herd, located on the coastal plain of the Arctic National Wildlife Refuge in Alaska. The international agreement for Porcupine Caribou conservation has been ineffective in addressing oil and gas issues because of political vagaries of US federal administrations, which have included not

selecting an Alaskan Gwich'in user to the international caribou conservation board, canceling international board meetings, and suspending international dialogue on habitat issues.

In response, the Porcupine Caribou Management Board of Canada has worked with its user communities using alternative methods to address the shortcomings of the international agreement, by initiating its own assessments on sensitive caribou habitat and distributing the products of that research directly to members of the US Congress. The Canadian caribou board also developed a relationship with the Canadian Consulate in Washington to monitor political activities and update local communities on legislative proposals and to communicate local concerns to the Canadian Foreign Affairs Ministry. The Canadian co-management board also raised private funds to facilitate the grassroots political campaign of Porcupine caribou user community members, which included training indigenous people to lobby in Washington and at public presentations across the USA. A central message of these effective campaigns has not been caribou science, but the mutual respect of people and wildlife of the North and the relationship between indigenous livelihoods and caribou conservation. Building on this experience, Gwich'in First Nations formed a political organization, Gwich'in International, which was later given observer status at the Arctic Council, an international governance body of state officials, indigenous peoples, and NGOs. Habitat protection for the calving grounds in Alaska remains a critical issue and continues to command the attention of the Gwich'in. Through the efforts of the Gwich'in and Canadian Porcupine Caribou co-management, the issue of oil development and caribou has also become widely recognized across the USA and internationally.

Co-management of Porcupine Caribou at the regional level has been effective at achieving consensus on a range of difficult issues and resolving them. For example, barter and trade policies have been established to provide legal backing for traditional systems of food sharing between families in Canada and Alaska to address export policies that created barriers for food exchanges. An offer by Asian buyers to purchase antlers for medicinal uses resulted in fear among local communities of wastage and disrespect for caribou. Eventually, a well-supported voluntary prohibition on raw antler sales (i.e., not art products) was formulated and endorsed by the Canadian Porcupine Caribou Management Board and the communities. Managing and achieving consensus on rules for hunting on Canada's Dempster Highway has been more difficult. Repeated efforts to implement rules for managing hunting from the highway have been unsuccessful because of intercommunity conflict and legal challenges, even when these rules were based on traditional knowledge. More successfully, co-managers, communities, and agencies have together developed a regional ecological monitoring program that draws on local knowledge, as well as science-based indicators. It has also created a strategic planning process that provides adaptive learning through the use of computer simulation models that inform policy choices, planning, experimentation, and reflection.

At the community level, local leaders have had to confront imposed legal regimes and negotiate new ones, learn to maneuver through and around bureaucratic institutions, while selectively maintaining and adapting traditional institutions in a rapidly changing social–ecological setting. Co-management of wildlife involving multiple communities, state agencies, international transactions is a difficult, often messy process. It is not a panacea. It does, however, provide an important mechanism of multiscale governance for building resilience and sustaining community livelihoods and resources.

Among the parties of a co-management arrangement, power relations are typically asymmetrical; that is, there rarely is a level playing field. As well, cultural differences among groups can create communication problems and limit understanding of each other's perspectives (Morrow and Hensel 1992, Kofinas 2005). The differences in approaches to knowledge production and legitimacy can also add to the difficulties (Berkes 2008). Yet many examples demonstrate that power sharing can yield better decisions, enhance adaptation, and contribute to social–ecological resilience (Pinkerton and Weinstein 1995, Armitage et al. 2007). There are many ways in which co-management can contribute to community livelihoods and social–ecological resilience:

- creating a regional forum for deliberations on resource-management issues;
- maintaining collaborative and systematic social and ecological monitoring;
- focusing research that draws on local, traditional, and scientific knowledge;
- evaluating sensitive habitat, protecting important habitat, and providing for community participation in impact assessments;
- developing strategic resource-management plans that are sensitive to community interests in economic development;
- overseeing appropriate policies for barter and trade of subsistence resources;
- guiding effective response polices for ecosystem disturbance, such as forest fires;
- building and using decision-support tools;
- achieving regional consensus and good compliance with co-management endorsed policies.

Case studies show that improved levels of social capital (i.e., trust) can emerge through the security of specified legal management rights for communities and or the ongoing interactions among leaders, resource professionals, and resource harvesters (Wilson and Raakjaer 2003). Factors that are particularly critical in ensuring the effectiveness of co-management for communities are that monitoring, research, and policy making be truly collaborative enterprises and resources rights be secure (Pinkerton 2003). In conditions of rapid change and novel problems, effective systems of cross-scale governance may prove critical in arriving at innovative solutions (Kofinas et al. 2007) and ultimately in supporting community and regional adaptation (Armitage et al. 2007).

Single-Species Management of Wild Resources

Single-species management of wild resources (IUCN Category IV; Box 6.1) often focuses on ecological controls rather than institutions that link conservation to community livelihoods. The use of local ecosystems as a source of food for personal consumption and as a connection to the land is not unique to hinterlands. People living in a broad range of rural to semi-urban environments pick berries, harvest mushrooms, catch fish, and hunt wildlife as culturally and nutritionally significant activities that link their livelihoods to local ecosystems. These activities are often responsive to formal and informal institutions that foster the conservation of these resources. For those wild resources such as berries and mushrooms in which harvest has little detectable effect on the future supply, local and traditional institutions tend to address access and use rights rather than the actual levels of harvest. In some countries, for example, rights to pick berries may be restricted to property owners. In other countries, particularly on public lands, families that have picked berries, harvested mushrooms, or caught fish in a particular location may return regularly to the same places, giving rise to informal, but predictable patterns of use. In still other cases, these resources may be treated as open-access resources to which all people have access.

The harvest of fish and wildlife is often regulated through additional formal (government) institutions because human overharvest can significantly reduce future availability. This formal management tends to focus on the maximum sustained production of particular species that are compatible with available habitat, natural mortality, and harvest rate. This single-species command-and-control approach to wildlife management (see Chapter 4) rests on the dubious assumption that other causes of variations in population density are unimportant or can be managed.

Because hunting pressure is usually more socially than biologically mediated, wildlife managers are challenged to manipulate hunting in ways that roughly mimic natural predation pressure, i.e., to increase hunting pressure when target populations increase and to reduce it when populations decline. In cases where potential harvest pressure may exceed the mortality rate that the population can support, management options might include shortened (or closed) seasons for hunting or fishing, reduced allowable harvest levels (e.g., through raffles or lottery for the right to hunt a particular species), catch-and-release fishing, etc. Alternatively, when population densities of the target species exceed the carrying capacity of the habitat, management options include increases in allowable harvest, focused harvest on females (which have a disproportionate effect on population growth rate), and economic incentives for harvest (e.g., bounty on predator species). Efforts to maintain a specific population target (maximum sustained yield) neglect the importance of wildlife harvest for community livelihoods. An alternative approach is to manage for resilience (see Chapter 2). Resilience-based wildlife management accepts a broader range of variability in wildlife populations as a normal state of the social–ecological system and places greater attention on community livelihoods as an integral component of the human–wildlife system. Many of these practices are similar to indigenous resource management described in previous sections. For example, habitat management that maintains a habitat mosaic of multiple successional stages is a cornerstone of wildlife management, just as in traditional resource management of the Ojibwa (Box 6.2).

Management of human harvest is challenging because of both biological and social uncertainties and societal pressures to manage populations in ways that deviate substantially from those that occur naturally. For example, rare species or individuals that are viewed as particularly desirable to maintain in the population (trophy individuals) are subject to more intense hunting pressure than they would likely experience under natural predation or harvest that responded to food needs.

Management is equally challenging when target populations become extremely dense. Deer in suburban areas of the northeastern USA, for example, are abundant because people have eliminated most potential predators, have not functioned as predators themselves, and created an abundance of favorable early successional habitat. Under these circumstances deer populations explode, become prone to wildlife diseases, and are more effective vectors of lyme disease (Ostfeld and Keesing 2000). The resulting disease risks have the perverse consequence of disconnecting people from nature by reducing human use of natural areas.

Synthesis and Conclusions

The conservation and maintenance of biodiversity at various levels are clearly social–ecological, rather than strictly biological, issues. Human actions are the principal causes of recent declines in biodiversity, but they may also be important pathways to potentially sustainable solutions. The establishment of protected areas is an important measure. But protected areas alone are not sufficient to meet the challenges of biodiversity conservation. We need to work with people, encouraging stewardship ethics and cultural connections to the land, to protect biodiversity everywhere, and not create artificial "islands" of conservation through protected areas.

The "fence-and-fine" approach to conservation that seeks to separate people from nature typically creates highly artificial, fragile systems that have little historical precedent and are unlikely to be resilient to future environmental, economic, and social changes. The case studies presented in this chapter show that the integration of livelihoods into a biodiversity conservation ethic has been an integral part of many traditional cultures. We need to support such relationships, build on them, and promote them as a component of innovative solutions to current conservation challenges.

Developing conservation programs in a social–ecological context is not easy because the inherent complexities of human–environment interactions, issues of power and politics, and uncertainties of future human behavior make precise planning impossible. However, these same complexities may be used to generate resilience that enables the system to adapt to unknown future changes. Active participation of local resource users in the design, implementation, and monitoring of social–ecological conservation efforts improves understanding, responsiveness to change, and compliance with rules. Under these circumstances, local engagement and innovation can become a source of resilience, by seeking solutions to emerging problems, as opposed to a conservation system that has inflexible rules and that seeks to exclude people.

Conservation that builds livelihoods for local people in ways that are compatible with conservation objectives seems crucial. Unless local people can meet their basic livelihood needs, they may have little choice but to conflict with conservation objectives. So the solutions often involve working with local communities in such a way that they can use some of the resources under a conservation scheme, creating a local stake in biodiversity protection. Although engagement of local people in conservation efforts does not always lead to desirable outcomes, generalizations are emerging as to the conditions that influence their likelihood of success or failure (Songorwa et al. 2000, Borgerhoff Mulder and Coppolillo 2005, WRI 2005, Berkes 2007).

In developed countries, the management of harvestable fish and wildlife also benefits from a social–ecological perspective that recognizes the importance of social determinants of harvest patterns. Hunting pressure, for example, may tend to increase for species that decline or become uncommon, particularly in areas where human densities are highest. This is precisely opposite to the natural predator–prey feedbacks that convey resilience to wildlife populations in less human-dominated landscapes. Under these circumstances, engagement of local communities in understanding the problems and seeking solutions to conservation challenges has the greatest likelihood of successful implementation.

In all cases, formal and informal rules affecting conservation of biodiversity and support of people's livelihoods will need to be evaluated and adapted, in some cases in subtle ways, and in other cases more radically. These adaptations may involve the renewal of traditional ways of relating to and managing resources. In other cases they may require a grand transformation of formal institutions to provide for better partnerships among communities, resource management agencies, and nongovernment organizations.

Most current conservation challenges have roots in processes occurring at many scales, including global climatic and economic changes, national and regional agendas, and local traditions and needs. Resilience-based approaches to conservation recognize the importance of scale, the inherent uncertainty of future conditions, and the need for multilevel management, as seen in the caribou conservation case. Resilience-based stewardship therefore requires the fostering of flexibility and innovation in a social–ecological context.

Review Questions

1. In comparing the conditions of the three case studies (Boxes 6.2, 6.3, and 6.4), identify common themes in conservation planning.
2. Can a conservation ethic be created in a society with little or no tradition of conservation?
3. What are some of the strategies that community leaders might use to ensure that community needs are seriously considered in the establishment of a protected area that includes community homelands?

4. What are the strengths and weaknesses of a national-park vs a community-based conservation approach in conditions of rapid social–ecological change? Debate the proposition that the *priority* for our biodiversity conservation efforts should be to establish people-free protected areas.
5. In what social–ecological conditions is each of the six classes of protected areas most appropriate?
6. What are the key institutions in each of the three case studies in the boxes? Are they local/traditional, governmental, or both?

Additional Readings

Anderson, M.K. 2005. *Tending the Wild: Native American Knowledge and the Management of California's Natural Resources*. University of California Press, Berkeley.

Bawa, K.S., R. Seidler, and P.H. Raven 2004. Reconciling conservation paradigms. *Conservation Biology* 18:859–860.

Berkes, F. 2008. *Sacred Ecology*. 2nd Edition. Routledge, New York.

Borgerhoff Mulder, M., and P. Coppolillo. 2005. *Conservation: Linking Ecology, Economics, and Culture*. Princeton University Press, Princeton.

Brechin, S.R., P.R. Wilshusen, C.L. Fortwangler and P.C. West (eds). 2003. *Contested Nature: Promoting International Biodiversity with Social Justice in the Twenty-first Century*. State University of New York Press, Albany.

Gomez-Pampa, A. and A. Kaus 1992. Taming the wilderness myth. *BioScience* 42:271–279.

Posey, D.A. (ed). 1999. *Cultural and Spiritual Values of Biodiversity*. UNEP and Intermediate Technology Publications, Nairobi.

Taylor, B.R. (ed). 2005. *Encyclopedia of Religion and Nature*. Thoemmes Continuum, London.

Turner, N.J. 2005. *The Earth's Blanket. Traditional Teachings for Sustainable Living*. Douglas & McIntyre, Vancouver; and University of Washington Press, Seattle.

7
Forest Systems: Living with Long-Term Change

Frederick J. Swanson and F. Stuart Chapin, III

Introduction

Human societies depend heavily on forests for resources and habitat in many parts of the globe, yet the global extent of forests has declined by 40% through human actions since people first began clearing lands for agriculture. Forests still cover about 30% of the terrestrial surface and support 70% of Earth's plant and animal species (Shvidenko et al. 2005). Societies residing in tropical forests alone account for half of the world's indigenous languages. People have lived in and interacted with forests for thousands of years, both benefiting from their services and influencing the dynamics of the forests in which they live. Forests are therefore most logically viewed as social–ecological systems of which people are an integral component. Forests are globally important because of their broad geographic extent and the great wealth and diversity of ecological services they provide to global and local residents, including substantial biological and cultural diversity.

The character of forests and societal interactions with them vary greatly across the globe (Plate 5, Fig. 7.1). Belts of dense tropical rainforests encircle the earth in the equatorial zone, bordered at higher latitudes of Africa and South America by extensive savannas of scattered trees. Boreal forests stretch across high, northern latitudes. Conifer or broadleaf and mixed forests are common in intermediate latitudes, especially in the northern hemisphere. Individual forest patches may be managed by a household, local community, small or large corporation, or local or national government agency. The social settings of forests range from urban forests within cities and small sacred forests adjacent to villages to extensive tracts of forest in remote wilderness areas. The portfolio of forest uses and the relative extent of human–forest engagement vary greatly from country to country, reflecting the complex histories of forestlands and societies.

Rapid global changes in biophysical and social factors (see Chapter 1) are particularly challenging to forest stewardship because many trees live longer than the professional lifetimes of people who manage them, making it challenging to develop and sustain institutions with

F.J. Swanson (✉)
Pacific Northwest Research Station, USDA Forest Service, Corvallis, OR 97331, USA
e-mail: fswanson@fs.fed.us

F.S. Chapin et al. (eds.), *Principles of Ecosystem Stewardship*,
DOI 10.1007/978-0-387-73033-2_7, © Springer Science+Business Media, LLC 2009

FIGURE 7.1. Patch clear-cutting that leads to plantation forests of native Douglas-fir in a matrix of 100- to 500-year native forest in the northwestern USA. Photograph by A. Levno, US Forest Service, photograph AEA-002 [online] URL: http://www.fsl.orst.edu/lter/data/cd‑pics/cd‑lists. cfm? topnav=116

time horizons that extend beyond the short-term motivations of individual decision makers. In this chapter we highlight four overlapping resource–stewardship challenges faced by forest managers throughout the world that are, in part, logical consequences of the long-lived nature of forest trees:

- **Sustaining forest productivity.** Institutions with a long-term view of the ecological conditions necessary to support forests and forest-dependent peoples are a foundation for sustainable harvest from forests. Given harvest intervals that are frequently 40–100 years or longer, large areas are required for forests containing the spectrum of age classes needed for a continuous flow of wood for human use. The large temporal and spatial scales required for sustainable forestry often clash with the short-term motivations and small-scale jurisdiction of forest managers. In order to meet these sustainability challenges it is critical to sustain social values and capacities for dealing with forest systems over long time periods.
- **Sustaining forest ecosystem services. Managing forests for multiple ecosystem services** involves strong tradeoffs among costs and benefits to different stakeholders, with choices having implications for multiple human generations. Forests provide a mul-

titude of ecosystem services, including fuelwood, timber products, water supply, recreation, species conservation, and aesthetic and spiritual values. Together these services generate a diverse and challenging mix of management objectives to meet multiple societal needs.
- **Sustaining forests in the face of environmental change. Managing forests under conditions of rapid change is challenging** because a forest stand is likely to encounter novel environmental and socioeconomic conditions during the life of the individual trees in the stand. Forest ecosystems across the globe face myriad threats from both intentional and inadvertent human impacts, including air pollution, invasive species, and, perhaps above all, global climate change.
- **Sustaining the forestland base. Forest conversion to new land uses** is a state change that is difficult and time-consuming to reverse, given the long regeneration time of forest trees and ecosystems. Historic and ongoing land use has converted vast areas of complex, native forests to agriculture, built environments, and plantations of a single or narrowly constrained set of species, often exotics. Under other conditions, large-scale agricultural abandonment or increased economic value of forests, as for carbon sequestration, can foster **reforestation** or **afforestation** (the regeneration of forest on recently harvested sites or planting of new forests on previously nonforested sites, respectively).

We next address the importance of social–ecological dynamics in forest planning. We then consider each of the above resource–stewardship issues in greater detail, showing both the challenges and the opportunities for sustainable forest stewardship. We conclude with some recommendations for sustainable stewardship of forests in the future.

Forests as Social–Ecological Systems

Because of the longevity of forest trees and forest crops, decisions made today for forest use and management have inevitable consequences for future generations operating in different environmental and social contexts. Societies often face a broad spectrum of forest-management objectives and associated approaches, ranging from maximum wood production with relatively short (decadal) rotations to development of old forest attributes on the multicentury time scale to simple retention of native forest and the species contained therein for conservation, recreation, spiritual, or other objectives. Furthermore, shifting legal, regulatory, and economic contexts of forestry can cause gradual or disruptive, abrupt changes in management, so social disturbances can be as important as ecological disturbances in shaping the future of forest resources.

This complexity of the societal context for forest management is confounded by a web of forces operating at local-to-global scales. Human population growth and sprawl of rural residential development into forest areas increase local demand for forest products while constraining the range of potential forest management tools, such as prescribed fire, that may be critical to sustaining the properties of native forests. Social forces may shift the emphasis of management objectives within the portfolio of ecological services provided by public or private land, such as from wood production to protection of species, with various intended and unintended consequences for society. Thus, social factors can trigger abrupt and profound changes in forest policy and management that can ripple across scales (see Chapter 5). Globalization of commerce, for example, may connect a market place in Europe to forestry operations in a distant part of the globe through forest certification and international agreements against illegal logging. **Forest certification** is a procedure for assessing forest-management practices against standards for sustainability so purchasers can support sustainable management, even over vast distances.

Changing societal and environmental factors are colliding and interacting. Global changes in climate, climate variability, and species invasions, much of them human-driven, make it increasingly difficult to ensure a predictable flow of desired quantities and diversity of services from forests. These global changes have impacts at local to regional scales, sometimes causing landowners to modify forests in expectation of climate-change effects not yet realized on the land. In some cases these threshold changes in system condition and capacity may be influenced by legacies of past management that affect the organization (e.g., age class distribution of forest trees or agency culture of past management) and vulnerability of the ecosystem to disturbance. The resulting uncertainties place a premium on social capital and adaptive capacity.

Supply and Use of Ecosystem Services

Forests provide many important ecosystem services to society both globally and locally over a range of time scales (see Chapter 2). At the global scale, the human conversion of forests to other land-cover types in the last two centuries has had important effects on the climate system (Field et al. 2007). For example, glacial–interglacial changes in forest cover in response to small changes in solar radiation contributed to massive shifts in global climate (Foley et al. 1994, Friedlingstein et al. 1995). As home to 70% of Earth's biodiversity, forests are important sources of ecosystem services ranging from food and medicines to cultural appreciation of their spiritual and aesthetic values. In addition to these global services, forests are home to human communities, whose local use includes harvest of goods; regulation of water and natural hazards; and recreation and cultural ties to the land. Different segments of society often place different priorities on these services, raising challenging questions about tradeoffs and synergies.

Forests are obviously *the* major source of timber and its products, including lumber, veneer, and paper. Despite the importance of lumber and pulp products, more than half (55%) of global wood consumption is for fuelwood, which is used locally; is not well characterized in economic summaries of forest products; and is the primary energy source for heating and cooking for 40% of the global population, particularly in developing nations and rural areas (Sampson et al. 2005). Nontimber forest products also have considerable economic and cultural importance. In the northeastern USA and Canada, maple syrup, produced from the sap of local trees, provides income to rural residents and contributes to local identity. In Alaska the economic value of blueberries and moose harvested by local residents each exceeds the value of harvested timber. About 25% of currently prescribed medicines originate from plant compounds that evolved as defenses against herbivores and pathogens (Dirzo and Raven 2003). Consequently, property rights to the genetic diversity of tropical forests are an issue that is contested between tropical nations, local indigenous peoples, and commercial bioprospecting firms (Kursar et al. 2006). The harvest and sale of bush meat to city residents in Africa is also a controversial topic because it creates both a source of local income and an extinction threat to many of the species that are harvested.

In some cases the products provided by forests are part of an agricultural rotation (**slash-and-burn** or **swidden agriculture**), in which small forest patches are cut, the land is cultivated, and forests regenerate, as people move to clear adjacent forest patches (see Chapter 12). This rotation has been sustainable for millennia, but increasing human population density is reducing the length of the forest rotation; below a rotation length of about 10 years, slash-and-burn agriculture no longer appears sustainable, causing a transformation to continuous cultivation of cash crops (Ramakrishnan 1992).

Many forestlands are managed for the critical ecosystem services of water supply and watershed protection (Rockström et al. 1999, Vörösmarty et al. 2005). Forested watersheds account for more than 75% of the world's accessible freshwater (Shvidenko et al. 2005). Given the increasing scarcity of freshwater relative to human demands (see Chapters 2 and 9), forests are likely to become increasingly important for their capacity to provide and purify freshwater. Water quality and quantity also affect other resources, such as fish, so an ecosystem approach is required for sustainable management. An important aspect of forest water-

shed management is therefore to meet water quality objectives by minimizing soil erosion, which also sustains soils as a base for terrestrial productivity, limits sediment accumulation in reservoirs, and prevents damage to downstream freshwater and marine aquatic habitats (see Chapters 2, 9, and 11). Watershed-focused forest management can also reduce the potential for landslides and snow avalanches, thereby protecting life and property on the hillslopes and valley floor below. Appropriate actions include distributing forest harvest and roads to avoid unstable areas, creating buffer strips of forest along streams, and adjusting the frequency, spatial pattern, and intensity of management actions to minimize their impact.

Forests and forestry affect the global carbon budget in ways that can either increase or decrease CO_2 concentration in the atmosphere, hence global warming. Forests account for about two thirds of terrestrial net primary production and half of the terrestrial carbon stocks (Shvidenko et al. 2005). Any increases in forest extent or a positive growth response to increasing atmospheric CO_2 or nitrogen deposition removes CO_2 from the atmosphere and reduces the potential climatic impact of fossil-fuel emissions (see Chapter 2). This capacity of forests to sequester carbon appears to be saturating (Canadell et al. 2007), indicating that we cannot count on forests to "solve the problem" of climate warming without more concerted efforts to reduce fossil-fuel emissions. Although increased forest extent sequesters more carbon, it also reduces albedo (short wave reflectance), especially in northern forests during the snow-covered seasons. The reduced albedo leads to greater absorption of solar energy and more heating of the atmosphere. In the tropics, the effects of carbon storage and cooling of the surface by high transpiration rates predominate over energy-exchange effects, so any reduction in deforestation or increase in forest regeneration tends to reduce the rate of climate warming. At high latitudes, the tradeoff is less certain; the greater atmospheric heating by forests due to their low albedo reduces the benefits of carbon sequestration, with the net effect on climate currently uncertain (Field et al. 2007,

Chapin et al. 2008). Therefore efforts to slow the rates of **deforestation**, the conversion of forest to a nonforested ecosystem type such as agriculture, are important to the global climate system, especially in the tropics.

Forests also provide important regulatory services at more local scales (see Chapter 2). Tropical forest pollinators, for example, are critical to coffee plantations, which show greatest fruit set and productivity adjacent to forests and forest fragments (Ricketts et al. 2004). Similarly, forests often harbor insect predators that reduce the likelihood of insect pest outbreaks (Naylor and Ehrlich 1997). As forests are cleared to support small-scale agriculture, people and their domestic cattle become exposed to diseases from forest animals. In this way, forest fragmentation reduces the capacity of forests to regulate diseases (Patz et al. 2005). In the northeastern USA forest disturbance and elimination of predators has led to dense deer populations that spread lyme disease, which in turn reduces human use of forests (see Chapter 6).

Cultural and spiritual values of forests have taken different forms in different parts of the world. The Druids, for example, held certain trees to be sacred deities in their Celtic religion. Patches of sacred forest, some only a fraction of a hectare, are an important part of life in many developing nations, including India (Ramakrishnan 1992), Madagascar (Elmqvist et al. 2007), and Benin and Togo (Kokou and Sokpon 2006; see Chapter 6). Forests also provide economically important cultural services such as recreation and tourism (see Chapter 2).

The existence of forests as reservoirs of biodiversity is important to many people and societies. Tropical forests alone house 50–90% of Earth's terrestrial species. Forested mountain landscapes are especially rich because the complex topography and steep environmental gradients create great habitat diversity. In some parts of the world large tracts of forestland have been reserved specifically for conservation of biological diversity, and rules for management of these reserves may, or may not, allow human use of other potential ecosystem services (see Chapter 6).

Sustainable Timber Harvest: A Single-Resource Approach to Forest Management

Evolving Views of People in Forests

Despite the huge ecological and cultural variations among forests as social–ecological systems, changes over time in social–ecological forest systems exhibit some striking parallels. Societal engagement with forests in Australia, Canada, and the USA, which share European cultural roots, for example, have exhibited a similar sequence of stages: "(1) use by hunter-gatherer societies, (2) exploitive colonization, settlement and commercialization, (3) wood resource protection, (4) multiple use management, (5) sustainable forest management or ecosystem management" (Lane and McDonald 2002), and finally (6) sustainable ecosystem stewardship (see Fig. 15.1). Transitions from one stage to the next have often occurred through a crisis triggered by biological forces (e.g., extensive insect outbreaks) or social events (e.g., law suits and court injunctions), or development programs sponsored by nonlocal agencies with interest in modernizing "primitive people" (Gunderson et al. 1995, Lane and McDonald 2002). These transformations can be important opportunities for innovation, because the crisis demonstrates that the current system is not working, making managers and the public willing to consider new alternatives (see Chapter 5). Although government controls on forestland and terminology for management systems may differ among countries, this general trajectory and the punctuated pattern of change have been quite similar among countries and biomes that range from drylands and freshwaters to cities and agriculture (see Chapters 8, 9, 12, 13, and 15).

However, not all societies engage with forests in this sequence. Developing countries, for example, sometimes find a path that draws on the ingenuity and knowledge of local people to meet pressing demands for both poverty alleviation and forest stewardship, providing a potential seedbed for new forest stewardship approaches that would benefit Western countries. However, indigenous and rural communities often face extremely difficult challenges in their efforts to use local forest resources because of governmental controls and corruption (Larson and Ribot 2007). In some very remote forests, however, population density may be limited by the capacity of forests to provide food, either through small-scale clearing and recovery of forests or through subsistence harvest of foods from the forests. These interactions limit both the human population that can be supported and the extent of forest harvest. Forests were harvested primarily to meet short-term, local human needs for resources provided by the forest rather than for commercial sale of wood products (see Chapters 6 and 12).

Expansion of agriculture and commercial trading of forest products led to more extensive forest clearing. This began in Europe in the Middle Ages and in eastern North America with European colonization and still characterizes many tropical and boreal regions. **Forest exploitation** is largely driven by demand for wood and depends on availability of labor, access, and markets in social contexts with limited local governance. Under conditions of illegal and even some corporate- and government-backed forest exploitation, little attention might be paid to whether forest practices are sustainable (Burton et al. 2003). For example, in the developing world forest exploitation is often fostered by unclear property rights, disempowerment of local institutions for managing common property, or government efforts to generate foreign exchange from wood exports (see Chapter 4). Under these circumstances, any unfavorable effects of forest harvest have relatively few consequences for those who make the decisions about the extent and nature of forest harvest.

As exploitation depletes forests, the importance of forest regeneration becomes more apparent, and there is a gradual transition to **maximum sustained yield** (MSY) of timber, rather than short-term profit as a guide for forest management. MSY sets a harvest level that does not exceed the expected annual growth increment (see Chapter 2). This leads to a harvest rotation system, in which forests are managed much like an agricultural crop. In this context, the value of the forests reflects both supply and demand. MSY is clearly motivated by efforts to manage sustainably, but with a narrow focus on wood supply and much less attention to other ecological services or to forest resilience to unexpected changes. Managing for MSY is challenging with long-lived species like trees and with uncertain variation in climate, pests, other disturbances, markets, and taxation systems. In many cases, estimates of future yield have been overly optimistic and probabilities of environmental consequences and disturbance have been underestimated, leading to harvest schedules that proved unsustainable over the long term. For example, climate warming may increase drought stress and the risks of fire and insect outbreaks to an extent that was not anticipated when plans for high levels of wood production were developed. Short-rotation production forestry on plantations is still guided by MSY, which provides a reasonable guide to sustainable timber production, if changes in slow variables such as soil fertility, disturbance, and pest populations are taken into account. Production forestry predominates where land is owned by forest companies or on public lands where wood production for short-term revenues is the primary objective for management. Even where wood production is the explicit management objective, silvicultural practices have been developed that enhance the delivery of other ecosystem services to society, as described below.

The shortcomings of MSY and public sentiment that public forests should serve more than industrial forestry interests pushed policy for public forestlands to embrace **multiple use management** that explicitly addresses a broad array of ecosystem services. In the 1990s growing global interest in species conservation, maintaining site productivity, and other ecological services, led to development of an **ecosystem-management** approach that emphasizes the well-being of the system as a whole, while capitalizing on natural ecological processes to do the work we wish to accomplish in the forest (Grumbine 1994). Even in the context of ecosystem management, crises may arise from influences such as extensive insect damage to forests or altered public opinion, resulting in environmental policy changes and new environmental regulations. Over the course of these changes in views and approaches to forest management, several trends are evident:

- Broader geographic and temporal scales are considered.
- More components of forests, hence more ecosystem services, are included as primary management objectives.
- Therefore, a greater variety of technical disciplines is engaged in planning and implementing forest management, and general professional oversight shifts from silviculturists and foresters to broadly interdisciplinary teams.
- In some cases adaptive management is implemented to monitor change, adapt plans based on new information, and, thereby, learn through doing.
- Research-management and science-policy ties are broadened and strengthened.

Current conditions of rapid environmental or social change have given rise to additional challenges that require an expanded vision of ecosystem management, which we term ecosystem stewardship (see Chapters 1 and 15). In this context, managers recognize that change is inevitable, although the nature and rate of change are generally uncertain. Under these circumstances, managers seek to respond to and shape changes and to be very judicious in identifying historical properties of the system to attempt to sustain into the future.

Despite these common trends, quite different forest landscape-management approaches are often evident even within a single forest region. In the Pacific Northwest (PNW) USA, for example, forestry practices range from an agricultural model (MSY) of 40-year cutting rotation on industrial lands to an 80- to 100-year rotation on government lands managed for timber production to wilderness, park, and other lands where no cutting occurs (Box 7.1). In some other countries, by contrast, laws designate narrowly prescriptive forestry management rules on most public and private lands. In yet other contexts ranging from communities in remote areas of developing countries to western government forestry, exploration of alternative future scenarios has been used to set a desirable future course of forest stewardship (Wollenberg et al. 2000). These contrasts suggest that there is no single "right" approach: different governmental jurisdictions, ownerships, and land allocations have different management objectives, hence management paradigms, approaches, and supporting science (see Chapter 4). Policy diversity and adaptive, learning social systems are sources of institutional and ecological resilience for an unknown future. Even if a "best policy" could be identified for today's conditions, other policies might prove more favorable as an uncertain future unfolds (Bormann and Kiester 2004) (Fig. 7.2).

Box 7.1. Blending Forest Production and Conservation in the Western USA

Development of forest issues and policies in the PNW of the USA (Fig. 7.2) over the past few decades presents dramatic examples of the challenge of conducting ecologically and socially sustainable forest management. The conflicting values people hold of PNW forests have been starkly framed — cut majestic, centuries-old forests or sustain local economies and families who for several generations have worked in the woods; save a cryptic owl species and iconic salmon or intensively manage highly productive forestlands for wood products that benefit a broad cross-section of society. Of course, the real issues are not such simple dichotomies, and societal complexities match or exceed ecological complexities. We consider this example of a region's quest for sustainable forestry in terms of the expanding breadth of multiple use objectives, the great diversity of livelihoods affected, and successes and failures in adaptive management.

Conflict over these forests grew out of three decades of intensive timber production on Federal forestlands commencing after World War II to supply wood for the postwar housing boom. Intensive timber harvest during this period focused on old growth (>200 years old) and cumulatively affected about 30% of the landscape by the late 1980s. Over the timber production era local communities grew dependent on livelihoods based on jobs in the woods and in the mills. Federal forestry agencies flourished by putting large volumes of wood into the marketplace, and the applied science community studied ways to enhance timber productivity and efficiency of logging systems. This pattern of natural resource system development follows the paradox framed by Holling (1995, p. 8): "The very success in managing a target variable for sustained production of food or fiber apparently leads inevitably to an ultimate pathology of less resilience and more vulnerable ecosystems, more rigid and unresponsive management agencies, and more dependent societies" — until some ecological or social disturbance triggers abrupt change.

That disturbance erupted as intense controversy in the late 1980s over the fate of old-growth forests, salmon, and northern spotted owl. Law suits hinged on protecting the northern spotted owl, "listed" under the Endangered Species Act, stopped all logging of forests in the 100,000 km^2 range of that species, extending from northern California to the Canadian border. To break the gridlock created by the resulting court injunctions that blocked timber cutting, President Clinton convened scientists to conduct a bioregional assessment (FEMAT 1993, Szaro et al. 1999) and craft a new plan with objectives of protecting species of critical interest and forming an interconnected network of old-growth forest reserves, while providing some flow of timber to local communities.

The resulting Federal lands policy, the Northwest Forest Plan (NWFP; USDA and USDI 1994), was a revolutionary departure from previous forest management in

the region, and set an example of ecosystem management with broader impact. The NWFP greatly expanded the scope of ecological considerations under the multiple use rubric. For example, hundreds of species were designated for special attention under protocols called "survey and manage." The geography of planning under the NWFP was based more on ecosystem considerations than political jurisdictions, spanning many Federal agency lands and aligned with watershed boundaries. The plan placed 80% of the land in reserves for terrestrial and aquatic species and reduced the timber harvest level by more than 80% of the level of the 1980s. Harvest of old-growth forest outside reserves was to provide a significant share of timber volume in the early decades of NWFP implementation. Ecological considerations strongly influenced the harvesting systems. For example, 15% cover of live trees and substantial amounts of deadwood were to be retained to meet ecological objectives in harvest units that would have been clear-cut in earlier logging systems. The NWFP incorporated adaptive management at several scales, including designation of ten Adaptive Management Areas (AMAs) covering 7% of the plan area. The NWFP also commissioned a region-wide monitoring program for change in forest cover, northern spotted owls, socioeconomic conditions, and other attributes, to support adaptations of the plan at the regional scale.

How is the NWFP working? Results of a regional monitoring program document successes and failures over the first decade of plan implementation (Haynes et al. 2006). Dire predictions of collapse of rural communities did not occur, although more isolated communities highly dependent of Federal timber did suffer and livelihoods dependent on jobs in the woods and mills declined markedly. A long-standing system of payment of timber revenues to counties in lieu of property taxes on Federal lands has collapsed, leaving counties with extensive Federal timber land without funds which they had depended on for roads, libraries, and schools. Federal timber harvest has not reached even the greatly reduced level projected by the NWFP because of procedural challenges by environmental organizations, limited funding of forestry agencies, and other issues. The harvest of old growth was limited because of strong public opposition. The regional monitoring program revealed that the extent of old-growth habitat increased because forest growth exceeded losses to harvest and wildfire. Ironically, although suitable habitat increased, spotted owl populations continued to decline, perhaps in part due to competition from the more aggressive barred owl expanding its range from the north and east. Adaptive management proceeded in two ways: (1) a regional monitoring program (Haynes et al. 2006) that reported findings likely to influence future plan revision and (2) the AMA program in which scientists, land managers, and in some cases local public groups undertook studies to test assumptions in the NWFP and explore other management approaches. However, the regional monitoring program faces uncertain funding, and the AMA program faced substantial institutional barriers to success (Stankey et al. 2003). Before the AMAs were a decade old most funding had disappeared, and the commitment to this learning process largely evaporated.

What does the future hold? Federal forests in the PNW will doubtless present profound surprises, as they have in the recent past. Three factors are in play—the forests, their social context, and environmental change. Is the present NWFP harvest level so low that it is socially unsustainable? There has been no great public clamor to increase logging in the Federal forests of the region, but it may develop for social (e.g., more revenue to local communities) or ecological (e.g., reduction of vegetation that could fuel fires) reasons. However, changes in the global market place and other social factors may someday trigger increased forest harvest. Will the barred owl displace spotted owls, reducing the legal motivation to sustain old-growth habitat? This now seems possible, but the spotted owl is no longer critical to protection of native forests, especially old growth. The public seems thoroughly committed to protection of the remaining old-growth forests. Climate warming, invasive species, and changing fire regimes are creating a very uncertain future for forest management. In short, the forces of change, both environmental and social, have great potential to trigger new convulsions. The region remains caught in Holling's paradox—in part because of limited success in developing sustained learning institutions (Holling 1995).

Forest Exploitation and Illegal Logging

Illegal logging dominates the timber production of some developing nations and accounts for up to 15% of timber production globally (Contreras-Hermosilla 2002, Sampson et al. 2005). More than half (50–80%) of timber production in Indonesia, Brazil (in 1998), and Cameroon, for example, is estimated to occur through illegal logging (Sampson et al. 2005), although precise estimates are seldom available. Illegal logging includes logging in protected areas, without authorization, or more than authorized; timber theft and smuggling; and fraudulent pricing and accounting practices (Sampson et al. 2005). It can focus on high-value tropical woods, depleting local diversity, or extensive forest clearing, causing land degradation. Illegal logging deprives governments and local communities of forest revenues, often strengthens criminal enterprises, and induces corruption among enforcement and other officials (Contreras-Hermosilla 2002, Sampson et al. 2005). It is analogous (and is a similar proportion of total harvest) to illegal and unreported fishing (see Chapter 10). In the former Soviet Union (FSU), illegal logging was estimated to account for about 20% of timber production, but this practice largely disappeared, when the collapse of the FSU eliminated subsidies for transportation from forests to processing facilities (Sampson et al. 2005).

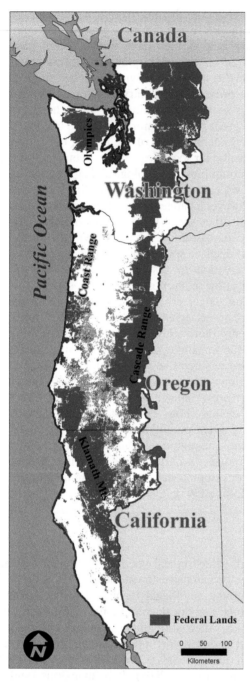

FIGURE 7.2. Shaded relief map of western Washington and Oregon and northwest California showing forested areas of coastal mountains and the Cascade Range and the outline of NWFP area covering the range of the northern spotted owl. Information from USDA and USDI (1994).

Just as in the forest exploitation phases of Europe and North America, illegal logging is driven largely by market value, access, and labor, with no regard for regeneration or sustainability. It often proliferates in response to multiple local, national, and global circumstances, therefore requiring multipronged responses. Commonly contributing factors include:

- Disempowerment of local institutions that provide for sustainable forest management. These could include a weakening of property rights or weakening of some of the factors that facilitate effective stewardship of common pool resources (Dietz et al. 2003, Agrawal et al. 2008; see Chapter 4).
- Political corruption that prevents local individuals or communities from benefiting from the timber sales or that leads to national policies with other social goals (e.g., forest clearing for oil palm plantations in Indonesia or meat production in Brazil) that subvert forest sustainability.
- High market prices that increase the benefits from illegal logging relative to the social costs and political risks.

Actions that might reduce the likelihood of illegal logging include the building of local social capital and livelihood options; bridging institutions at local-to-national scales; clarification of property rights; international aid that is tied to changes in the incentive structure for illegal logging; or global forestry certification programs that encourage sustainable management. The engagement of local forest users in decisions about rules for management of protected areas generally results in greater compliance and oversight of others than in cases where rules are imposed by higher authorities (Nagendra 2007, Agrawal et al. 2008). At the scale of international trade, on the other hand, European purchasers of South American wood would like to know that it came from forests managed sustainably. Forest certification currently covers a negligible proportion of timber production, primarily in the temperate zone, with variable verification, but ultimately could substantially reduce forest exploitation, especially for high-value timber species with specialized markets. International accords on forest conservation provide potential avenues to address both illegal logging and forest certification.

Timber Production from Natural Forests

Natural forests are forests that have regenerated naturally, in contrast to **production forests** in which trees are planted, often in regularly spaced patterns of a single species (Sampson et al. 2005). Logging of natural forests accounts for most (65%) of global timber production and is a significant source of local employment and livelihoods. In this chapter we use **timber** synonymously with **industrial roundwood**, which includes sawlogs and pulpwood (and the resulting chips, particles, and wood residues). Throughout the world, individuals, families, or local enterprises harvest most timber, providing both local income and cultural connections to forestlands. Although timber production continues to increase globally in response to global increases in consumption, the rate of this increase is slowing and is largely met by increases in production forestry. Until recently, the growth in timber production was approximately balanced by increased labor productivity (1.45% annually in the USA, for example), as labor-saving technologies and infrastructure expand, leaving a roughly constant employment in forest production (Sampson et al. 2005). This apparent stability hides huge national and regional shifts in timber harvest. For example, harvest declined radically in the FSU due to economic collapse and in the western USA due to shifts in forest policy (Box 7.1). Conversely, harvest from natural forests is increasing in many tropical countries, in part through illegal logging. Given the long-lived nature of forest trees, uncertain environmental change, and shifting markets, how can natural forests be managed more sustainably? The answer appears to be context-specific and differs substantially between developing and developed nations.

Rural communities in many developing nations depend strongly on the products of nearby forests, especially in the tropics and subtropics (Bawa et al. 2004). In some of these countries, such as Costa Rica, Mexico, and Nepal, large-scale reforestation is occurring either coincident with or following deforestation. In other countries, such as China, deforestation is the predominant trend (Nagendra 2007). Even within a country the balance between forest cutting and reforestation can be regionally variable. In Nepal, for example, all forests are nationally owned but are managed differently, depending on the land-tenure status of local residents. In national forests and parks that attempt to exclude local residents to meet conservation goals, deforestation is the predominant trend as a result of illegal logging, whereas areas that are managed by communities or by individual leaseholders show net reforestation (Fig. 7.3; Nagendra 2007). These trends toward reforestation are strongest where local residents participate in monitoring the patterns of forest use. These patterns of community forestry are more sustainable under conditions of clearly specified land-use rights and active monitoring, consistent with research on successful management of commons (see Chapter 4). Community engagement does not always lead to sustainable forestry, however, in part because several factors create significant misfits of institutional and ecological conditions (Brown 2003). Factors that can undermine sustainable forestry include inability of local residents to exclude nonresidents from forest harvest (e.g., illegal logging) or to participate meaningfully in defining the rules or resolving conflicts related to forest use (see Chapter 4). Globally there is a trend toward decentralization of forest-management authority (Agrawal et al. 2008), providing opportunities for concerted efforts to strengthen community-based management.

Sustained timber production from natural forests involves both the first cutting of native forest and the subsequent cuttings of forests that regenerate after the initial harvest. If reforestation is managed to sustain a near-natural mix of native species; biotic structures, such as standing and down deadwood; and ecological processes, such as a natural fire regime, successive forest rotations may provide many

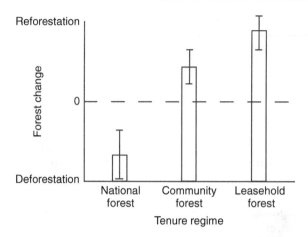

FIGURE 7.3. Relative extent of deforestation and reforestation of government-owned forests in Nepal. National forests are managed by the government, with little community engagement or protection, resulting in open-access use by people living nearby. Community forests are managed by local communities through community forestry user groups. Lease-hold forests are small patches of forest that are generally highly degraded and available to households in regions of high poverty. Leaseholders often replant their own forests and harvest from nearby national forests. Adapted from Nagendra (2007).

of the same services as the initial forest. In ecological terms, this forestry system contrasts sharply with management of plantations of exotic species. In some parts of the world first cutting of natural forests remains the dominant practice—in Siberia, for example, where old-growth forests are extensive and forest regrowth is slow, and parts of the tropics, where ecological, social, and economic factors have constrained reforestation. Even in places where native forests were converted to agriculture and then monocultures of native or exotic tree species, such as forest plantations on public lands in the Southeastern USA, some forests are being managed for more natural biological conditions of species and processes combined with the intent of maintaining a flow of forest products from the land. In these cases forestry has both wood production and forest restoration objectives.

A common theme in forestry in natural systems is the need to understand the ecology of the system. This is the gist of ecosystem management (Grumbine 1994). What limits and capacities does the native ecosystem impose? How can native ecosystem processes help us efficiently work toward our management objectives? For example, many native ecosystems contain plant species with the capacity to fix

nitrogen, which can contribute to soil fertility. Intensive plantation forestry may preclude these nitrogen-fixing species by eliminating the stages of forest development in which they prosper. More natural and diverse ecosystems may help retain the roles of nitrogen-fixing species in the forest system, resulting in greater, sustained site productivity.

The practice of sustained production from natural forests therefore requires development of tightly integrated ecological and social dimensions of forestry practices. Adaptive management (see Chapters 4 and 10; Walters 1986) provides a context for bridging the contrasting workplace cultures of science and management. Scientists like to ask questions and challenge ideas; land managers tend to work together and seek public license to operate by using "best management practices." But, if scientists, forest managers, and communities can work together in an adaptive management framework, it may be possible to develop management systems that are scientifically credible, socially acceptable, and adaptive to changing social and environmental circumstances (Box 7.2). Some government agencies, such as Forestry Tasmania and the US Forest Service have both management and research branches, which have the potential benefit of cultural proximity, but the disadvantage of too much familiarity and colle- giality, which may limit objectivity and critical analysis. Another research-management part- nership model is the alliance of state forestry with academia, such as extension programs in the USA for forestry and agriculture. However, these tend to be simple consultations rather than sustained programs of adaptive manage- ment. In very limited instances the forest indus- try has had in-house research programs that may set the stage for a sustained adaptive man- agement program. A critical issue in all such cases is the institutional capacity to sustain adaptive management programs over the long time spans appropriate to trees, cropping rota- tions, and institutional change.

Box 7.2. The Many Roles of Scientists in Management and Policy

Scientists involved in forestry stewardship are among the many people whose liveli- hoods are deeply engaged with forest systems—such as loggers, heavy equipment operators, mill workers, forestry and regula- tory agency staff, and environmentalists, to name but a few (see Chapter 3). Among these groups scientists have a unique and difficult challenge in their need to produce new knowledge, interpret monitoring and research findings, and in many cases com- ment on policy alternatives.

To display some of the complexity of types of livelihoods and other engagements with forests, we briefly examine scientists' roles in shaping, supporting, and even attacking for- est management and policy (Swanson 2004). It is useful to view these roles in the context of stages of forestry management and pol- icy paradigms in the panarchy framework of Gunderson and Holling (2002), who envision natural resource management as a series of cross-scale, interacting adaptive cycles punc- tuated by periodic crises (see Chapter 1). However, some roles of scientists are contin- uous, such as conduct of basic science, despite periodic crises and regardless of the manage- ment approach of the day. The role of sci- entists can be broadly grouped as operating inside the system in support of the prevailing management paradigm and those operating outside as shadow networks that attempt to influence decision making through the pro- duction of knowledge and critique of pro- posed management and policy options.

Recent forestry conflicts reveal the quite varied roles of scientists over the course of development of a resource exploitation man- agement paradigm, through the crisis of its demise, and into establishment of the next set of policies (Box 7.1, Fig. 7.4). These roles may take the forms of:

1. Exploitation phase

 - "Applied scientists" working in sup- port of the prevailing paradigm study- ing methods to enhance productivity and management efficiency. Informa- tion exchange is termed "technology transfer," implying one-way flow of information in the form of technology. Government and industry are likely to fund this science work.
 - "Canary in the mine" scientists iden- tify species or other ecosystem compo- nents or processes that are perceived as imperiled by the management paradigm of the day. This is a risky, public, and generally unfunded role.

2. Conflict phase

 - Scientists may conduct bioregional assessments of the history and cur- rent conditions of the ecosystems and resources in question to set the stage for resolution of conflict (Szaro et al. 1999).
 - Scientists may participate in develop- ment of broad-scale management plans that set new policy intended to resolve past conflict. Bioregional assessments may be used as a foundation for the new plan.

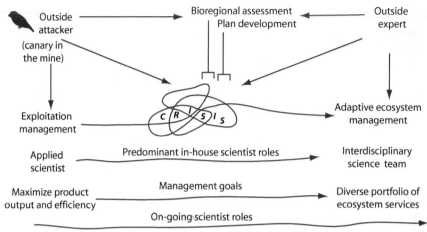

FIGURE 7.4. Schematic representation of the varied and changing roles of scientists during a transition from a predominantly exploitation phase of forest management through a conflict phase to a postconflict ecosystem-management phase. This paradigm shift involves important changes in the roles of in-house (agency or industry) scientists from leadership by applied scientists, whose goal is to improve productivity and economic efficiency, to interdisciplinary teams of basic and applied scientists who may have pivotal roles in bioregional assessments and development of new management plans. These scientists together with land managers may assume responsibility to design and implement adaptive management. Scientists outside of agencies also play key roles—most frequently as outside attackers in the exploitation phase or as outside experts in the postconflict phase. Ongoing roles for scientists at all times include conducting basic science and serving as sources of expert opinion (e.g., blue-ribbon committees). Adapted from Swanson (2004).

3. Postconflict phase

- "Adaptive management roles" have scientists working in support of the new management paradigm. In current terminology that may mean participation in adaptive management programs with scientists and land managers working in close partnership to develop new management approaches that have scientific and operational credibility. This is quite different than the old "technology transfer" model of scientist and manager relations. The adaptive management relationship between scientists and managers can feature two-way exchange of views and shared learning. A strong base of adaptive management can set a stage so that when the policy window opens, innovative, grounded approaches are ready to be spread via a shift in policy (see Chapter 5).

4. Science roles at any time

- "Long-term basic research" is a continuing role of scientists, whose findings may be useful in policy and management at some later point.
- "Outside attackers" criticize the prevailing management paradigm and may work for interest groups or with personal motivations. Attackers may operate at any stage of policy and management, but are especially significant in triggering crises and bringing down an existing paradigm. Attackers of a new management approach may come from the ranks of those who benefited from the earlier management paradigm.
- "Episodic venues" for science input to policy makers occur via participation in processes such as "blue ribbon" commissions on special topics intending to inform the public and the policies.

- "The isolated individual" is a single scientist or small group with limited support operating in areas with little science infrastructure, including developing countries. In this case an individual may play many of the above roles to try to improve stewardship of forests and other natural resources.

Scientists engaging in natural resource issues, especially leading up to, through, and shortly after a major policy convulsion, can find professional life very challenging. Education programs generally do not train students for the social and political turbulence that can come with these roles. Many dilemmas about professional conduct arise in the context of public policy disputes. Some scientists play multiple roles during their careers or even simultaneously, such as government scientists who may be a dispassionate provider of science-based information in that government role, but work clandestinely on behalf of an interest group in the off hours. Scientists confront many tough personal choices, such as advocate aggressively for a policy outcome in the immediate crisis or hold back to maintain a reputation of objectivity, hoping this to be an asset in future crises.

Thus, scientists have many decisions to make about their conduct. Even advocating use of science in natural resource decision making may be a contentious proposition in some circumstances. Young scientists in particular have a great deal at stake — credibility with the public and peer groups, continued and future employment prospects, even self-esteem. Attitudes about appropriate roles of scientists differ among scientists, land managers of science information, interest groups, and the informed public, although these groups do endorse the role of scientists to help integrate their knowledge with other information (Lach et al. 2003). Attitudes about appropriate roles of scientists vary from one societal context to another (Lach et al. 2003). Decision making and policy advocacy by scientists is more accepted in Europe than in the USA, for example; and in the USA these actions may be more accepted for academics than for government scientists, who are viewed as sources of objective scientific information to decision makers, who may be elected officials or their appointees possessing little or no scientific background. Therefore, decisions about professional conduct as a scientist are very dependent on institutional context, potential impacts on career, and the environment in both the near term and the long term. This issue has been debated at length (Fischer 2000, Weber and Word 2001), especially in the aftermath of highly disputed cases. University programs in natural resources and related fields should provide training in values, professional conduct, and environmental ethics, but ultimately the resolution rests with the scientist or natural resource professional.

Production Forestry

Intensive production forestry on plantations accounts for an increasing proportion of the world's timber supply, about 35% in 2000, with the proportion expected to increase to about 44% by 2020. As in the case of intensive agriculture (see Chapter 12), an important benefit of production forestry is that a substantial proportion of the global demand for timber can be met on a relatively small land base, reducing pressures for harvest of natural forests that have high conservation, watershed, and recreational values. Production forests currently account for only 5% of forestlands, but 35% of timber production. Production forests are intensively managed, regularly spaced stands, often monocultures of *Pinus* or *Eucalyptus* where they are exotics. They are managed with agricultural approaches that maximize the efficiency of wood production.

Although management practices vary geographically, production forests can provide many important ecosystem services in addition to the wood for which they are explicitly managed. Water and energy exchange, for example, may not differ substantially between production and natural forests. Production forests have high rates of carbon gain, but their role in carbon sequestration depends on the size of dead wood carbon pools left on the forest site and the lifetime of wood products from these forests. Wood used in construction, for example, may have as long a lifetime as a solid form of carbon as it would in the forest, whereas paper products might decompose and return their carbon to the atmosphere quite rapidly. Old-growth forests, in contrast, have rates of production that are only slightly higher than rates of decomposition and therefore sequester carbon relatively slowly (Schulze et al. 2000). These forests have high standing crops of carbon, including in the dead wood component, which gets released to the atmosphere upon cutting and conversion to intensively managed young forests (Harmon et al. 1990).

Sustainable management practices that maximize both forest production and other ecosystem services generally also sustain slow variables, including the retention of soils and their capacity to supply water and nutrients, provisioning of high-quality water in streams and groundwater, and many of the aesthetic and cultural benefits that foster public support for forests and forestry (see Chapter 2). Production forests can provide employment for local residents, enabling people to remain in rural areas and sustain their cultural connections to the land. However, not all production forests are managed in an environmentally sound way. As with intensive agriculture, some production forests receive large applications of fertilizers and pesticides, with similar detrimental consequences for the environment and local communities (see Chapters 9 and 12).

The most consistent losses of ecosystem services with production forestry are associated with reduced biodiversity. Production forests are typically single-aged, single-species stands that maximize the efficiency of planting, thinning, and harvesting of trees, but lack the structural and biotic diversity of natural forests. In regions dominated by production forestry or plantations of trees such as oil palms, much of the regional diversity has been lost, especially those species that require the structural complexity, long periods for dispersal, or decaying wood typical of old-growth forests.

Restoring biological diversity within highly managed landscapes is one of the greatest challenges faced by production forestry, but in some cases it can be accomplished to a significant extent without greatly sacrificing economic efficiency. The most simple and cost-effective way to restore diversity is to facilitate processes that naturally generate and maintain ecosystem and landscape diversity. These processes include periodic disturbance, such as fire — perhaps approximated by harvest patterns that create habitat heterogeneity within and among stands; maintenance of landscape patterns that provide corridors for spread of native species and barriers to spread of early successional exotic species; and protection of rare habitats to which some species are restricted (Hannah et al. 2002, Chapin et al. 2007, Millar et al. 2007). In general, if a diversity of habitats and environments is created and sustained, the appropriate organisms will find and exploit them, if they are locally available.

Within-stand heterogeneity and diversity can be increased by silvicultural practices such as retention of some live trees, dead standing trees, and decaying wood at the time of harvest, and protection of key habitats and buffer strips between managed stands (Larsson and Danell 2001, Raivio et al. 2001). A spectrum of tree ages also fosters diversity, because wood-decaying fungi, insects, and bird predators found in late-successional forests are quite different than species found in younger stands (Chapin and Danell 2001). A flexible rotation schedule can augment stand-age diversity. Allowing some trees to die and produce dead wood is critical to maintaining trophic diversity. As coarse woody debris becomes more available on the landscape, it will likely be colonized by wood-decaying taxa. In some forestry cultures, dead wood was removed to protect crop trees — to remove logging slash to reduce fire danger and ease planting, and to remove habitat for pest organisms that might attack live crop trees (e.g., removal of dead wood under forest hygiene laws in parts of Europe). In a manner similar to shifting attitudes about fire—from enemy to potential collaborator—forest managers have come to see dead wood as integral to ecosystem management rather than a threat to the forest. In summary, a spectrum of forest-management approaches exist from highly "agricultural" forest plantations to plantations that are managed with substantial structural and biological diversity to natural forests that are managed largely for timber production to natural forests that are never harvested. This blurs the distinction between natural and plantation forests.

Economic and cultural diversity can be just as important as ecological diversity to sustainable forestry, even in production forestry operations. Economic subsidies to initiate new types of businesses, such as the harvest of nontimber forest products (e.g., berries, guided hunts, medicinal herbs, edible fungi, greens for floral displays), the production of local crafts, and promotion of recreation can generate a wide variety of economic options, increasing the likelihood of long-term economic vitality, regardless of the social and economic events that may occur (Chapin et al. 2007). Alternatively, payment to local forest owners to provide ecosystem services such as berries and wild game could alter the incentive structure for forestry planning in rural areas. If local users strengthen their personal and cultural connections to the land, they will have a stronger long-term commitment to sustaining its important qualities.

Multiple Use Forestry: Managing Forests for Multiple Ecosystem Services

Given the breadth of ecosystem services provided by forests, there is a long history of efforts to balance exploitation of forests for wood production and their value in providing many other services. In some cases this balancing act has been fraught with conflict (Box 7.1); in other cases societies have been very accepting of

tradeoffs to meet multiple objectives. Management of forests for multiple objectives was well established long before the term "**multiple use**" gained formal definition in contexts such as the US Multiple-Use Sustained-Yield Act of 1960. Many indigenous societies traditionally managed forests for multiple uses (see Chapter 6) and recognized that the gifts received from forests (i.e., ecosystem services) entailed a responsibility to manage the forests in ways that sustained the capacity of forests to continue providing these services. Community forestry, which has deep roots in indigenous cultures, has recently experienced a resurgence of interest in many societies, leading to establishment of NGOs and governmental programs to provide support (Gunter 2004). The case of moving from the management of forests for timber to multiple uses requires at least two steps: (1) recognition by forests managers (whether they are in a government agency, a company, a forest community, or a family) that forests provide multiple services with both synergies and tradeoffs among the provisioning and use of these services and (2) fostering a sense of community responsibility to sustain the capacity to provide multiple services over the long term.

During the 1990s public pressures and conflicts forced forestry from an emphasis on resource extraction to a focus on stewardship of the ecosystem as a whole. The resulting **ecosystem-management** paradigm emphasizes forest practices for multiple services by capitalizing on and sustaining native ecosystem processes (Grumbine 1994, Szaro et al. 1999; see Chapter 2). As a consequence, in many jurisdictions the extent of forest reserved from cutting has been increased; the intensity and frequency of cutting have declined; and historical disturbance regimes have been increasingly used to guide future management.

Given the inevitable tradeoffs among ecosystem services, multiple use management is challenging and benefits from regular engagement and open communication among multiple stakeholders. Sometimes acceptable solutions are possible by managing different portions of the landscape using different but complementary approaches.

Managing Forests Under Conditions of Rapid Change

Global Change

Rapid changes in environmental and social drivers of forest dynamics increase the uncertainties of how to provide long-term forest stewardship. Environmental changes occurring at local to global scales shape the forest environment in ways that alter options for sustainable forestry. For example, air pollution, such as that resulting in acid rain and nitrogen deposition, affects large regions, such as northeastern North America and Eastern Europe, with net effects across the gradient of pollution ranging from stimulation of tree growth by nitrogen addition to severe growth reduction due to leaching of essential cations from soils (Driscoll et al. 2001). Resulting changes in soil properties, such as organic matter content and pH, have effects that will likely last for centuries. Introductions of exotic species such as earthworms or forest pests have in some cases directly and radically altered entire ecosystems. For example, the spread of exotic insects (e.g., defoliators, bark beetles) and pathogens (e.g., root-rot fungi) can cause range-wide extirpation or profound suppression of key forest trees, with impacts that ripple through the affected ecosystems (Ellison et al. 2005). Notable examples in the USA include the chestnut blight in the early twentieth century and the hemlock wooly adelgid and Sudden Oak Death Syndrome that are current, acute management concerns. Perhaps the most daunting driver of forest change will come from global climate change, which may cause migrations of species across the landscape, reshaping forest communities and making them vulnerable to pests, pathogens, and other disturbance agents at unprecedented scales. Some system responses to climate change and/or land use history may be gradual and others abrupt. Threshold system changes can have critical effects on forest systems by mechanisms including pine beetle outbreaks, permafrost thaw, and altered disturbance regimes. Some threshold changes may shift forests to other ecosystem types (Folke et al. 2004), thus eliminating forest-specific ecological services. Not all forest changes are negative. Abandonment of agricultural lands has led to forest expansion in portions of Europe and eastern North America. Combinations of rising CO_2 and modest nitrogen deposition can stimulate forest production, which can have both positive (e.g., carbon sequestration) and negative effects (e.g., reduced species diversity; Chapin et al. 2002) .

The world is now hyperlinked via near-universal, instantaneous communications and intricate economic networks. Public attention to the well-being of forests has grown over recent decades as a result of controversies over deforestation in the tropics, global loss of biodiversity, and protection of old-growth forests in temperate regions. In this context growing attention to sustainable forest stewardship and combating illegal logging appear poised to make substantial advances.

Disturbance Management

The paradigm for managing forest disturbances has changed dramatically in recent decades. Through much of the twentieth century "disturbances" were viewed as bad, so fires and insect outbreaks were fought with military zeal. Training and practice of this dimension of forest management was termed "forest protection." When forests are viewed as a crop, disturbances are a simple loss of capital and investment, so this single-minded management response was logical. Mid-twentieth-century practices, such as extensive use of persistent chemicals like DDT, were gradually revealed to be very damaging to the environment, so they were replaced by pesticides more targeted to the "pest" species. But even these pesticides, as well as native biocontrol agents like *Bacillus thuringiensis*, a bacterium that attacks the blood of Lepidoptera, such as gypsy moth, raised challenges because it can kill desirable, nonpest relatives of the target species. Thus, as the appreciation of ecosystem complexity grew, so too did the difficulty of combating pests and other disturbances.

Furthermore, practical experience and ecological studies have shown that disturbances are integral to the well-being of many types of forest ecosystems and their provisioning of ecosystem services (Gunderson and Holling 2002). Attempts to suppress native disturbance processes have in some cases resulted in even more explosive disturbance events when they escape that suppression—extremely intense fires may feed on an unnatural buildup of fuels; floods that breach dikes may create unusual havoc when the dikes slow drainage of the inundated area; insects feeding on stressed, vulnerable forest stands and landscapes may spread rapidly (Holling and Meffe 1996). Consequently, there has been an accelerating shift from attempting to exclude fire and other disturbance agents from forests to adopting management systems that seek to retain roles of these processes within accepted limits. For example, frequent, low-intensity prescribed fire is now an important management tool in systems where ground fires occurred naturally (Schoennagel et al. 2004). In some cases a legacy of fire suppression has permitted forest fuels to build to unnatural levels that pose the threat of stand-replacing fires in systems where that was not characteristic. In these circumstances mechanical reductions of woody biomass may be necessary before reintroducing fire to the forest. In other cases managers attempt to control forest vigor by thinning overcrowded stands, with the goal of keeping insect outbreaks within limits. However, these forestry practices are generally expensive and require extensive implementation to be effective at the landscape scale, so they have been applied primarily in targeted situations, such as the "wildland–urban interface" where high-value homes are threatened by wildfire (Radeloff et al. 2005).

Historic disturbance regimes are increasingly used as a reference point to guide management of future landscapes (Perera et al. 2004). This approach is based on the premise that native species and ecological processes might best be maintained by retaining ecological structures (including whole landscapes) and disturbance processes in a seminatural range of conditions to which native organisms are well adapted.

Several provinces in Canada (Perera et al. 2004) and forestry organizations in the USA (e.g., Cissel et al. 1999), for example, have explored this approach to forestland management by studying historic wildfire patterns over time and space and then developing management plans with frequency and severity of harvest treatments that emulate the wildfire regime. Although this concept has been extensively discussed in the literature (Burton et al. 2003, Perera et al. 2004), no examples have been implemented extensively, in part because of the long time periods required to produce an imprint on forest landscape structure where cutting rotations approach or exceed 100 years.

In addition, many challenging questions have been raised concerning the use of the **historic range of variability** to guide future management (see Chapter 2): What period of history do we use as a guide? What aspects of historic disturbance regime do we incorporate in the management plan (e.g., frequency, severity, spatial pattern of disturbances)? Climate change presents many challenges that are difficult to anticipate. Millar et al. (2007) present the options for employing adaptive strategies in these cases, but we know of no cases where actual practices are in place on the ground.

The challenge of using an historic approach brings into focus the question: What does "conservation" mean in a world with so much change? Given the dynamic state of the environment, an iterative approach seems prudent:

1. Identify the climatic and other forms of environmental change that are most likely to occur and the likely ecological responses, both gradual and abrupt/threshold;
2. examine and implement management approaches that may confer social and ecological resilience in the face of anticipated change;
3. design an assessment process that provides a basis for adaptive change, as learning occurs; and
4. be prepared to adjust objectives from cultivating resilience to transformation, if environmental or social change is so great that an alternative trajectory may provide greater

ecological and social benefits in the face of inevitable transformation.

This sequence of steps required for adaptive management (see Chapters 4 and 8) may take many decades. In many cases, desirable and undesirable outcomes are strongly shaped by social forces, as described in the next section.

Forest Conversion to New Land Uses

Conversion of forests to other land uses and potential for reversion to forest are critical determinants of future forest extent and the services they provide to humanity. The history of forest clearing is deep, but the process is accelerating in response to a wide range of social and economic drivers. Forest clearing for shifting agriculture has caused cyclic change in the past, although increasing population pressure may lead ultimately to deforestation. Other forms of forest clearing may be immediate and irrevocable.

Tropical deforestation, both legal and illegal, to make way for agriculture, such as production of beef and biofuels, has been an issue of critical concern because of the profound social and environmental consequences. All ecological services that are distinctively provided by forests are lost. Tropical deforestation involves cross-scale interactions in driving local change and the large-scale feedbacks to the climate system, as described earlier.

An important form of forest conversion may occur due to the intimate mixing of forest and other land uses, even where the landscape remains largely forested. For example, low-density residential development in forest landscapes is common in many parts of the USA, creating conflicts when forest processes, such as wind-toppling of trees, wildfires, and wild animals (e.g., cougars drinking from children's wading pools) impinge on a suburban life style. In these circumstances, some ecological services provided by forests may be partly maintained (e.g., carbon sequestration, water regulation, habitat for some groups of forest species), but others are lost (e.g., wood production, fire, and

habitat for species considered threatening to people).

Not all forest conversion is irrevocable. For example, waves of land-use change began in New England when the "soft" deforestation for agricultural development took place in the late eighteenth and early nineteenth century (Foster and Aber 2004). Industrialization, access to superior farmlands in the Midwest, and other factors led to farm abandonment and reforestation largely by natural reseeding. Now, however, the sprawl of urban and rural residential development is a new wave of "hard" deforestation; the pavement and dwellings will be more difficult to return to forest. New conservation strategies are being put forth to identify the best of the remaining forestland tracts in a regional design intended to provide large habitat patches and dispersal corridors between them (e.g., the Wildlands and Woodlands program in Massachusetts and neighboring states— http://harvardforest.fas.harvard.edu/wandw/).

Changes in subsidies and taxation systems can precipitate shifts in property ownership and attendant shifts in commitment to sustainable management. In the USA, forest companies that have a long-term stake in forest productivity are being taken over by Timberland Investment Management Organizations (TIMOs) and Real Estate Investment Trusts (REITs) that seek very short-term (fraction of a cutting rotation) financial return in part by selling forestlands for residential development (Fernholz 2007). In the eastern US conservation NGOs are partnering with timber companies to purchase development rights that protect forestlands from development and allow small-scale logging to continue (Ginn 2005). Similarly, in some developing nations, debt-for-nature swaps have allowed foreign debts to be forgiven in return for conservation protection of lands that might otherwise be cleared. The long-term outcomes of these emerging transactional mechanisms controlling forests are far from clear—will they lead to more sustainable forest-management practices or to a serious departure from sustainable forestry? TIMOs and REITs appear to encourage unsustainable practices, whereas the debt-for-nature program may protect forests from clearing. The intermediate case of conservation easements on private lands may prove to be a compromise that provides both ecological and social benefits. Management of public lands seems to be on the pendulum swing between reserve and rather intensive management, as, for example, the New Zealand bifurcation of lands to native forest without management and plantations largely of exotic species (mainly *Pinus radiata*). Only time will tell the outcome of these alternative approaches to governance.

Change in national policies to shift energy production from fossil fuels to forest-generated biofuels will intensify forest management, as currently being considered in Sweden, or lead to clearing of native forests for oil palm plantations, as is occurring in Indonesia.

Synthesis and Conclusions

These many approaches and challenges to forestry lead us to seek an expanded view of forest stewardship for tomorrow's forests, drawing on the large bodies of practical and scientific knowledge of forests and associated communities and also calling for greatly enhanced adaptive capacities to face the changing societal and environmental conditions. Despite their social–ecological diversity, forests face similar challenges throughout the world, suggesting some general strategies for sustainable forest stewardship. The implementation of these strategies must be context-specific, taking into account local, national, and global drivers of change.

- Institutions governing forest stewardship must allow planning for the long term by promoting practices that maintain the social and ecological capital required for multiple generations of trees and forest users, whether they are in developed or developing countries (see Chapter 1). In the absence of such institutions, positive social outcomes are most likely to result from attention to capacity building and adaptive governance that would foster development of institutions and policies that might lead to favorable social–ecological transformations to an alternative state.

- Sustainable timber production requires sustaining the slow variables that maintain the productive potential of ecosystems, including soil fertility, a disturbance regime to which local organisms are adapted, and both the ecological and the socioeconomic diversities to maintain future options.
- Effective stewardship to sustain multiple ecosystem services provided by forests requires a clear understanding of the controls over these services and the synergies and tradeoffs among them. Although it is unlikely that all these services can be maximized simultaneously, careful planning that engages stakeholders can lead to well-informed choices that address multiple needs and concerns.
- Given that social and environmental conditions are quite likely to change and unknowable surprises will occur, forest stewardship should foster flexibility to respond to these changes through adaptable governance and the fostering of biological and socioeconomic diversity that provides the seeds for multiple future options (see Chapter 5).
- Given the frequent mismatch between the requirement for long-term vision for forest stewardship and the short-term motivations and path dependence of decision options, forests are quite prone to social–ecological transformations, suggesting the benefit of planning that considers multiple alternative states and their relative social and ecological benefits. Think outside the box.

Although none of these strategies is unique to forests, the longevity of forest trees makes the importance of the long view particularly apparent and could therefore inform ecosystem stewardship in a broad range of social–ecological systems.

Review Questions

1. What are some of the challenges to forest stewardship that result from the long-lived nature of trees? What other systems face similar stewardship challenges?

2. What are the major stages of development of management practices that have characterized forestry practices in many parts of the world? Based on this understanding, how might policy makers avoid detrimental phases and foster sustainable forest stewardship in regions that are just beginning to develop forest resources?

3. What common and distinctive challenges do communities in developed and developing countries face in exercising local influence on forest management and ecological services received from forests?

4. How can production forests be managed in ways that provide multiple ecosystem services within a management framework geared to maximizing wood production?

5. What institutions and social practices are most likely to foster long-term sustainability of forest values that reflect inevitable trade-offs among multiple ecosystem services?

6. What roles do scientists play in natural resource decision making? What challenges do they face in each of these roles?

Additional Readings

Brown, K. 2003. Integrating conservation and development: A case of institutional misfit. *Frontiers in Ecology and the Environment* 1:479–487.

Ellison, A.M., M.S. Bank, B.D. Clinton, E.A. Colburn, K. Elliott, et al. 2005. Loss of foundation species: Consequences for the structure and dynamics of forested ecosystems. *Frontiers in Ecology and the Environment.* 3:479–486.

Folke, C., S. Carpenter, B. Walker, M. Scheffer, T. Elmquist, et al. 2004. Regime shifts, resilience, and biodiversity in ecosystem management. *Annual Reviews in Ecology, Evolution, and Systematics* 35: 557–581.

Foster, D.R., and J.D. Aber. 2004. *Forests in Time: The Environmental Consequences of 1,000 Years of Change in New England.* Yale University Press, New Haven.

Ginn, W.J. 2005. *Investing in Nature: Case Studies of Land Conservation in Collaboration with Business.* Island Press, Washington.

Lane, M.B. and G. McDonald. 2002. Towards a general model of forest management through time: Evidence from Australia, USA, and Canada. *Land Use Policy* 19:193–206.

Millar, C.I., N.L. Stephenson, and S.L. Stephens. 2007. Climate change and forests of the future: Managing in the face of uncertainty. *Ecological Applications.* 17:2145–2151.

Nagendra, H. 2007. Drivers of reforestation in human-dominated forests. *Proceedings of the National Academy of Sciences* 104:15218–15223.

Sampson, R.N., N. Bystriakova, S. Brown, P. Gonzalez, L.C. Irland, et al. 2005. Timber, fuel, and fiber. Pages 585-621 *in* R. Hassan, R. Scholes, and N. Ash, editors. *Ecosystems and Human Well-Being: Current State and Trends.* Millennium Ecosystem Assessment. Island Press, Washington.

Szaro, R.C., N.C. Johnson, W.T. Sexton, and A.J. Malk, editors. 1999. *Ecological Stewardship: A Common Reference for Ecosystem Management.* Elsevier Science Ltd, Oxford.

8
Drylands: Coping with Uncertainty, Thresholds, and Changes in State

D. Mark Stafford Smith, Nick Abel, Brian Walker, and F. Stuart Chapin, III

Introduction

Drylands cover 40% of the terrestrial surface (Table 8.1, Plate 6) and are characterized by high ecological and cultural diversity. Although they are, by definition, of low productivity, they have been a source of biotic, social, and scientific innovation. A third of the global biodiversity hotspots are in drylands, with a diversity of large mammals in savannas, high diversity and endemism of vascular plants in shrublands, and a high diversity of amphibians, reptiles, birds, and mammals in deserts. Succulence, the CAM photosynthetic pathway, and camels' tolerance of changing blood water content are examples of biological innovations that arose in drylands. Drylands are culturally diverse and account for 24% of the world's languages (Safriel et al. 2005). Traditionally, many social groups moved both seasonally and in response to prolonged droughts (e.g., Davidson 2006). The need to cope with harsh conditions and repeated episodes of scarcity

have given rise to strong cultural traditions such as the invasive effectiveness of the Mongols, the rule base for several major religions, and traditional ecological knowledge backed by powerful sanctions as in Aboriginal cultures in Australia. In ecology, attention to the extreme conditions represented by drylands has helped create paradigm shifts of wider relevance, such as the development of disequilibrium concepts and state-and-transition models (Westoby et al. 1989, Vetter 2005) that were important drivers of the development of resilience theory (Walker 1993). Such concepts are especially pertinent as humans prepare for climatic change.

Today, about a third of the world's population (two billion people) live in drylands (Table 8.1), about half of them dependent on rural livelihoods. Water availability strongly constrains both biological productivity and human development. Drylands receive only 8% of the world's renewable fresh water supply − 30% less per capita than the minimum considered essential for human well-being and sustainable development (Safriel et al. 2005). Dryland populations tend to lag behind those in other parts of the world on a variety of economic and health indices, even controlling for "ruralness" (MEA 2005a, b), with higher infant

D.M. Stafford Smith (✉)
CSIRO Sustainable Ecosystems, PO Box 284, Canberra ACT 2602, Australia
e-mail: mark.staffordsmith@csiro.au

F.S. Chapin et al. (eds.), *Principles of Ecosystem Stewardship*,
DOI 10.1007/978-0-387-73033-2_8, © Springer Science+Business Media, LLC 2009

TABLE 8.1. Key features of four dryland subtypes. Information from Safriel et al. (2005).

	Hyper-arid	Arid	Semi-arid	Dry subhumid	Total
Aridity index[1]	<0.05	0.05–0.20	0.20–0.50	0.50–0.65	
Area (million km^2)	9.8	15.7	22.6	12.8	60.9
Share of globe (%)	6.6	10.6	15.2	8.7	41.3
Total population (million)	101	243	855	910	2109
Share of globe (%)	1.7	4.1	14.4	15.3	35.5
Land-use (% of subtype)					% of all drylands[2]
Rangelands	97	87	54	34	65
Cultivated	0.6	7	35	47	25
Urban	1	1	2	4	2

[1] Ratio of precipitation to potential evapotranspiration
[2] Does not add to 100% as 8% is taken up in other categories, notably inland waters.

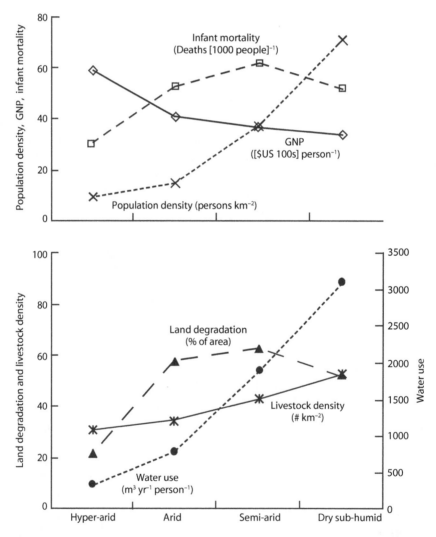

FIGURE 8.1. Some vital statistics across the dryland aridity gradient: top, GNP per capita (excluding OECD countries), infant mortality (excluding OECD countries), population density (developing countries only); bottom, proportion of dryland subtype that is degraded (%, 1990 estimates, excluding "low" category), livestock density (developing countries only), and water supply per capita. Information from Safriel et al. (2005).

mortality, severe shortages of drinking water, and much lower per capita GNP (Fig. 8.1). Dryland populations are among the most ecologically, socially, and politically marginalized populations on Earth (Khagram et al. 2003). Given that 10–20% (depending on measures) of drylands are "desertified," their populations are seen as some of the most vulnerable to the increased frequency of drought events expected under climate change (MEA 2005b, Burke et al. 2006). Rapid population increase coupled with livelihoods at risk creates the potential for further water shortages and land degradation, most likely causing movement of environmental refugees and flow-on effects to other biomes of the world.

In many ways, drylands represent the epitome of management challenges emphasized in this book. They are social–ecological systems that were traditionally well-adapted to uncertainty and variability, and many dryland civilizations persisted for millennia. However, in part through attempts to reduce this variability, many of the recent ecological and social changes have reduced the social–ecological resilience of drylands. The social context and

the climate under which these social–ecological systems evolved have changed and will continue to change. What were some of the original sources of resilience in drylands? How have these changed, and how can we restore or generate new sources of resilience under the altered social and climate contexts? In some cases, drylands have undergone dramatic and apparently irreversible degradation. What changes in social–ecological slow variables (see Chapter 1) have caused these thresholds to be exceeded? How can we reduce the likelihood of this occurring in the future?

Dryland Features and Functioning

The key attributes of drylands can be summarized as unpredictability, resource scarcity, sparse populations, remoteness, and "distant voice" – the **drylands syndrome** (Reynolds et al. 2007). Remote regions of Australia clearly illustrate this syndrome (Fig. 8.2; Stafford Smith 2008), although these attributes are important in most drylands to a greater or lesser

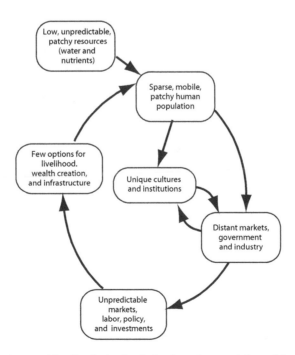

FIGURE 8.2. Factor interactions and feedbacks in the dryland syndrome. Adapted from Stafford Smith (2008).

TABLE 8.2. Gross domestic product (GDP) and area of Australian rangelands used by different sectors in 1999. Information from Fargher et al. (2003).

Sector	Rangeland contribution to Australian GDP (% of GDP)	Area of rangelands occupied (% of area)
Mining	2.4	0.04
Tourism	0.4	Not available (small)
Pastoral leasehold	0.2	50

extent. Reynolds et al. (2007) applied resilience and systems theory to drylands with these properties to develop a **drylands development paradigm (DDP)**, a conceptual framework consisting of five principles for analyzing past and future policy and management (see Table 2 in Reynolds et al. 2007):

P1. Social and ecological systems are coupled in drylands to form complex adaptive systems (see Chapter 1); this is particularly important in drylands because livelihoods are tightly coupled to the biophysical environment and vulnerable to market and policy processes that are beyond local control. These systems have no simple target equilibrium point for resource users and policy makers to aim for. Tracking change is thus more difficult and important in drylands than elsewhere.

P2. Although drylands are highly variable in time, these dynamics are determined by a few slow variables (see Chapter 1) that are regulated by feedbacks. Identifying and monitoring slow social and ecological variables is particularly important in drylands because high variability in fast variables masks the fundamental change indicted by slow variables. The limited suite of key variables and processes makes the complex system tractable.

P3. Thresholds in key slow variables define different states of social–ecological systems. Soil loss from a landscape that supports an arable farming community, for example, may shift the system to another state that can only support livestock, when the threshold of soil loss is exceeded. Thresholds are particularly important in drylands because there is usually little capacity to

invest in recovery from an undesirable change in state. Where calls must be made on outside agencies, the transaction costs of doing so are high in drylands due to relative remoteness. Such thresholds contribute to the unpredictability of drylands.

P4. Dryland social–ecological systems are hierarchical, nested, and connected through social networks, communications, and infrastructure to the broader nation in which they are located. This is particularly important in drylands because their typically low primary productivity supports only sparse populations that tend to have weak political influence in remote seats of governance. Consequently, they receive lower levels of services and infrastructure than national averages. Cross-scale linkages are important but usually weak in drylands, thus requiring particular institutional attention.

P5. Local ecological knowledge is critically important for maintaining the functions and coadaptation of dryland social–ecological systems. This is particularly important in drylands because experiential learning and innovation tend to occur more slowly in these highly variable systems. In addition, there is often less formal research than in mesic regions. Consequently, the development of hybrid scientific and local knowledge systems is vital for local management and regional policy in drylands.

In this chapter we use the DDP principles, which are derived from resilience theory for environments driven by variability and scarce resources, to analyze and seek solutions to drylands problems.

Physical Functioning

Drylands span a gradient from arid to semiarid to dry subhumid areas. Precipitation is scarce and variable (Middleton and Thomas 1997) along the entire gradient. Drylands are defined climatically by their low **aridity index** (<0.65) – the ratio of precipitation to potential evapotranspiration (Table 8.1) – resulting from high air temperatures, low humidity, and abundant solar radiation.

Water is thus the factor that most directly limits biological productivity and human population density. Water's greatest societal significance is to support forage production for pastoralists at the drier end of the gradient and to support agriculture at the moister end of the gradient. Not only are rainfall inputs more-or-less low in drylands, but they are generally less predictable than in other systems. Rainfall variability increases as mean annual rainfall decreases, as latitude decreases, and as the effect of the El Nino-Southern Oscillation (ENSO) increases (Nicholls and Wong 1990), thus delineating the great midlatitude belts of deserts in both hemispheres with added impacts from ENSO, especially in the southern hemisphere drylands. Even where rainfall itself is more reliable, other climatic elements (e.g., extremely cold winters in Kazakhstan or Patagonia) can create other sources of interannual variability from the perspective of human use.

Climate is coupled to vegetation cover at a regional scale. Loss of vegetation through land use increases albedo, which lowers surface temperature, reduces convective uplift (and in some cases advection of moist marine air), and reduces monsoon rains (Foley et al. 2003b, Xue et al. 2004). In contrast, controlled grazing, afforestation, and irrigated agriculture in the northern Negev of Israel led to a 10–25% increase in rainfall (Otterman et al. 1990), suggesting multiple effects of management on dryland climate.

Rock type determines parent material and thus affects spatial patterns of plant-available nutrients (PAN). Parent material affects soil texture, and thus soil permeability and soil moisture storage capacity, as well as soil depth. Soil moisture and nutrient levels can be generalized at broad scales. The productivity, structure, composition, and functioning of ecological communities (Cole 1982) and the resilience of soil–vegetation systems can be partly explained in terms of PAN and plant-available moisture (PAM; Frost et al. 1986, Walker and Langridge 1997), both of which act as slow variables, with thresholds that can be crossed if over-cropped or overgrazed. In general (Walker et al. 2002):

- The ratio of woodland to grassland (a slow variable) increases with increasing rainfall;
- So does the likelihood of increases in undesirable shrubs when a site is grazed;
- The productivity of a site is highest where PAN and PAM are both high, permitting faster rates of recovery after grazing. However, this says nothing about the propensity of a site to change to a new state.

Biological Functioning

Soil and vegetation characteristics strongly influence water retention and use (d'Odorico and Porporato 2006). Because natural rates of erosion from wind and water are high, soils tend to be shallow and therefore have low soil moisture storage capacity, a key slow variable. This determines vegetation productivity as well as influencing the ratio of water infiltration to runoff. Most dryland soils also have low organic matter and low aggregate strength, which make them highly erodible. When tillage or grazing reduces ground cover below a critical threshold of between 30 and 40%, soil loss from intense winds and rainfall can reduce soil depth (thus soil moisture-storage capacity) below a critical threshold that can preclude recovery of vegetation productivity and consequent ground cover (Fig. 8.3). This negative effect on soil moisture is reinforced because a decrease in vegetation cover reduces the rate of water infiltration. Losses of mineral soils are accompanied by losses of clay and organic matter, so cation exchange capacity, a slow variable that determines PAN, is also reduced. Seeds may also be transported off the landscape by water flows, reducing the capacity of vegetation to recover.

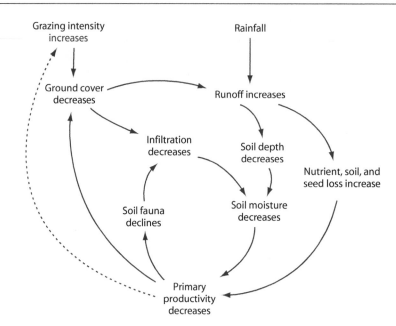

FIGURE 8.3. Grazing, aspects of landscape function, and feedbacks among these. Changes in ground cover (a fast variable with fluctuations driven mainly by interannual rainfall variability but affected by grazing) can cause soil depth (a slow variable) and consequent soil moisture storage capacity to cross thresholds below which the rest of the cycle moves into a new state characterized by impoverished soil fauna, reduced nutrients and seed banks, and long-term loss in productivity. There is a negative feedback to grazing intensity, but management interventions generally weaken this control by maintaining stock numbers.

Infiltration and soil moisture storage are also influenced strongly by soil invertebrates that create macropores and support nutrient cycling. Decreases in soil moisture resulting from grazing can cause declines in soil fauna with negative effects on nutrient availability and soil moisture. This further reduces vegetation productivity and increases vulnerability to further soil loss (Fig. 8.3). Thus a state change can be induced as thresholds of PAN and PAM are crossed.

Toward the drier end of the climatic gradient there is insufficient PAM to support more than very sparse vegetation cover if the land surface is flat. However, even slight topographic variations redistribute water (and seeds and nutrients) to produce patches of vegetation at all scales. Soil–water–nutrient–plant–animal processes have been integrated in the concept of **landscape function** (Ludwig et al. 2005). Landscape function is the capacity of a landscape to regulate nutrients and water, con-

centrating them in fertile, vegetated patches where soil biota maintain nutrient cycles and water infiltration and where vegetation cover impedes surface water flow, retains nutrients and seeds, maintains infiltration, and protects against erosion (Fig. 8.4). The resulting vegetation structures, such as **brousse tigré** (banded landscape patterns) and perennial grass and shrub "islands" (Ludwig et al. 1999, Rietkerk et al. 2004), are self-organizing expressions of landscape function. These support biota that could not occur if the same resources were evenly distributed across the landscape.

Loss of landscape function, commonly evident from the loss of vegetation patterns and homogenization of landscape cover, can drive a landscape from a productive state across a threshold to a state of low productivity (Fig. 8.3). The key control variable is vegetation cover at ground level. Ground cover on cropland is affected by crop type, sowing method, fertilizer level, and residues. On rangelands it is

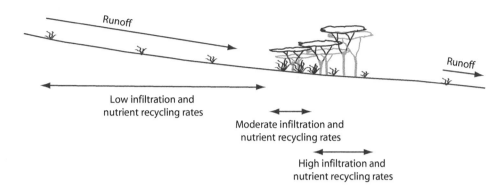

FIGURE 8.4. Functioning of a "brousse tigré" or banded shrub landscape when it selforganizes to capture water, nutrients, litter, and seeds. Modified from Tongway and Ludwig (1997).

affected by fire and grazing. As cover declines, there is increased loss of water, soil, nutrients, organic matter, and propagules from the system (Ludwig et al. 2005). There is a critical threshold level in landscape function below which, even if all livestock are withdrawn or fields are left fallow, the process continues to an irreversibly degraded state, or with a slow, hysteretic return path (Rietkerk et al. 2004). In this case we see landscape function as a critical slow variable with a threshold that separates alternate states.

Pringle and Tinley (2003) have extended the landscape function of drylands to a catchment scale that integrates a geomorphological framework. Water erosion caused by grazing or cropping can incise a landscape and lower its base level (a slow variable), for example, by incising natural levees and riparian sills. This can increase drainage from, and reduce soil moisture in, the landscape above the incision. Edaphic wetlands can be transformed to a new state characterized by scrub, for example. The decline in PAM and PAN reduces primary productivity. The management implication is that the problem needs to be addressed at catchment scale.

Socioeconomic Functioning

More than in most regions of the world, drylands are characterized by a majority of livelihoods that still depend on natural and cultural resources, through grazing, dryland agriculture, tourism, and mining. In this section we focus mainly on grazing and dryland agriculture, with a brief discussion of other uses.

Long-established social systems in drylands have coadapted with the unpredictability, resource scarcity, and large extent of rangelands. Compared to mesic areas, and a few major desert cities notwithstanding, three social and economic features of dryland systems are pervasive: the human populations of drylands are usually more sparse, mobile, remote from markets, and distant from the centers (and priorities) of decision makers. It is consequently more costly to deliver services, and institutional arrangements devised in other regions may be dysfunctional when imposed on drylands (Stafford Smith 2008).

Systems of property rights and associated reciprocal obligations between individuals or social groups reduce conflicts over scarce forage and water by enabling movements of people and stock. Such arrangements are slowly changing social variables that have been critical to traditional human adaptation to drylands. However, some rangeland regions have been privatized, leading to long-term trajectories of changes in systems, driven by increasing populations of stock and humans, expanding commercial opportunities, alienation of land, blockage of traditional migration routes, and sometimes veterinary disease control fences (Fig. 8.5). These processes alter the cost-benefit trade-off of subdivision and enclosure

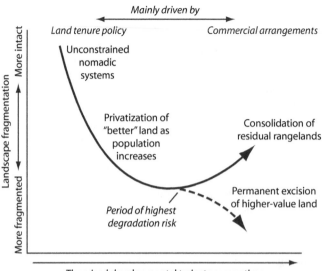

FIGURE 8.5. Hypothesized general trajectory of land fragmentation, potential excision for permanent agriculture, and consolidation in drylands. The depth of the curve and the degree to which the dashed or solid part to the right predominates depends on a variety of factors including productivity (more excision at the more productive end of the spectrum), population density, and political stability (more excision for higher population and political stability). Modified from Behnke (2008).

of common resources (Behnke 2008). In developing countries, subdividing land and assigning it to individuals, families, or small groups or alienating land for irrigation schemes is often driven by development policies devised in distant capitals. It disrupts previously nomadic cultures and resource-use strategies. Traditional strategies are commonly focused around key land units where landscape function is robust, and forage and surface water are available during dry periods (Scoones 1995). Such land is commonly selected for privatization because its higher productivity, unlike that of surrounding land, justifies capital investment (Behnke 2008). However, there are thresholds, such that, if too much of the key land resources are removed from the system, a nomadic strategy on the residual land becomes unsustainable. Cheap and readily available weapons such as assault rifles have further transformed resource access disputes in some dryland regions (Carr 2008). Where previously they were resolved through negotiations or with spears, they may now become much bloodier feuds as in, for example, Darfur in Sudan, and the Horn of Africa among many other regions.

Today pastoralism and rangeland use still dominate land use in hyperarid and arid (desert) portions of drylands, but agriculture now covers a greater area than rangelands at the moist end of the drylands gradient (Table 8.1). Cities are also concentrated in regions with greater water availability, where they occupy 4% of the land area (Safriel et al. 2005). Consequently, human population density is sevenfold greater at the moist end of the dryland moisture gradient (71 persons km^{-2}) than at the dry end (10 persons km^{-2}; Fig. 8.1). Other uses are also important. Mining occupies a trivial proportion of the area of the drylands, but its economic importance can hugely exceed that of pastoralism and agriculture. Tourism is also important, as, for example, in Australia (Table 8.2). In developing countries tourism based on wildlife has a history at least a century long, relying on protected areas from which local people are wholly or partly excluded (see Chapter 6).

Causes of Degradation and Their Impacts

The central issue in drylands is the same as in other systems – the sustainable production of outputs valued by humans. In drylands these outputs are milk, meat, hides, grains, timber, biodiversity conservation, aesthetic, recreational and tourism experiences, and hunting trophies, as well as minerals. Sustainability implies that the net production of values is maintained indefinitely (see Chapter 1). The features of the drylands syndrome – unpredictability, resource scarcity, sparse populations, remoteness, and distant voice (Reynolds et al. 2007) – create particular problems in achieving this goal compared to other systems. Geist and Lambin (2004) provide a comprehensive analysis of 132 cases of major change in different dryland systems worldwide, identifying six clusters of fundamental biophysical or social drivers that underpin the proximate causes of dryland degradation – climatic, demographic, technological, economic, policy and institutional, and cultural processes.

Climatic – Climate is changing at a variety of temporal and spatial scales, from the short- and medium-term variability discussed earlier, through regional changes due to deforestation and other land-use changes, to global climatic change. The past 50 years have already shown a tendency toward drying trends in many dryland areas of the world, with these trends expected to become more pronounced in the next 50 years (Burke et al. 2006). Variability is also likely to increase. Both drying and increased variability can cause degradation if resource uses and institutions become unsuited to the climatic regime.

Demographic – The human population is increasing more rapidly (18.5% between 1990 and 2000; Safriel et al. 2005, p. 645) in drylands than in any other major biome. However, drylands worldwide can be broadly split between those (mostly in developing nations) where there is strong population growth and consequent pressures on the resource; and those (mostly in developed nations) where drylands tend to become depopulated (notwithstanding the development of major cities such as Phoenix and Las Vegas in the USA, and Dubai in the Gulf States that are metropolitan islands set in arid hinterlands). Drylands are also a major source of migration.

Technological – A whole series of innovations continually alter production dynamics, so that the balance between management and environment continually needs updating. These include physical technologies such as more powerful boring rigs that can dig deeper for water, new types of fences, new and cheap weapons used in resource-access disputes, and more accessible four-wheel drives for tourists, as well as biotechnologies such as higher yielding or pest-resistant crop cultivars and new breeds of grazing animals. There are also communications and transport technologies that link drylands to the opportunities and threats of globalization.

Policy and institutional – Medium- and long-term shifts in political structures and geopolitical allegiances (e.g., loss of the Soviet Union, growing strength of China and India, development of the Convention to Combat Desertification, and other multilateral arrangements) alter the pressures on drylands. Changing social processes and institutional structures (e.g., methods for delivering aid, increasing roles of NGOs, network development such as "Desert Knowledge" in Australia [Stafford Smith et al. 2008]) provide opportunities and challenges.

Economic – Changing market needs and access, as well as world trade agreements and bilateral arrangements, mostly associated with a broad suite of globalization effects, reach into drylands profoundly, as exemplified by the huge surge in mining in deserts worldwide (e.g., Chile, Australia) as a result of demand from China.

Cultural – Positive cultural trends in developed nations such as rising demand for tourism, greater appreciation of cultural and biological diversity, and the opening of premium markets such as Fair Trade deliver opportunities to dryland areas, at the same time that advances in communications technologies increase the imposition of a dominant culture from developed nations into remote areas, causing an

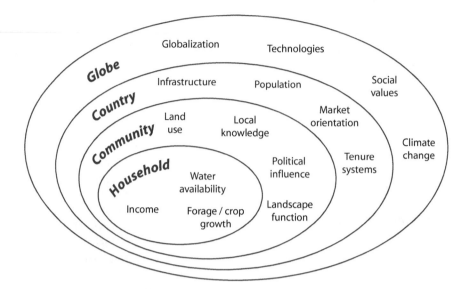

FIGURE 8.6. The effect of scale on the role of processes as external drivers, internal controlling (slow) variables, or fast variables representing immediate goods and services. Factors outside any ellipse are generally slow (but not necessarily controlling) variables, whereas those inside are relatively fast.

unprecedented rate of loss of local languages and culture.

Whether a factor such as population is a driver or a controlling (slow) variable depends on the scale of analysis (Fig. 8.6). The following sections explore some major system changes, usually driven by multiple factors in combination, and often creating both risks of degradation and new opportunities. The discussion aims to highlight slow variables and thresholds.

Expanding Aridity

The drying trends of the past 50 years and increased human use of freshwater, made possible in part by improving borehole technology, have combined to cause groundwater tables to drop in most areas, a widespread problem, for example, on the Indo-Gangetic Plain of India. For a given technology, there is a threshold below which water extraction is not worthwhile or is not feasible, possibly rendering some areas unsuitable for human occupation. The chronic water shortage that charac-

terizes drylands is expected to become more pronounced, by 11% per capita at the moist end of the gradient to 18% at the dry end of the gradient for the period 2000–2010 (Safriel et al. 2005) due to a combination of population increase, land cover change, and climate change.

A key climatic feature of drylands is variability of rainfall. Ellis (1994) proposed that there is a threshold in the **coefficient of variation** (CV – standard deviation divided by the mean) of annual rainfall of 30% above which it is more useful to think of a system as disequilibrial (that is, perpetually buffeted away from any equilibrium), than as an equilibrium system experiencing occasional disturbance. Increasing variability resulting from climatic change may cause this threshold to be crossed in regions that are currently managed as equilibrial systems. Adaptation to avoid land degradation in such cases may require resource users to shift to an opportunistic production strategy that is spatially more mobile and that can accommodate large fluctuations in livestock numbers and outputs. On the other hand, it may become a rational alternative to vacate lands that, through climatic change, have lost productive potential and have become increasingly risky to use. Policies to address the risks of land degradation, inequity, and impoverishment could include incentives to leave the region, radically reduced stocking densities, or shifts to alternative occupations within or outside the affected region.

Biodiversity Loss

Biodiversity is believed to be a key foundation of the resilience of social–ecological systems because biodiversity augments response diversity and the maintenance of functions that may become important if circumstances change (see Chapter 2; Elmqvist et al. 2003). Biodiversity is also valued for its own sake, its commercial or subsistence utility (for example, tourism and recreation), future values that are not yet known, or its support of cultural practices, including traditional ones.

Biodiversity conservation policies and management attempt to restore or maintain functioning examples of selected ecological communities and the processes and biota they contain. Processes include the dynamics and dispersion of populations, trophic and other interactions, and the evolutionary processes that generate diversity. Successful conservation of biodiversity requires the maintenance of sufficient areas of well-functioning land, representing all the biological units and processes deemed important, including the ecological interactions within and among the conserved areas (Sarkar et al. 2006). This is an ideal, and policies and management are usually pragmatic rearguard actions against threats from habitat loss, pests and weeds, increased human use, and other causes. The best that can be achieved in most circumstances is to make and keep each managed area:

- large enough (a slow variable with a threshold) to maintain the viability of sedentary populations;
- sufficiently well connected (slow variable with a threshold) to other conservation sites

to enable movements of mobile species and recoveries from local extinctions.

Connectivity depends on distance between similar sites, as well as the suitability of the land between the sites for dispersal by mobile species. The majority of drylands, particularly those that are grazed rather than cropped and not too fragmented, are often in seminatural condition, so the opportunities to combine conservation with moderate land use are greater than in dramatically transformed agricultural landscapes.

With increasing population and appropriation of primary production, humans in drylands threaten biodiversity by removing habitat, heavy grazing of plant species, blocking dispersal paths, killing wild species to eat or sell, changing fire regimes, and grazing their animals in competition with wild species. For example, there is a long history of conflict between wildlife conservation and indigenous uses of African drylands. Conservation areas were established by colonial governments, and indigenous users were formally excluded in most cases, although poaching and other uses continued (Abel and Blaikie 1986). Today conservation areas continue to be supported by postcolonial governments because of the foreign exchange they earn from tourism (see Chapter 6). There are thresholds in the slow variables of size, representativeness, and connectivity that affect the success of biodiversity management. Reconsolidation of fragmented land, sometimes under local initiatives, sometimes promoted by government policy, has often been necessary in order to incorporate animal populations within an ecologically viable unit (Abel et al. 2006). Where the movements of wildlife valued for tourism (e.g., wildebeest, zebra) occur at a very broad scale, incentive schemes are being established to encourage landholders not to block migration routes with fences or fields.

On the other hand, the establishment of conservation areas can cause the loss of key resource areas or migration routes for hunter-gatherers or agro-pastoralists. As with wild species, the accessible areas of key resources and the connectivity among them are both slow variables with thresholds that, if exceeded, threaten the viability of particular social–ecological systems. Some of the traditional uses by hunter-gatherers that are disallowed in conservation areas may be entirely compatible with conservation and can be restored through co-management arrangements between government and indigenous residents, as in some Australian national parks (see Chapter 6).

In sparser rangelands, the number and placement of artificial water points are critically important to the management of forage for pastoralism, such that more water points enable the landscape to be grazed more evenly. Grazing intensity by domestic stock is highest near the water and declines as a power function with distance, affecting both landscape function and biodiversity (Ludwig et al. 2004). Improvements in borehole technologies and cheaper water distribution using plastic pipe have caused rapid water point development in Australia. Consequently, fewer areas are distant from water, leaving less habitat in which grazing-sensitive species can persist (James et al. 1999, Landsberg et al. 2003). In this way, slow developments by individual pastoralists can cause a regional threshold to be reached suddenly, causing widespread biodiversity losses because no water-remote refugia remain. Parallel issues occur with respect to borehole provision in the Sahel.

The establishment and management of regional biodiversity conservation systems are further complicated by actual and potential climatic changes. In the absence of climatic change a resilience-based approach might attempt to maintain a conserved area or system of conserved areas within a particular basin of attraction (see Chapters 1 and 5). Climatic change causes the basin of attraction to shift so that the functionality of the conserved area is no longer possible in its current location. When basins of attraction shift along a climatic gradient, policies and management must adapt to these changing conditions or disappear entirely. Alternatively, triage may encourage the reallocation of resources to achieve the feasible, rather than futile, attempts to maintain species and ecological communities that cannot adapt.

Intensification and Abandonment

Population increase leads to intensification and fragmentation (with increasing demands for services), whereas population declines lead to abandonment or consolidation (coupled with withdrawal of services; Fig. 8.5). These different trajectories create distinctive interactions with policy issues in drylands of developed and developing nation (Campbell et al. 2000).

The size of the human population in a region strongly influences the area cropped and the numbers of stock. One threshold that is crossed as population (a slow variable) increases is the number of people who can be supported by pastoralism alone. Once crossed, cropping or mixed farming may become obligatory since it can support more persons per unit area than can pastoralism (Ruthenberg et al. 1980). Thus population increase can lead to sedenterization of human populations that had previously been mobile (often also promoted by governments to enhance the provision of services such as education, health and energy, and perhaps control of the population). Drylands of intermediate aridity are the most susceptible to degradation thresholds; this is because they can support opportunistic cropping, but, being economically and ecologically marginal, their low productivity does not justify the establishment of secure property rights and investment in sustainable cropping (Safriel et al. 2005; Fig. 8.1). Cropping has a much larger negative impact on landscape function than grazing because it exposes soils to erosion and temperature increase, loss of soil structure, and nutrient loss by runoff and harvesting (Hiernaux and Turner 2002), so landscape function declines. The traditional method of maintaining landscape function, controlling weeds, and accumulating soil moisture, is fallowing, which declines when there is excessive demand for land under pressure of population (Ruthenberg et al. 1980). Mixed farming and manuring offers some remediation, or fertilizer can be purchased if crops are grown for sale, but with attendant risks of acidification and soil structural decline. More intensive cropping, as practiced at the wetter end of the gradient, tends to have less impact because its higher productivity is more likely to justify high labor or capital inputs for soil conservation (Ruthenberg et al. 1980).

In drylands of developed countries, declining populations lead to the withdrawal of key services, such as health, education, and financial services. Each service is likely to have a threshold of population size below which provision is deemed by politicians or public servants to be not cost-effective. Potential policy responses include triage, in which public funds are not invested in communities that are not viable, but citizens are assisted in moving to more socially and economically viable locations, and consolidation of services in selected settlements. These trends parallel patterns observed in remote arctic and boreal regions where increasing costs associated with climate change and/or rural depopulation threaten the viability of rural indigenous communities (Chapin et al. 2008). In developing countries where populations are increasing, such thresholds for service provision are likely to be reversed, moving from cost-ineffective to cost-effective as demand grows. The extent to which the services are provided to a region, whether in a developed or a developing country, will depend largely on the political influence of the region (Sandford 1983, Abel et al. 2006).

Population increase provides more labor for land management, promotes innovation, and can foster cooperation, so degradation is not inevitable (Tiffin and Mortimore 1994). Social networks also increase with population growth, perhaps enhancing the potential to support those affected by drought and facilitating the exchange of information about resource availability in times of scarcity. Technological changes such as public electricity supplies may have positive impacts on woodland resources around villages through reducing firewood collection. They also allow access to television and telephones that bring new commercial, political, and social influences to remote places.

Loss of Local Community Governance Structures

Governance (Ostrom 2005; see Chapter 4) includes:

- the formal and informal social rules (e.g., property rights) and norms (e.g., equity) that guide the behavior of individuals and groups toward one another and their access to and use of resources and services;
- the political system that makes and changes the rules and the policies through which they are implemented; and
- organizations and social networks that implement and monitor compliance with the rules.

Governance operates across levels from local (e.g., a social network that implements rules about reciprocal obligations during drought) to regional, national (e.g., national policy on drought support for drylands), and global (e.g., attempts to establish carbon trading to mitigate climatic change). Interactions of governance rules and organizations across scales are critically important to the resilience of dryland regions and the well-being of its occupants (Abel et al. 2006). Cross-scale networks link individuals, groups, and organizations in remote regions to metropolitan governments so that calls for infrastructure, services, and drought relief are heard. The response depends on the political influence of the remote region, which, because of its remoteness, distant voice, and sparse population, is often weak. The problem can be exacerbated by differences in the ethnicity of the members of government and the public service, on one hand, and the ethnicity of populations in the dryland region, on the other (Sandford 1983).

Government interventions, coupled with the globalizing effects of communications and transport technologies, can disrupt regional and local governance processes. Early in the trajectory represented by Fig. 8.5, formalization and centralization of governance tends to weaken traditional governance that evolved to cope with spatiotemporal variability in rainfall (Sandford 1983). It commonly involves privatization of group property rights; provision of water points, clinics, and schools; and increased rule of law. The net benefits of these changes depend on many details, but the process of change is often disruptive. For example, collectivization was imposed in the USSR on some nomadic peoples and livestock ownership appropriated by the state. This policy failed because the collectives were sedentary and nonviable in the face of spatiotemporal variability. Sedenterization policies, widely applied in the Middle East and Africa in the 1970s and 1980s (Sandford 1983), also commonly failed for similar reasons. In some African rangelands, tribal land was excised for small collective groups of pastoralists, but these "group ranches" were usually too small to accommodate spatiotemporal variability. In the Americas, Australia, and Africa, drylands were taken from indigenous peoples and privatized by invading Europeans. In some regions, there has been a successful effort to rebuild group property rights or at least some regional peer group governance of resources (e.g., LandCare in Australia; the Desert Community Initiative in Niger).

Privatization (fragmentation) of communal land, commonly following a US ranching model, does not necessarily increase productivity per unit area, measured in physical or monetary terms, nor the number of people that land can support (Abel 1997). In Australia state-legislated subdivision of initially extensive holdings created units that lacked both the financial and the ecological resilience of larger holdings (Young 1985). Policies now promote reconsolidation, but even large units are still below the threshold of property size (a slow variable) needed to encompass broad-scale spatiotemporal variations in forage. In such cases **agistment**, which is widely practiced in Australia, has become an important means of redistributing livestock at regional and interregional scales (McAllister et al. 2006). In the practice of agistment, private properties that have too many animals for the available forage in a particular year will transport stock to other properties where the imbalance is in reverse, and the senders pay the receivers. In other years and in the long run the imbalance and the payments may be reversed, thus reducing pressures on areas where rain has not fallen. Other enterprises have achieved the same effect by owning land in climatically different regions and moving stock among them. Both approaches use commercial arrangements to mimic traditional systems of reciprocal obligations and mobility that have been practiced by Aboriginal hunter-gatherers (e.g., Keen 2004) and pastoralists in Africa and Asia (Sandford 1983, Alimaev and Behnke 2008).

Disruptive Subsidies

At national and international levels a common response to spatiotemporal variation has been policies directed to drought relief. In Australia pastoralists have received support for animal feed and debt and tax relief during droughts. Many of these measures create perverse incentives to retain stock in poor times and cause damage to the land (Stafford Smith 2003). Today some of these measures persist, although emergency relief is more directed toward transport subsidies to support moving animals, and some better tax instruments that allow producers to build up financial reserves in good years that are not taxed until they are used. In developing countries famines have triggered international food and medical aid. While the food has saved lives during the emergency, its long-term benefits are questionable. In many cases the problem is not lack of food at a national scale that leads to famine, but inability of drought-struck people to pay for it. In such cases it would be better to subsidize the purchase of food from farmers within the nation and thus stimulate investment and production (see Chapter 12).

Subsidy of drylands can be ecological as well as economic. The intensively used parts of the landscape (often the majority at the moister end of the drylands spectrum) generally depend on imported subsidies of nutrients and chemical pest and weed control to maintain their productivity. Subsidies can make these areas more productive when they function well, but highly vulnerable if these subsidies are withdrawn (as in times of war or economic recession). Also, fertilizer use can lower pH (a slow variable) below a critical threshold, or irrigation can increase soil salinity above a threshold so that yields and ground cover decline. Because most dryland crops are annuals, soil is exposed to erosion and the risk of irreversible loss of landscape function if poorly managed. Pesticides can eliminate predators and foster resistance in pests, both changes reducing the self-organizing capacity of the system and reducing its resilience. The policy implications are clear, but ethically and politically difficult, because a commitment to high-input development can increase shorter term welfare substantially, but at the cost of longer term vulnerability. When populations are increasing, policy makers may feel they have no choice but to support intensification (see Chapter 3). However, the resilience of a human–ecological system stems largely from its capacity to self-organize and recover from a disturbance. Although rebuilding this capacity at times requires access to external resources, excessive subsidy can reduce the incentives and capacity to self-organize. Cross-scale subsidies, whether of nutrients or through drought relief, increase vulnerability at the local scale and, if feasible, should end when self-organization becomes apparent. If many regions are subsidized, this can reduce the resilience of the system as a whole (Abel et al. 2006).

Commercialization and Diversification

All regions are affected by the increasing connectedness of the world, but globalization creates particular threats and opportunities for drylands. Communications and transport technologies provide dryland inhabitants access to new markets. Thus aboriginal art is sold across the Internet from remote Australia, animal products are marketed internationally, Botswana beef is sold to the European Union, and nature-based tourism and hunting experiences attract international customers. All bring new money into remote communities. With such changes, of course, comes a market orientation, which in developing countries often brings social, cultural, and technological threshold changes.

Commercialization can disrupt traditional cultures by transforming the social relations of production. For example, a shift from the production of milk for subsistence, which is traditionally managed by women for their families, to meat for sale, which is commonly managed by men, requires different herd structures and stocking rates. This can alter family nutrition and vulnerabilities to market fluctuations. Alternatively, traditional cultures that are unwilling or unable to participate in markets may lose political influence and become marginalized. Likewise, producing crops for sale often requires the use of fertilizer and pesticides to raise yields above those needed for subsistence. This increases dependence on input markets which are themselves vulnerable to fuel price increases and wars. It is evident that many thresholds of potential concern can arise from entry into new markets, which wise policy makers would consider carefully.

Changes from complex mixes of livestock and other bush products in a subsistence society to a market orientation toward maximum production of one or two saleable products like beef can have significant flow-on effects to the ecosystem as livestock forage needs become focused, and complementary mixes of browsers (goats, camels, and browsing wildlife) and grazers (cattle, sheep, and grazing wildlife) are replaced by herds of one commercial species.

New markets can diversify income sources and enhance resilience. Income generated on- or off-farm from nonagro-pastoral activities such as a service industry (tourism, for example) may be more stable or vary asynchronously with an agro-pastoral income that is driven by rainfall. If an alternative or supplementary income is from a source not dependent on regional rainfall, such as mining or tourism, it enables farmers and pastoralists to survive drought and rebuild assets afterwards. Such diversification is not new, for many traditional societies in arid lands are accustomed to using wildlife during times of scarcity.

In some regions cropping and livestock production by pastoralists are being replaced by employment in tourism, conservation, specialist native products, and mining. In eastern and southern Africa, for example, commercial livestock production has in some cases been replaced by or is managed alongside commercial production of wildlife for sport hunting, meat production, or tourism. In some cases this reverses a colonial process in which pastoralists and indigenous hunter-gatherers were excluded from the then newly established national parks. A reduced focus on livestock production may mean increases in woody vegetation and other landscape-scale changes with beneficial implications for landscape function, carbon storage, and energy feedbacks to climate.

Case Studies

To illustrate the issues raised by an analysis of resilience in different dryland systems, we now present four case studies (locations shown in Plate 6) that provide examples from developed and developing nations (capturing differing population trajectories, production goals, and connectivity with markets and governments) in different environments, and at a variety of different scales.

A Crop–Livestock System in the Southern Sahel, Western Niger

Mixed crop–livestock systems in western Niger illustrate the strong interdependence of the social and biophysical subsystems and ways in which increasing pressure of use can lead to the crossing of thresholds in both (Fernández et al. 2002, Hiernaux and Turner 2002). About 15% of this arid region (300–500 mm annual rainfall) is rangeland on nonarable soil, and the remainder is either active cultivation or fallow land on sandy, infertile soils. The crops are mostly staple cereals for subsistence with some other crops, and livestock are mainly cattle, sheep, and goats with a few donkeys, camels, and horses used for draft.

The evolved system of land use is quite complex, involving two kinds of households with different livelihoods: a village household (VH, about 70% of the people) and a camp household (CH, about 30%). VHs have primary rights to arable land. CHs have mostly livestock and undertake wet season **transhumance** (sea-

sonal movements) to northern areas for livestock forage. The VH have fewer livestock, because they cannot maintain them year round on the available forage. The CH livestock use the fallow and cropped lands in the dry season, providing fertilizer in the form of manure.

Soil fertility is a critical limiting factor for crop production. It declines under cropping and in the absence of manure from livestock; 3 out of 8 years must be in fallow to maintain long-term fertility. This translates to a minimum of 3 out of every 8 hectares in fallow at any one time. This "3/8" value is a biophysical threshold in the system. Below that value, fertility declines; above it the system's soil fertility is self-sustaining.

Finally, in the rangeland component of the system, soils have a higher clay content. Under heavy grazing they are prone to erosion and compaction, as perennial grass cover gives way to a sparse cover of small annuals, constraining the number of animals and time spent grazing on rangelands. Three indices of sustainability can be defined for this system:

1. Soil fertility is sustained unless land is farmed more than 5 years out of 8. If manure or other sources of nutrients are added, less fallow is needed.
2. Grazing is sustained unless there is insufficient palatable herbage to meet intake needs throughout the season.
3. Household economy is sustained if the aggregate production of crops, dairy products, and livestock exceeds the minimum threshold need for these by a household.

The mixed economy of farmers and camp pastoralists is viable as long as none of these thresholds is exceeded, which requires cooperation among the two types of households. If VHs leave too little fallow land in order to increase crop yields, then soil fertility declines and camp pastoralists are forced to spend more time on rangelands, moving the system toward two thresholds of unsustainability. Alternatively, if lands are left fallow for too long, soil fertility increases but total harvest declines, leading toward a threshold of economic unsustainability. Interventions that enhance the system's adaptive capacity might reverse this trend, for example, through access to inorganic fertilizers, crop diversification, agro-forestry, and increasing farmers' husbandry skills (human capital).

An important factor in the dynamics of this system is the strong linkage between the household and the community scales. Available forage for livestock depends directly on community livestock density and reciprocity agreements that determine this. In the dry season, the VH depend on the CH for livestock manure, and the CH depend on VH for access to crop and fallow land. The system as a whole depends strongly on the continued access by CH to grazing resources outside the area.

Learning from a Century of Degradation Episodes in Australia

Eight well-documented degradation episodes associated with major droughts have occurred in different regions of Australian rangelands between 1880 and 1990 (McKeon et al. 2004). A metaanalysis of seven of these (Stafford Smith et al. 2007) illustrates the importance of focusing on slowly changing state variables in environments where fast variables are very poor indicators of their slow counterparts. The analysis also enriches our understanding of ways in which local environmental knowledge links human and environmental subsystems at multiple scales.

The Australian rangelands constitute three quarters of the Australian land mass, about half of which is grazed by cattle or sheep under leasehold pastoral systems. Most productive lands are subject to high climatic variability on multiple time frames, including intra- and interannual fluctuations driven in part by ENSO events with return times of about 4–8 years and by longer term drivers (e.g., the Interdecadal Pacific Oscillation) with mean return times of about 20 years (White et al. 2003). An analysis of social (process and policy) and biophysical (particularly climate and pasture condition) drivers (McKeon et al. 2004; Fig. 8.7) revealed the following syndrome that was common to all events:

- [–10 years] Good prices, good rains, animal numbers increase; human expectations rise. Decline in pasture condition causes reduced resilience in short dry periods.
- [0 year] Onset of severe drought and/or market decline; stock retained in hope of rain/price increase, sometimes encouraged by perverse policy incentives.
- [+2–3 years] Extended dry period reveals decline in pasture condition more dramatically as pasture production collapses; further damage, big stock losses.
- [+5–7 years] Eventual relief through rains, but a more-or-less permanent decline in pasture condition becomes evident in subsequent short dry periods. Local learning by individual pastoralists does occur, often resulting in reduced stocking rates. Formal government inquiries report on events, but this occurs too late for intervention in the present drought.
- [+20 years] Pastoralists turn over; memory loss in pastoral and policy communities occurs over the next 20 years.

Both industry and government organizations failed to recognize key slow variables and their thresholds. Pastoralists responded to pasture growth (and good prices) rather than land condition (and market trends); governments responded to actual drought events rather than climatic cycles or the adaptive capacity of industry. Land condition in many cases had crossed a critical threshold before the drought, but this was only evident once the severe drought episodes had begun. By this time it was too late.

There were evident cross-scale influences affecting pastoral decision-making, in particular those of markets being driven by national and global changes and the influence of policy in creating perverse incentives to retain livestock in drought times (Stafford Smith 2003). The influence of scale is particularly important when we consider learning. Almost all episodes were followed by clear evidence of some learning by individual pastoralists. However, there was little evidence that this learning persisted from one drought event to the next, because the local knowledge appeared to be *too* local (Stafford Smith et al. 2007). The long return

FIGURE 8.7. Linking the social and ecological aspects of an Australian pastoral grazing system. Redrawn from Stafford Smith et al. (2007).

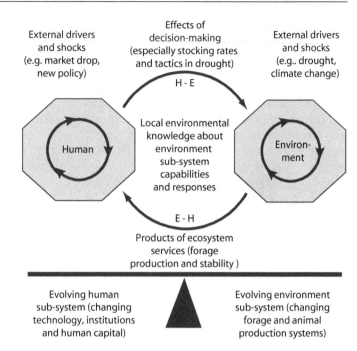

External drivers and shocks (e.g. market drop, new policy)

Effects of decision-making (especially stocking rates and tactics in drought)

External drivers and shocks (e.g.. drought, climate change)

H - E

Human

Local environmental knowledge about environment sub-system capabilities and responses

Environ- ment

E - H

Products of ecosystem services (forage production and stability)

Evolving human sub-system (changing technology, institutions and human capital)

Evolving environment sub-system (changing forage and animal production systems)

time of major drought events, on the order of 20 years, was of the same magnitude as the longevity of a pastoralist's managing lifetime, so the inevitable turnover of managers meant that their learning was lost by the time another major drought occurred. 'Local' learning in this case needed to occur at a regional scale, with a large enough community involved and with the support of policy and research so that it would persist over many decades.

This case study constitutes a "natural experiment" of eight regional degradation episodes that occurred under broadly similar policy and social settings over a century in the Australian rangelands. Long-term impacts on pasture production of significance to household livelihoods were documented in events that were triggered by drought but really driven by overoptimistic human expectations of the environment and markets in the years preceding the drought. Both managers and governments were responding to the wrong variables, and governments sometimes exacerbated the situation with perverse policy incentives that caused pastoralists to hold onto stock even longer. Most importantly, this case study illustrates that the crit-

ical scale for learning (DDP P5) may not be very local, where the controlling variables (in this case, particularly drought) are operating at larger scales in time and space. Encouragingly, after a century, Australia has been moving toward a more regional alliance between industry, research, and government to deliver peer-supported "safe carrying capacity" estimates to pastoralists, monthly drought alerts from the scientists, and hopefully a better policy context from government (Stafford Smith et al. 2007).

Immigration and Off-Site Impacts in Inner Mongolia

Between 1949 and 1997 cross-scale interactions have caused a pronounced co-evolution between the human and the environmental subsystems and influences of Inner Mongolia (Jiang 2002). The population of the Uxin *banner* on the Mu Us Sandy Land on the Ordos Plateau has tripled, the cash income per capita has risen 50-fold, pastureland has been privatized to the household, and irreversible changes in household consumption and market-oriented aspirations have occurred in the population. During that period, one policy aimed at making the existing livestock production hazard-resistant and able to support the growing population through forage shrub plantations and irrigated cropping. This policy succeeded to the extent that the region is now self-sufficient in winter forage, and stock mortalities have dropped from 8–17% to 1.5%. The irrigated cropping now produces mostly cash crops but often remains in proximity to plantations that help to protect against sand movements. Irrigated cropping practices originated with in-migrating Chinese Han people but are also spreading to the indigenous Mongolians. During the same period, a second policy aimed at stabilizing sand dunes, the mobilization of which was the main indicator of desertification in the region. The North China Revegetation Program sought to plant trees, shrubs, and grasses on the dunes, with periodic injections of effort from the central government in Beijing, which is affected by sandstorms from the region.

Unfortunately, the irrigation appears to be drawing down the shallow water table in the region. The landscape is increasingly polarized and consolidated into irrigated areas with high cover and production, and sandy areas with a declining water table and worsening sand movement. Intermediate grasslands are declining in productivity and extent. Human aspirations for production in the region will not decline. The issue is now therefore whether the local understanding of the water table-mediated connection between higher overall production and increasing area of mobile sands will develop to the point that action on the trade-off in values can be taken. Regardless of the social and environmental drivers at the household level, there are emerging regional impacts that must now be addressed. The latest efforts in revegetation were driven by Beijing as a result of sandstorms affecting the capital that might affect China's reputation during its hosting of the 2008 Olympics – international expectations playing through a national government to policies that are implemented at regional and local levels.

As Jiang (2002: 182) notes in her analysis of the effects of cultural change on desertification, "Aiming at a moving target, the dryland ecosystem does not exhibit recovery but transition toward a new state." Just as dryland ecosystems are disequilibrial, so too are their embedded social–ecological systems. Uxin has essentially crossed mixed social and environmental thresholds and been transformed into a new type of system (Walker and Meyers 2004), perhaps not yet in balance. The option of recrossing the thresholds back to the original regime no longer seems possible or desirable: the new system has reduced livestock deaths and improved household economies (Fig. 8.7). With this change in state have come fundamental changes in system functioning. At one time these revolved around grazing and its local impacts on grass cover, which at times became too low, causing the mobilization of sand. Today, the more important functionality is groundwater for irrigation and the impact of water withdrawal on the water table for trees and shrubs. With the change in system has come a change in the controlling slow variables. Previously the important variable to monitor was land condition for pasture production (and sand mobilization) but today water table levels are more important in the new system state. Drivers contributing to the change include population increases, a change from collective ownership when pastureland was distributed to households in 1984, and the cultural and technological changes associated with irrigated agriculture, all of these ultimately driven by policies from another scale. Survival in the future depends on whether the community can develop new local knowledge and solutions for groundwater-mediated landscape linkages.

Managing Under Uncertainty in the Kruger National Park, South Africa

The Kruger National Park (KNP) illustrates the integration of all of the issues discussed above in the multiscaled planning and management of a major multiple-use dryland region (Biggs and Rogers 2003, du Toit et al. 2003). KNP spans 22,000 km^2 of land in northeast South Africa. Across its area, mean rainfall ranges from 300 to 700 mm yr^{-1}. At least three extreme droughts have occurred between 1981 and 2000. The high heterogeneity of its rocks, landforms, and soils is reflected in the diversity of the structure, composition, and functioning of its ecological communities (du Toit et al. 2003) and its resulting value for the conservation of biodiversity. The Park is famous for its large populations of elephant, buffalo, hippopotamus, rhino, lion, leopard, cheetah, and antelope, but it also conserves an array of other biota. However, this diversity is also a consequence of its resource scarcity and sparse human populations, because the area is relatively unproductive for farming and supports trypanosome-bearing tsetse fly that make it unsuitable for livestock. Its ecological communities are consequently relatively less simplified by humans than those in more productive areas. What is now the KNP was nonetheless established on land and among cultures that have seen waves of social–ecological changes during which the area was subject to different burning, resource use, and disease regimes that changed the structure, composition, and functioning of its ecosystems. The stone-age San hunter-gatherer peoples were overrun around 400 AD by iron-age farming and trading cultures that exported gold, ivory, and other wildlife products. Afrikaaner and British colonizers hunted the area heavily from the 1830s until rinderpest, introduced by Europeans to Africa, drastically reduced the numbers of susceptible large herbivore species in 1896. Formal protection of wildlife in what is now the KNP began in 1902. Protection from hunting and recovery from rinderpest enabled the recovery of large herbivore populations. The KNP was established in 1926 (du Toit et al. 2003).

Management and use of this relatively remote area has been strongly influenced by national policies. In 1948 the National Party was elected to govern South Africa by a white electorate and remained in power until 1994. Its apartheid regime attempted to run the economy and society with races segregated. Its policies applied to the KNP too, which was run partly to foster the cultural identity of Afrikaner people. Physical and psychological separation of nonwhite peoples from the KNP remains a problem today, though a de jure mul-

tiracial democratic society was established in 1994 (du Toit et al. 2003). The future of the KNP is tied by multiple cross-scale political, cultural, and economic links to the future of the evolving South African nation. It is now surrounded by a variety of land uses, including towns, mining, agriculture, forestry, and wildlife utilization. It depends on water from catchments that support large human populations. Its mobile wildlife populations, including elephant, move across its permeable boundaries into neighboring Zimbabwe and Mozambique. It is linked to international markets through a major tourist industry, and neighboring mines export minerals and use water.

The KNP has inherited abundant opportunities as well as multiple problems from its cultural and ecological past. At the national scale, the importance of biodiversity conservation and tourism has to be considered alongside the extremes of poverty, health, and educational opportunities of the population as a whole. Around the KNP, traditional users of the land that became the Park press claim to historical rights. Competition for water is already being addressed by explicit recognition in South African water policy of the transpiration of water from each land use, and the rights to water of the users. This has consequences for river flow in the Park and its aquatic ecosystems. Within the Park the dynamics of plants, animals, and fire drive large and often unpredictable changes in the composition, structure, and functioning of ecological communities. KNP's policy makers, managers, and users thus face high levels of uncertainty stemming from the unpredictability of the social–ecological system at several scales. This case study describes the adaptive framework that guides their management strategies.

South Africa's history is consistent with Adaptive Cycle theory (Gunderson and Holling 2002), in which the pre-1994 racist society had become increasingly rule-bound, hierarchical, and resistant to new ideas. At a regional scale, the KNP-management system reflected a similar pattern. The rest of this case study is drawn from Biggs and Rogers (2003), who describe how the management system subsequently adapted. They describe the science-management system of the KNP in the years shortly before democratic and social changes of 1990 as:

- treating nature as linear and predictable;
- compartmentalizing knowledge along disciplinary grounds and around particular problems;
- insular and autocratic in its decision-making;
- neglecting biodiversity and heterogeneity in its management.

Since then the science-management system has moved in directions that acknowledge the rapid change and high uncertainty at the national scale, the multiple policy objectives of a multiracial and economically heterogeneous society, and its interaction with a dynamic and heterogeneous ecosystem. Here we describe the current science-management approach – called Strategic Adaptive Management - under six headings: ecological theory; objectives hierarchy; thresholds of potential concern; monitoring and adaptive decision making; communities of practice; and outcomes and challenges.

Ecological Theory. The concept of hierarchical patch dynamics is the scientific basis for management interventions in the structure, composition, and processes of the KNP's ecological communities. The ecological system as a whole is seen as a set of nested and interacting patches. This is consistent with panarchy theory (see Chapter 1), with its nested but interlinked subsystems, each operating at different temporal and spatial scales. Management interventions in KNP are made at particular times, particular spatial scales, and in particular ways to maintain or enhance this spatial and temporal heterogeneity, both because it is considered fundamental to ecological functioning, and because of the intrinsic and market values of the biodiversity that it fosters. Gillson and Duffin (2007) show how pollen analyses spanning 5,000 years can guide thresholds of probable concern (TPCs – defined below) for vegetation structure, thus influencing TPCs for the Strategic Adaptive Management of elephant densities and fire. That said, the dynamics of the patches are not well understood, many pro-

cesses are nonlinear and often unpredictable, and the ecological system as a whole is in constant flux. Lack of ecological predictability reinforces the need for adaptive management (see Chapter 4) that would in any case be required to address the social, economic, and political uncertainties in the broader social–ecological system (Fig. 8.8).

Vision Statement and Objectives Hierarchy. A vision statement for the KNP, developed with public participation, is to "...maintain biodiversity in all its natural facets and fluxes and to provide human benefits in keeping with the mission of the South African National Parks in a manner which detracts as little as possible from the wilderness qualities of the KNP" (Biggs and Rogers 2003, p. 60). A hierarchy of objectives has been developed to link the abstract vision to management activities at a hierarchy of scales. Main groups of objectives are atmospheric, aquatic, terrestrial research, terrestrial management, terrestrial monitoring, alien species management, human benefits, wilderness, and integrated environmental management.

Thresholds of Probable Concern. TPCs are management goals that together define current views about the spatiotemporal heterogeneity conditions of ecosystems (Biggs and Rogers 2003). TPCs are upper and lower levels in selected indicators. When a level is reached, or modeling predicts it is likely to be reached, reasons are investigated, and management action is applied, if necessary. Alternatively, the level of the TPC may be reset if it is thought to have been wrongly set. TPCs are hypotheses about the limits of acceptable change in ecosystem structure, composition, and function. As hypotheses, they are necessarily modified, replaced, or discarded as new evidence dictates (Fig. 8.8). As a set, they form the multidimensional envelope within which variation of the ecosystem is acceptable under the vision statement. In terms of resilience theory (see Chapter 1), it is thus a practical definition of a basin of attraction for the ecological elements of the social–ecological system (Table 8.3).

Monitoring and Adaptive Decision-Making. The TPCs indicate what management actions should be done, where, and when. The choice

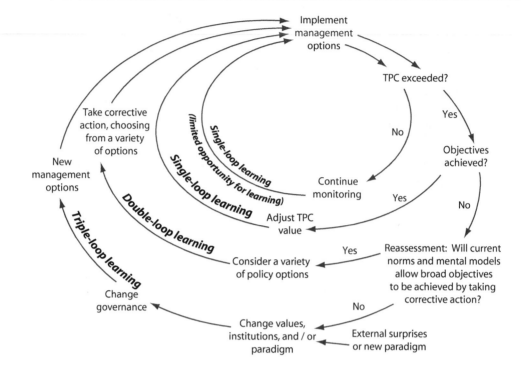

FIGURE 8.8. Strategic approach to river management that integrates indicators, endpoints, and values in the Kruger National Park. Thresholds of probable concern (TPCs) define the acceptable levels of heterogeneity. If the system remains within these limits, monitoring continues. If management objectives are met despite the TPC being exceeded, the threshold value is changed, but monitoring continues. If the TPC is exceeded and management fails to meet its objectives, a more fundamental reassessment of action is made, acknowledging uncertainties about outcomes. The TPCs also define what is monitored and where, so that managers can learn whether or not their actions maintained the ecological system within the TPCs, and ment occurs, leading to a modification or invention of new management strategies. Sometimes this reevaluation leads to an entirely new paradigm or new understanding, especially if the process has been triggered by unexpected events or is considered in the context of a new paradigm. These more fundamental reevaluations often require a change in governance, involving new sets of actors (Biggs and Rogers 2003). See also Fig. 5.1 on single-, double-, and triple-loop learning.

whether the TPC levels, or indeed the TPCs themselves, were appropriate.

Communities of Practice. Sharing of knowledge among managers, stakeholders, and scientists is fundamental to the implementation

TABLE 8.3. Examples of thresholds of probable concern (TPCs). Adapted from Biggs and Rogers (2003).

Trigger or lead-up	TPC exceeded or expected to be exceeded
Silt release episode from upstream mining operation	Turbidity, resulting in fish kills
Fires started by humans	Area burned by fires lit by humans exceeded TPC
River sedimentation	Modeled geomorphic and riparian species indicators
Occurrence of an alien plant species along rivers	New occurrence of an alien with a high index of potential threat.

of this approach (Biggs and Rogers 2003), supporting the principle of local ecological knowledge as critical in maintaining the functions and coadaptation of social–ecological systems in drylands. Sharing of views and knowledge in the KNP takes place within self-organizing communities of practice, which are networks that link across organizational, disciplinary, and interest-group boundaries, and between research, practice, and management (see Chapter 4). In addition to breaking down such boundaries, they enhance collective and individual learning. Formal research processes are thus linked to the local knowledge of managers and stakeholders in a dynamic learning system that adapts to new problems, opportunities, and knowledge.

Outcomes and Challenges. According to Biggs and Rogers (2003), the application of Strategic Adaptive Management in the KNP has resulted in, for example:

- More proactive behavior – for example, actions taken now to forestall sedimentation of rivers, even though these effects are not predicted to reach unacceptable levels for many years hence;
- Stronger links to external stakeholders – for example, use of poverty-relief programs to clear alien species and address the external cause of river sedimentation (mining);
- Acknowledgment and use of conflicting views about the causes of and remedies for TPCs being exceeded – for example, in the management of rare Sable Antelope;
- Revision of fire policy when it was realized that the frequency of unintended fires would exceed its TPC regardless of management action.

Among the challenges of applying Strategic Adaptive Management, Biggs and Rogers identify:

- Need to interpret and sift multiple sources of information, of varying quality and often conflicting, and synthesize it for monitoring and for setting or resetting TPCs;
- Interactions with programs that operate outside the Strategic Adaptive Management framework, such as veterinary control of dis-

eases that affect both wildlife and livestock, with which park managers are legally obliged to comply;
- Keeping the TPCs to a number that achieves the KNP mission and can be monitored and responded to with available resources;
- Difficulty of knowing when monitoring data are signaling an impending crossing of a TPC threshold.

The KNP's Strategic Adaptive Management system has continued to learn and develop since publication of the du Toit et al. (2003) book on which this case study was based. It is an important model for policies and management for biodiversity conservation in general, as well as in drylands, and the concept of TPCs is being advocated and applied in Australia at least.

Synthesis

Thanks to their underlying features of an unpredictable climate and resource scarcity, drylands tend to possess sparse populations, be remote from markets, and distant from centers of governance. The analysis of issues in drylands is therefore most likely to be adequate if carried out through the lens of the five principles of the Dryland Development Paradigm (DDP – Reynolds et al. 2007). Our case studies have illustrated the importance of each, in particular:

- The coevolutionary nature of social and ecological systems, such that system collapse principally occurs when this relationship becomes dysfunctional, not just because there is change (DDP P1);
- The need to focus very carefully on the appropriate slow variables and their thresholds in order to determine the state of this coevolutionary system as a matter of particular importance in variable environments (DDP P2, P3);
- The massive effect that cross-scale interactions can have on dryland systems that are usually particularly poorly equipped to deal with these because of their distant voice (DDP P4); and,
- The vital importance of the right shared mental models in the form of local

knowledge at a variety of scales for maintaining the functionality of the coupled system – particularly important in drylands where variability slows down experiential learning (DDP P5).

These lessons are true for all levels of engagement in drylands, from local management, through regional research efforts, to national and international policies. The marginal nature of productivity in drylands usually means that they also end up being politically marginal. The challenge for dryland inhabitants is to accept this as inevitable and focus on managing it by creating strong links to the outside world that can be activated readily in times of need. By contrast, the challenge for (well-meaning but distant) policy makers is to pay particular attention to the form of institutional arrangements in drylands, to support local governance where this is possible, to create adequate links to other regions, and to put extra effort into hearing the articulation of needs from those regions that otherwise get swamped by demands of more heavily populated areas.

In a number of dryland regions around the world an unfortunate outcome of the drylands syndrome is that cross-scale effects coupled with inherently low adaptive capacity has resulted in changes to undesirable, but resilient, system regimes. The efforts of people to survive and prosper have led to declines in ecosystem productivity; further efforts seem to make things worse, in terms of both ecosystem health and human well-being. For these regions, the problem is no longer how to increase resilience, but how to increase transformability (see Chapter 5). The need is to facilitate transformation from the kinds of systems they now are to some other kind of system. This may entail changing the ways people make a living, developing new "products" (goods and services) and operating at different scales. There are no clear rules for how to do this, although recognition of the dryland syndrome is a useful prerequisite for effective action. It requires strong leadership and vision, a high level of social capital to get agreement on what may be difficult or painful changes, and most likely financial and other support from higher scales. Transforma-

tion and transformability are emerging as critical areas for research within the broad area of resilience science, with particular relevance and poignancy for marginalized peoples in drylands.

Review Questions

1. Why are drylands more likely than many other systems to undergo changes in state?
2. What social changes in the last century have reduced the resilience of many pastoral societies?
3. What are some of the major causes of desertification, and how do these interact? What policy interventions might reduce the likelihood of desertification?
4. Why has it been difficult for landowners in Australian drylands to learn how to cope with drought?
5. What key attributes of drylands need to be considered when analyzing policy or management responses to drylands problems?
6. Which principles of resilience theory are most important in drylands, and why?
7. Why is an analysis across scales particularly important in drylands?
8. What is the role of local knowledge in managing dryland social–ecological systems?

Additional Readings

Biggs, H.C. and K.H. Rogers. 2003. An adaptive system to link science, monitoring, and management in practice. Pages 59–80 in J.T. du Toit, K.H. Rogers, and H.C. Biggs, editors. *The Kruger Experience: Ecology and Management of Savanna Heterogeneity*. Island Press, Washington.

Galvin, K.A., R.S. Reid, R.H.J. Behnke, and N.T. Hobbs, editors. 2008. *Fragmentation in Semi-Arid and Arid Landscapes: Consequences for Human and Natural Systems*. Springer, Dordrecht.

Geist, H.J., and E.F. Lambin. 2004. Dynamic causal patterns of desertification. *BioScience* 53:817–829.

Reynolds, J.F. and D.M. Stafford Smith, editors. 2002. *Global Desertification: Do Humans Cause Deserts?* Dahlem University Press, Berlin.

Reynolds, J.F., D.M. Stafford Smith, E.F. Lambin, B.L. Turner, II, M. Mortimore, et al.

2007. Global desertification: Building a science for dryland development. *Science* 316: 847–851.

Safriel, U., Z. Adeel, D. Niemeijer, J. Puigdefabregas, R. White, et al. 2005. Dryland Systems. Pages 623–662 *in* R. Hassan, R. Scholes, and N. Ash, editors. *Ecosystems and Human Well-Being: Current State and Trends. Millennium Ecosystem Assessment.* Island Press, Washington.

Walker B.H. 1993. Rangeland ecology: Understanding and managing change. *Ambio* 22(2–3) 80–87.

Walker, B.H., and M.A. Janssen. 2002. Rangelands, pastoralists and governments: Interlinked systems of people and nature. *Philosophical Transactions of the Royal Society of London, Series B* 357:719–725.

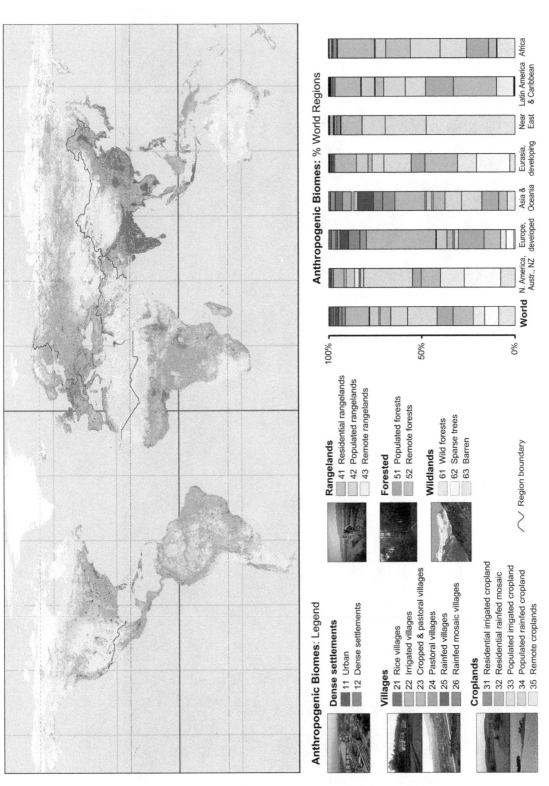

Anthropogenic Biomes: Legend

Dense settlements
11 Urban
12 Dense settlements

Villages
21 Rice villages
22 Irrigated villages
23 Cropped & pastoral villages
24 Pastoral villages
25 Rainfed villages
26 Rainfed mosaic villages

Croplands
31 Residential irrigated cropland
32 Residential rainfed mosaic
33 Populated irrigated cropland
34 Populated rainfed cropland
35 Remote croplands

Rangelands
41 Residential rangelands
42 Populated rangelands
43 Remote rangelands

Forested
51 Populated forests
52 Remote forests

Wildlands
61 Wild forests
62 Sparse trees
63 Barren

∿ Region boundary

Anthropogenic Biomes: % World Regions

World — N. America, Austr., NZ — Europe, developed — Asia & Oceania — Eurasia, developing — Near East — Latin America & Caribbean — Africa

Plate 1. Anthropogenic ecosystems of the world. Human activity has fundamentally altered both the nature of Earth's ecosystems and the way they are conceptualized. Reprinted from Ellis and Ramankutty (2008).

Plate 2. Global map of city lights. This night view of the planet illustrates the magnitude and extent of human use of Earth's resources. Reprinted from NASA website.

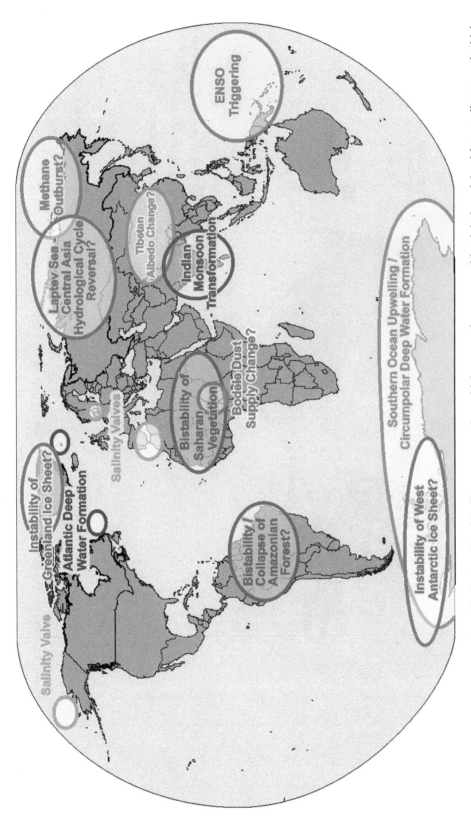

Plate 3. Tipping points that could radically alter Earth's climate system, but which are not incorporated in global models of future climate and which would constitute an important surprise. Reprinted from IPCC (2007c).

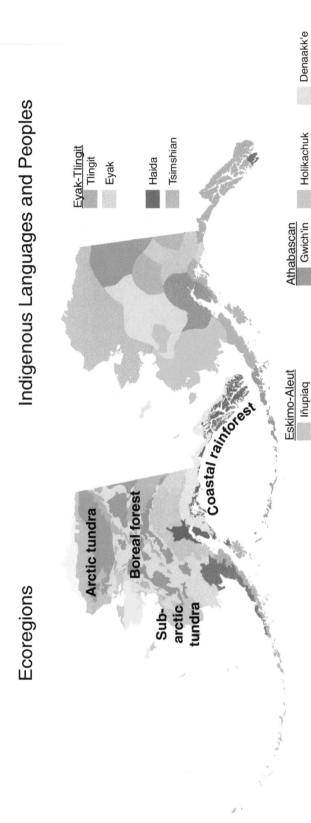

Plate 4. Maps of ecosystems (left) and cultural (linguistic) groups (right) in Alaska. The close correspondence between these maps demonstrates the tight linkage between ecosystems and society. Redrawn from Gallant (1996) and Krauss (1982), http://agdcftpl.wr.usgs.gov/pub/projects/fhm/ecoreg.gif.

Plate 5. (Top) Viewed from a lookout, the cover of this forest in the Western Ghats, India, appears dense and healthy. However, this forest is far from a "wild" forest. (Bottom) When observed at the ground level, much of the forest consists of a mosaic of multispecies plantations. (Photo: Fikret Berkes).

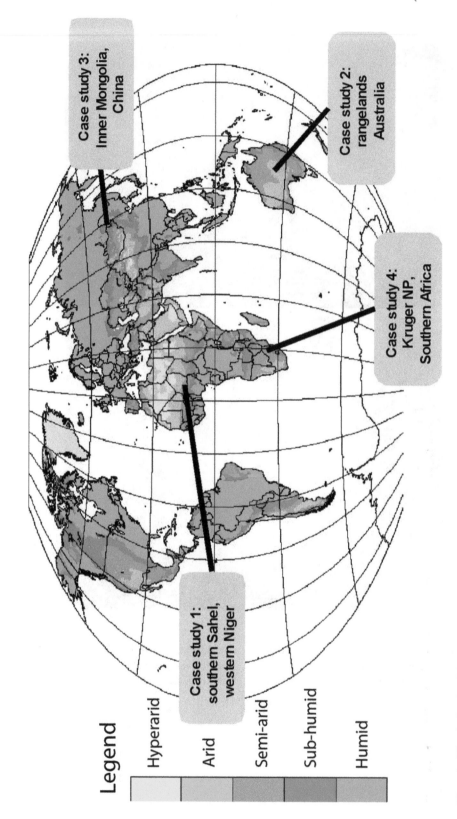

Plate 6. The drylands of the world, showing locations of the Chapter-8 case studies. Redrawn from Leemans and Kleidon (2002).

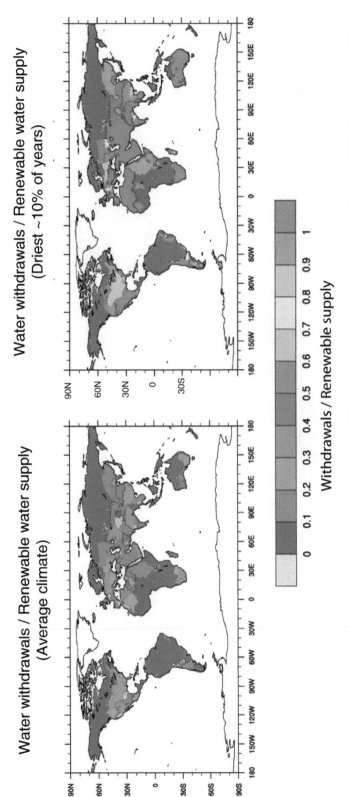

Plate 7. Water withdrawals for the year 2000 compared to long-term (Left) in the driest ten years (Right) between 1901 and 1995. This analysis highlights the susceptibility of certain regions to water scarcity. From Foley et al. (2005).

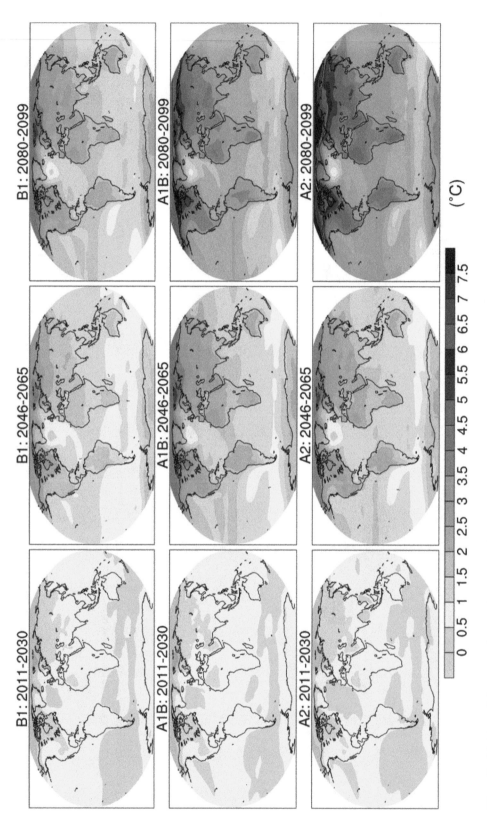

Plate 8. Multi-model mean of annual mean surface air temperature change relative to 1980–1999 for three scenarios of future human CO_2 emissions: (top) B1, which assumes a strongly reduced emissions; (middle) A1B, which assumes the current rates of emission continue (bottom) A2, which assumes that energy consumption continues to increase on its current trajectory ("business-as-usual"). Reprinted from (IPCC (2007a)).

9
Freshwaters: Managing Across Scales in Space and Time

Stephen R. Carpenter and Reinette Biggs

Introduction

Freshwaters include groundwater, rivers, lakes, and reservoirs. These ecosystems represent about 7% of earth's terrestrial surface area. Although aquatic and terrestrial ecosystems appear clearly separate to the human eye, groundwaters, lakes, and rivers are in fact closely connected to terrestrial systems (Magnuson et al. 2006). Climate, soils, and water-use characteristics of terrestrial plants affect infiltration of water to groundwater and runoff to surface waters. Terrestrial systems contribute nutrients and organic matter to freshwater systems. Rivers in flood fertilize their valleys. Terrestrial organisms are eaten by aquatic ones, and vice versa. A natural unit for considering coupled terrestrial and aquatic ecosystems is the **watershed**. Within a watershed, ecosystems are closely linked through flows of water, dissolved chemicals, including nutrients and organic matter, and movements of organisms. Thus watersheds are natural units of analysis for freshwater resources.

People are closely connected to freshwaters of the watersheds where they live. Historically, people depended on water for drinking, washing, agriculture, and transportation. Rivers and lakes have been important routes for travel and commerce for thousands of years. Today, people also use freshwaters at unprecedented scales for industrial processes as well as irrigation of agriculture for food production. Freshwaters are therefore critical in supporting modern society. Given the close connections between human and ecological aspects, management of freshwaters is likely to be most effective if they are treated as social–ecological systems. Freshwaters exemplify several features of complex systems that make them fascinating to study and difficult to manage: spatial heterogeneity of flows, mixtures of fast and slow variables, and thresholds (Box 9.1).

S.R. Carpenter (✉)
Center for Limnology, University of Wisconsin,
Madison, WI 53706, USA
e-mail: srcarpen@wisc.edu

F.S. Chapin et al. (eds.), *Principles of Ecosystem Stewardship*,
DOI 10.1007/978-0-387-73033-2_9, © Springer Science+Business Media, LLC 2009

Box 9.1. Key Principles for Freshwaters

Like other social–ecological systems discussed in this book, freshwaters support human activities and are altered by human action. Freshwater social–ecological systems provide particularly good examples of three principles that occur frequently in this book: spatial heterogeneity and flows across boundaries, fast and slow turnover times, and thresholds.

Spatial heterogeneity and flows across boundaries. Groundwater, lakes, and rivers within a watershed are interconnected, exchanging water, nutrients, organic matter, and organisms. Human action alters these connections in many ways. People change land use and land cover, thereby altering flows from land to freshwater. People change water movements with channels, levees, and dams. People extract water for irrigation and industrial and household uses, and people harvest aquatic organisms such as rice or fish. Many of the changes in freshwater ecosystems and their services are a result of human alteration of spatial flows of water, solutes, and organisms.

Fast and slow turnover times. Turnover time is the amount of time that it takes to replace the pool of a material in an ecosystem, if the pool size is steady over time. If a reservoir holds 1 km^3 of water, and the annual inflow is 0.5 km^3, then the turnover time of water in the reservoir is 2 years. Turnover times vary widely among the components of freshwater ecosystems. Water turnover times range from millennia in groundwater to years in lakes to days or min-

utes in streams. Phosphorus turnover times range from centuries in soil, to decades in sediments, to years in fish, to milliseconds for phosphate in surface waters. Freshwater food webs also span a wide range of turnover times, from years for fish biomass to about a day for plankton biomass. Differences in turnover times sometimes lead to interesting nonlinearities in ecosystem dynamics such as alternate stable states of aquatic communities, shifts in nutrient limitation, alternate states of water quality, and trophic cascades (Carpenter and Turner 2000, Carpenter 2003).

Thresholds. Sharp changes in flows, feedbacks, or the state of a system occur at thresholds. For example, a light rain on an agricultural field will be absorbed by the soil. As the intensity of rainfall and water saturation of the soil increases, a threshold is exceeded, causing some of the water to runoff as surface flow, carrying soil particles and nutrients into streams or lakes. Some important thresholds for freshwater ecosystems affect dominance of primary production by rooted plants versus phytoplankton, clear water versus turbid water, outbreaks of invasive species or waterborne diseases, and large changes in relative abundance of harvested fishes (Scheffer et al. 2001, Carpenter 2003). Thresholds also occur in human behavior and institutions for ecosystem management (Brock 2006). Thresholds are important because system behavior on one side of a threshold is a poor guide to system behavior on the other side of the threshold.

Sustainability of freshwater social–ecological systems involves many issues, of which three are paramount:

1. People use freshwater intensively. In many regions, the rate of water use by people exceeds the rate of supply by the hydrologic cycle. Extractive use of water competes with in-stream uses of water to meet human needs (pollution dilution, transporta-

tion, fish and game, recreation) and ecosystem needs (water for support of terrestrial ecosystem services).

2. Freshwater ecosystems are frequently degraded by changes in chemical or biological drivers. Pollution, especially runoff from agriculture and urban areas and sewage discharge, causes **eutrophication** (nutrient enrichment) and sanitation-related disease. In terms of the numbers of people impacted

globally, degraded water quality is as severe a problem as insufficient water supply. Freshwater ecosystems, like oceanic islands, also tend to have strong internal ecological feedbacks because changes in species traits strongly influence ecosystem dynamics. Thus species invasions and overfishing can have strong effects on freshwater ecosystems.

3. Competition for scarce water resources can lead to conflicts and requires management of tradeoffs. Many kinds of institutions have developed to manage freshwater, depending on local ecosystem conditions, the social and historical context, and the state of knowledge about the system. Climate change and rapidly-rising human demand for water are increasing the pressures on these institutions as well as ecosystems.

The chapter begins by describing freshwater resources, their use by people, and the ecosystem services that freshwaters provide. Next we describe the degradation of freshwater resources and the drivers of degradation, such as land use, chemical pollution, habitat loss, and species invasion. We then turn to challenges of managing freshwater resources that occur commonly in case after case – conflict between upstream and downstream users, human versus environmental flows, managing for heterogeneity, and equity across human generations. Such conflicts can be addressed by institutional processes, such as markets, trade in virtual water, conservation, technological innovation, or rescaling of management institutions. The chapter closes with a short summary.

Human Use of Freshwater Resources

Water is an essential and nonsubstitutable resource for people and ecosystems. This critical resource is, however, rare and patchily distributed on earth. Only about 2.5% of all water on the planet is freshwater. Less than 1% of Earth's freshwater occurs in lakes, rivers, reservoirs, or groundwater shallow enough to be tapped at affordable cost (Vörösmarty et al. 2005). Some freshwater occurs as soil water

available to plants, but most of the remaining water occurs as ice or deep aquifers that are not available to people or other organisms (Fig. 9.1).

Freshwater supports human societies and ecosystems in two main forms: "blue water" and "green water" (Falkenmark and Rockström 2004). **Blue water** is what we typically think of when we consider water resources: liquid water in rivers, lakes, reservoirs, and groundwater aquifers. Blue water is important for a host of essential services, including drinking water and sanitation, food production through irrigation, transport, and energy production. Blue water also supports a diverse array of aquatic ecosystems. **Green water** is the moisture in the soil that supports all nonirrigated vegetation, including rainfed crops, pastures, timber, and terrestrial natural vegetation. Green water flows via evaporation and transpiration exceed blue water flows in rivers and aquifers (Fig. 9.2).

Available Freshwater Resources

Flows of freshwater are more important than storages when considering the freshwater available for use by humans and ecosystems (Fig. 9.1). For example, the total amount of blue water stored in all the world's rivers is about 2,000 km^3, much less than the annual withdrawal of 3,800 km^3. A more appropriate measure of water availability is the 45,500 km^3 that is discharged annually by the Earth's rivers (Oki and Kanae 2006). Water that is replenished by rainfall or snowfall can be regarded as a renewable resource. Renewable water sources are available on a sustainable basis provided the rate of extraction does not exceed the replenishment rate, and adequate water quality is maintained. Water quality is usually defined by its desired end use. Water for drinking, recreation, and habitat for aquatic organisms, for example, requires higher levels of purity than water for hydropower. Nonreplenishable stocks of freshwater such as fossil groundwater aquifers are nonrenewable in the same way as oil resources.

Freshwater replenishment rates vary substantially among ecosystems and over time.

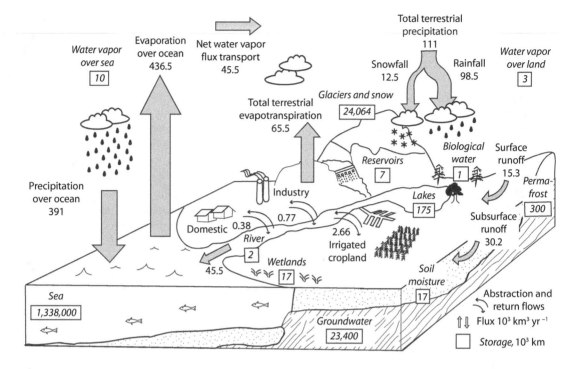

FIGURE 9.1. Global hydrological fluxes (1,000 km³ yr⁻¹) and storages (1,000 km³). The direct groundwater discharge, which is estimated to be about 10% of total river discharge globally, is included in river discharge. Adapted from Oki and Kanae (2006).

The water of a lake in the moist tropics may be replaced several times per year, whereas accessible groundwater of a dryland steppe may be replaced only once every few decades. The sustainable rate of extraction for two equal-size bodies of water will therefore be very different in these two systems. Additionally, many drier parts of the world experience substantial interannual variability in precipitation. What may be a replenishable rate of extraction in a particular year or decade may not be sustainable in another. Such variability poses particular challenges for water resources policy and management because it demands high levels of adaptability. Lastly, not all renewable water is available for people or ecosystems. Flood waters may flow away before the water can be used, or groundwater may become polluted and unusable. Different watersheds therefore tend to each have their own particular set of management challenges.

Globally about 3,800 km³ of blue water is withdrawn annually for human use. Although this represents only 10% of the maximum renewable freshwater resource (Oki and Kanae 2006), high variability in water supplies over space and time means that extraction in many regions exceeds renewable supplies (see Plate 7). Overextraction has resulted in dramatic decreases in the extent of several large aquatic ecosystems, including the Aral Sea, the Mesopotamian Marshes, and Lake Chad (Fig. 9.3). Approximately 700 km³ of the total annual withdrawal comes from groundwater, and much of this is extracted from fossil aquifers or at rates substantially greater than the rate of recharge. To stabilize the spatial and temporal variability in water supplies, humans have substantially altered the flows of freshwater by constructing dams and reservoirs and through interbasin transfers. At any point in time, the world's man-made reservoirs now hold three to six times the total amount of

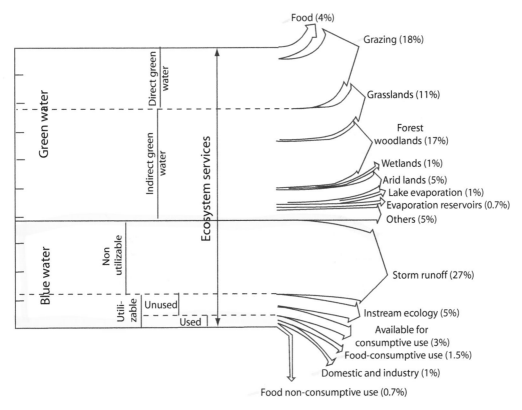

FIGURE 9.2. Blue-water and green-water flows used to support global ecosystem services, based on data in Rockström et al. (1999). Direct green water is the water directly used by food-production systems; indirect green water benefits society through other ecosystem services.

water in all the world's rivers (Vörösmarty et al. 2005). The total flow of water being diverted without return to its stream of origin in Canada alone was 140 km³ in the 1980s, more than the mean annual discharge of the Nile (Vörösmarty et al. 2005).

Society's management of freshwater is largely organized around the different human uses of freshwater. While this may be practical to some degree, it means that different agencies are commonly responsible for managing different parts of the hydrological cycle. For example, water utilities are usually responsible for providing drinking water, agricultural departments for providing water for agriculture, and environmental departments for maintaining aquatic ecosystems and water quality. Management of freshwater systems therefore tends to be fragmented.

Freshwater for Drinking and Sanitation

Access to adequate water for drinking and sanitation is a basic human need, and was declared a human right by the UN in 2002. The minimum drinking water requirement for human survival is 3–5 l per person per day, depending on climate (Gleick 1996). Drinking water must be sufficiently clean to prevent water-related diseases such as cholera. The major determinant of clean drinking water is access to water for sanitation. About 1.7 million people, primarily young children, die each year from diseases associated with inadequate sanitation and hygiene (Vörösmarty et al. 2005). If water for basic sanitation, hygiene, and food preparation is taken into account, each person needs at least 20–50 l of clean water each day to

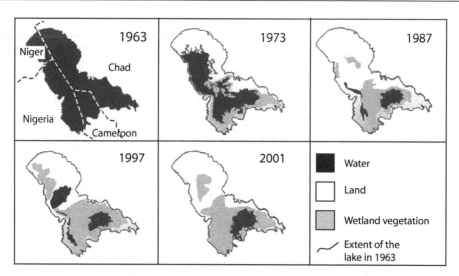

FIGURE 9.3. Lake Chad was once one of Africa's largest bodies of fresh water, similar in size to North America's Lake Erie. Lake Chad has decreased dramatically to about 5% of its original size due to growing human demand, especially for large-scale irrigation projects, combined with a series of droughts. Most of the change in Lake Chad happened over the 15-year period from 1973 to 1987. The drier climate and reduced size of the lake has compromised the irrigation farmers, fishermen who depended on the lake fishery, and pastoralists whose cattle grazed in the area surrounding the lake. Some of these people have moved to cities or other regions, where they have often joined the ranks of the jobless or come into conflict with host communities. Redrawn from http://www.grida.no/climate/vitalafrica/english/14.htm

survive. The minimum requirement varies due to differences in climate, culture, technology, and lifestyle. Providing basic freshwater needs depends both on availability of water in the ecosystems where people live and on human knowledge and infrastructure to extract, transport, and clean the water.

Much of humanity still lives with inadequate water and sanitation supplies, a major factor affecting well-being and limiting development. For each dollar invested in improved water supply and sanitation, an estimated return of $3–34 can be expected on account of averted deaths, reduced health care costs, days gained from reduced illness, and reduced time spent obtaining and transporting water (Hutton and Haller 2004). Nevertheless, in 2004, approximately 1.1 billion people (17% of the world's population) lacked access to clean water, and 2.6 billion (40% of the global population) lacked access to improved sanitation (Vörösmarty et al. 2005). Many governments are too poor or dysfunctional to make the expensive infrastructural investments required to deliver these basic

services. This is particularly true where local water supplies have become exhausted or polluted, and water supply costs have risen dramatically. For example, in parts of Bangladesh groundwater supplies have become contaminated with arsenic. Arsenic occurs naturally in many soils, but becomes toxic when exposed to the atmosphere when groundwater tables are lowered through excessive extraction. As discussed in the section on pollution, developed and developing countries alike also face increasing problems with chemical pollution of drinking water from industrial and agricultural waste (see Chapter 12).

Freshwater for Food Production

Irrigation is by far the largest user of blue water, accounting for 70% of all blue water withdrawals. These withdrawals support 40% of the world's crop production on 18% of the world's cropland (Cassman et al. 2005). Advantages of irrigated agriculture are that crop yields are higher than for rainfed crops, and production is less susceptible to the vagaries of climate. The expansion of irrigated agriculture together with dramatic improvements in transportation has helped reduce the incidence of severe famines and malnutrition despite large increases in human populations (see Chapter 12). Expanding irrigated agriculture in very poor parts of the world such as Africa presents one of the largest opportunities for addressing poverty (Shah et al. 2000). However, as discussed later, irrigation is also a major cause of freshwater and environmental degradation. Furthermore, in some areas irrigation withdrawals far exceed replenishment rates. The sustainability of irrigation practices in many parts of the world is therefore of concern (Foster and Chilton 2003, Vörösmarty et al. 2005).

Blue water is also critical in providing food in the form of fish. Inland fisheries are of particular importance in developing countries because fish are often the only source of animal protein. However, overharvesting in combination with habitat degradation and invasive species pose substantial threats to freshwater fisheries (Finlayson et al. 2005). Several well-studied aquatic ecosystems such as the North American Great Lakes showed dramatic declines in native fish populations in the twentieth century. These declines have been related to a combination of factors, including the establishment of invasive species such as sea lamprey and alewife, overharvesting, severed migration routes, and stocking of exotic sport fish such as Pacific salmonids. In recent years, the production of inland fish has become dominated by aquaculture, which is the fastest growing food production sector in the world. Although aquaculture may add a source of protein to food supplies, this benefit must be balanced against adverse environmental effects. Aquaculture facilities are foci for disease, concentrate pollutants that cause eutrophication, and may increase harvest of wild fishes used as food (Naylor et al. 2000).

Green water plays a major role in food production. Almost 40% of the world's terrestrial surface is now used for food production: 16 million km^2 for croplands (82% of which is rainfed), 18 million km^2 for managed pastures, and 16 million km^2 as rangeland. Evapotranspiration is estimated to be 7,600 km^3 yr^{-1} from crop-

land and 14,400 km^3 yr^{-1} from permanent grazing land used for livestock production (Oki and Kanae 2006). Together, cropland and grazing land therefore account for one third of total terrestrial evapotranspiration. About 40% of the world's population depends directly on these agricultural systems for their livelihoods, with the proportion increasing to over 60% of the population in poorer parts of the world.

Other Services Supported by Freshwater

Besides water and food, freshwater ecosystems provide a host of other services that sustain modern societies. The most important of these are ecological processes that maintain the environment and resources on which people depend. Aquatic ecosystems such as lakes and wetlands play an important role in regulating water flow. They help attenuate floods, recharge groundwater, and maintain river flow during dry periods by releasing water stored during wet periods. Such hydrological regulation reduces the need for expensive engineered flood control and water storage infrastructure. The vegetation in many inland waters traps sediments, nutrients, and pollutants such as heavy metals. This helps maintain water that is of adequate quality for drinking and irrigation without the need for expensive water treatment. Trapping sediments and breaking down pollutants also reduces degradation of downstream habitats that are important for fish production. Wetlands, and peatlands in particular, store exceptionally large quantities of organic carbon per unit area and thereby contribute to climate regulation (Finlayson et al. 2005, Cole et al. 2007). The value of the regulating services provided by functioning natural aquatic ecosystems is often not recognized until they are degraded or destroyed.

By supporting aquatic and terrestrial ecosystems, freshwater provides important nonfood products such as fiber, construction timber, and energy. Nonfood crops such as fiber (e.g., cotton for clothes and textiles), biofuels, medicines, pharmaceutical products, dyes, chemicals, timber, and nonfood industrial raw materials account for almost 7% of the world's harvested crop area (Cassman et al. 2005). The area used for biofuel production could grow substantially as petroleum supplies decline or nations seek to decrease their dependence on petroleum-producing nations. In poorer parts of the world such as sub-Saharan Africa, over 80% of the population depends on fuelwood and charcoal from natural vegetation for domestic cooking and heating. Peatlands, a type of wetland, have been mined extensively for domestic and industrial fuel, particularly in Western Europe. Today peat mining for use in horticulture is a multimillion dollar industry in Europe (Finlayson and McCay 1998).

Blue water directly supports nonfood sectors of modern society by providing water for industry, electricity generation, and transport. About 21% of blue water withdrawals globally are used for industrial purposes, which account for 32% of global economic activity (CIA 2007). Freshwater is also important for electricity production, both through hydropower (17% of global electricity production) and in nuclear and coal power plants (16 and 66% of global electricity production, respectively) (EIA 2006). Blue water in rivers and lakes has long played a major role in transportation. Although cargo traffic has substantially declined due to aviation, railroads, and trucking, cargo shipping remains important in areas such as the North American Great Lakes, the Rhine River in Europe, and the Yangtze River in China.

Freshwater ecosystems provide many cultural and recreational ecosystem services. People are drawn to water. Lakes, rivers, or wetlands are associated with sacred sites or religious activities in many cultures and have inspired artists. Many freshwater ecosystems are today protected as National Parks, World Heritage sites, or Wetlands of International Importance (Ramsar Sites). Income generated by recreation and tourism associated with freshwater can be a significant component of national and local economies. This includes activities such as water sports, recreational river cruises, and recreational fishing. The educational value of freshwater systems, and wetlands in particular, is closely associated with recreation. For example, approximately 160,000 people visit a 40-ha wetland complex in the heart of London each year. Similar to many

such centers around the world, this site offers an educational exhibition center and activities within a recreational setting with boardwalks, hides, and pathways (Finlayson et al. 2005).

Degradation of Freshwater Ecosystems

The Millennium Ecosystem Assessment (2005a) concludes that "It is *established but incomplete* that inland water ecosystems are in worse condition overall than any other broad ecosystem type." Freshwater ecosystem condition has been degraded by draining of wetlands, fragmentation of rivers through construction of dams and canals, pollution by fertilizers and toxic chemicals, invasion by harmful exotic species, and overfishing (Fig. 9.4).

In preindustrial times, transfers across watershed boundaries were few, and flows between watersheds were usually small compared to flows within watersheds. Flows between watersheds occurred in the form of critical nutrients (in dust) and microorganisms moving through the atmosphere and mobile animals such as insects, birds, and mammals moving among watersheds. In modern times, people have dramatically increased the rates of flows between watersheds by moving water, fertilizers, agricultural products, and people and by increasing the airborne transport of nutrients and contaminants. These changes have contributed to alterations in the ecological processes within freshwater ecosystems and have expanded the impacts of human activities to far beyond the particular watershed in which people live.

Draining, Fragmentation, and Habitat Loss

Extensive areas of wetlands have been drained to make way for agriculture or urban development. By 1985, losses of wetland area since the industrial revolution were 56–65% in North America and Europe, 27% in Asia, 6% in South America, and 2% in Africa (MEA 2005a). Rates of loss have declined steeply in many rich nations and are rising in some poor nations. Degradation of watershed vegetation,

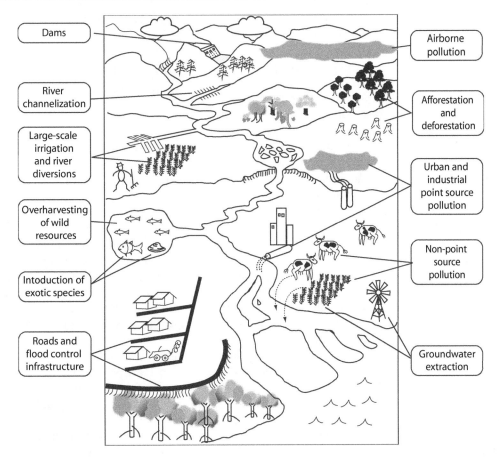

FIGURE 9.4. Ways in which human use affects the water cycle and freshwater ecosystems. Modified from Finlayson et al. (2005).

shorelands, and wetlands increases the probability of damaging floods and reduces the natural storage of water for release during dry periods (MEA 2005a). Floods typically affect the poorest people most severely, though even rich countries suffer devastating floods that are exacerbated by degradation of wetlands and terrestrial vegetation (Kundzewicz and Schellnhuber 2004). For example, large-scale clearing of savanna in Brazil was associated with a 28% increase in wet-season flows (Costa et al. 2003).

Impervious land cover, such as buildings, roads, or packed earth, prevents infiltration of precipitation into soil water or groundwater. Instead, most of the precipitation runs off the land as surface flow. Consequently, water inputs to streams and lakes are more variable and have higher concentrations of sediment, phosphorus, and contaminants that are associated with sed-

iment particles. Agriculture and some urban land uses (such as fertilized lawns) increase the nutrient content of runoff. Reduced infiltration contributes to declining groundwater tables, so that groundwater accounts for a smaller proportion of water input to streams and lakes. The net effect is more variable water levels, poorer water quality, and loss of springs and small streams that are crucial habitat for many plants and animals, including spawning fish.

Dam construction has fragmented about 40% of the world's major river systems. The number of large reservoirs has increased from about 5,000 in 1950 to about 45,000 in 2000 (Vörösmarty et al. 2005). This estimate excludes many small impoundments created for local use by farmers or small municipalities. While dam construction has contributed to economic development, food security, and flood control, it has also had substantial environmental, social, and human health impacts. Dams have had enormous effects on the water cycle and hence on aquatic habitats, suspended sediment, carbon fluxes, and waste processing. In some cases, substantial amounts of water are transferred among watersheds, altering the water balance of extensive areas. Large dam projects often involve the displacement and resettlement of people, sometimes causing enormous social disruption. In addition to losing their livelihoods and cultural and historical heritage, resettled communities often suffer from a marginalized status and cultural and economic conflicts with the communities in which they are settled.

Loss of habitat through draining and fragmentation is the principal driver of biological change in freshwater ecosystems. Habitat loss can be caused by changing climate or land use or by directly engineering waterways for hydropower, transportation, or shoreline stabilization. Some habitat losses are obvious. Dams and levees, for example, prevent movement of fishes and are an important factor in loss of fish species from river networks (Rahel 2000). Other habitat losses are subtle or indirect. Clearing of trees and brush along shorelines decreases the numbers of fallen trees in lakes, thereby decreasing habitat for fishes and their

prey and ultimately decreasing fish production (Sass et al. 2006).

Eutrophication, Salinization, and Chemical Pollution

Many freshwater supplies are degraded by eutrophication, chemical pollution, and salinization. Eutrophication leads to cloudy water, blooms of cyanobacteria (some of them toxic), oxygen depletion, fish kills, and taste and odor problems that impair the use of water for drinking. In some regions of the world, sewage discharge is a major cause of eutrophication. In addition to degrading water quality, inadequate sewage treatment is an important cause of human disease. Most of the world's wealthier countries have adequate facilities for treating sewage before it is discharged to surface waters. However, as sewage discharges came under control, eutrophication in many places remained stable or increased because of nonpoint pollution (Box 9.2). **Nonpoint pollution** is nutrients or toxins scoured by runoff from agricultural or urban lands. Nonpoint pollution accounts for more than 80% of the nitrogen and phosphorus inputs to the surface waters of the USA and is the major source of pollutants to surface waters in many countries (Carpenter et al. 1998b).

Box 9.2. Managing Lake Mendota

When Wisconsin's legislature in 1836 selected a location for the capital city, Madison, they chose a gorgeous site on an isthmus between two lakes (Mollenhoff 2004; Fig. 9.5). A reporter visiting from Massachusetts noted that he could "see the drifts of white sand far down to the transparent depths." By the 1870s, the fertile savannas of the watershed had been converted almost entirely to agriculture, and the formerly white lake bottoms were covered with a thick layer of black soil eroded from the uplands (Lathrop 1992a). By the 1880s, when E.A. Birge arrived to launch America's first limnology program at the University of Wisconsin, the lakes were plagued in summer by algae blooms and fish kills. Carp were introduced

and exacerbated the lakes' water-quality problems. Eutrophication and carp led to changes in rooted aquatic plants and loss of some fish species that depended on these plants for habitat (Magnuson and Lathrop 1992). Other fish species were introduced to the Madison lakes from rivers or lakes that dried out during the deep drought of the 1930s (Magnuson and Lathrop 1992). Starting the late 1940s, intensification of agriculture and development brought further deterioration of water quality (Lathrop 1992a). During the 1950s, most of the deep-water benthic invertebrates disappeared from Lake Mendota, the largest of the lakes (Lathrop 1992b). Thick blooms of cyanobacteria were common in summer.

FIGURE 9.5. Lake Mendota. http://mrsec.wisc.edu/Edetc/NSECREU/2007%20N4.jpg.

Management of the Madison lakes has gone through a series of phases (Carpenter et al. 1998, 2006b, Carpenter and Lathrop 1999). During the first half of the 1900s, management focused on treating algal blooms with direct applications of herbicides. Nuisance rooted plants were harvested or treated with herbicides. By the 1950s, public concern with water quality evoked proposals to decrease sewage inflow to the lakes. By 1971, all sanitary sewage was diverted from the lakes. However, water quality did not change much. Phosphorus runoff from agriculture, construction sites, and towns, combined with phosphorus recycling from sediments, maintained high levels of nuisance algae in the Madison lakes. The focus of management shifted to runoff, or nonpoint pollution. During the early 1980s, efforts to manage nonpoint pollution failed because of low participation by farmers. In 1987, the food web of Lake Mendota was manipulated to increase grazing on phytoplankton and improve water quality (Kitchell 1992, Lathrop et al. 2002). However, water qual-

ity remained highly variable and was especially poor in years of high runoff. In the late 1990s, a massive program was initiated to cut nonpoint pollution of Lake Mendota by half (Carpenter et al. 2006b). At the time of writing, this project was still underway. Though the outcome is unknown, it seems likely that many of its goals will be met.

Meanwhile, the Madison area faces ongoing changes in water quality and supply. Climate is changing, with unknown but potentially large consequences for precipitation, evapotranspiration, and runoff. Development has increased the amount of impervious surface in the watershed and thereby decreased infiltration of water to soils and increased runoff. As a consequence, water levels in the lakes have become more variable. New invasive species, such as zebra mussel or silver carp, could become established and alter water quality. Thus management of the Madison lakes is always a work in progress. New challenges and (one hopes) new solutions are likely to continue in an ongoing cycle.

Overuse of fertilizers and high densities of livestock are major causes of nutrient runoff and eutrophication. Phosphorus and

nitrogen are two of the main plant nutrients supplied by fertilizers and have particularly large impacts on environmental quality and ecosystem services. In pre-industrial times, inputs of phosphorus to earth's terrestrial ecosystems were 10–15 Tg yr^{-1} (Bennett et al. 2001). By 2000, mining of phosphate rock and increased weathering due to land disturbance by people had increased these inputs to 33–39 Tg yr^{-1} (Fig. 9.6). Most of the mined phosphorus is used as fertilizer to grow crops. The phosphorus in crops is fed to domestic animals or people, which recycle the phosphorus in waste or accumulate it in their bodies. Phosphorus applied to soils as fertilizer or manure can wash into aquatic systems and cause eutrophication. Globally, the fluvial transport of phosphorus through freshwater ecosystems to the sea is about 22 Tg yr^{-1} at present, versus only 8 Tg yr^{-1} before the advent of industrialized agriculture. Human additions to the phosphorus cycle have therefore almost tripled the amount of phosphorus annually cycled through the world's ecosystems.

About 20% of irrigated croplands worldwide suffer from **salinization**, the accumulation of salts to the point that soils and vegetation are degraded. The main cause of salinization is inputs of salt dissolved in irrigation water from

rivers or aquifers. When plants take up irrigation water from the soil, the salts are mostly left behind. In the absence of sufficient drainage to leach the excess salts deeper into the soil profile, salts accumulate in the topsoil over time and result in reduced crop yields. A secondary cause of salinization is waterlogging. Waterlogging occurs when aquifers are unable to take up the excess irrigation water that has drained deeper down into the soil profile. This causes the groundwater table (containing salts leached from the topsoil) to rise to within the crop rooting zone. Water tables may also rise as a consequence of clearing trees for agriculture, especially in arid regions such as Australia (see Chapter 8). The replacement of deep-rooted trees by shallow-rooted crops can dramatically reduce the amount of water taken up and transpired, resulting in rising groundwater tables.

Chemical pollution of surface and groundwaters is of increasing concern in developed and developing countries alike. Exposure to heavy metals, long-lived synthetic compounds, and toxic substances has been linked to a range of chronic diseases, including cancer,

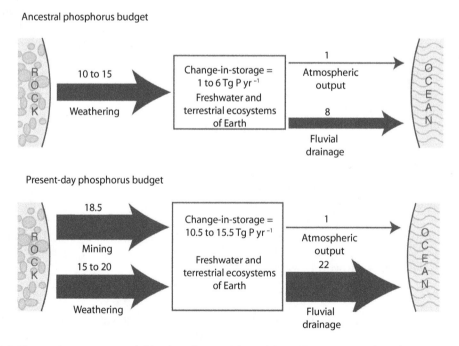

FIGURE 9.6. Human impacts on erodable phosphorus. Adapted from Bennett et al. (2001).

lung damage, and birth defects. Some pollutants may persist in aquatic systems for decades after they have been found to be harmful and banned. Cleaning water of these compounds is usually very expensive and in some cases impossible.

Invasive Species and Overharvesting

Habitat loss interacts with other drivers of biological change, including eutrophication (see above), species invasions or introductions (Kolar and Lodge 2000), and harvesting (Post et al. 2002). Species invasions often lead to secondary losses of species that are eaten by the invader or lose habitat due to expansion of the invading population. **Trophic cascades** are changes driven by changes in top predators that affect lower trophic levels, primary producers, bacteria, and nutrient cycles (Carpenter and Kitchell 1993). Cascades in aquatic ecosystems can be caused by overfishing or by invasions of top predators. Trophic cascades change not only food webs, but also primary production, sedimentation rates of nutrients and carbon, and rates of release of greenhouse gases from freshwater ecosystems. Often the effects of species invasions, fish harvesting, and nutrient input interact, leading to massive changes in lakes or rivers (Box 9.3). Such changes are difficult to reverse or can be reversed only slowly. More often, the changes lead to new and different ecosystems as well as new kinds of relationships between people and freshwaters.

Box 9.3. Lake Victoria: The Nile Perch "Gold Rush"

Lake Victoria in East Africa is the world's second largest freshwater lake. It is the site of one of the most rapid and extensive evolutionary radiations of vertebrates known, with most of the over 500 endemic species of haplochromine cichlid fish having evolved in the last 15,000 years (Stiassny and Meyer 1999, Kocher 2004). It is also the site of one of the most recent vertebrate mass extinctions on the planet. Up to half the endemic cichlids, some of them never described, are believed to have gone extinct during the 1970s and 1980s. The mass extinction is mainly linked to the population explosion of the Nile perch (*Lates niloticus*) in the mid-1980s, although other factors such as overfishing and eutrophication also played a role (Kaufman 1992). The predatory Nile Perch was introduced to Lake Victoria in the 1950s and 1960s by the British colonial administrators in an attempt to develop a productive commercial fishery in the lake and address the declining native subsistence tilapia fishery (Pringle 2005b).

The establishment of the "perch regime" has had profound impacts on the lake's ecology and people. The species communities, food web, and ecological processes in the lake have been radically transformed. Increased numbers of people in the watershed have contributed to large-scale deforestation and the eutrophication of the lake. These ecological changes may have made the lake more vulnerable to invasion by the South American water hyacinth, which occurred in the 1990s and negatively impacted drinking water quality, fishing, tourism, and transport (Balirwa et al. 2003) – but has subsequently been successfully controlled. The perch regime also brought significant benefits. The large, fleshy perch is highly suited for commercial export and transformed the regional subsistence economy into a cash-based economy linked to global markets. This change undoubtedly provided valuable employment opportunities, raised the living standards of people living around the lake, and earned valuable foreign exchange for the bordering countries (Reynolds and Greboval 1988, Pringle 2005a).

Today the lake region faces substantial challenges in simultaneously addressing poverty and environmental degradation. There are signs that the Nile perch fishery is in decline. Population growth coupled with few formal employment opportunities and poor environmental enforcement have resulted in intense fishing pressure. One result of the declining perch population is that some of the cichlids that were believed extinct have recently been recorded again. On the other hand, globalization of the perch fishery means that lakeside communities now compete with consumers in the West, and recent price increases limit local access to perch as a food and protein source

(Fig. 9.7). Many people also depend on the jobs centered on the Nile perch export industry. Transforming the current regime into one that is ecologically and economically sustainable presents a considerable challenge. No "magic bullet" solution is likely to exist; rather concerted effort will be needed on a number of fronts, such as developing alternative livelihood options and improving fisheries monitoring, regulation, and enforcement. Experience elsewhere suggests that these are most likely to succeed if local fishermen and riparian communities are actively involved in developing and implementing solutions to address the challenges they face (Dietz et al. 2003).

FIGURE 9.7. Fish landing on Lake Victoria, Jinja, Uganda. Photograph by Jim Kitchell.

Challenges in Water Management

Declines in water supply or water quality often lead to conflict among competing users.

These include competition among people or between people and ecosystems for scarce water, as well as political conflict over mitigation of water quality. Such conflict can lead to nonsustainable use of water supplies,

development of costly alternative water supplies or purification methods, limitations to
economic growth and development, pollution
and public health problems, international disputes in transboundary river basins, and political and civil instability (Vörösmarty et al.
2005). In this section, we address four challenges that frequently arise in freshwater management: upstream/downstream flows, human
versus environmental flows, managing for heterogeneity, and cross-generational equity.

Upstream/Downstream Flows

Water extraction and pollutant discharges in
upstream areas clearly have direct impacts
for downstream users. They affect both water
quantity and quality, as well as ecosystem services such as coastal fisheries that are important
in many large river systems. These conflicts are
similar to those that often exist between multiple users in a particular river section or around
a particular lake.

The Watershed Trust Fund of Quito,
Ecuador, illustrates a potentially successful
partnership for maintaining water supplies
and conserving biodiversity in the watershed
(Postel 2005). Organized with support from
The Nature Conservancy and the US Agency
for International Development, the trust fund
pools the demand for watershed conservation
among downstream users, including municipal
users, irrigators, industries, and hydroelectric
utilities. In 2004, the trust fund mobilized more
than half a million dollars for conservation
of the watershed. An important aspect of the
trust fund is the merger of water supply goals
with conservation of biodiversity. This combination expands the pool of participants and
funders.

Upstream/downstream conflicts can be especially difficult to deal with in freshwaters that
are shared across international boundaries.
There are 261 international rivers, whose watersheds cover almost half the total land surface of the globe. In addition, many groundwater aquifers and large lakes and reservoirs
are shared between two or more countries.
Water has been a source of cross-border ten-

sion, for instance in the Middle East, southern
Africa, the US–Mexican border, and parts of
Asia, and there has been much talk of potential
"water wars." However, history suggests that
the strength of shared interests usually induces
cooperation rather than inciting violence. In
the twentieth century 145 water-related treaties
were signed, while there were only seven minor
water-related skirmishes (Wolf 1998). Such historical analyses suggest that war over water
is seldom strategically rational, hydrographi-
cally effective, or economically viable. Rather,
cooperative water regimes established through
international treaties have tended to be impres-
sively resilient over time, even between hostile
nations that have waged conflicts over other
issues. For example, the 1960 Indus Waters
Treaty between India and Pakistan has survived
major conflicts between the two nations.

Human versus Environmental Flows

Providing water for domestic, agricultural, and
industrial use, while also meeting the water
requirements of aquatic ecosystems is a central
challenge in water resource management. As
discussed earlier, maintaining aquatic ecosystems is important for the provision of a host
of ecosystem services, such as flood protection,
maintaining fisheries, and regulating waterborne diseases. It can also be argued that
aquatic organisms and ecosystems have intrinsic value in and of themselves irrespective of
their value to human society. Some people
believe that humans have an ethical responsibility to limit their use of freshwater resources so
that other organisms can live and thrive (Singer
1993, Agar 2001).

One response to human–environmental conflicts practiced in parts of Australia, Europe,
New Zealand, North America, and South
Africa has been to specifically allocate water
for environmental flows (Box 9.4). Environmental flows refer to the quantity, quality, and
timing of water flows considered sufficient to
protect the dependent species and the structure and function of aquatic ecosystems (King
et al. 2000, Dyson et al. 2007). Flow variability is important because different benefits are

provided by high and low flows. Low flows, for example, can exclude invasive species, while high flows may provide spawning cues for fish and replenish floodplains. Environmental flow requirements can range from 20 to 80% of mean annual flow, depending on the river type, species composition, river condition objectives (e.g., pristine, moderate modification, minimum flows) (Vörösmarty et al. 2005). The volumes of flow required indicate a high degree of potential conflict between direct human uses and the maintenance of freshwater ecosystems.

Box 9.4. Water Policy in South Africa

The South African National Water Act introduced in 1998 is regarded as a landmark in international water policy (Postel and Richter 2003). The new legislation defines the freshwater resource as river ecosystems rather than water, in recognition of ecosystem service values. The Act gives priority to basic human needs and the needs of aquatic ecosystems through establishment of a "Reserve." Instead of private ownership, the legislation is based on the legal principle of public trust, whereby the government holds certain rights and entitlements in trust for the people and is obliged to protect those rights for the common good. The Act treats the hydrological cycle holistically, recognizing that surface and groundwater flow are connected and also linked to land use. The new policy promotes participatory decision-making, where allocations are based on interest-based negotiations rather than water rights.

The "Reserve" is one of the most innovative aspects of the Water Act and has two parts: the basic human needs reserve and the ecological reserve. The basic human needs reserve provides for 25 l per person per day for essential drinking, cooking, and sanitation needs. The ecological reserve is defined as the water quantity and quality required to protect the structure and functioning of aquatic ecosystems in order to secure sustainable development. This two-part Reserve has priority over all other uses and is the only water guaranteed as a right. International obligations are provided for once the Reserve has been met. Only after these needs are satisfied, is the remaining water allocated to other uses such as irrigation and industry.

Another innovative aspect of the national water legislation is that it is regarded as a tool for transforming post-Apartheid South Africa into a socially and environmentally just society (Funke et al. 2007). To achieve this, the legislation is based on four key principles: decentralization, equitable access, efficiency, and sustainability. The decentralization principle makes provision for people to participate in decision-making processes that affect them and for governmental functions to be delegated to the lowest appropriate level. In accordance, Catchment Management Agencies (CMAs) and Water User Associations are being established for all the major watersheds in the country. The CMA governing boards must represent all major stakeholders and are mandated to develop detailed catchment management strategies for their watersheds through cooperative approaches. The principle of equitable access is provided through the public trust doctrine and an administrative licensing system that regulates the extraction of water for all nondomestic uses. To ensure efficiency, the social, economic, and environmental benefits and costs of competing water uses have to be evaluated. To facilitate this process, provision is made to appeal against licensing decisions, and economic instruments such as pricing and subsidy programs are used. In particular, in order to provide the "lifeline" basic human needs water supply of 6,000 l per household per month free of charge to all households, a sliding-scale pricing scheme has been adopted whereby users pay progressively more per unit of additional water use. The two parts of the Reserve are the pillars that provide for the interlinked social, ecological, and environmental sustainability of the country's freshwater resources. The Reserve thereby provides for two constitutional rights guaranteed to all South Africans: the right to enough water to meet basic needs and the right to a safe environment that is sufficiently protected to ensure socioeconomic development and ecological sustainability.

Habitats, Fisheries, and Managing for Heterogeneity

When relatively undisturbed, freshwater ecosystems have enormous heterogeneity. Sizes and shapes of waterbodies and drainage networks are similarly variable (Downing et al. 2006). Time series of discharge from undisturbed rivers vary across a wide range of timescales. Freshwater organisms range in mass over about 18 orders of magnitude, from bacteria ($\sim 10^{-12}$ g) to the largest freshwater fishes ($\sim 10^6$ g). Thus great variability – in habitat size and shape, in flow, and in organism size – is characteristic of freshwater ecosystems.

Human intervention tends to reduce structural variety of freshwater ecosystems by engineering shorelines, altering sizes of ecosystems using dams or drains, and harvesting species in certain size classes (such as larger fishes). While the role of structural variance in the function of freshwater ecosystems is not well understood, interventions that alter habitat shape or the biotic size structure tend to cascade, affecting many species as well as ecosystem processes such as net ecosystem production, ecosystem respiration, gas exchange with the atmosphere, and nutrient supply ratios. The capacity to absorb disturbance and maintain ecosystem processes seems to be related to the variability in habitat configuration, flow regime, and biota of freshwater ecosystems.

Maintenance of variability is crucial for managing freshwater ecosystems, but most management practices are designed for stabilization. Paradoxically, uniform application of such practices can create severe instabilities. If fisheries of a lake district are managed with a "one-size-fits-all" policy that stabilizes management targets for all lakes, there is greater risk that spatial cascades of collapse will spread contagiously across fisheries of many lakes (Carpenter and Brock 2004). In contrast, policies that allow local managers to set incentives or regulations on a lake-by-lake basis, taking account of conditions in neighboring lakes, tend to maintain a patchwork of fisheries that are variable yet persistent. In other words, heterogeneous policies that emerge from local decisions foster resilience of the fisheries across the lake district as a whole (see Chapter 1).

Humans often demand stability in freshwater ecosystem services. For example, people want the supply of freshwater, protection from floods, the capacity to carry boat traffic, or the fish production of freshwaters to be predictable and stable. Yet this expectation of stability conflicts directly with the heterogeneity that creates resilience of freshwater ecosystem services. Designing institutions that accommodate the heterogeneity of resilient freshwater ecosystems and the needs of people for freshwater ecosystem services is one of the greatest challenges in ecosystem management.

Cross-Generational Equity and Long-Term Change

Human-caused changes to aquatic ecosystems can have long-lasting consequences. Many species invasions or extirpations are difficult or impossible to reverse. Water bodies may remain eutrophic for decades or longer, even if excess nutrient use ceases, because of the slow turnover of phosphorus in enriched soils and efficient recycling of excess phosphorus within lakes and reservoirs. Altered flow regimes due to levees and dams may also last for decades, and species or habitats lost to fragmentation may be lost permanently. Depleted aquifers may take decades to replenish, and polluted groundwaters may remain unusable for hundreds or thousands of years. Thus, actions taken in the present may affect the condition of freshwater ecosystems for many generations in the future. There is a conflict between present uses and future options for use of freshwater ecosystems.

Some economic analyses use discounting to compare the value of natural resources in the present and the future. Conceptually, the **discount rate** is similar to the inflation rate, whereby a given amount of money in the future is worth less than the same amount of money in the present. Discounting often seems like a reasonable tool for evaluating projects when the stakes and uncertainty are modest, and there is general agreement about the benefits and costs of the project.

In many cases, however, proposed uses of water resources involve high stakes, unknown and potentially irreversible outcomes, and considerable controversy about the best way to proceed. Such decisions involve judgments of values or ethics that are not well represented by simple discounting formulations or cost-benefit methods (Ludwig et al. 2005). In polarized political contexts, each interest group may have its own preferred "objective" analysis, and the diversity of analyses adds to the overall uncertainty. Interestingly, if there is broad uncertainty about the discount rate then on average discounting tends toward zero, thereby placing present and future generations on the same footing (Ludwig et al. 2005). Thus, moral judgments about future risk assume a central role.

Institutional Mechanisms for Solving Conflicts

While water scarcity often generates conflict, it can also provide an opportunity to develop institutions or technologies that enhance cooperation and build resilience. Responses to water scarcity are many and varied and depend on social and historical context, local ecosystem conditions, water infrastructure, and the state of knowledge. Different types of responses tend to be employed depending on the phase of the adaptive cycle in which the local social–

ecological freshwater system finds itself (see Chapter 1). For example, markets and virtual water trade tend to be employed most often in the growth phase, while conservation responses feature prominently in the conservation phase. Technological innovation and decentralization usually have their biggest impacts in the renewal phase.

Globally, there has been a shift in emphasis from increasing water supply (mainly by building dams and reservoirs) to reducing demand through conservation or improved technology. At a global scale, this may indicate a shift from the growth to the conservation phase in the way humanity interacts with freshwater resources.

Markets and Benefit–Cost Comparisons

Economic studies often show substantial benefits from conservation of aquatic ecosystems. New York City avoided $6 billion in capital costs and about $300 million per annum in operating costs for water-purification facilities by spending $1.5 billion in watershed protection over a 10-year period (NRC 2000). When the floodplains of northeastern Nigeria were evaluated for dams and irrigation projects, researchers compared the economic benefits of the proposed projects with the benefits of the intact floodplains for agriculture, fuelwood, and fishing. When used for these purposes, water had a net economic value about 60 times greater than its value in the irrigation projects (Barbier and Thompson 1998). Thus economic analysis favored conservation of the floodplain.

In theory, markets could allocate water resources efficiently if values of water could be properly monetized. Yet experiences with water markets have been mixed. In sub-Saharan Africa, markets for drinking water have failed because private investments have been lower than expected, and in many cases users have been unable to meet payments (Bayliss and McKinley 2007). The solution may lie in public utilities or in more effective partnerships of development agencies and private industry. Successful water markets involve tiering of water rates, so that basic household uses of water are inexpensive while heavy water users such as industry or agriculture pay higher rates (Postel 2005).

Nutrient flows that degrade water quality can also exhibit remarkable diseconomies. For example, farming with a net yield of $4 million per annum causes losses to society of more than $50 million per annum due to damages to water quality of Lake Mendota, Wisconsin (Carpenter et al. 1999 and Box 9.2). Market mechanisms may be useful in solving such imbalances, though often they are addressed by regulation or litigation instead.

"Cap-and-trade" has been used to create markets for nutrient discharge or water withdrawals. A maximum tolerable nutrient discharge or water withdrawal (the "cap") is established by a regulatory authority. The multistate commission responsible for the Murray–Darling River basin in Australia set a cap on withdrawals to mitigate degradation of the river system (Postel 2005). To create a market, the regulatory authority issues marketable credits for pollutant discharge and water withdrawal (the "trade"). In North Carolina in the USA, state officials set a cap for phosphorus and nitrogen discharges into Pamlico Sound and allowed for trading of nutrient credits among polluters. A major problem with cap-and-trade is that it does not allow for natural variation in the capacity of ecosystems to provide services. In some years the Murray–Darling River may have abundant flows, and in dry years the flow may be insufficient to meet the cap. In some years, Pamlico Sound may flush frequently and be able to dilute larger nutrient inputs; in other years the estuary may not flush at all, and the cap may lead to dangerous algal blooms and deoxygenation. Making markets flexible enough to cope with ecological variation is a challenge with cap-and-trade.

Virtual Water and Trade

Virtual water is the volume of freshwater needed to produce a specific product or service (Fig. 9.8). While water requirements can be calculated for any product, virtual water requirements have mainly been explored for crop and livestock production. There is a vast mismatch between the weight of agricultural commodities and the virtual water required for their production. For instance, producing 1 kg of grain requires 1,000–2,000 kg (liters) of water, and 1 kg of beef requires an average of 16,000 kg of water (Hoekstra and Chapagain 2007). In water-scarce regions, transporting water over long distances to meet these large demands is usually too expensive. However, water-scarce regions can offset their water demand for food production by importing products that require large volumes of water for their production from water-rich regions. Such "virtual water trade" is estimated to be about 1,000 km^3 yr^{-1}, although not all of it is to compensate for water shortages (Allan 1998, Oki and Kanae 2006).

Virtual water trade encompasses more than trade in water. In terms of agricultural production it involves transfers of crop nutrients. For

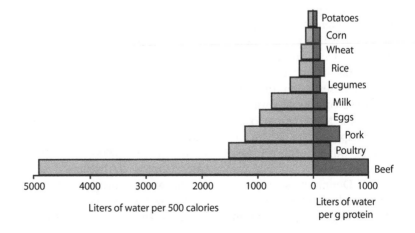

FIGURE 9.8. Virtual water, the liters of water needed to produce 500 calories of food or 10 g of protein. Data from Renault and Wallender (2000).

instance, Africa depends on a net import of more than 30 million tons of cereal crops annually, accounting for a net inflow of around 1.5 million tons of nitrogen, phosphorus, and potassium to the African continent each year. The need for food imports is partially a result of the depletion of local soil nutrients in Africa, which results in low crop yields and inadequate local food production. Soil nutrients in much of sub-Saharan Africa have progressively been depleted because most farmers cannot afford or do not have access to commercial fertilizers, and production demands have outstripped the capacity of traditional soil fertility practices, such as fallow periods or applying animal manure, to replenish soil nutrients (Cassman et al. 2005; see Chapter 8).

While food importation may offset local water and soil limitations, it also displaces the impacts of crop production. The impacts of excess fertilizer runoff and biodiversity impacts due to habitat loss are borne by the crop-producing country. In other regions, movement of food to feed livestock leads, through manure, to high soil nutrient levels and eutrophication of downstream water bodies (Bennett et al. 1999, 2001). On the other hand, policies aimed at food self-sufficiency are likely to intensify present-day patterns of water scarcity and associated conflicts. Crop production under marginal or poorly managed conditions can also have disproportionately large environmental impacts.

Conservation

Although aggregate global withdrawals continue to increase, a major feature of global water use is that per capita use rates have been declining, from around 700 m^3 yr^{-1} in 1980 to about 600 m^3 yr^{-1} in 2000 (Vörösmarty et al. 2005). Conservation, or increased efficiency of water use, offers many ways of mitigating water shortages. Most cities lose 20–50% of their water supplies to leaks in the distribution system or other forms of waste (Postel 2005). Leakage is highly variable among cities; Copenhagen, Denmark, loses only 3% of its water supply to leakage. Cities can reduce their water loss through engineering improvements and

public education. For instance, Boston in the USA decreased water use by 31% from 1987 to 2004 despite substantial economic growth and a stable population size (Postel 2005).

Food production offers the greatest opportunities for increased efficiency. Irrigation diverts 20–30% of the world's available water resources, but inefficiencies in distribution and application mean that only 40–50% of that water is used in crop growth. Irrigation water is often wasted because governments subsidize water supply so that farmers have little incentive to conserve. Yet technology exists to make irrigation far more efficient, for instance through drip irrigation systems (Postel 1999). Production per unit water can also be improved on rainfed lands that are not irrigated. Methods vary with local conditions. Changes can involve the timing of seed sowing or rooting depth of crop varieties, intercropping to create more shade on the soil surface, cultivation to promote infiltration of water, mulching, or weed control (see Chapter 12). Changes in diets can also affect water usage. Diets high in meat demand a lot of water, because a unit of animal production requires many units of plant production, as well as the water consumed by the animal through its lifetime (Fig. 9.8).

Nutrient runoff, the major cause of water quality degradation, can also be addressed by conservation measures. Nitrogen and phosphorus that leave a farmer's field in runoff are a loss to the farmer and to agricultural production and also degrade water quality. Manure disposal is a major factor in overfertilization of croplands (Bennett et al. 1999, 2001). If manure applications were adjusted to crop demand, waste of nutrients, and damage to water quality would be avoided. Also, reduced demand for meat would decrease animal production and thereby decrease nutrient flow through manure.

Terrestrial restoration projects can improve both water flows and water quality. Forest conservation plays a key role in maintaining the water supply for New York city (NRC 2000). The Working for Water project in South Africa is a very successful land-management program that seeks to simultaneously address pressing environmental, economic, and social issues. About 20,000 previously unemployed people are hired per year to manually clear invasive alien plants; 60% of the jobs are reserved for women, and HIV/AIDS awareness training and child care facilities are included (Magadlela and Mdzeke 2004). The impetus for the project was the finding that many invasive alien plants use substantially more water than native species, and that clearing stands of alien plants could substantially enhance streamflow in water-stressed regions (Gorgens and van Wilgen 2004).

Technology

Technological advances to improve water availability are likely to have their biggest impacts on the largest water user, irrigation. **Drip irrigation** is a method that minimizes the use of water and fertilizer by allowing water to drip slowly to the roots of plants, either onto the soil surface or directly onto the root zone. Drip irrigation methods range from high-tech, computerized systems to low-tech and relatively labor-intensive methods. **Hydroponics** is a method of growing plants using mineral nutrient solutions instead of soil. Water use can be substantially less, and soil-borne diseases and weeds are largely eliminated. However, hydroponic systems require greater technical knowledge and more careful monitoring than conventional soil-grown crops.

Genetic engineering of crops may have important implications for future use and impacts on freshwater resources. Genetically modified crops have been developed to increase resistance to pests and harsh environmental conditions such as droughts, as well as to improve shelf life and increase nutritional value. Genetic engineering can therefore play a role in reducing water demand, the need for fertilizer and pesticides, as well as the amount of land needed for agriculture. However, genetically engineered crops and other organisms are also associated with environmental risks, many of which are as yet unknown.

Traditional supply-side technological measures, such as interbasin transfers and building dams and reservoirs, are today regarded with some reservation (WCD 2000) but nevertheless remain important, especially in devel-

oping countries. An estimated 1,500 dams are under construction worldwide, and many more are planned. Interbasin transfers are also likely to remain an important mechanism for alleviating regional water shortages (Vörösmarty et al. 2005). Knowledge about the ecological impacts of dams and about ecological functioning can help design and manage engineering structures so that they are less harmful. For example, many dams are now built with fish ladders that facilitate natural fish migration. Flows from reservoirs can also be managed to mimic natural flow patterns by allowing for higher and lower releases at different times of the year.

Desalinization, converting saltwater into freshwater, is currently the most costly means of supplying freshwater and is highly energy-intensive. Nevertheless by 2002 there were over 10,000 desalinization plants in 120 countries, supplying more than 5 km^3 yr^{-1} of freshwater. About 70% of the installed desalinization capacity is in the oil-rich states in the Middle East and North Africa (UN 2003). While its use may be difficult to justify for high-consumption activities such as irrigation, investments in desalinization technology are likely to improve efficiency and bring down costs, creating a potentially important source of freshwater, at least for domestic use (Gleick 2000). Adequately managing brine waste from the desalinization process to protect nearby coastal ecosystems remains an unresolved issue requiring special attention (Vörösmarty et al. 2005).

Rainwater harvesting, the collection and storage of rain water from roofs and other surfaces can be practiced through traditional methods or using modern technology. Traditionally, rainwater has mainly been harvested in arid and semiarid areas, providing water for drinking and domestic use, livestock, small-scale irrigation, and as a way to replenish ground water levels. Rainwater harvesting can be particularly beneficial in urban areas. It augments cities' water supply, increases soil moisture levels for urban greenery, facilitates groundwater recharge, and mitigates urban flooding. In New Delhi, India, for instance, it is now mandatory for multistoried buildings to have rooftop rainwater-harvesting systems (Vörösmarty et al. 2005).

Many technologies are available for improving access to adequate sanitation facilities, some of which require very little or no water and little capital investment (Gleick 1996). Conventional high-volume flush toilets use as much as 75 liters per person per day, but efficient designs that require less than 10 liters per person per day are now available. Composting toilets and improved pit latrines require no water other than for hand washing and no connection to a central sewer.

Freshwater ecosystem restoration is an active area of research and development (Cooke et al. 2005). Although the term "restoration" seems to imply a return to past conditions, restoration most often involves the conversion of an ecosystem to a more desirable condition that has some elements of the past but is in many respects a new kind of ecosystem (see Chapter 2). Technologies are well-developed for mitigating eutrophication by a range of methods, including changes in hydrology, in-lake chemical interventions, and manipulation of aquatic food webs. Wetland construction, habitat improvements, removal of harmful dams or levees, and foodweb management (through managing fisheries) are also practiced in various ways around the world. Each freshwater restoration project is in some respects unique, and different combinations of methods are appropriate for different circumstances.

Decentralization and Integration of Management

It is increasingly apparent that the complexity and specificity of issues that characterize freshwater systems and watersheds require management approaches that are more collaborative and integrated than in the past. In contrast to conventional resource management that tended to be highly centralized, many regions are now adopting freshwater management structures that are far more decentralized and directly involve users (e.g., Box 9.4). Research on natural resource governance has shown that, when users are genuinely engaged in deciding on the rules that govern their use of the resource, they are far more likely to follow the rules and

monitor others than when an authority simply imposes the rules (Dietz et al. 2003; see Chapter 4). Such self-enforcement and monitoring, through formal and informal local institutions, can be particularly important in poorer countries that do not have the resources to police and enforce regulations.

Users' direct involvement in designing the resource-management rules also encourages greater creativity and sensitivity in the rules, which increases their effectiveness. For example, treaties around shared international waterways often show great sensitivity to local conditions. In the boundary waters agreement between Canada and the USA that allowed for greater hydropower generation in the Niagara region, both states affirmed that protecting the scenic beauty of the falls was their primary obligation. The treaty guarantees a minimum flow over the famous Niagara Falls during the summer daylights hours, when tourism is at its peak. In the Lesotho Highlands Treaty, South Africa helped finance the hydroelectric/water diversion facility in Lesotho, one of the world's poorest countries. South Africa acquired rights to drinking water for Johannesburg, while Lesotho receives the power generated in addition to substantial annual water payments from South Africa (Wolf 1998). Although the management rules designed for a particular place may incorporate elements from other places, it is clear from the great diversity of management contexts that there could never be a one-size-fits-all "management recipe."

Most freshwater systems have clear hierarchical structures: smaller watersheds nested within larger watersheds. Management is often most effective if carried out at multiple levels, via hierarchical or polycentric institutions (McGinnis 1999; see Chapter 4). In this way decision-making processes at different levels partially overlap, conferring some flexibility in scaling management appropriately to the patterns of ecosystems. Polycentric governance systems therefore provide some autonomy at each level but also a degree of consistency across levels. Polycentric institutions cut across scales, involving individuals, local communities, municipalities, and central government in managing ecosystem services from the level of a village to a watershed and thereby mediate between global and local knowledge. Bridging organizations that link across scales, such as South Africa's Water User Associations (Box 9.4), help to balance the roles of local autonomy and cross-scale coordination.

Conflicts in freshwater use are also contributing to the emergence of an alternative model for the relationship between science, management, and society (Poff et al. 2003). This model emphasizes the need for partnerships between scientists and other stakeholders in developing water-management goals. It also emphasizes the need for new experimental approaches that advance understanding at scales relevant to whole-river or whole-lake management. In this model, existing and planned management policies can be seen as opportunities to conduct ecosystem-scale experiments, in accordance with the ideas of adaptive management (Walters and Holling 1990; see Chapter 4).

Summary

Freshwater resources are embedded in terrestrial ecosystems and must be understood as parts of interactive landscapes. Moreover, freshwater resources, including the quantity and quality of the water itself as well as fish and wildlife, interact strongly. Freshwater is a nonsubstitutable resource, in the sense that people have finite needs for drinking, washing, sanitation, and growing food, while both terrestrial and aquatic ecosystem processes depend on adequate supply and quality of freshwater. In a world of expanding human populations and associated demand, along with a changing climate that alters patterns of precipitation and evaporation, there is increasing pressure on freshwaters and the many resources that they support.

Human interactions with water tend to be characterized by sector-specific decisions and unavoidable tradeoffs. The condition of freshwater ecosystems has been compromised by the conventional sectoral approach to water management and, if continued, will constrain progress to enhance human well-being (Vörösmarty et al. 2005). For example, flow

stabilization through dam construction can severely degrade aquatic habitats and lead to losses of economically important fisheries. It is clear that substantial inconsistencies will develop between major development and sustainability strategies, such as the Millennium Development Goals, the Convention on Biodiversity, and the Kyoto Protocol, if they do not become better integrated.

Problems of water, food, health, and poverty are highly interlinked. This is particularly true where freshwater is scarce, and the local economy is too weak to provide the infrastructure for safe domestic water supply or to allow large-scale imports of food. On the other hand, if access to safe drinking water can be secured through appropriate investments in infrastructure, public health conditions improve, the potential for industrial development increases, and the time devoted to collecting water can be spent on more productive work or educational opportunities. Addressing poverty-related problems in very poor nations may require a threshold level of infrastructural investment to enable the economy to develop and grow.

Spatial heterogeneity, multiple turnover times, and thresholds characterize freshwater resources. Because of these complexities, freshwater resources are vulnerable to **regime shifts**, or large-scale changes maintained by new sets of feedbacks. Some regime shifts are harmful, such as eutrophication, salinization, or loss of fish and wildlife stocks. Other regime shifts are desirable, such as the deliberate destabilization of eutrophication during lake restoration. Resilience of freshwaters is inversely related to the amount of change necessary to cause a regime shift. In practice, resilience occurs at multiple scales within a watershed. Regime shifts in a particular subsystem may cascade to other subsystems through movements of water, organisms, or people. Thus management of watersheds involves the breakdown or enhancement of resilience at multiple scales.

So far, success in managing freshwater resources has been mixed. Many methods and technologies exist for managing freshwaters, and even more are under development. Freshwater-management problems are somewhat individualistic; each particular problem seems to require a unique combination of approaches. Moreover, spatial connectivity means that all freshwater resources within a watershed are connected, and these connections must be accounted for. Decentralized management with institutions nested at several spatial scales seems to be necessary for successful management of freshwater resources. Such networks of institutions seem to be capable of adapting as circumstances change, and adaptation is crucially important because of the directional changes in climate, land use, and human demand for freshwater resources. It is not easy to create and integrate such networks of institutions, yet it is being done in many places around the world. Freshwater management demands resilience thinking. Therefore management of freshwaters continues to be an incubator for novel approaches to transform the relationships of people and nature.

Review Questions

1. What ecosystem services are connected to freshwaters, both directly and indirectly?
2. How do changes in terrestrial ecosystems alter freshwater ecosystems?
3. What are thresholds, and why are they important in ecosystem management?
4. How have people responded to conflicts over freshwater? Which responses build resilience of regional social–ecological systems, and which responses erode resilience?
5. What are the implications of a single regional institution, polycentric institutions, and decentralized management for freshwater resources?

Additional Readings

Falkenmark, M., and J. Rockström. 2004. *Balancing Water for Humans and Nature: The New Approach in Ecohydrology*. Earthscan, London.

Finlayson, C.M., R. D'Cruz, N. Aladin, D.R. Barker, G. Beltram, et al. 2005. Inland water systems. Pages 551–583 *in* R. Hassan, R. Scholes, and N. Ash, editors. *Ecosystems and Human Well-Being: Current State and Trends. Millennium Ecosystem Assessment*. Island Press, Washington.

Magnuson, J.J., T.K. Kratz, and B.J. Benson, editors. 2006. *Long-Term Dynamics of Lakes in the Landscape*. Oxford University Press, London.

Naiman, R.J., H. Décamps and M.E. McClain. 2005. *Riparia*. Elsevier Academic Press, Amsterdam.

Postel, S.L., and B. Richter. 2003. *Rivers for Life: Managing Water for People and Nature*. Island Press, Washington.

Vörösmarty, C.J., C. Leveque, C. Revenga, R. Bos, C. Caudill, et al. 2005. Fresh water. Pages 165–207 *in* R. Hassan, R. Scholes, and N. Ash, editors. *Ecosystems and Human Well-Being: Current State and Trends. Millennium Ecosystem Assessment*. Island Press, Washington.

Walters, C.J., and S.J.D. Martell. 2004. *Fisheries Ecology and Management*. Princeton University Press, Princeton.

10
Oceans and Estuaries: Managing the Commons

Carl Walters and Robert Ahrens

Introduction

Most of Earth (71%) is covered by oceans, whose fisheries provide an important protein source for people, both locally and globally. Fisheries also contribute economically to the well-being of coastal communities and other sectors of society. The demand for marine fish continues to increase because of human population growth, migration to coastal zones, increased income and health concerns that strengthen a luxury-seafood market, and growth of fishmeal-based aquaculture, and livestock production. About 20% of the most food-deficient countries export fish to provide foreign exchange and to service their national debt—a further motivation to increase fish harvests (Pauly et al. 2005). Fishing capacity has increased substantially to meet this demand, primarily through development of large industrial-scale fleets that can harvest fish that were once too remote, deep, or dispersed

for efficient commercial harvest. As a result, the annual global fish catch has more than tripled in the last 50 years. Since about 1985, however, global landings appear to have declined, despite continued increase in fishing capacity, suggesting that past harvest levels are not sustainable.

About half (53%) of marine fish and invertebrate harvest comes from coastal (continental shelf) areas that occupy <10% of the global ocean, particularly along steep coastlines that have high upwelling rates (Peru current off South America, Benguela current off Africa). With the exception of upwelling areas, most fisheries production is from high-latitude (subarctic, e.g., Bering Sea and North Atlantic) areas, and from near the mouths of major rivers (like the Mississippi) that provide estuarine rearing-environments for juvenile fish and high nutrient delivery rates to coastal areas. Tropical oceans tend to have much more diverse, complex, and less-productive fish communities than temperate and subarctic oceans. Most tropical fish harvest comes from very large but highly dilute **pelagic** (water column) ecosystems, mainly from tuna fisheries.

Severe overfishing has occurred in most of the major ocean production regions (Christensen et al. 2003, Jackson et al. 2001). A recent analysis of global fishery catch statistics

C. Walters (✉)
Fisheries Centre, University of British Columbia, Vancouver, BC V6T 1Z4, Canada
e-mail: c.walters@fisheries.ubc.ca

F.S. Chapin et al. (eds.), *Principles of Ecosystem Stewardship*,
DOI 10.1007/978-0-387-73033-2_10, © Springer Science+Business Media, LLC 2009

indicate about 25% of global fisheries have **collapsed**, i.e., a 90% or greater reduction in catch from peak historic levels (Mullon et al. 2005). Notable exceptions to stock depletions include the Bering Sea and Gulf of Alaska, where very conservative harvest policies have restricted fishing over most of the history of the fishery. In some regions like the North Atlantic, the overfishing problem has not been solved and is apparently getting worse. In a few other high-production regions, like the Peru upwelling system, early (1960–1980) periods of overfishing have given way to apparently more conservative and sustainable harvest management.

In this chapter we explore the interactions between people and fish, focusing primarily on industrial-scale fisheries that account for most of the global harvest. Chapter 11 focuses on smaller scale community-based fisheries that have tight linkages to the livelihoods of many coastal communities. Quite different controls over social–ecological dynamics characterize the extremes of this continuum of fisheries types.

Social–Ecological Structure of a Fishery

Marine fisheries are unusual as a social–ecological system because one process (harvest) dominates social–ecological interactions. Other processes (e.g., pollution, habitat modification, and diversity effects on ecosystem processes) vary in importance and future unknown changes in climate and perhaps ocean circulation may have important consequences. This provides an unusual opportunity to use a systems framework focused on a single dominant interaction. In this section we briefly describe the unique features of fish and fishermen that are important in understanding their interactions.

Most fish species have a basic life history pattern that exposes them to a complex set of ecological interactions, including human impacts that alter coastal and estuarine systems. Fish typically start out life as very small eggs, fol-lowed by a larval stage in which juveniles are widely dispersed and depend on small marine organisms (lower trophic levels, phytoplankton, and zooplankton) for food. This start at the bottom of the food web generally occurs in localized spawning areas to which older fish have migrated, for example, areas of high production (e.g., estuaries) and/or relatively low predation risk. As juveniles grow, they typically exhibit complex **ontogenetic habitat shifts** (i.e., shifts related to stage of development) to locations where they can utilize larger food organisms and where increased body size makes them better able to both forage efficiently and move rapidly to avoid predation. These areas include reefs, rocky bottom areas, sea grass beds, and salt marshes. An important policy implication of ontogenetic habitat shifts is that **marine protected areas (MPAs)** cannot easily be designed to cover the full life cycle of the fish.

Since ontogenetic habitat shifts generally involve use of shallow coastal waters and/or estuarine areas, many fish species spend at least some critical parts of their life cycle in those marine habitats most vulnerable to human impacts from habitat modification (e.g., dredging and filling of mangrove swamps) and eutrophication. But remarkably, long-term studies of fish (and commercially important invertebrates like shrimp) recruitment patterns, where recruitment is measured as net number of juveniles surviving to harvestable ages, have rarely shown temporal patterns of recruitment decline that are obviously correlated with known historical habitat alterations. We will review one case study (oysters in the Cheasapeake Bay) where an obvious linkage between recruitment and human coastal habitat alteration can be demonstrated, but most such "obvious" linkages have been difficult to demonstrate empirically. There are at least three potential explanations for the general lack of demonstrated linkage between habitat modification and population dynamics: (1) overwhelming impact of fishing as a control over population dynamics, (2) insufficient observation and/or experimentation to demonstrate habitat effects, and/or (3) habitat modification may be less important than some managers have assumed. One management implication is that careful management of one process (fishing) can have enormous influence on fish stocks and their role in sustaining ecosystem services in coastal and marine environments. We return to the issue of management experiments later.

The biophysical structure of most productive coastal habitats is maintained through a dynamic balance of structure-building processes like sedimentation and growth of sedentary organisms that offer structure (mangroves, sea grasses, reefs), and erosive processes associated with water movements (currents, waves). These processes result in slow-variable changes in structural patterns, punctuated by rapid changes during violent physical events like hurricanes. For example, sediment delivery to the coast has declined globally by about 10% because sediment capture by reservoirs exceeds increases in erosion (Agardy et al. 2005). The ecological consequences of this slow change can be quite abrupt (e.g., 1,500 km^2 of Louisiana wetlands lost during hurricane Katrina in 2005).

Localization of spawning and juvenile nursery opportunities further complicates fisheries management, because fish populations tend to be divided into complex metapopulations with many local subpopulations, where natural selection has led to many adaptations to local habitat conditions (e.g., differentiation in spawn timing, movement patterns of juveniles, growth and maturation schedules). Such complex spatial organization inevitably leads to subpopulation differences in vulnerability to human harvest or, alternatively, a compensatory increase in productivity in response to fishing-induced increases in mortality. Fishing typically causes at least some loss of spatial population structure (the most vulnerable and sensitive subpopulations are typically overharvested) even when overall population productivity and harvest appear to be sustainable. For this reason harvest statistics often provide inadequate warning of incipient stock depletion.

There has been much speculation about the possibility of complex ecological dynamics and multiple stable states in marine ecosystems (Walters and Martell 2004). However, the empirical evidence from long time-series (50–100 years) of harvest data most often indicates relatively simple dynamic change, punctuated for some species by dramatic collapses apparently due to overfishing. Data from tropical and pelagic fisheries show no hint of multiple equilibria. But for very productive, relatively low-diversity ecosystems like the Bering Sea and the northern continental shelves of the Atlantic Ocean, there may be at least two stable states, a **pelagic** (water-column-dwelling) **dominance state,** where the fish community is dominated by small **planktivores** (herring, sprat, sardines, etc., that eat plankton), and large **piscivores** (fish that eat fish) are relatively rare, and a **benthic** (bottom) **dominance state** where large benthic piscivores (cod, large flatfish, pollock) are very abundant, and planktivores are reduced through high predation rates. For example, the Bering Sea has been moving into a benthic dominance state, since initiation of industrial fishing, with growing populations of Alaska pollock, cod, and arrowtooth flounders (NRC 1996, 2003). Over the same period, the Scotian shelf off eastern Canada has been moving in the opposite direction, with declines in large piscivores like cod and increases in small pelagic species (Choi et al. 2004, Frank et al. 2005, Zwanenburg et al. 2002).

The world's largest fisheries involve species with highly dispersive and migratory life histories (like tunas, cods, and small pelagics), but most of biological diversity of fish communities is in coastal areas with complex bottom structure (rocky bottoms, coral reefs). Thus there is a big difference between sustaining fisheries production in overall biomass terms, versus protecting biological diversity as a conservation objective. In California, for example, a large network of MPAs (mandated by the California Marine Life Protection Act) offers effective protection to less than 20% of the harvestable biomass, primarily in long-lived benthic species like rockfish that have very low sustainable harvest rates (10% per year or less; Hilborn et al. 2006).

Sociocultural systems and their management institutions are closely linked to variation in patterns of fish production and biodiversity. Where there is high biomass, productivity, and mobile fish stocks (subarctic, temperate, and upwelling systems and the tropical open-

TABLE 10.1. Comparison between small-scale (artisanal) and large-scale (industrial) fisheries in Norway. Information from Alverson et al. (1994).

Fishery characteristic	Large-scale fishery	Small-scale fishery
Number of fishermen employed	2 million	>12 million
Fishermen employed per $1 million invested in fishing vessels	5–30	500–4,000
Capital cost per fisherman	$30,000–300,000	$25–2,500
Annual fuel consumption	14–19 million tons	1–3 million tons
Fish caught per ton of fuel	2–5 tons	10–20 tons
Annual catch for human consumption	29 million tons	24 million tons
Annual catch for industrial products (fishmeal and oil)	22 million tons	almost none
Discarded bycatch	16–40 million tons	almost none

ocean tuna fishery), most fish are harvested by industrial fleets with efficient gear, relatively low employment, and well-developed assessment and regulatory agencies and commissions (Table 10.1). Conversely, where fish populations are less productive and/or less mobile, for example, in diverse tropical ecosystems or in temperate invertebrate (e.g., shellfish and lobsters) fisheries, more localized fishing communities tend to develop (see Chapter 11). In these situations fishermen utilize a diverse collection of simpler technologies, rely more upon local institutional arrangements to prevent overfishing and generally only cause severe depletion in local areas near communities where fishing activities are based. We discuss these socially diverse types of fisheries later in the chapter, with social dimension discussed in Chapter 11. The cases where local institutions fail to prevent overfishing mainly involve industrial fishing operations that have moved into tropical coastal areas through agreements with local governments (often involving corruption and payoffs to local officials) to fish in their **exclusive economic zones** (**EEZs**; Bonfil et al. 1998). This has happened, for example, with European fleets using the coast of Africa.

The major institutions that govern industrial-scale fishing are (1) regulations that constrain fishing pressures to protect stocks and govern allocation of stocks among competing groups of fishermen and (2) subsidies and market forces that influence the behavior of fishermen within this regulatory framework. Each nation regulates fisheries within 200 miles of its coast (its EEZ). International bodies (e.g., the International Whaling Commission) and treaties (e.g., the

Convention on the Conservation of Antarctic Marine Living Resources, CCAMLR) provide similar protection to some stocks in international waters. The effectiveness of international efforts is seriously limited by illegal, unreported, and unregulated (IUU) fishing and by the choice by some nations not to participate in treaties (e.g., Norway with respect to whaling). Fishing pressure can be reduced by restrictions on the number of vessels (e.g., individual transferable fishing quotas (ITQs) that are assigned property rights to a proportion of the annual catch), types of gear, or times and places where fishing is allowed.

Subsidies are a major factor contributing to overfishing. These subsidies are large (globally about 50–100% of the value of the landed catch, Christy 1997) and are intended to reduce economic hardships to fishermen when stocks decline or alternatively to increase national competitive advantages in industrial fishing. About 90% of subsidies go to industrial fishermen, who tend to have greatest political influence (Pauly et al. 2005). The types of subsidies vary among countries, for example, unemployment benefits in Canada, tax exemptions in the USA, and payment of fees to gain access to foreign fishing grounds by the European Union. Many food-deficient developing nations sell fishing rights at low prices to other countries to provide foreign exchange because they lack the fishing capacity to exploit offshore stocks. Other subsidies include cheap fuel and vessel buyback programs to reduce fishing capacity in overcapitalized fisheries. Vessel buyback programs often fail to achieve their objective because fishermen invest in new boats with technologies that enable them to exploit depleted stocks more effectively (Pauly et al. 2005). All types of subsidies enable fishermen to continue fishing under circumstances when declining stocks have reduced profits below levels that would permit them to fish in an unsubsidized fishery.

In order to understand the overfishing issue, it is useful to think of a "fishery" as a complex adaptive system (see Chapter 1) consisting of three main subsystems or components:

1) a collection of targeted organisms (stocks) that may range from a single genetic stock of one species to a complex assemblage of species living in an arbitrarily defined area of institutional management responsibility and pursued by
2) a collection of fishermen, which again may range from a single fleet of similar fishing vessels to a complex assortment of fishing gears operated by a multinational set of fleets, motivated by economic and/or recreational interests and regulated to some degree by
3) a management authority, which is usually a public or international agency/commission with some legal mandate to monitor the stocks and fishing, and to regulate fishing activity (input control approach) or catches (output control approach) with the dual aims of assuring ecological sustainability and reasonable or fair allocation of benefits among competing fishermen. Local informal institutions serve this role in some coastal fisheries.

Such systems are essentially a predator–prey interaction, somewhat moderated in its dynamics by restrictions imposed by regulations, incentives, or other institutional arrangements. These three subsystems are embedded within larger ecosystems that sustain (and sometimes threaten) the target species, larger economic and social systems that generate demand for fish, and larger institutional systems that define and constrain the powers of the regulatory authority and sometimes create perverse subsidies for fishermen as part of social programs aimed at maintaining employment and economic well-being. As target stocks

vary over time in response to natural factors and fishing impacts, fishermen respond through changes in investment in new gear and technology, fishing activity or effort, and spatial choices about where to fish. Regulatory authorities are faced with the complex problems of (1) inferring how much of measured changes in outputs (catches, fisher success rates) and abundance are due to human-induced versus natural changes and (2) varying fishing restrictions over time accordingly. Further, in complex multifleet fisheries, regulatory authorities often become preoccupied with issues of allocation among competing fleets (or recreational versus commercial users) and lose sight of conservation or sustainability objectives.

Regulatory authorities typically exhibit a decision-making pathology known as **indecision as rational choice** when faced with evidence of stock decline (Walters and Martell 2004). In such circumstances, decision makers hear two types of empirical assertions. On one hand, they hear scientific assessments purporting to demonstrate the decline, generally presented in the form of complex assessment reports with arcane terminology and much hedging about problems with inadequate data and population models. On the other hand, they hear simple (and hence compelling) assertions from fishing representatives about how there are still plenty of fish and scientists are often wrong. So faced with a hard choice between causing immediate economic and social harm by imposing more severe regulations (reduced fishing time, lower quotas, etc.), versus gambling that the scientists are wrong and that the decline will reverse itself naturally if no action is taken, it is quite rational for decision makers to choose the gamble and to defer fishery reductions for as long as possible. Such time delays and thresholds in the regulatory subsystem cause dynamic instability in the fishery system as a whole.

In the following examples we describe how fisheries dynamics vary over a range of situations. We begin with relatively simple cases in which the fish–fisherman predator–prey interaction is not substantially impacted by management, yet can lead to persistently productive systems. Then we look at cases in which persistence has not occurred, mainly because of particular ecological spatial dynamics interacting with an inadequate management subsystem. Finally we examine a few cases in which management has been relatively successful despite ecological and fishing dynamics that could easily produce collapse and review a few key characteristics of such success stories.

Why Fisheries Should Not Collapse: Shrimp and Tuna Fisheries

If fisheries are viewed as dynamic predator–prey systems, there are bioeconomic mechanisms that could prevent collapse without intensive management intervention. It is not self-evident that fishermen in search of profits will always drive fish stocks to collapse unless prevented from doing so by effective institutional arrangements. If profitability of fishing decreases as the predator (fleet) population grows, the predator recruitment (new investment) rate may decrease and mortality rate may increase, leading to a rough balance or **bioeconomic equilibrium** even in the absence of effective regulation (Clark 1985, 2007).

There are now hundreds of time series of data showing how fish populations respond to increased fishing pressure. If reproductive and mortality rates showed no compensatory change, increased mortality due to fishing should accelerate declines toward extinction, because fishermen first remove the largest, oldest, most fecund fish. However, most exploited fish populations do not show this simple rapid population decline but instead show strong, compensatory increases in juvenile survival rates (from egg to the age of first capture by fishing). Consequently, the net recruitment of fish to the harvestable population remains nearly independent of spawning population size until the spawning population is greatly reduced. Juvenile survival rate typically increases at least threefold as stock size declines, and many fish (like cod and flatfishes) exhibit increases of 50-fold or more (Myers et al. 1999, Goodwin et al. 2006). Increased juvenile survival is often accompanied by compensatory improvements in body growth rates. Even when fish populations are also affected by environmental changes (like ocean regime shifts), these compensatory responses imply that there can be a rough balance between biological production and fishery removals, if fishing mortality rates decline at low fish population sizes due to loss of economic incentives to keep fishing.

The open-access character of many fisheries (i.e., where fishermen pursue fish and no one claims ownership of the stocks) is often assumed to be the main cause of fishery collapse. In other words, in the absence of effective regulatory restraints, individual fishermen have no incentive to restrain their catches, because any fish so saved can be taken by other fishermen. However, from a bioeconomic perspective, the real issue is whether there are situations where the expected decline in profitability (decreasing income relative to fishing costs) fails to reduce fishing pressure as stocks decline.

There are a few excellent examples of fisheries that have apparently reached bioeconomic equilibrium and remain highly productive without effective regulation of total fishing effort or fleet size. One such example is one of the most valuable fisheries in the USA, the Gulf of Mexico shrimp fishery (Fig. 10.1a). Regulations in this fishery are primarily closed seasons and areas, aimed at protecting shrimp until they reach marketable body sizes. The total number of vessels allowed to pursue shrimp during open seasons and in open areas has never been effectively controlled, except through the bioeconomic process of declining profitability as more vessels enter the fishery and compete for about the same total biological production each year. This apparent bioeconomic equilibrium does not, however, imply that the Gulf of Mexico *ecosystem* has reached an equilibrium. There are continuing, slow declines in several fish (like Atlantic croaker and marine catfishes) that are killed in large numbers by the trawling or discarded as bycatch. Over the long term, disruption of the benthic habitat by trawling or declines in other species could impact the sustainability of the shrimp fishery itself.

The total fishing effort or catches of the major pelagic tuna fisheries have also never

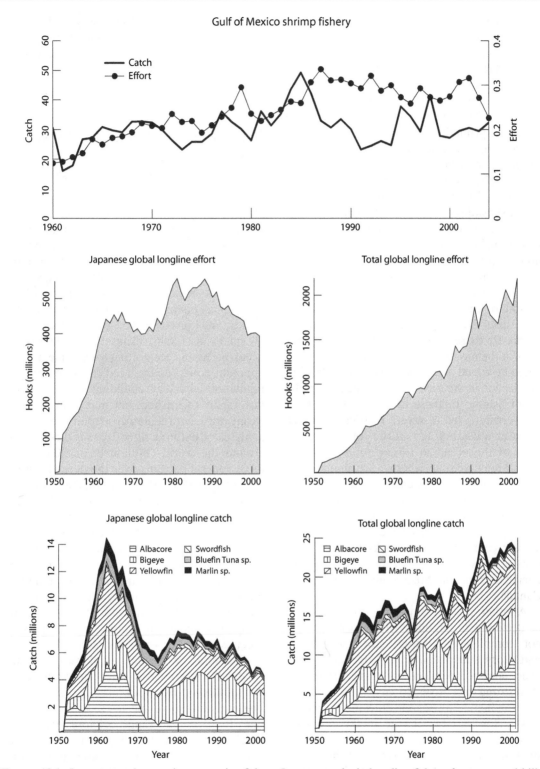

FIGURE 10.1. Long-term changes in two major fisheries. The Gulf of Mexico shrimp fishery has reached a "bionomic equilibrium" where stock sizes and catches have remained at healthy levels despite lack of effective control on the fishing fleet size. The Japanese pelagic longline fishery for tunas and billfish has also reached an apparent bionomic equilibrium, but global longline effort and catch continue to grow apparently due to economic subsidies that a few nations have provided.

been effectively regulated. For these fisheries, at least some national fishing fleets appear to have reached bioeconomic equilibrium (e.g., Japan, Fig. 10.1b), while others continue to grow (Fig. 10.1c) despite slow declines in profitability as measured by catch per unit fishing effort. The continued growth in effort has occurred primarily in fishing fleets of those nations that provide large financial subsidies to maintain fishing (e.g., Taiwan and South Korea), effectively shifting the bionomic equilibrium to lower or even zero fish abundance. Other subsidies seek to prevent economic hardship to fishermen at times when stocks are too low to make a profit also altering the bioeconomic equilibrium.

There are also a few obvious examples of fisheries that have been sustained over long periods of time, most notably the Pacific salmon in British Columbia and Alaska, where the stocks are extremely concentrated at spawning (e.g., at the mouths of salmon rivers) and hence could be wiped out without fishermen "seeing" the decline in the form of decreasing profitability of fishing. In these cases, the blatant risk of overfishing led to severe restriction on fishing over a century ago, in the form of closing most of the ocean to fishing for most of the year. Many salmon fisheries are managed by allowing only brief, intensive fishery openings at river mouths, which each take limited (and well-known) percentages of the stocks. Similar regulatory systems have been developed more recently for other stocks that form vulnerable spawning aggregations, most notably Pacific herring and Atlantic herring off eastern Canada.

Another reason why the "invisible hand" of declining profitability with stock size cannot completely substitute for an effective regulatory subsystem is that declining abundance and competition among fishermen creates strong incentives to invest in technologies for increasing search efficiency. This creates a slow-variable change in the predator–prey parameters and a decline over time in the stock size needed for the average fisherman to catch enough to just meet costs. Technological change tends to cause fisheries to slowly "ratchet down" (decrease stock sizes) over time

(Ludwig et al. 1993, Hennessey and Healey 2000), with episodes of apparent stability interspersed with punctuated declines, as new technologies spread through fishing fleets.

Why Some Fisheries Collapse Anyway: Atlantic Cod and Herring Stocks

Despite the possibility of bioeconomic balance or equilibrium even without regulation, reviews of trends in fishery catch statistics show that a growing percentage of the world's fisheries (roughly 25% as of 2006) have "collapsed" in the sense that catches have declined to 10% or less of peak historical levels (Mullon et al. 2005, Caddy and Surette 2005, Worm et al. 2006). Worm et al. (2006) claim that collapses are occurring at an accelerating rate. Some of these apparent collapses have led to more stringent regulations aimed at rebuilding stocks that were historically overfished, but many have not led to any apparent regulatory response. Statistical analyses of patterns of collapse in 1,519 stocks around the world (Mullon et al. 2005) show three general patterns over the decade prior to the collapse (Fig. 10.2): (1) "plateau" where a period of stable catches is followed by rapid, apparently **depensatory** (accelerating) decline; (2) "smooth" where catch declines more or less steadily; (3) "erratic" where there is at least one strong peak in catches in the decade preceding collapse. Caddy and Surette (2005) argue that these patterns can be largely explained through direct impacts of fishing, without invoking more complex changes in trophic interactions or other aspects of ecosystem structure.

The most worrisome collapses are the "plateau" pattern cases, where apparent stability is followed by rapid, accelerating decline. These cases involve species (e.g., cod and herring) that show a "range collapse" pattern in which the density of fish (fish per unit area) remains stable or even increases as the total stock declines because of decreases in the area occupied by the stock (MacCall 1990). Fishermen are generally able to detect this pat-

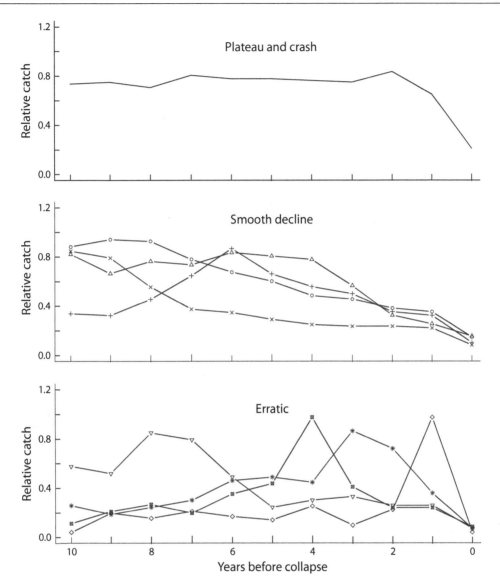

FIGURE 10.2. Selected examples of three patterns of temporal change in catch over the last decade before collapse occurred, for various fish stocks that have collapsed in the last 50 years as classified by Mullon et al. (2005) from analyses of over 1500 catch time series. The most worrisome pattern is the "plateau", where catch statistics (and often other indicators of stock status and trend) show little or no warning of impending rapid collapse. Modified from Mullon et al. (2005).

tern and concentrate their fishing activity in the remaining occupied range area. The net effect of this ability to find the remaining fish is to cause the fishing mortality rate to increase rapidly even if the total fishing effort declines somewhat as the stock collapses. In fisheries terminology, we say that the **catchability coefficient** (fishing mortality rate generated by one unit of fishing effort) increases as stock size decreases; this effect has been recognized as a major management risk for many years for fisheries where control of fishing effort is seen as the main way to prevent overfishing (Paloheimo and Dickie 1964, Harley et al. 2001, Shertzer and Prager 2007). For cod off Newfoundland, the effect was to cause a dramatic

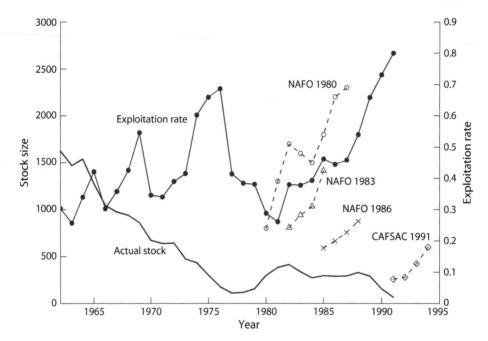

FIGURE 10.3. The recent history of the Newfoundland "northern" cod stock. Beginning in the 1950s, foreign fishing fleets caused rapid stock decline. After Canada assumed management of the fishery in the late 1970s (under extended jurisdiction over coastal waters), stock assessments and projections (trajectories labeled "NAFO" and "CAFSAC") grossly overestimated stock size and recovery rate, leading to development of a Canadian fishery that eventually decimated the stock. Note how annual exploitation rates increased greatly during the last few years of the collapse, as fishing effort remained high while the range area occupied by the stock decreased and fish became more concentrated in remaining fishing areas. Modified from Walters and McGuire (1996).

increase in total fishing mortality as the stock collapse proceeded (Fig. 10.3; Atkinson et al. 1997, Rose et al. 2000, Swain and Sinclair 1994).

Stock assessments often fail to detect range collapse for species like cod because population trends are estimated from trends in relative abundance along with catch and size–age composition statistics, without considering the area occupied by the stock (Walters and Martell 2004). The resulting stock assessments overestimate stock size when a stock begins to decline, which in turn leads to a delay in implementing needed catch and effort reductions. The stock estimates and projections of future population growth can be grossly incorrect (Fig. 10.3), right up to the point where fishermen can no longer take even the allowable catch and the fishery must be closed completely. This occurred in the Newfoundland cod case. Management decisions based on scientific assessments that did not consider distortions in relative abundance trends due to range constriction ultimately collapsed the fishery (Walters and McGuire 1996).

Even when scientific assessment errors have not contributed to the problem, two key decision-making pathologies have characterized rapid collapses of range-collapsing species for which it remains economical to keep fishing until stock size is very low. First, there is a **dynamics of denial**, in which fishermen argue that scientific assessments must be wrong because they are still catching plenty of fish (and if there is a decline, recovery will occur naturally due to favorable environmental circumstances). Political decision makers are not blind to such arguments; after all, fishermen are the ones with firsthand knowledge of what is happening on the fishing grounds. Second, there is a kind of ecological brinksmanship or **inde-**

cision as rational choice (Walters and Martell 2004). Decision makers when faced with certain economic pain if they take decisive action, or a chance of natural recovery (or underestimates of stock size by scientists) if they do not, will make the quite rational choice to gamble on the chance of natural recovery rather than face the certainty of economic harm. No one can blame decision makers for taking such gambles, especially considering how frequently scientists have been wrong in claiming "the sky is falling." A good example of such claims is the highly publicized assertion by Worm et al. (2006) that there may be global fishery collapse within 50 years; such claims have aimed to foster public support for conservation but may end up having just the opposite result by further increasing mistrust in scientific assessments.

The points of the previous paragraph emphasize that decision-making processes in fisheries harvest management are rarely done by, or even dominated by, scientists in the first place. There are rare situations, like the Pacific Halibut Commission and the Pacific Salmon Commission, where independent management authorities have been empowered to set harvest limits supposedly based purely on scientific evidence, but even in such cases there is political influence through the composition of decision-making boards and commissions. In most developed fisheries, decision-making falls ultimately on the shoulders of politicians who hear evidence from other stakeholders besides scientists and are empowered to balance ecological and social/economic concerns. There is gross misunderstanding about this point in the scientific community, with most attention being focused on ecological changes (such as depensatory decreases in productivity at very low stock size) without much thought about why dangerously high harvests were allowed in the first place. This disciplinary myopia is evidenced for example in a review by Longhurst (2006) claiming that the classical "theory of fishing" has failed. In fact that population dynamics theory correctly predicted most patterns of fishery decline, but its predictions have been widely ignored by decision makers who have engaged in the pathology of indecision as rational choice.

This institutional structure of industrialized fisheries contrasts with informal institutions often found in local fisheries (see Chapter 11).

Range collapses and associated stock-assessment errors are not the only causes for fishery collapse, especially in cases of slow collapse where no bioeconomic equilibrium (or at least decline in fishing effort) becomes evident as the stock becomes very low. Many stocks have continued to collapse simply because they have been "caught in the crossfire," i.e., are incidental or more sensitive species taken along with more productive or valuable ones in multispecies fisheries. This has been a problem in both the large pelagic fisheries, where fishing is targeted on tunas but incidentally generates high mortality rates of larger (and less productive) billfishes and dolphins, and the multispecies benthic fisheries where long-lived, slow-growing species like rockfish have been depleted while fisheries have continued to be supported by more productive cod, flatfish, etc. There are also some cases where environmental change and habitat loss have contributed to apparent collapses, for example, cod in the Baltic sea where the habitat periodically becomes unfavorable for them, and Pacific salmon species that depend on stream habitats impacted by cumulative effects of logging and urban development. There are cases like the tuna longline example in the previous section where fishing effort has been driven to artificially high, eventually nonsustainable levels by outright government subsidies aimed at maintaining fisheries employment. Finally, there are a few species that bring such high prices that fishing remains economical even after abundances have been driven very low (e.g., bluefin tuna and abalone).

Slow-Variable Dynamics and Shifting Baselines: Chesapeake Oysters

Changes in critical slow variables can cause degradation of a fishery. One of the saddest stories in renewable resource management is the decline of oysters in the Chesapeake Bay. These valuable creatures were spectacularly abundant when the region was first settled, occurring in large, inshore reefs where oysters grew atop reefs created from the accumulated shells of their ancestors. Early fisheries depleted live oysters from these accessible reefs, and the shell reefs themselves were "harvested" to provide sources of lime and bed material for roads and other construction. By the mid-1800s, the fishery had moved offshore, using dredges from sailboats. From that time onward, the fishery has shown a progressive decline (Fig. 10.4; Kennedy and Breisch 1983, Rothschild et al. 1994), to the point that it is now almost extinct. An obvious slow variable in this system is the availability of hard bottom materials, especially oyster shells themselves, as settlement sites for new recruits. This nonliving "biomass" would have declined somewhat over time with reductions in live oyster populations, even if shell had not been actively mined for other uses.

The need to maintain and restore shell reefs was recognized even in the 1800s, and attempts to regulate the fishery led to colorful "oyster wars" by the 1880s. Yet despite such attempts to regulate the fishery in the face of relatively simple dynamics of the loss of recruitment habitat, there was an apparent lack of learning; commons decision-making was defied (e.g., oyster wars). In the 1900s the decline was punctuated by disease invasions. Recognition of the need to stop dredging and removing shell came early, and efforts to haul shell back into bay that began during the 1920s continue today. Today in the Cheasapeke there is much public investment in large, complex scientific monitoring programs and plans for restoration. The studies have shown, for example, that high chlorophyll (phytoplankton) levels in the Bay are due partly to eutrophication from surrounding cities and farmlands, but also to loss of the massive water-filtering function provided by the oyster

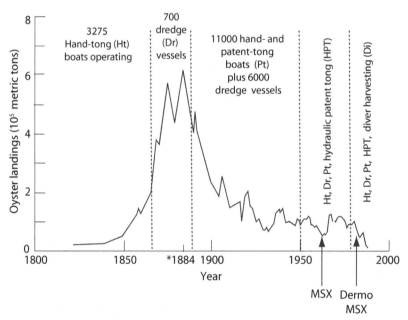

FIGURE 10.4. Development and decline of the Chesapeake Bay oyster fishery. The long, slow decline is most likely a slow-variable effect of loss of oyster shell bottom habitat (where most successful settlement of juvenile oysters occurs), due to mining the shell and loss of "recruitment" to the shell population caused by harvesting of live oysters with- out returning their shells to the beds from which they were harvested. During each successive time period (A to E), new harvest methods and additional harvest pressure were added. MSX arrows refer to times of exotic disease outbreaks. Modified from Rothschild et al. (1994).

beds, which, prior to exploitation, probably fil- tered the entire Bay's water mass at least once per week.

Today, as outside observers, we see a system in which there has been huge ecological impact of oyster harvesting, combined with massive sci- ence efforts that appear to be leading nowhere. How could this have happened? Given the sedentary oyster stock, why was there never an institutional move to local governance with incentives for local communities and fishermen to protect and restore their resources? Did peo- ple expect the problem to correct itself? Did they simply ignore the long-term trend, i.e., was there a **shifting baseline syndrome** as defined by Pauly (1995)?

Coral reef systems have been lost in some coastal areas due to slow-variable increases in nutrient loading and sedimentation from agricultural and urban activities (Brodie et al. 2005, Bellwood et al. 2004) in combination with overfishing (see Chapter 11), which depletes stocks of parrotfishes and other herbivores that would otherwise prevent algae from overgrow- ing the corals (Bellwood et al. 2004, Hughes et al. 2007). Such changes may have con- tributed to development of dynamic instability in trophic interactions. For example, the crown- of-thorns starfish has exhibited coral-destroying outbreaks since the 1950s on Great Barrier Reef of Australia. These outbreaks may have been a natural part of this system for at least 7000 years (Walbran et al. 1989), but this evi- dence has been disputed (Keesing et al. 1992). Models of the outbreak dynamics look much like the fast–slow variable mixed dynamics of forest–insect outbreaks that motivated early ideas about resilience, with slow coral recov- ery leading under some conditions to increas- ing larval survival of the starfish and eventu- ally to outbreak dynamics where the starfish spreads widely across reef systems. These con- ditions for improved larval survival may involve slow changes in nutrient loading. In other cases, warming climate has been a slow-variable change that leads to bleaching of corals (loss of their symbiotic alga), death of corals, and, in extreme cases, loss of reef structure that is crit- ical in supporting the fishery, just as in the case of Chesapeake oysters.

Resilience in Biologically Variable Systems: Pacific Salmon and Herring

Appropriate management policies have made some of the most vulnerable fisheries highly resilient. As noted above, Pacific salmon and herring fisheries have prospered over most of the last century, despite dramatic spatial con- centration and high vulnerability to overfish- ing as fish move into spawning areas. Bla- tant risk of overfishing led as early as the late 1800s to management systems involving clo- sure of large areas to fishing for most of each year, with fishing concentrated in small areas for brief periods. This management approach represented essentially a reversal of the idea of MPAs, by creating small exceptional areas for fishing, instead of closing areas to fishing. Also early in the development of this very restrictive approach to management, agencies began to adopt **feedback policies** for coping with environmental fluctuations, in the form of rules for varying harvests with changes in stock size. The earliest such rules were very sim- ple, so-called **fixed exploitation rate** rules: catch the same fixed percentage of the stock each year, and leave the rest to spawn. Later, **fixed escapement rules** became popular, in which the management goal is to allow the same fixed number of fish to spawn each year. Opti- mization studies (see review in Walters 1986) showed that yield should be maximized by fixed escapement policies, but less variable and socially more acceptable patterns of yield over time should arise from fixed exploitation rate policies.

Resilience is the capacity of a system to cope with unexpected changes in its environ- ment. In resource-management contexts, this means the capacity of management systems to respond effectively to unplanned, usually sur- prising changes in ecological and resource user dynamics (Walker et al. 2004, Gunderson et al. 2006).

A few fisheries management systems have demonstrated high resilience, coping effec- tively (and continuing to support relatively high biological production and healthy dependent communities) on time scales of up to a century despite wildly variable fish stock sizes, periods of unintentional overfishing, unexpected destructive events such as catastrophic die-offs of fish associated with blockage of migration paths, and extreme changes in fishing technologies and allocation among license holders and gears. These highly successful management systems have mainly involved species like Pacific salmon and herring that are highly concentrated in space, relatively easy to monitor (and to overfish), capable of supporting specialized local communities, and so obviously needful of careful management and protection so as to preclude much debate among stakeholders and politicians about the basic need for management.

Most of the cases of high resilience have involved species with complex spatial population structures, with many local stocks or genetic races of fish contributing to overall production. This allows fishermen flexibility to dampen variation by moving among species and areas. There are successful management systems for stocks with simpler ecological structure, e.g., Pacific halibut and Australian shrimp and rock lobster fisheries, but these cases also involve spatially complex fish life histories (migration, dispersal) and fishing patterns and species that have either been relatively easy to monitor (shrimp) or have some natural protection from overharvesting due to biological characteristics (e.g., movement of some fish into deeper waters that do not attract much fishing effort, use of juvenile nursery areas where fishing is either not practical or is banned deliberately to protect the juveniles).

Such cases are at an opposite extreme from cases like the Chesapeake oyster fishery, which has been in decline for over a century and has shown virtually no capability to respond effectively to regulatory needs or to unexpected disturbances like appearance of exotic disease organisms. Fortunately, such extreme management failures are also rare.

Between these extremes are a variety of fisheries that have shown some capability to cope with "routine" natural variability but have failed when faced with persistent changes in productivity. Good examples are the cod and herring fisheries of the north Atlantic, where periods of low recruitment led to stock size declines that were too rapid for available monitoring data and stock-assessment methods to detect and where the need for reductions in harvest occurred more rapidly than was considered socially acceptable. In these cases, low resilience results from a complex stew of rapid ecological change, inadequate monitoring, and incentives for denial of the need for adaptive response.

Social Sources of Resilience in Fisheries Systems

Much of the previous discussion has focused on industrial-scale fisheries. However, there is a continuum of fishery types, including industrial fleets that catch and process fish; small-to-large privately owned commercial vessels that sell their catch to markets or processors; charter boats and sport fishermen; individual subsistence fishermen who consume or share what they catch; and various combinations of these categories. Small-scale fishermen are more likely to use inshore fisheries, where it is easier to monitor the status of stocks, fishing effort, and catch compared to the open ocean where at least parts of many stocks are too distant from fishing villages to be worth pursuing. In many cases, local institutions (both formal and informal) have developed to regulate the fishery for the long-term benefit of local users. As in the industrial fishery, some of these arrangements have been successful and others have failed to sustain stocks (see Chapter 11). Even in the case of failures, however, these are usually local in extent, because of the local nature of the fisheries. There has been considerable interest in these local fisheries, because success often occurs without top-down governmental regulation. There is no simple formula to success of these local fisheries, but comparisons of dozens of case studies show that certain features enhance the likelihood of success (see Chapters 4, 9, and 11; Schlager and Ostrom

1993, Pinkerton and Weinstein 1995). Considering the spectrum from small-scale to industrial fisheries, common denominators that have demonstrated high resilience are observed.

In most cases, a degree of localized management control is present. Such control, from government regulatory representatives or locally evolved institutions, has developed some degree of cooperation between stakeholders to design mutually acceptable monitoring programs and rules for responding to change. Such monitoring programs are capable of rapidly detecting change as it occurs and are carried out in a manner that is easily understood by stakeholders (i.e., do not involve complex scientific surveys and arcane stock-assessment modeling procedures). Management objectives are clear and relatively simple. Focus is on a few key performance measures and management actions are aimed at improving those measures, often with little regard for collateral damage to other ecological factors like bycatch/discard species.

Flexibility in economic opportunity for stakeholders to target different fishing locations, stocks, and jobs without catastrophic loss in income is observed in systems with high resilience. Complex spatial structure in the ecological production system, so that disturbances and irreversible losses in local productivity are not felt immediately throughout the managed system is also a common trait. Finally, when components of the fishery are overfished, there is a willingness to entertain and implement actively adaptive, experimental management policies. Such policies are aimed at rebuilding historically overfished stock components (e.g., Walters et al. 1993) and evaluating impacts of artificial stock-enhancement (hatchery) programs.

Note that this list does not include factors that ecologists have sometimes incorrectly pointed to as possible requirements for successful ecological management, such as high biological productivity (rapid population growth rates after disasters) and predictability or stability of population sizes over time. Further, note that not all fisheries with these characteristics have been managed successfully; for example,

extremely valuable abalone (*Haliotis* spp.) populations satisfy most of the criteria, but have been decimated along most of the Pacific coast of North America and Australia by illegal fishing (poaching).

There are disturbing signs of decreasing resilience in some management systems that were historically successful (e.g., Pacific salmon in western Canada). Indicators of an erosion of resilience in fisheries management include:

1) complication of management objectives, particularly in relation to protection of weak stocks and biodiversity that have resulted in severe, contentious, and economically disruptive regulatory changes (e.g., large area closures);
2) diversion of management agency resources (people, funds) resulting in the deterioration of necessary monitoring-assessment-enforcement (core adaptive management activities) into other activities;
3) adding complexity to stakeholder involvement and decision-making processes aimed at meeting the more complex objectives which appear necessary but do nothing to improve the management system;
4) adopting ecologically trivial or risky projects like habitat restoration and hatchery production when such activities detract form core activities, or no framework is put in place to determine if such activities are a benefit;
5) complication of scientific research and modeling activities aimed at developing more precise predictive models (which have been extremely costly, especially oceanographic research, and almost completely ineffective; Myers 1998).

The basic result of increasing institutional complexity has been to generate much verbiage about planning, but no improvement in essential management activities and in some cases even an apparent loss of recognition of what those essential activities are, i.e., a basic loss of understanding of priorities for effective management.

Actively Adaptive Policies in Fisheries and Coastal Zone Management

Active adaptive management (see Chapter 4) is essential to evaluate the success of ecosystem approaches to fisheries management. The concept of actively adaptive or experimental management of renewable resources had its origins in case studies of fisheries, when we began to encounter management options and questions for which historical data and process-based modeling were incapable of providing definitive answers to managers and when we started to suggest that the most efficient way to resolve many uncertainties might be to try management alternatives within an experimental framework (Walters and Hilborn 1976, Walters 1986). In those early case studies, we examined simple policy questions with correspondingly simple management objectives in mind, for example, "would increases in salmon spawning escapements result in increases in long-term abundance and harvest, and would those increases be great enough to compensate for losses in harvest associated with increasing spawning escapements in the first place?"

Curiously, adaptive management is now widely advocated as an approach to environmental management in general but has only rarely been used in the design of fisheries policies. There have been a few experimental tests of policies involving increased spawning runs of salmon, tests of simple habitat enhancement and restoration measures such as lake fertilization, and assessments of impact of fishing on tropical continental shelf and coral reef fish communities (Sainsbury 1988, Sainsbury et al. 1997, Campbell et al. 2001). There are cases like the California Marine Life Protection Act where an adaptive approach to management of MPAs has been legally mandated (i.e., required by law), but little indication as yet whether such mandates will be respected in practice in terms of (1) sound experimental design principles used to select experimental areas and (2) adequate investment in monitoring programs capable of detecting differences between pro-

tected and nonprotected areas as these differences change over time.

So the simple fact is that actively adaptive management is not widely used, either in fisheries or in coastal restoration management in general (Thom 2000). One excellent review site for coastal restoration studies (http://www.nemw.org/restoration.htm) only mentions adaptive management for two case studies (San Francisco Bay and Florida Everglades), where very little is actually being done in the field. For two other cases where very large-scale restoration tests are in fact being conducted (flow and sediment restoration in the Mississippi River Delta, oyster bed restoration in Chesapeake Bay), the tests are not even labeled as adaptive management experiments.

Infrequent past application of adaptive management to fisheries does not lessen its relevance to current issues. In response to widespread fisheries declines, a wide range of new policies is being implemented, often with unknown consequences. For example, fisheries management agencies are rapidly moving from traditional single-species harvest-management approaches to broader ecosystem management (Corcoran 2006), largely in response to political pressure from conservation interests. The resulting policy initiatives will cause immediate economic hardship for some traditional fisheries stakeholders, but the ecological outcomes are grossly uncertain, and there is almost no historical experience from which to build. These initiatives range from developing networks of MPAs, to requirements for use of more selective fishing gear, to direct control or culling programs for unwanted species, to establishment of new legal arrangements of fishing rights intended to create incentives for fishermen to cooperate in sustainable management. Available ecosystem models warn us that some of these initiatives could backfire badly, particularly those involving protected areas (see, e.g., Walters et al. 1999 or Fig. 11.10 in Walters and Martell 2004) and culling of unwanted species (Yodzis 2001). Even the outcomes of "obvious" management improvements aimed at reducing bycatch of nontarget sensitive species (e.g., bycatch reduction devices for shrimp trawling) are indicated by ecosystem models to be highly uncertain, due to our inability to predict how the fish community as a whole will respond to the changes in mortality rates and competition-predation regimes that they will cause (Walters 2007). It is not that ecosystem management is a bad idea; rather it is important to treat this shift in management paradigm as an experimental opportunity from which managers can learn.

Several authors have reviewed reasons for the lack of widespread implementation of adaptive management (Walters 1997, 2007, Allan and Curtis 2005, Schreiber et al. 2004; see Chapters 6–8). Reasons range from the high cost of monitoring programs implied by complex spatial experimental designs, to institutional barriers created by inappropriate incentive systems for government employees, to lack of individual leadership in coordinating and pushing forward the activities of various groups (scientists, managers, stakeholders) whose expertise and participation is needed in complicated management analyses and implementation. But despite much analysis and discussion, the only point on which experienced people can agree is that development and implementation of an effective adaptive management plan requires a very large personal investment of time and effort by at least one key person in a leadership position, and such key people rarely step forward and accept such an arduous challenge.

Overview of Dynamics and Policy Options for Sustainability

Careful attention to processes controlling supply and demand for fish provide a framework for assessing fisheries sustainability. A very simple graph can be used to summarize the main ideas presented in this chapter (Fig. 10.5). Over time, the dynamics of fishery development create a shrinking domain of stability for sustainable management. As harvest grows, stock sizes and spatial stock structure are inevitably eroded, so the safe catch that can be taken without eroding or endangering future options for sustained harvesting decreases over time. This shrinkage may be exaggerated by slow changes

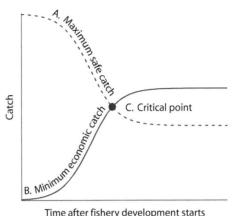

FIGURE 10.5. Loss of resilience and adaptability of fisheries as they develop is associated with convergence of two curves, one representing the maximum harvest that can be safely taken without impairing future production (Curve A) and one representing the minimum harvest that must be taken to prevent economic disruption (Curve B). If the two curves cross (Point C, where minimum economic catch exceeds maximum safe catch), the fishery becomes unsustainable from both ecological and economic perspectives, fishery managers are faced with a pathological situation where they must choose between risk of ecological decline or economic hardship for fishermen, and there may be either accelerating ecological collapse (if the stock shows collapse in its range) or at least partial economic collapse.

in habitat capacities as well. The growth in harvest creates increasing economic commitment, in the sense of an increasing minimum catch that must be taken in order to sustain economic and social functioning of the fishery. That minimum tolerable catch increases particularly quickly in technology-intensive fisheries, where participants must typically finance capital development through borrowing, which in turn creates a minimum catch necessary to meet loan repayment (interest) requirements.

Provided stock sizes are maintained high enough, and harvest level is low enough not to cross point C in Fig. 10.5, there is considerable ecological and economic flexibility (resilience) to carry out management experiments aimed at testing ecological potential, to cope with unpredictable variations in ecological production due to environmental factors, and to invest in the development and maintenance of stable and profitable markets for fish products. But once a fishery has substantially passed point C, so that maximum sustainable catch is substantially less than minimum required economic catch, decision-making dynamics become dominated by concerns about economic collapse, risk taking comes to dominate harvest-regulation choices, and the fishery is bound to collapse unless economic factors prevent continued overharvesting as stock sizes decline (i.e., unless there is a bioeconomic equilibrium at nonzero stock size).

A wide range of strategies have been recommended for managing fisheries so as to avoid point C in Fig. 10.5. These can be roughly divided into two categories, production-side and economic demand-side, corresponding to management of the two curves in Fig. 10.5.

Strategies for production-side management mainly involve **precautionary policies** that have been developed by highly risk-averse fisheries scientists and other ecologists, where stock assessments and harvest regulations are deliberately chosen to be conservative enough to assure that the safe harvest curve remains high over time. A second production-side strategy, which we might call the **engineering policy approach**, attempts to shore up the safe catch curve through engineered habitat management and artificial propagation (hatchery) systems. That second strategy has been used primarily with Pacific salmon and has been widely criticized as a no-win option, where natural production systems are likely to be replaced by expensive and possibly unsustainable artificial production. The replacement effect occurs because of two mechanisms, failure to reduce harvest impact on natural stocks when enhanced stocks continue to attract fishing, and competitive impacts of enhanced stocks on survival and growth rates of naturally produced fish (Hilborn 1992, Hilborn and Eggers 2000).

Recommended strategies for demand-side management mainly revolve around assignment of property rights or other institutional constraints on fishing effort, along with regulatory approaches that discourage overinvestment in fishing capital and create incentives for fishermen to advocate rather than oppose production-side limits on harvesting. Tactically, such systems may be as simple as license limitation programs that cap the number of fishing vessels and gear deployed by each vessel (to limit fishing effort), or individual transferable/vessel fishing quotas (ITQ or IVQs) that limit either fishing activity or total catch per owner (see Chapter 11). Once in place, such rights often lead quickly to demands by fishermen for cooperative science and assessment programs, and even for management experiments, to provide information needed to protect their investments. In some cases where fishermen have taken over the financing of research and assessment (essentially, adaptive management) programs, spending has increased compared to what public agencies were willing to spend. So there is considerable hope for the future of fisheries research and management despite the rather dismal history implied by many of the case studies discussed earlier.

Summary

Overfishing is the primary cause of recent unsustainable changes in marine fisheries. This results from both increased demand for fish and increased capacity to catch fish. About 25% of the world's fish stocks appear to be producing only a fraction (10% or less) of historical catches. Subsidies to industrial fleets drive most of the overfishing that has occurred. These are intended to maintain national capacities to reap profits from international fisheries and to soften the economic hardships to individual fishermen when a fishery collapse occurs. Subsidies prevent the fishery from reaching a bioeconomic equilibrium in which fishing pressure declines in response to declining stocks. With subsidies, fishermen continue to fish even when stocks are depressed to levels at which stock recovery is unlikely to occur. In localized fisheries in which industrial fishing has not developed, the status of the fishery depends largely on changes in slow variables, including pollution, habitat structure, and biological interactions related to biodiversity (e.g., in coral reefs), and local institutions that govern fishing pressure. Fisheries management is moving into a new era of ecosystem management involving multispecies management; establishment of MPAs; regulation of bycatch and habitat disruption (e.g., by trawling); and enhancement of fishery production (through hatchery and habitat enhancement programs). Given the emergent conditions, adaptively managed fisheries programs of experimentation are important to evaluate the currently unknown potential of these programs to enhance fishery sustainability and resilience.

Review Questions

1. Describe the two main ways that spatial stock structure changes when a fish stock is reduced by fishing. When fishermen are capable of following these changes, what happens to their perception of available fish abundance (and to fishery-based indices of abundance used by scientists) as stock size declines?

2. Numerical abundance of most fish is limited to at least some extent by restricted habitat use at juvenile life stages, which in turn is driven by selection for habitat choices and behaviors to reduce predation risk. Such limitations on abundance tend to weaken trophic interaction (predator–prey) linkages. Fishing down top predators tends to result in simple decreases in natural mortality rates of their prey, and fishing on species in the middle of the food web tends not to cause large changes in abundances of species higher in the web. Does this mean that good single-species assessment and management is enough to insure sound ecosystem management?

3. The restricted habitats used by most fish species are in coastal areas, including shallow reefs and estuaries. These areas are highly vulnerable to impacts of human activities ranging from eutrophication to outright habitat destruction through physical modification, freshwater flow modification, etc. What are the key monitoring technologies that can be used to document coastal

changes, and what policy instruments (e.g., zoning) can be used to prevent them?

4. In a controversial review of impacts of fisheries on biodiversity loss and ocean ecosystem services, Worm et al. (2006) correctly assert that many multispecies fisheries involve specific targeting behavior by fishermen, such that fishermen may "switch off" a given species as its abundance declines. They assert further that this switching behavior may allow the species some chance to recover. Explain why this second assertion is wrong in situations where overall fishing activity remains unregulated. Do depleted stocks recover, or are they simply kept down by reverse switching whenever recovery starts to become apparent to fishermen?

5. A fisheries management system (fish + fishermen + management agency) may fail, leading to stock collapse at least for range-collapsing fish species, because scientific assessments are not heeded by the management agency or because the assessments are commonly biased upward by the same factors that make it difficult for fishermen to see that stock size is declining. What does this say about the need for detailed spatial data in fisheries monitoring programs?

6. Consider a developing fishery for some sedentary species that may play some key role in the food web, such as a sea urchin that may have major impact on development of kelp communities and the fishes that use them. Instead of the usual approach of allowing fishermen to pursue them over a broad area under overall harvest regulations and/or individual quotas, develop an alternative management plan based on territorial use rights (TURFs) assigned to individual fishermen, with each right holder fully responsible for monitoring and regulation of harvests within one TURF. What opportunities would this create for organizing the allocation of TURFs by the management agency over time so as to create a large-scale adaptive management experiment to test for undesirable ecosystem impacts before such impacts become widespread?

7. Slow variables that can undermine management systems based on myopic models and

data range from fishery-induced habitat loss (e.g., oyster shell beds) to erosion of spatial stock structure to technological changes that reduce the cost of fishing enough to make it economical to drive stocks to very low levels. Formal monitoring programs (ecological and economic) seldom provide quantitative data over long-enough periods to document such slow-variable changes. Where do scientists need to look for evidence of the impact of such variables, e.g., historical accounts, "traditional knowledge" of older fishermen, etc.?

Additional Readings

Bellwood, D., T.P. Hughes, C. Folke, and M. Nyström. 2004. Confronting the coral reef crisis. *Nature* 429:827–833.

Clark, C.W. 2005. *Mathematical Bioeconomics: The Optimal Management of Renewable Resources.* 2nd Edition. Wiley-Interscience, New York.

Clark, C.W. 2007. *The Worldwide Crisis in Fisheries: Economic Models and Human Behavior.* Cambridge University Press, Cambridge.

Hilborn, R., and C.J. Walters. 1992. *Quantitative Fisheries Stock Assessment and Management.* Chapman-Hall, New York.

Jackson, J.B.C., M.X. Kirby, W.H. Berger, K.A. Bjorndal, L.V. Botsford, et al. 2001. Historical overfishing and the recent collapse of coastal ecosystems. *Science* 293:629–638.

Pinkerton, E.W., and M. Weinstein. 1995. *Fisheries that Work: Sustainability Through Community-Based Management.* The David Suzuki Foundation, Vancouver.

Walters, C.J. 1986. *Adaptive Management of Renewable Resources.* McGraw-Hill, New York.

Walters, C.J., and S.J.D. Martell. 2004. *Fisheries Ecology and Management.* Princeton University Press, Princeton.

11
Coastal Marine Systems: Conserving Fish and Sustaining Community Livelihoods with Co-management

Evelyn Pinkerton

Introduction: What Is a Coastal Marine System and Why Is It Important?

Community contributions to fisheries management are important to the sustainability of coastal ecosystems, including the people who most depend on fisheries. A majority of the world's population lives along coastlines. People are an integral component of coastal marine systems because of what they do to either conserve or damage marine resources. Just as we think of ecological systems as having structure, function, and patterns of interrelationship and change, so too are human populations usually organized into communities that have institutions (laws, rules, customary practices) and measurable effects on coastal marine systems. Human communities and the coastal marine zone that they most strongly impact are thus logically viewed as an integrated social–ecological system (see Chapter 1).

Anthropologists have documented the impact of small-scale preindustrial societies on natural systems in a subfield called **cultural ecology** or **human ecology**. They think of human communities as having adapted to the natural resources in their immediate environment in a manner that has produced sustainable rates of human use. For example, studies of hunter-gatherers concluded that these small, mobile groups tended to limit their populations to about 60% of the carrying capacity of their adjacent local resources (Lee and DeVore 1968). In other words, given fluctuations in resource abundance, human societies adopted patterns of use that were conservative enough to allow them to survive during times of scarcity.

But human communities did far more than passively adapt to what was available to them. They also significantly altered the landscape over time. For example, they burned patches of grassland and forests to encourage vegetation for their own use and to attract animals for hunting purposes (Lewis 1989; see Chapter 6). In the coastal zone, they burned wetlands, contributing to the raising of land levels and thus reduced flooding, an important function when sea levels were rising (Kirwan 2008). They systematically fished certain species at

E. Pinkerton (✉)
School of Resource and Environmental Management, Simon Fraser University, Burnaby, BC V5A 1S6, Canada
e-mail: epinkert@sfu.ca

F.S. Chapin et al. (eds.), *Principles of Ecosystem Stewardship*,
DOI 10.1007/978-0-387-73033-2_11, © Springer Science+Business Media, LLC 2009

certain times and places in a manner that clearly affected marine populations and food webs. So **it is more appropriate to think of a coastal marine system as a case of mutual adaptation and social–ecological coevolution**. We have much to learn from the ways in which traditional human communities coexisted with and actively managed adjacent marine resources. There are few coastal areas without a long history of human use in which these types of interactions occurred, so it is inappropriate to think of "pristine coastal ecosystems, untouched by humans," because these were rare indeed.

Instead, what should impress us is the enormous diversity of species and sustainable patterns of livelihood that developed in coastal zones before the industrial revolution in now-developed countries and the colonization of developing countries (Swezey and Heizer 1977, Johannes 1978, 1981, McCay and Acheson 1987, Berkes 1988, Cordell 1989, Anderson 1996, Walter et al. 2000). The industrial revolution and colonization introduced outsiders and more intensive fishing methods by groups that did not always live adjacent to the marine resources they fished. The introduction of industrial processing capacity, alongside the stripping away of local rights to protect local resources and livelihoods, led in many cases to the overfishing and collapse of at least some local stocks. Yet in other cases, local systems were resilient enough to survive, raising fascinating questions. Maritime anthropology has focused on what made the preindustrial livelihoods sustainable and what happens when these old fishing communities become part of the modern world – the industrial economy. What changes and what remains the same? How do they evolve? How do they survive intense external pressures that threaten to destroy local resources? What kind of histories can these local coastal systems tell? What can they teach us about principles for sustainable use? What is the role and importance of resource dependence, place orientation, ownership, and resilience to resource stewardship?

Increasingly, interdisciplinary scholars have become interested in the survival of these traditional mechanisms in contemporary societies because they represent a reservoir of institutional strategies for solving certain problems (Schlager and Ostrom 1993, Wilson et al. 1994). To support the survival of small-scale traditional institutions, there is an interest in **co-management** arrangements that provide for power sharing between local communities and agencies in resource management (Pinkerton 1989a, 2003), and **adaptive co-management** as arrangements for building the capacity of contemporary communities to design institutions capable of working at multiple levels, learning from experience, and responding to change (Berkes et al.1989, Armitage et al. 2007, Plummer and Fennell 2007) (see Chapter 4). This chapter examines the conditions that facilitate the success of these arrangements in the context of coastal marine fishing communities.

Problems of Privatizing Coastal Marine Systems

The rise of **neoliberalism** (the state acting to advance capitalist markets) and neoclassical economic paradigms and models for managing resources (focusing on the benefits sought by individuals and the market as the best regulatory mechanism) has created problems for sustainably managing many coastal marine systems. In such paradigms, the allocation of private rights to individuals and the free transferability of such rights via the market are considered to be the ideal. Such systems are assumed to deliver efficiency and even to create incentives for conservation of the resource (NRC 1999). Starting with New Zealand and Iceland in the 1970s, many countries have experimented with privatization of access and withdrawal rights to marine resources through **individual transferable quotas (ITQs)**. Quota-based systems were usually designed to solve problems in state regulation, but were implemented without recognizing the possibility of community-based regulation. In many cases this approach led to the transfer of fishing rights away from a diverse array of coastal communities (or even out of a country), separating place from access, and removing the possibility of capturing the benefits of co-management or adaptive management as described above and below. Above all, ITQ systems can represent a loss in flexibility and adaptability in the management system, the hallmarks of resilient systems (deYoung et al. 1999). Flexibility and adaptability are compromised with these policies because property rights to capture a specific amount or percentage of fish leads to ever more costly investments in quotas, even while it lowers the cost of fishing through concentration of fewer fishermen on fewer vessels. Especially problematic for the flexible management of ecosystems is the fact that quota rights are usually based on ownership of single-species licences and do not create incentives to attend to habitat linkages or species interactions. Privatization has also created problems with ecological sustainability and equity; some analysts find that quotas do not score well on efficiency – usually their most prominent claim (Tietenberg 2002, Pinkerton and Edwards 2009). The conservative Organization for Economic Cooperation and Development in a study of ITQs worldwide found many cases of stock failures, higher enforcement costs or problems, underreporting of catch, and data degradation (OECD 1997). Agrawal (2002) and Baland and Platteau (1996) also found that community institutions were not necessarily any less efficient than privatization. However, ITQs can be appealing to government agencies because they solve an important problem for fisheries management agencies: in times of declining agency funding, they allow agencies to use quotas-holders to pay for the cost of stock assessment and monitoring of catch (such as 24-h surveillance cameras on board vessels, and dockside monitoring at delivery sites on land). The first generation of quota holders can afford to pay these and other management costs, because they have been gifted with a valuable public good that they are then allowed to sell at what the market will bear. The results have been mixed, in many cases resulting in overfishing and struggles over the public's (and even government's) right to the catch data and research of quota holders.

In cases where quotas are unavoidable because they are the accepted approach, communities and cooperatives have sometimes adapted the model by creating **community quotas** (Langdon 1999, 2008). In such cases, rights are usually transferable among individual members of a specified organization, but not at speculative prices and not out of the organization. Hence they are neither fully individual rights nor are they fully transferable via the market. In less-formalized arrangements, fishermen may develop an arrangement to share their quota with other community members in times of resource abundance (Loucks 2005).

The reliance on a privatization model such as quotas to solve problems highlights the more fundamental issue of overreliance on single-species management and a single disciplinary approach. Many resource-management situations have shown the need for multidisciplinary teams. Ever since Peter Larkin's famous dictum **that "fisheries management is about managing people, not fish,"** there has been recognition of the need to integrate social sciences into problem solving. At least half the problems are caused by difficulties in managing what people do, and the failure to understand how human societies function culturally, sociologically, politically, psychologically (Wilson 1998). Unfortunately, current rates of social–ecological change confound efforts to understand the nature of the problem. Still, it is collaborative and multiscaled efforts, such as adaptive co-management that offer the best opportunities for meeting the challenges of sustainability.

International Recognition of Local Fishing Communities

There has been a growing recognition of the importance of coastal marine systems since they were acknowledged internationally by the **1992 United Nations Conference on the Environment and Development**, whose initiative on coastal communities was followed by the 1994 **UN convention on the Law of the Sea**, and the **1995 UN Food and Agriculture Organization's Code of Conduct on Responsible Fisheries**.

Chapter 17 of the Rio Declaration from the 1992 UN Conference on Environment and Development (paragraphs 17.79, 17.81, and 17.93) stated that governments should "take into account traditional knowledge and interests of local communities, **artisanal fisheries** [i.e., small-scale mixed commercial and subsistence fisheries] and indigenous people in development and management programs." The Code of Conduct went on to state that resource managers should "implement strategies for sustainable use of marine living resources, taking into account the special needs and interests of small-scale artisanal fisheries, local communities, and indigenous people to meet nutritional and other development needs." Finally, it noted that governments should "recognize the rights of small-scale fish workers and the special situation of indigenous people and local communities, including their rights to utilization and protection of their habitats on a sustainable basis." The 1995 FAO Fisheries Code of Conduct on Sustainable Fishing insisted on "transparency in data collection and in decision-making practices, to ensure that independent analysis of data and assumptions is possible by credible third parties, and that the public owners of the resource have access to basic information." The document also emphasized the importance of measures that prevent excess fishing capacity and ensure that the exploitation of the stocks remains economically viable and that the biodiversity of aquatic habitats and ecosystems is conserved (FAO 1995).

These international agreements and codes illustrate the growing recognition that the sustainability of coastal communities is critical to the resilience of the costal marine social–ecological system as a whole. Through the recognition of the rights of fishermen, the legitimacy of their local knowledge, and their meaningful role in the data collection, research, and decision making, resource governance is more likely to be responsive to change and provide for the livelihoods of those who most depend on the marine coastal system.

These international agreements and codes do not, however, explain the conditions that are necessary for this vision of community participation to actually occur. This chapter looks at the key factors that allow the vision to become a reality, by asking the following questions: Under what conditions will coastal communities play a positive role in fisheries and

coastal management? Under what conditions might we expect communities to play a negative role, damaging these same systems? How can community involvement in coastal systems governance address the increasing pressures from nonsustainable fishing practices by large nonlocal industrial fleets and from many forms of development and pollution in the coastal zone such as oil and gas development, tourism, and unplanned urban sprawl? How can involvement of communities contribute to the flexibility and responsiveness of coastal marine systems in conditions of directional change?

The pages that follow explore a variety of situations in different parts of the world that illustrate both the conditions allowing sustainable outcomes in coastal marine systems and the specific benefits that are associated with this. In so doing, critical challenges to sustaining community livelihoods of coastal systems will be explored.

General Conditions Under Which Coastal Communities Conserve Marine Resources

If international bodies believe that coastal communities can make such a difference in conserving coastal marine resources, we need to know more about *when* we can expect this to occur and when we should be concerned that coastal communities might plunder these resources. It is useful to consider the impact of five conditions discussed below that social scientists believe are good predictors of conservation by coastal communities. Because resource governance emerges in each situation through its own path-dependent set of conditions (see Chapters 1 and 4) and because all situations are complex, no specific set of conditions is necessary or sufficient to achieving sustainability. Yet identification of key conditions under which coastal communities make a positive contribution to fisheries conservation can help groups at all scales to identify the possible actions needed to sustain coastal marine systems.

Condition 1: Communities Have Strong Access Rights to Local Marine Resources

Think of it this way: If you are relatively poor and have a historical dependence on your local marine resources, but you now have no right to fish, what are you likely to do? Will you poach as a way of taking fish away from the outsiders who do have the rights? Will you decide to develop the coastal zone for other purposes that provide economic opportunities for you? Will you support oil and gas development, intensive finfish aquaculture, wind farming and ferry terminals in tidal flats, cruise ship dumping, supertanker traffic, and many other industrial uses for the coastal zone that can negatively impact fish habitat, because at least these will provide some economic activity in your community? **There must be enough local livelihoods in fishing that are perceived as sufficiently important by locals to counteract possible temptations to develop coastal zones for other purposes.** Since fishing is often perceived as an occupation of last resort, or as poverty alleviation in developing countries, a local constituency willing to stand up and speak for protection of the fish is often required to prevent development with no regard for fish habitat. If local communities have no economic incentive to fight for these values, there may be little standing in the way of fish habitat destruction in the estuaries, bays, and river mouths that are often critical nursery areas for juvenile fish. Fishing rights holders in distant urban centers usually have little knowledge of the importance of such areas and do not depend on any one area for their livelihood, so they are not likely to be a strong voice for protection.

Similarly, it is the **rural fishermen catching fish at a smaller scale than industrial fleets who are most likely to notice changes in stocks related to fishing**, as is illustrated in the following example. The Catholic priest Thomas Kocherry in India organized two successful nation-wide strikes of fish sellers and the small fishermen who were their spouses, because the poorest and smallest fishermen observed that the species that were once abundant in their catches were disappearing due to the growth of an offshore trawl fleet. As a result, India banned foreign trawlers from its territorial waters (Thalenberg 1998). In this case, the small fishermen had no formal legal rights to access fish that could stand up against the large industrial fleets. However, through the organized action of Kocherry, they were able to exercise de facto rights to catch their share of the fish and to prevent outsiders from taking it first. Without the action of these small fishermen, it is likely that the trawl fleet would have overfished or extirpated many species before it was evident that their actions had this effect and before the Indian central government would have been willing to take any action. If the small fishermen in the coastal zone had had no fishing access rights at all, there would have been no nationwide strikes to ban foreign trawlers. **Holding rights to fish is thus the first and most fundamental condition that must exist for coastal communities to play a role in conservation.**

Condition 2: Communities Have Strong Rights to Participate in Management Decisions

It takes far more than informal access rights, exercised through strikes and protests, to produce conservation, however. To really make a difference, communities must have a bundle of rights to participate in the management system at multiple levels, from data collection to policy development (Pinkerton 2003. Table 11.1 illustrates a hierarchy of rights, some far more powerful than others, that would allow a coastal community to participate meaningfully in management and assert its vision of conservation. On a continuum from weaker and smaller-scope rights to stronger and larger-scope rights, these involve rights to data collection, data analysis, planning the timing and location of the fishery, fishing methods, allocation of catch, exclusion of outsiders or nonrights holders, protection of habitat, monitoring of harvest and habitat, creation of the rules by which all these decisions are made, enforcement of the rules, coordination of harvests and other competing uses, returning optimum value

TABLE 11.1. Hierarchy of rights in fisheries-management decision making, based on Pinkerton and Weinstein (1995).

Type of management right	Specific right
Lower order rights	Data collection
	Data analysis
Higher order rights	Plan timing and location of fishery
	Rule-making re fishing methods
	Allocation of fishing opportunity among rights holders
	Enforcement of fishing rules
	Defining who has fishing rights
Broader rights affecting other actors and users of marine space	Rule-making re fish habitat protection
	Enforcement of habitat-protection rules
	Coordination of fishing and other competing uses of marine space
	Returning optimum value to fishermen
Highest level rights	Fisheries policy development
	Identification of key problems, issues
	Creating a vision of what fishery is desired, goals of management

to fishermen, policy development, identification of key problems, issues, and goals of management, creating a vision of what kind of fishery is desired.

Two of the oldest examples of coastal communities developing this range of fishing rights are discussed below: the US western Washington State treaty tribes and the coastal municipalities of the Philippines. Both of these systems evolved from long-term precolonial de facto management by local communities and became de jure systems when laws or court decisions clarified the nature of the rights in the 1970s. Both systems continued to evolve, clarifying and extending the rights of coastal communities over the next four decades, as summarized in Table 11.2 and briefly discussed below.

The 1974 "Boldt Decision" (*US* v *Washington*) was a US Federal Supreme Court decision (based on the court's interpretation of earlier treaties) that recognized the right of Indian tribes in Western Washington to up to a 50% share of the salmon that passed their customary fishing sites, mostly located on rivers, bays, and estuaries. Tribes had been unable to *exercise* rights to a share of fish recognized in earlier court decisions, because the harvest was managed by the state government in such a way that few salmon remained by the time this migratory species reached the marine and riverine

territories in which the tribes could legally fish. The state managed the nontribal commercial and sport fisheries so that all but approximately 5% of the salmon were harvested elsewhere. Judge Boldt reasoned that only by recognizing the tribes' right to *participate in planning and regulating the entire harvest* (which he called "concurrent management") would their allocation right ever be exercised. In other words, the treaty's promise that the tribes would share in access to the fish alongside the citizens of the territory (Condition 1) was unrealizable without a second and higher-level right being granted: the right to participate in management decisions about how the harvest would be conducted. The treaty tribes invented the term "co-management" to describe this situation and gradually over the next two decades began to assert the full range of rights so that they shared management authority with the state of Washington in all aspects of management. Thus we can note that, in the hierarchy of rights, the higher-level more powerful rights (such as harvest management) do not automatically accrue simply because a lower right (such as access to an allocation of fish) exists.

The Washington Department of Fisheries (WDF) originally resisted the exercise of the tribes' right to co-manage, beginning with tribal demands for access to WDF's stock abundance

TABLE 11.2. Fishery management rights held by Washington tribes and Philippines municipalities and Fishery and Aquatic Resource Management Councils (FARMCs).

Rights	Washington tribes	Philippines municipalities and FARMCs
Data collection on fish stock abundance	de jure	No
Data collection on fishermen's gear, vessels	de jure	de jure for registration of gears and vessels by municipality (not fish)
Data analysis	de jure	No
Plan timing and location of fishery	de jure	de jure, local fisheries ordinances, approved by state
Rulemaking re fishing methods	de jure	de jure, local fisheries ordinances, approved by state
Allocation of fishing opportunity among rights holders	de jure	de jure only in that local fishing cooperatives have preferential access and marginal fishermen's organizations are exempt from fees
Monitoring and enforcement of fishing rules	de jure	de jure, shared with state, municipal-based interagency law enforcement teams
Rule-making re fish habitat protection	de jure	de jure, shared, municipal involvement in designation and management of marine protected areas
Enforcement of habitat protection rules	State	de jure, by state and/or FARMC in their own protected areas
Coordination of fishing and other competing uses of aquatic space	Shared with state; de facto participation in Regional Management Councils	de jure, assist in the preparation of development plans
Returning optimum value to fishermen	No	No
Fisheries policy development	de facto	de jure
Identification of key problems, issues	de facto	de jure, identification of problems in particular areas and strategies to address them
Creating a vision of what fishery is desired	de facto	de jure, mandated to FARMCs as method of rural development
Creation of financial capacity to manage	de jure (federal funding)	de jure, municipalities levy taxes for management and partner with NGOs and international development agencies to deliver basic services, build capacity, develop livelihood projects; FARMCs depend on municipalities for budgets and divide fishing violation fines with them

data, catch data, and participation in the very definition of conservation. In this case, defining "conservation" meant deciding how much salmon should not be fished, but rather allowed to spawn in each stream in order to reproduce the next generation. (The tribes tended to advocate lowering the catch and allowing more salmon to spawn.) The WDF did not want to reveal the paucity of its stock abundance data and the level of uncertainty surrounding its analysis and decision-making about the allowable catch. Furthermore, it did not trust the tribes more than any other fishermen to report their catch accurately, especially since some tribes had asserted their treaty rights (see Cohen 1986) through illegal fishing for decades. The final negotiation of a formal co-management agreement (see Chapter 4) that resolved these problems, among others, took 10 years and resulted in a complex power-sharing relationship in which the state and tribes agreed to work jointly on every aspect of data gathering, data analysis, and harvest planning, and eventually played complementary and mutually supportive roles. The tribal–state relationship was less a delegation or decentralization of powers than a complex division of powers and a collegial collaboration in problem-solving as presumed coequals. The two parties eventually developed a high level of trust and learned to make the best use of limited funding by sharing (and sometime agreeing to tradeoff specializations) in virtually every aspect of management and every stage of planning, from international negotiations to collecting data on indicator streams. The sharing and improvement of data gathering and data analysis through mutual accountability provided the foundation on which trust was built in harvest planning (Pinkerton 2003).

The court had not envisaged all the complexities of co-management; so many of the other rights were recognized only through negotiations after intense political struggles, or continued use of the courts and its extensions. Through the repeated exercise of their harvest management right, the tribes gradually learned what other rights were necessary for making this core co-management function operable. As soon as harvest co-management protocols were agreed to in 1984, a new set of issues around co-management emerged. These included habitat protection, regional planning, setting broader policies at a higher level, and international allocation (interception) agreements. The co-management system set the stage for a complex multistakeholder exercise in watershed analysis and eventually for the most challenging exercise in complex collaboration ever attempted, involving federal agencies regulating endangered species protection and water quality under federal statutes. **Thus harvest regulation emerged as only a small part of what would eventually be involved in co-management** (Pinkerton 1992, 2003, Ebbin 1998, Singleton 1998).

A similar evolution of rights process occurred in the Philippines, which also had had a system of village jurisdiction over coastal resources before Spanish colonization in the seventeenth century imposed state control of fishing and gradually eroded traditional village rights (Pomeroy 2003). The decreasing catch by small-scale fishermen in the 1970s led the Philippines government to increase fishing effort by equipping them with vessels and gear to venture further from shore, a strategy that ultimately resulted in lowered catches. Beginning in 1991 with the Local Government Code, which devolved key governance functions from the national to municipal level (covering areas 15 km from shore), the national government began to actively promote devolution of management authority as part of a development and poverty-reduction initiative. Then in 1995, the Fishery and Aquatic Resource Management Councils (FARMCs) were set up and eventually brought under national legislation as government consultative bodies to allow greater participation of fishermen's organizations in management. These advisory and monitoring bodies work mainly at and across three levels of governance (municipal, regional, and national) and assist in the formulation and implementation of national fisheries policy. Advisory and monitoring bodies are considered public–private partnerships between government, nongovernmental organizations (NGOs), and private industry. In this case "private industry" consist of commercial fishing ventures (6% of fishermen), processors, and aquaculture (26% of fishermen), while NGOs consist of organizations of fishermen participating in municipal fisheries (65%). Since 80% of the dietary protein in rural areas comes from fish, food security is even more important than livelihoods in municipal fisheries. The participation by these groups in **civil society** (engagement with informal and formal institutions) was deemed to be a critical element of poverty alleviation, a key goal of the legislation, and hence, fishermen and fish workers make up one third to two thirds of all FARMCs. These bridging organizations are most active in law enforcement and pollution control (Benjamin et al. 2003).

Initial and subsequent devolution of decision making to communities through the implementation of FARMCs was further supported by other aspects of the process that (1) simplified procedures for national approval of local fisheries ordinances that had to comply with federal legislation and policy; (2) strengthened enforcement of fisheries laws through municipal-based interagency law-enforcement teams; (3) allowed the identification of specific problems in particular areas and the development of strategies to address them; (4) mandated the protection of coastal fish habitat; (5) permitted consolidation and coordination of the services and resources of local governing units for common purposes; (6) allowed municipal governments to levy taxes and fees for resource management that had once been allocated to the federal government; (7) allowed municipalities to partner with NGOs and international development agencies to deliver basic services, build capacity, or develop livelihood projects; (8) mandated the rights of local fishing cooperatives to have preferential access and, in the case of marginal fishermen's organizations, exemption from fees; (9) mandated that the fisheries management councils at national and municipal levels would assist in the preparation of development plans; and (10) mandated municipal involvement in the designation and management of marine protected areas (Pomeroy 2003, Pomeroy and Viswanathan 2003).

In practice, the FARMCs and municipalities have been actively involved so far mostly in the creation of protected areas, the protection of habitat, and in enforcement of regulations (Benjamin et al. 2003). One island municipality in the Philippines has used the protected area strategy effectively to rehabilitate coral reefs and ban use of destructive fishing technology (cyanide, explosives, fine mesh nets) introduced by new arrivals to this area in the 1970s. By 1988, living coral cover had declined to an average 23% and catches had also drastically declined. With the help of a local NGO, in 1989 the municipality created an inner no-take sanctuary and an outer reserve allowing only nondestructive technology, then proceeded to cooperatively create ordinances

for fishing, provide a boat and equipment to patrol coastal waters, and monitor fishing activities closely. The local daily average catch per fisherman went from 3 kg in 1988 to 10 kg in 1998, while living coral reef cover increased from 23 to 57%, accompanied by increased fish species diversity (Pomeroy and Viswanathan 2003). This example illustrates a frequent finding that when local authorities have a central role in the design of marine protected areas (MPAs), and the regulations are well-respected and well-enforced, local fishermen directly benefit from increased fish abundance and strongly support the MPA. In this case, the design of the no-take and fishing zones, the banning of introduced technology that local fishermen understood to be destructive, and effective monitoring and enforcement were key to the success of the MPA.

Another case demonstrates that flexibility through adaptive co-management can also contribute to the success of MPAs. In that case, Cinner et al. (2005) found that locally-made regulations for a MPA invoked high rates of compliance because there were adaptive periodic openings and closures practiced explicitly to meet social goals, including providing food resources at a time of social significance, rather than to fulfill nonlocal notions of conservation or resource management. The authors concluded that "The reality in many developing countries is that fully closed reserves often suffer from low levels of enforcement, monitoring, and compliance and are thus, in many cases, ineffective at either conserving marine resources or enhancing fisheries. In contrast, we found that adaptive periodic closures increased fish biomass and average fish size . . . . When the proper socioeconomic factors are found, and large, permanently-closed reserves are unrealistic or have repeatedly failed, adaptive periodic closures may be a more viable conservation strategy" (Cinner et al. 2005).

But sometimes the forces of habitat degradation, global warming, and overfishing threaten to overwhelm an MPA's efforts to the extent that adaptation at another scale is called for. Australia offers an example of a transformation in MPA governance through reconceptualizing the need for greater and more interconnected no-take areas *after* a marine park was long established. The Great Barrier Reef Marine Park (GBRMP), the largest coral reef system in the world, went from 5 to 33% no-take areas in response to a drastic decline in species and coral health, combined with greater understanding of the importance of connectivity of larvae and interactions between reef and nonreef habitats for maintaining the resilience of the entire ecosystem (see Chapter 5). The focus of the rezoning was on protecting biodiversity and maintaining ecosystem function and services, rather than on maximizing the yield of commercially important fisheries. Key to the rezoning was internal reorganization of the reef-management agency to build support and understanding for the MPA, to encourage agency scientists to think beyond their individual sample sites or specialized expertise, and for all groups to collectively reach a bioregional perspective on the GBR as a whole. Also key was extensive public education, two rounds of extensive public consultation, innovative interactive forms of information exchange, and strategic building of public and legislative support. As a protected area the size of California, this site in Australia set a new standard in mobilizing public support at a scale commensurate with the scale at which the agency believed the park ecosystem should be viewed (Olsson et al. 2008).

Table 11.2 illustrates the similarities in the bundles of management rights held by the co-managing coastal communities in Washington State and the Philippines and also suggests some of the differences in the situations of developed and developing countries in their efforts to enact co-management. The role of government in providing technical support, credit, marketing assistance, or protective legislation is especially important to co-management in developing countries such as the Philippines (Chevalier and Buckles 1999, Pomeroy and Berkes 1997), although this is often the case in developed countries as well. Economic development and the protection of the large percentage of poor or marginal fishermen is a more central concern of governments in developing countries. It is worth noting, however, that external or nonfishing NGOs such as universities, international bodies, or advocacy groups may be the chief sources of legitimation for the co-management relationship in developing countries such as the Dominican Republic (Stoffle et al. 1994, Chevalier and Buckles 1999), not necessarily government, even though government must eventually provide protection for co-management to work. But dependence on legitimation by external bodies such as NGOs can be unhelpful or even a disadvantage if a developed country NGO does not have adequate experience with the community in a developing country it is attempting to assist, and if there is no internally generated effort and strong desire to develop co-management. If external NGOs offer funding to communities to experiment with and adopt co-management processes, **goal displacement** may occur (see Chapter 4). A community may discontinue all efforts toward co-management after the external funding is exhausted (Hara and Nielsen 2003). In this situation, the community's goal in the co-management project becomes to secure funding rather than to develop co-management relationships and management rights that would contribute to conservation and livelihoods. In such a case, the community may have lacked the capacity, vision, or will to take on the responsibilities of co-management (as discussed under Condition 4), and the mere provision of funding and legitimation by external NGOs was not sufficient to overcome the absence of these conditions and allow the development and assertion of management rights. In Chile, most of the existing MPAs are sponsored and administered by external private organizations (Fernandez and Castilla 2005). Superimposing co-management policy, in the form of territorial use rights for fishermen over an existing traditional community-based natural-resource management system in Chile, eroded trust in the community and reduced adaptive capacity and resilience in resource-stewardship (Gelcich et al. 2006).

All of these findings underline the overwhelming importance of management rights held by communities as a condition necessary for their ability to implement conservation and sustainable fishing rates. As noted above in the Philippines and in the Dominican Republic examples, the ability of communities to ban destructive technology is a particularly important example of how these rights can be implemented to the benefit of conservation. Also noted is the fact that institutions for co-management evolve through time, and active adaptation of these arrangements and the processes they employ can strengthen their performance. In the following discussion, these rights are considered co-management rights (given that government will play some role in management, even if communities take the lead on many issues) and are assumed to be a basic underlying condition that enables other conditions to occur.

Condition 3: The Nature of the Resource Must Lend Itself Well to Co-management Arrangements

So far the discussion has focused on the nature of the governance arrangements between the coastal community and the level of government charged with the legal responsibility to manage fisheries, usually the national government. But the biophysical and logistical characteristics of the fish matter a great deal to sustainability because of the difficulty or ease they create in management (see Chapters 4 and 10 about the fit between institutions and ecosystems, and the problems of misfits). For example, species that are immobile such as shellfish or that remain year-round in nearshore waters, lend themselves well to one of the oldest and most common forms of local management, **Territorial Use Rights Fisheries** or **TURFs**, as they were famously named by FAO economist Francis Christy (1982). A TURF is adjacent to a coastal community so that fishing activity can be monitored and locally made rules can be easily enforced with low transaction costs. The community exercises either de facto (informal) or de jure (formal and legal) ownership over its local "territory." Many of the fish species governed by municipalities and FARMCs in the Philippines were immobile or less-mobile species, and it was thus possible to exclude

other fishermen from taking them. However, the fact that the western Washington tribes could co-manage salmon, a highly migratory species, shows that mobility does not make local management impossible, merely more difficult.

In addition to resource mobility, Agrawal (2002) identified seven other resource characteristics that can affect the ease or difficulty of local management: (2) resource size (affecting ease of monitoring and enforcement), (3) boundary clarity (affecting ease of exclusion and governance), (4) riskiness of resource flows (affecting willingness to invest in management), (5) scarcity and value (affecting how important the resource is economically, which in turn affects willingness to invest), (6) salience (affecting how culturally important and valuable a resource is considered to be), and (7) resource storability (affecting the logistical and technical possibilities for holding or processing after catch). To this list, Pinkerton and John (2008) added (8) resource visibility (affecting ease of monitoring; 9) resource spoilability (affecting how long a resource can be held after being caught without significant loss of value), and (10) resource defendability (the location of resource concentrations in relation to possibilities for exclusion, monitoring, and enforcement). A historical or dynamic dimension should be added to this list, because a number of studies suggest that, if a stock is badly overfished by outsiders or newcomers, locals may abandon hope of conserving and participate in finishing it off (Befu 1980, McGoodwin 1994). In other words, the resource must be sufficiently abundant to not trip off the feeding frenzy that occurs when everyone decides it is not possible to conserve the resources and turns attention instead in getting a share of what will soon be gone.

Condition 4: The Characteristics of the Community Must Lend Themselves Well to Co-management

Integrating the work of many scholars, Agrawal (2002) also identified eleven characteristics of a community that is well-adapted to self-management or co-management. (1) relatively small size (not too large to work together effectively), (2) clear membership rules (who is a community member and who can fish), (3) shared norms of behavior, (4) trust because of experience of past success in working together (social capital), (5) appropriate leadership (connected to both traditions and possibility of innovation), (6) interdependence (need for support of other community members to have a viable livelihood and participation in a social group), (7) homogeneity of identities and interests, heterogeneity of endowments (a variety of skills and capacities possessed by community members), (8) low level of poverty, (9) high level of dependence on resource, (10) residents located close to resource, (11) low or gradually changing demand, (12) access to conflict resolution.

To this list I add eight other characteristics related to local culture and capacity (Pinkerton 1992, 1991, 1989): (13) a shared sense of history and cultural continuity (leading individuals to care about more than their immediate self-interest and to identify themselves with a historical tradition and with future generations), (14) shared values that are sufficiently clear to grant moral legitimacy to a management system which reflects these (e.g., sense of fairness in access to the resource by users, stewardship), (15) ability of key individuals (leaders) to articulate a broad, holistic vision regarding sustainability, including ecosystem values, (16) political will to work consistently toward the vision, (17) willingness and ability to develop local skills and capacity, (18) an energy center, dedicated person, or group who applies consistent pressure to advance the process, (19) political attitudes that allow strategic alliances and collaboration, or ability to contain conflict for the sake of larger goals, (20) local and regional bodies that can coordinate planning processes, act as information depots and clearinghouses, mediators, regional consensus builders, policy developers, and local human and financial resource mobilizers and organizers.

Communities lacking some of these characteristics will still be able to co-manage, but the more of these assets they lack, the more difficult will be their task.

Condition 5: The Nature of the Community's Relationship with Outside Groups and Government Must be Favorable, i.e., Strong Enough to Promote its Attempt to Claim Power in Governance

There are also 17 characteristics that enable communities to mobilize sufficient political sympathy and power to implement power sharing with government agencies (Pinkerton 1992, 1993). A first set of characteristics emerge from communities' capacities to express their political aspirations effectively: (1) ability to articulate local ecological knowledge and local understanding of what works in fishing regulations in a manner understandable to natural scientists and government regulators, (2) ability to identify community interests with the public interest in sustainability, (3) ability to demonstrate that radical reform is necessary and not being addressed.

A second set of characteristics pertain to civil society and allow communities to mobilize sufficient power to roughly balance the power of groups that have captured the government agency in the past: (4) access to old and new public forums of debate and dissemination of opinion, including issue networks, as a source of ideas, technical information, alternative models and complementary resources, linking community to government agencies, universities, NGOs, and credible organizations, (5) a social movement in the larger society, in the form of new and expanding organizations, new and expanding forms of political expression, (6) access to sources of power such as legal advice, strategy advice, legislative bodies, public boards, (7) access to logistical and financial resources.

A third set of characteristics concern existing institutions that can be accessed by the community to press its case for claiming management rights: (8) sympathetic courts and supportive legal precedents (re attitude toward local involvement in decision-making, tribal rights, etc.), (9) protective or permitting legislation, (10) an appeal body to assist with local equity questions and principles, (11) an adequate scale of planning to control enough variables, i.e., the political boundary to which rights apply.

A fourth set of characteristics concerns the nature of the management agencies with which communities have to negotiate power-sharing: (12) agencies have hands-on experience working with fishermen, including personal relationships and common understandings, (13) agencies grasp the importance/value of local knowledge and are willing to work to combine it with natural science, (14) agencies are willing to work with communities to experiment through adaptive co-management with what level of fishing impact is sustainable, (15) agencies are willing to negotiate with communities and together with them experiment and do pilots to get started, (16) pilot agreements end up being formalized, legalized, and multiyear.

Finally (17) market and product form development is concomitant with the supply (so that communities can successfully sell their fish).

Benefits of Involving Communities in Coastal-Management Systems

The contribution of coastal communities' involvement in fisheries management depends on the specific ecological, economic, and social objectives. We saw in the Philippines example that some of the national objectives of fisheries access allocation in developing countries may be primarily social: to provide economic and food-security opportunity to coastal communities with few alternative livelihoods, and with long traditions of dependence on fishing, giving preference to the most needy and dependent groups. In developed high-latitude countries such as Norway and Canada, government policies for the allocation of fishing opportunity have historically been a way to encourage and stabilize settlement on the most northern coastline. Coastal fishing communities serve multiple national interests by increasing a country's ability to claim jurisdiction over adjacent waters (thus controlling shipping and oil drilling) and the ability to support national defense bases

of operation or reconnaissance. Governments therefore sometimes allocate fishing opportunity in the form of fishing licenses or favorable policies toward fishing/farming communities for purposes that have little or nothing to do with conservation or economic efficiency (Jentoft 1993, deYoung et al. 1999).

Whatever multiple purposes may be served by fisheries management, most federal fisheries legislation in developed or developing countries is likely to identify the primary goal to be conserving fish stocks and managing them sustainably. Let us therefore consider what the contribution of coastal communities might be to this particular goal.

As human understanding of ecological processes grows more complex, we increasingly appreciate the importance of long-term monitoring, not only for understanding food webs, species-habitat linkages, and trends in abundance in particular locations, but also for understanding and anticipating specific impacts of regime shifts caused by social or environmental change such as global warming (see Chapters 2 and 4). Without community involvement in and concern for monitoring, it would not be possible to document or grasp in much depth the nature and extent of change. In addition to global processes that may be reflected in long-term monitoring, there are also local processes that have their own particular patterns. Documentation of these local processes can help decision makers take advantage of more precise time- and place-specific knowledge so they can integrate analyses of different scales (Neis and Felt 2000, Kofinas 2002, Degnbol 2003). Structured and informal/traditional community-based monitoring is also still used by some indigenous communities to control local use, and to curtail it in times of overuse (see Chapter 6). Historically, tribal chiefs and heads of families within a given territory had the authority to allow access to specific resources, such as shellfish beds, and to call a harvesting moratorium if they deemed that the local stocks needed to recover (Turner et al. 2000). I was recently told by a British Columbia Stolo First Nations chief that his grandfather could identify five different stocks of salmon in their territory by examining their belly scales. His grandfather told him that his great grandfather would identify all the salmon stocks in their traditional territory (stocks that were adapted to spawn in different tributaries or creeks). This kind of detailed local knowledge allowed very specific management actions to be taken in response to detailed observations of different stock condition.

A major challenge for fish-management agencies and scientists is how to support and connect with community willingness to contribute to monitoring, often as volunteers because of their concern for and connection to the resource. In Washington State, Professor James Karr at the University of Washington was able to mobilize hundreds of citizens who live on reaches (sections) of salmon streams to monitor water quality and other conditions. Using simple straightforward indices that are easy to monitor, Karr and his colleagues demonstrated that citizen science can help a great deal in documenting very specific trends in particular areas (Fore et al. 2001, CCDCD 2004; see Chapter 4). Similarly, fishermen in Nicaragua were willing to monitor local stock condition and report in to a research team a great distance away. "After the fieldwork was over they went on recording the water level during different months of the year and mailed self-made charts to the scientist. No one had asked them to do this, but they apparently had the feeling that the study should somehow continue…" (Fischer 2000b).

An even more valuable contribution is made when a community can undertake its own research and when communities and researchers work together to link monitoring at different scales with research to co-produce knowledge on social–ecological change. In several cases these research programs have involved observations by local fishermen that were unnoticed by scientists. For example, two indigenous communities in the Broughton Archipelago between northern Vancouver Island and mainland British Columbia, Canada, noticed degradation in the condition of their clam beaches and in the clams themselves. Because of the timing of these changes, local fishermen suspected that the changes resulted from the presence of salmon farms that had proliferated in the area for a couple of decades (Schreiber 2003, 2006). They

noticed an increase in sludge worms (*Capitela*) and macro algae matting on the beach that apparently contributed to the beaches becoming smelly and unable to drain normally. (These phenomena are normally associated with ecological stress due to an overabundance of nutrients.) They also found blackened clams with a bad taste. The value of multigenerational local knowledge in a case like this provides a baseline of what is "normal." These conditions had never been observed before (Heaslip 2008). The communities were eventually able to get the attention of a research scientist at the Department of Fisheries and Oceans, and engaged the agency in a collaborative study by scientists, communities, and the salmon farmers to look at the sediments on particular clam beaches identified by the community as good candidates. The study is examining the different isotopic signatures in sediments on clam beaches to identify different sources of materials. The joint study will test the hypothesis of the communities that some of the carbon, metals, and nutrients so identified will be found to originate from salmon farm feed (Weinstein 2006, 2007). Since salmon farms on the British Columbia coast have been highly controversial, and the controversy has consumed vast financial and time resources, a study that establishes or denies the suspected linkages will be an important contribution to science and public policy. In this case, the co-production of knowledge involves systematically testing a hypothesis generated by the community that would have been unlikely to gain serious attention by researchers without community initiative.

A similar and ongoing study in Kitimat on the northern coast of British Columbia measures the effect of pulp mill effluent on the oolichan run (*Thaleichthys pacificus*, a major subsistence food of the local indigenous community) via sensory evaluation methodologies in which the panelists determine if the oolichan is tainted or not. The human sensory abilities of taste and smell are qualitative measures that have been adapted as a standard scientific approach to detecting contamination in foods. It is widely recognized that the human palate is extraordinarily sensitive to various kinds of "off" tastes and odors, and that properly established testing protocols can produce consistent, replicable results.

In the Kitimat situation, standard analytical chemistry would not have been helpful because there could be up to 2,000 different chemicals in pulp mill effluent, most of which have not been formally identified. Because the tainting propensity could come from any one of these or any number acting in combination, it was not practicable to proceed with analytical chemical identification of compounds. Previous studies have measured the tainting propensity of pulp and paper mill effluent (Shumway and Chawick 1971, Brouzes et al. 1978, Gordon et al. 1980) as a means to determine uptake of contaminants and the potential for biological effects. Furthermore, the sensory methods go to the heart of the matter by determining whether the food is "off" and does not concern itself with what specific chemical is causing it. Consultants for both the indigenous community and the pulp mill agreed at the start that sensory testing was appropriate. Sensory tests can employ expert evaluators, trained panelists, or naïve panelists depending on the purpose and objective of the evaluation. In this case, all sides agreed to use a panel made up solely of the local indigenous community, on the grounds that only people from the local community are sufficiently familiar with the normal taste of oolichans to be able to tell when something is off.

Assessment of tainting with pulp and paper mill effluent is a very sensitive indicator of pollution effects. And because of their extraordinarily high fat content, oolichans are particularly susceptible to becoming tainted from exposure. Therefore, the ability of the mill to eliminate tainting of oolichan would also greatly reduce the contaminant loading and pollution effects in the receiving environment in general. Thus, the Kitamaat indigenous community's right and ability to protect a traditional food source translates into a broader benefit for adjacent resources that do not enjoy the same protection.

Coastal communities also contribute to conservation by the very nature of their mode of fishing. The small-scale local fishermen who live in these communities are the most adaptable part of the commercial fleet. They have low overhead costs and flexibility to shift to other resources when conservation requires fishing closures. Unlike the highly invested large industrial fleet that exerts political pressure to keep fisheries open because of their high overhead costs, the small-boat fleet does not require a huge volume of fish to support it and can adapt flexibly to radical shifts in abundance. Large highly capitalized vessels are efficient at times and in places when mass production is called for, and risks and fuel costs are relatively low (such as "mop up" fisheries during high abundance times or large concentrations of migratory fish that must be taken quickly). Small vessels have a different kind of "efficiency" in that they are well-adapted to the risk and uncertainty prevalent in many modern fisheries (deYoung et al. 1999). These small fishing vessels are also "efficient" in that they are used in coastal communities to procure food fish, firewood, wild game, and supplies and to provide transportation. Loss of commercial fishing opportunity means the loss of a boat and thus the loss of the means to survive (Pinkerton 1987). It is important in viewing the role of coastal systems in supporting coastal populations to recognize that the food-security role they play in the larger economy (providing for the basic food and livelihood needs of coastal residents, enabling them to care for their own needs and not be dependent on government or other forms of social assistance) is enabled by their small-scale commercial fishing activity. In turn, the recognition by coastal communities of their dependence on the local system for their livelihood is a driver in their being candidates for conservationists. The argument here, as should be clear from the conditions outlined in the previous sections, is not that coastal communities are automatic conservationists, but that they are predisposed by many conditions and circumstances to be so, and should be seen as an important form of human and social capital that can aid in management.

The contribution of volunteers in coastal communities to the restoration of fish habitat and the enhancement of fish runs is underappreciated and little understood. Unlike the offshore marine fisheries described in Chapter 10, nearshore and riverine coastal fisheries are heavily dependent on healthy habitats, and highly vulnerable to upland uses such as logging, farming, and industrial development that cause soil erosion or release toxins that especially affect nursery and spawning areas. There are some 10,000 volunteers in fish habitat restoration, protection, and stock enhancement in British Columbia. In a study of the reasons why people volunteer for this work, Justice (2007) found that the closer the volunteers lived to the resource, the more likely they were to perceive the urgency required to rehabilitate the resource and also the more time they had available to participate. The primary reason people volunteered was to "put something back," hoping to improve resource outcomes. Secondary reasons for volunteering were to find camaraderie and appreciation among fellow volunteers for doing something worthwhile. Volunteer participation in decision-making forums, where it was perceived to have an effect, encouraged further volunteerism (Justice 2007). This research suggests that the perception of the importance of the resource, not only to livelihoods but also to identity, is a factor in producing volunteerism.

This research reminds us that the capacity of a resource system to mobilize human energy is one of its most important characteristics. Peter Senge (1990), in researching successful learning organizations, identified three conditions for effective organizational function. People in the organization must believe in the work; they must feel part of a team; they must feel that their individual contribution is valued. Under these conditions, people are likely to "go the extra mile" to solve problems and facilitate effective function, creating a working environment in which extra energy and creativity are produced. Bandura (1982) likewise found that when people experience **self-efficacy** (the sense of having an effect on events through one's efforts, having the power to make a difference) in their work, they are able to mobilize more energy. If we imagine coastal communities as places where people, if they feel empowered, are likely to contribute to things that are important to them and that they care about, then we can see important possibilities for the role these communities can play in resource

management through their knowledge, skills, vision, and energy. In an ideal world in which many of the conditions identified in the last section were met, coastal communities would form a political constituency with a vision of abundant coastal resources that would dedicate itself to working for this vision.

Conclusions

This chapter has shown how coastal communities have coevolved with coastal ecosystems as parts of linked social–ecological systems. The diversity of species and the sustainable patterns of livelihood that developed in coastal zones both in pre- and postindustrial times are products of a number of conditions that permit coastal communities to achieve conservation of adjacent marine resources and to remain resilient to change. Social scientists have found that five general conditions are good predictors of conservation by coastal communities: (1) strong community access rights to local marine resources, (2) strong community rights to participate in management decisions, (3) resource characteristics that lend themselves well to co-management arrangements, (4) community characteristics that lend themselves well to co-management arrangements, and (5) community relationships with outside groups and government strong enough to support its attempt to claim power in governance. Since there is not a single formula for success, but rather a multiplicity of possible combinations of favorable conditions, the listing of conditions can encourage those involved to look for windows of opportunity. Where these conditions exist, there are strong possibilities for coastal communities to contribute to the ecological, economic, and social objectives of fisheries management, and ultimately, to the resilience of the coastal social–ecological system.

Communities can uniquely contribute to long-term monitoring and research of global and local trends, model a flexible and adaptable way of fishing, and contribute volunteer energy toward the restoration of fish habitat and the enhancement of fish runs. These activities position communities to ban destructive gear and technology that may negatively impact fisheries. Because of their relationship to local aquatic life, coastal communities are in a constant process of learning and responding to their environment. As components of larger, linked, and multilevel systems of governance, they can act as the tentacles that give a first warning about change and can generate hypotheses about the linkages between different components of the system as these respond to change.

Review Questions

1. What is an appropriate way to conceptualize the relationship between coastal communities and marine ecosystems? Why?
2. Under what conditions will coastal communities play a positive role in fisheries and coastal management? Under what conditions might we expect communities to play a negative role?
3. What is the role and importance of resource dependence and place orientation on fisheries conservation?
4. What can the involvement of coastal communities contribute to the ecological, economic, and social objectives of fisheries management?
5. What role can coastal communities play in monitoring, research, and policy making? What are some of the potential benefits and hazards of their involvement in these processes?
6. How might communities resist external pressures that threaten to destroy local resources?
7. How can co-management contribute to adaptation of small-scale coastal communities that are dependent on fisheries?
8. What lessons can be drawn from the various histories of community involvement in MPAs?

Additional Readings

Agrawal, A. 2002. Common resources and institutional stability. Pages 41–85 in NRC. *The Drama of the Commons*. National Academy Press, Washington.

Degnbol, P. 2003. Science and the user perspective: The gap co-management must address. Pages 31–50 in D.C. Wilson, J.R. Nielsen, and P. Degnbol, editors. *The Fisheries Co-Management Experience: Accomplishments, Challenges and Prospects*. Kluwer, Dordrecht.

Heaslip, R. 2008. Monitoring salmon aquaculture waste: The contribution of First Nations' rights, knowledge, and practices in British Columbia. *Marine Policy* 32:988–996.

Holm, P., B. Hersoug, and S.A. Ranes. 2000. Revisiting Lofoten: Co-managing fish stocks or fishing space? *Human Organization* 59:353–364.

Neis, B., and L. Felt. 2000. *Finding our Sea Legs. Linking: Fishery People and Their Knowledge with Science and Management*. Institute for Social and Economic Research, St. Johns, Canada.

Pinkerton, E.W. 2003. Toward specificity in complexity: Understanding co-management from a social science perspective. Pages 61–77 in D.C. Wilson, J.R. Nielsen, and P. Degnbol, editors. *The Fisheries Co-Management Experience: Accomplishments, Challenges and Prospects*. Kluwer, Dordrecht.

Pinkerton, E.W., and L. John. 2008. Creating local management legitimacy. *Marine Policy* 32: 680–691.

Pinkerton, E.W., and M. Weinstein. 1995. *Fisheries that Work: Sustainability Through Community-Based Management*. The David Suzuki Foundation, Vancouver.

Schlager, E., and E. Ostrom. 1993. Property rights regimes and coastal fisheries: An empirical analysis. Pages 13–41 in T.L. Anderson and R.T. Simmons, editors. *The Political Economy of Customs and Culture: Informal Solutions to the Commons Problem*. Rowman and Littlefield Publishers, Lantham, MD.

Wilson, J. 2002. Scientific uncertainty, complex systems, and the design of common-pool institutions. Pages 327–360 in NRC. *The Drama of the Commons*. National Academy Press, Washington.

12
Managing Food Production Systems for Resilience

Rosamond L. Naylor

Introduction

Managing food production systems on a sustainable basis is one of the most critical challenges for the future of humanity, for the obvious reason that people cannot survive without food. Ecosystem health is both a "means" and an "end" to resilient crop and animal production. Being fundamentally dependent on the world's atmosphere, soils, freshwater, and genetic resources, these systems generate some of the most essential ecosystem services on the planet. They are also the largest global consumers of land and water, the greatest threats to biodiversity through habitat change and invasive species, significant sources of air and water pollution in many locations, and major determinants of biogeochemical change from local to global scales (Matson et al. 1997, Vitousek et al. 1997, Naylor 2000, Smil 2000). The inherent interplay between human welfare, food production, and the state of the world's nat-

ural resources underscores the need to manage these systems for resilience—to anticipate change and shape it in ways that lead to the long-run health of human populations, ecosystems, and environmental quality.

The production of crops and animals for human consumption epitomizes the social–ecological connection developed throughout this volume. This chapter differs from most other chapters in that food production systems are, by definition, human-created. The tight coupling of social and ecological processes for food production has been complete for at least 10,000 years since the early domestication of crops and animals (Smith 1998). Few "wild" systems exist today, in which humans hunt and forage for food. A major exception is the fisheries sector, but even there, the fish species in greatest demand for human consumption have long been enhanced through hatchery production, and **aquaculture** (fish and shellfish cultured in confined systems) is rapidly becoming the dominant source of global fish supplies (Naylor et al. 2000, Naylor and Burke 2005, FAO 2006).

Food production systems are also unique in terms of their economic, institutional, and cultural contexts. Markets link demand and supply of food commodities throughout the world. This linkage is strong within commodity groups

R.L. Naylor (✉)
Woods Institute for the Environment and
Freeman-Spogli Institute for International Studies,
Stanford University, Stanford, CA 94305, USA
e-mail: roz@leland.stanford.edu

F.S. Chapin et al. (eds.), *Principles of Ecosystem Stewardship*,
DOI 10.1007/978-0-387-73033-2_12, © Springer Science+Business Media, LLC 2009

(e.g., the wheat market in China is coupled economically to the wheat markets in the USA, Argentina, and Australia), between commodity groups (e.g., an increase in the price of maize causes the quantity demanded for substitute food crops, such as wheat, to rise), and between sectors (e.g., the demand for maize as a livestock feed or feedstock for bio-ethanol affects the price and demand for direct consumption of maize as a food grain; Naylor et al. 2007c). In almost all cases, agricultural and food markets tend to be influenced heavily by policy involvement within countries. Culture also plays a role in the design and management of food production systems. In the words of Anthelme Brillat-Savarin (*circ.* 1825), "Tell me what you eat, and I will tell you who you are."

How food production systems are designed, managed, and redesigned throughout the world depends on a myriad of social and ecological factors, such as soil type, climate, water availability, pests and pathogens, genetic advances, economic incentives driven by market forces and policy, and cultural influences including tastes, traditional practices, and urbanization. A key question to be addressed before any discussion of sustainable management can begin is: What are we trying to sustain? Food supplies and adequate nutrition per capita over time? The environmental quality of farm systems and ecosystems affected by food production practices? The cultural integrity of farming communities? Food quality, diversity, and safety? In a directionally changing world, with continually rising demand for agricultural products driven by population and income growth pressing on a finite land base, tradeoffs among these sustainability goals often occur. These tradeoffs are likely to become more acute in the future as the demand for biofuels adds to the already large and growing global demand for food and animal feed and as climate change limits agricultural productivity growth in certain locations.

The Agricultural Enterprise in Perspective

What makes food production systems interesting—and challenging—to study is their wide diversity throughout the world.

Maize is produced with high-yielding, hybrid seeds in mechanically dominated systems on thousands of hectares in the USA, Australia, and South Africa. Maize is also produced with local seeds on small plots of sloping land by farmers in Central America, and by poor farmers on marginal lands in East Africa using few external inputs apart from family labor. Rice is grown in irrigated, lowland fields in China; it is also grown in deepwater, flooded systems of Bangladesh and Thailand, on dryland, hillside plots of Cambodia and Laos, and in integrated rice–fish ponds in Malawi. It is not uncommon to see rice grown on quarter-hectare plots in rotation with vegetables and cassava in Indonesia and on thousand-hectare plots with no rotation apart from the winter fallow in the USA. Hogs are raised in confinement in large industrial systems in Mexico and in small numbers in backyard pens, and even inside houses, in many Asian countries. Some agricultural regions are dominated by cash crops for export; others are devoted to staple crops for domestic consumption. Large private companies play a major role in the development of agricultural technology, trade, and policy of most industrial nations, whereas agribusiness involvement directly in farming is minimal in the world's poorest countries. Some countries like the USA, France, and Japan subsidize agricultural production heavily, while many developing countries tax it.

These examples simply illustrate the diversity in agricultural systems seen throughout the world and should not be considered as the typical agriculture of any particular region. For example, industrial livestock systems are also found throughout Asia, and high-productivity lowland rice is grown in Bangladesh, Thailand, Cambodia, and Laos. The USA grows staple crops both for export and for domestic consumption of food, feed, and fuel. Laos exports cassava through Thailand to China for feed and fuel. There is a message to this madness: lessons on resilience and sustainability from one location do not necessarily apply to other areas with similar biophysical systems. And lessons are not easily transferred across scales of production. The global food production system is complex.

Intensification

Despite the heterogeneity in food production systems worldwide, there are important aggregate trends worth noting. The rising global population—now at 6.5 billion people and headed toward 8 billion or more by the middle of the twenty-first century (UN 2008)—coupled with steady urbanization and increasing demand for animal protein with income growth, has created the need for more high-yielding production systems. Over the past 40 years, roughly 80% of the growth in agricultural output has resulted from **intensification**; that is, the move toward high-yielding crops with adequate water availability, soil quality, and nutrients (including synthetic fertilizers) to achieve significant increases in yields (Conway 1997, Evans 1998, MEA 2005a). The pastoral lifestyles and ecosystem use described in Chapter 8 on drylands are examples of **extensification,** and they stand in contrast to the highly modified food production systems discussed in this chapter. Shifting cultivation for subsistence still represents a large share of global cropped land—particularly in the tropics—but many of these systems are experiencing declining productivity with human population pressure (Box 12.1). The need to intensify all systems is likely to increase over time.

Box 12.1. Shifting Cultivation

Food production systems have experienced an evolution from extensive to intensive cultivation in many parts of the world as a result of continued population growth on a limited land base. Yet extensive slash-and-burn production systems still exist on about 1 million ha of the earth's land surface, and account for 22% of all agricultural land in the tropics (Giller and Palm 2004, Palm et al. 2005). In these systems, land is typically cleared by burning and cultivated for several cycles until there is a significant loss in soil fertility (see Chapters 2 and 7). The land is then fallowed in order to recuperate its natural vegetation and nutrients. If the land is given sufficient time to recover, it can be cleared and cultivated again after a number of years in a sustainable fashion. However, if the land is farmed too long and is not given sufficient time to recover, excessive nutrient loss causes yields to decline and forces farmers onto more marginal lands. Under these circumstances, the ecosystem may not revert back to a mature forest, but instead develop into grassland that is much less productive biologically (Conway 1997). For this reason, shifting cultivation is often blamed as a primary cause of detrimental land-use change, most notably of tropical deforestation (Amelung and Deihl 1992, Myers 1993, Rerkasem 1996, Ranjan and Upadhyay 1999).

Shifting cultivation is designed to be resilient through time rather than space (Fox et al. 1995). In other words, when a farmer clears, cultivates, and fallows land for sufficient time, the land changes – in fact its ecological structure may change dramatically – but as time passes, it returns to alternative stable states of high resilience. Soon after the land is cleared, conditions are generally good for planting; there are abundant nutrients from the ashes of burned vegetation, and stressors such as weeds and pests are low. The land can be used to farm productively for a few years before stressors become prohibitive, at which point it must be abandoned in order to permit forest regrowth. Forests have natural stabilizing components such as a diversity of flora and fauna (that prevents sudden population swings) and long-living trees (that retain nutrients and maintain productivity). Given sufficient time to recuperate, forests generally represent a resilient system with productive soils, ample water, and resistance to pests, all of which are important for productive farming activities (Ewel 1999).

The key to resilience in these systems is timing. It is thus important for managers to understand the determinants of reduced fallow periods or expansion into marginal land. Growth in the human population dependant on slash-and-burn agriculture is the leading culprit of such change; more food is demanded from the system, and there is simply not time to wait for full forest recuperation unless crop yields can be improved with some form of productive intensification. Other factors are also important; for example, land privatization can disrupt a sustainable clear–cultivate–fallow cycle, and access to new markets can provide incentives to farm for profit rather than for subsistence, perhaps resulting in less resilient monocultures (Bawa and Dayanandan 1997, Amelung and Diehl 1992, Angelsen and Kaimowitz 1999). The introduction of intensive and diverse agroforestry systems is one avenue for resilience-based management in the face of continued population growth (Palm et al. 2005). The establishment of institutions and policies that support a transition from degraded, extensive systems toward more productive, intensive systems is essential as population pressures erode the existing land base (Fig. 12.1, taken from Palm et al. 2005).

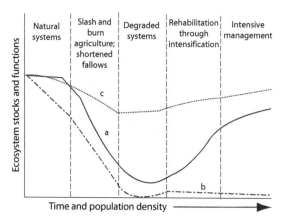

FIGURE 12.1. Land use intensification pathways and changes in stocks of natural capital such as carbon and nutrient stocks, biodiversity, and other ecosystem services, with time and increasing population density in the tropics (Palm et al. 2005). Line *a* represents the usual pattern of land degradation and eventual rehabilitation when the proper policies and institutions are in place, line *b* represents the continued state of degradation that can occur in the absence of appropriate policies and institutions, and line *c* represents the desired course where there is little degradation of the resource base yet improved livelihoods are achieved. Redrawn from Palm et al. (2005).

Investments in irrigation, improved crop cultivars, and animal breeding have resulted in impressive growth in global food production—around 260% between 1961 and 2003 (FAO-STAT 2007). Cereal output grew by 2.5-fold during this period, and poultry, pork, and ruminant production rose by 100, 60, and 40%, respectively. However, these changes have occurred at the expense of natural ecosystems. During the past half-century, intensive food production systems have contributed substantially to anthropogenic emissions of greenhouse gases (particularly methane and nitrous oxide) and pollution mainly in the form of nitrates, nitric oxide, and a wide range of pesticides (Tilman et al. 2002). The three largest staple crops—maize, rice, and wheat, which provide 45% of the calories consumed by the human population—are grown on almost half of the total arable cultivated land and account for over half of the synthetic nitrogen fertilizer applied to agriculture worldwide (FAOSTAT 2007, Cassman et al. 2003).

On a global basis, agricultural lands extend over 5 billion ha (34%) of the earth's terrestrial surface (Fig. 12.2). Agricultural land area expansion averaged 7.3 million ha per annum on a net basis between 1990 and 2005, after taking into account urban expansion, conversion of rural lands to nonagricultural uses, and degradation (FAOSTAT 2007). The rate is falling over time; between 2000 and 2005, net land area expansion was only 2.6 million ha per annum. Although rates of area expansion have been declining over the past quarter century, growth often occurs in natural habitats with

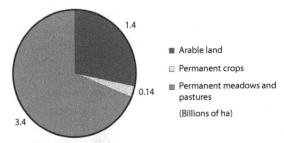

Arable land

Permanent crops

Permanent meadows and pastures

(Billions of ha)

FIGURE 12.2. Global agricultural land use. Agricultural area refers to: **arable land**—land under temporary crops, temporary meadows for mowing or pasture, land under market and kitchen gardens, and land temporarily fallow (less than five years). The abandoned land resulting from shifting cultivation is not included in this category. Data for arable land are not meant to indicate the amount of land that is potentially cultivable; **permanent crops**—land cultivated with crops that occupy the land for long periods and need not be replanted after each harvest, such as fruit trees, cocoa, coffee, and rubber; this category excludes land under trees grown for wood or timber; and **permanent pastures**—land used permanently (5 years or more) for herbaceous forage crops, either cultivated or growing wild (wild prairie or grazing land). Information from FAO-STAT Database 2006. FAO, Rome. 12 Nov 2006 (FAOSTAT accessed Nov 27, 2007).

high biodiversity value, such as the Amazonian rainforest (Myers et al. 2000). Given the limits on net land area expansion, maintaining yield growth in food and feed production will be essential in order to meet the expected global demand increase of 70–85% between 2000 and 2050 as the human population and incomes continue to rise (MEA 2005a). Even greater growth will be needed if significant cultivated area is used for biofuels and if climate change reduces productivity in many regions.

The availability of water through rainfall and irrigation is another important determinant of yield growth and agricultural productivity throughout the world. Irrigated crop systems account for about 40% of global food production, but less than 20% of the world's cultivated land is irrigated (Gleick 2002). Irrigated systems are abundant in Asia, whereas

over 90% of Sub-Saharan African agriculture is sown on rainfed lands. Most of the "easy" irrigation investments were made in the second half of the twentieth century; the annual rate of increase in irrigated area is currently below 1% due to costs and to environmental and social protests (FAO 2004, Khagram 2004). Unfortunately, many irrigation systems were built without proper drainage systems, and now an estimated 20% (45 million ha) of irrigated land suffers from salinization and waterlogging (Ghassemi et al. 1995, Postel 1999, 2001). Roughly 70% of the available surface water withdrawn for human activities globally is used for agriculture (Postel 1999, Gleick 2002; see Chapter 9). The diversion of surface water for irrigation often comes at the expense of surrounding natural ecosystems.

Hunger

Despite growth in productivity of food systems in many parts of the world during the past 40 years, over 800 million people still live in chronic hunger (Chen and Ravallion 2007, MDG 2007; see Chapter 3). A set of Millennium Development Goals was adopted by the United Nations General Assembly in 2000, which included halving the world's undernourished and impoverished by 2015. Today, more than halfway through this target period, virtually no progress has been made toward achieving the dual goals of global hunger and poverty alleviation. Even worse, most of the gains made toward these goals since 2000 have been erased by the world food crisis that emerged in 2008 (Economist 2008). The world food crisis has been characterized by extraordinarily high agricultural commodity prices, low grain stocks, and restrictions on cereal exports by several major trading countries. The persistence of global hunger among such a large number of people is particularly disturbing in light of the widespread economic and technological progress experienced in many parts of the world and the growing problem of obesity in industrial and middle-income countries, most notably the USA.

The Challenge Ahead

The challenge for the twenty-first century is thus clear: to develop food production systems in a way that will support rural incomes, enhance yield growth, utilize inputs efficiently (particularly water and added nutrients), minimize environmental impacts, and provide healthy diets for the human population (Conway 1997, Power 1999, Tilman et al. 2002, Robertson and Swinton 2005). Given the outlook for future global food demand, there is no question that the productivity of food production systems must continue to increase in order to achieve any definition of sustainability. But it is not at all clear that the goal should be to maintain or augment productivity on existing systems in all cases. Instead, the goal should be to redesign systems to promote genetic and crop diversity, stability in the face of future shocks (e.g., climate change and associated pest–predator impacts), and **food security** as defined by access to affordable food for all people at all times (see Chapter 5). In order to achieve this goal, efforts should be directed toward both small-scale farming systems that primarily meet local and regional demands, and large-scale surplus systems that meet national and global demands (Robertson and Swinton 2005). A framework of structural dynamics for the agricultural sector is presented below to help conceptualize the opportunities for—and constraints on—advancing resilience-based management of food production systems.

Structural Dynamics of Food Production Systems

Resilience-based management of crop and animal production requires a focus on the dynamics of both demand and supply. Humans play a dominant role in these dynamics, fundamentally shaping the agricultural landscape in an effort to meet local, regional, and global demands for food, animal feed, and fuel. At a broader scale, humans shape the agricultural landscape in order to generate incomes, employment, and in some cases ecosystem services (e.g., pollination, pest control, watershed management) that have positive impacts on the agricultural system itself and on surrounding areas. Human behavior plays a more intrinsic role in the dynamics of food production systems than is the case for most other ecosystems. Agriculture is by definition a human artifact—starting with the initial selection of genetic material and encompassing genetic manipulation, breeding, the deployment of new crop varieties and animal breeds, farm management, and food processing to meet consumer demand (Evans 1998, Smil 2000, Pretty 2002, Manning 2004).

Determinants of Demand and Supply

Growth in demand is a function of population increases, per capita income growth, urbanization, and cultural preferences. Per capita income growth and urbanization, in particular, typically lead to diversification of diets. Two empirically based rules in the agricultural development field—"Engel's Law" and "Bennett's Law"—have held up well over space and time (Timmer et al. 1983). **Engel's Law** states that as incomes grow, the share of household income spent on food in the aggregate declines. Although food quality rises with income growth, there are fundamental limits of food intake (the law of the stomach), which lead households to spend an increasing share of incremental income on nonfood items such as education, housing, health, and material goods and services. Engel's Law holds over time (the share of household income spent on food declines as countries develop) and over space (households in poor regions spend a greater share of their income on food than do households in wealthier regions). **Bennett's Law** states that the caloric intake of households is dominated by starchy staples at low levels of income, but is characterized by a diversified diet of fruits, vegetables, and animal products with income growth.

Urbanization creates another shift in demand; households tend to eat food that is easier to prepare at home or that is sold at restaurants and food stalls. The demand for meat products often rises in urban settings, as does the demand for products such as cooking oil used by street vendors. The process of urbanization and suburbanization also supports the rise in national and multinational super-market chains and other large retail operations (e.g., WalMart and Cosco), which favors pro-ducers who are connected to these marketing chains and certain products over others (e.g., farmed salmon over wild salmon because the former can be supplied consistently in large volumes throughout the year; Eagle et al. 2004, Reardon and Timmer 2007). Finally, income growth leads to higher fuel energy demand, including demand for motorized fuels, which creates new linkages between the energy and the agricultural sectors through biofuels when fossil fuel prices are sufficiently high (Naylor et al. 2007c).

The ability of crop and animal production to meet these various demands depends crit-ically on factor market conditions in individ-ual locations—that is, on the dynamics of labor, land, and credit markets—and on the state of infrastructure (e.g., roads, irrigation networks) and natural capital (water availability, climate, soils, genetic resources). Individual production activities are aggregated to create regional or national supplies, which are either consumed domestically or traded internationally. Tech-nology also plays an important role region-ally and globally in terms of genetic manipu-lation (e.g., improved crop cultivars and ani-mal breeds), labor-saving mechanization, and tools to maximize input use efficiency (e.g., nitrogen sensors, machines designed to incor-porate residues in low-till systems, drip irri-gation technology). Producers throughout the world have a complicated set of decisions to make each season, particularly given inherent uncertainties in output price, weather, and pests and pathogens at the beginning of each pro-duction cycle, and the long payoff times for many technological investments. Most farm-ers strive to maximize expected profits given a set of constraints that includes, for exam-ple, factor availability, agro-climatic conditions, technology, and infrastructure (Timmer et al. 1983). Other farmers—particularly in very poor regions—strive to minimize variability in pro-

duction systems that are used mainly for subsistence.

The overall impact of agricultural develop-ment on the environment is determined by the **IPAT Law** (Ehrlich and Holdren 1974). This law states that the impact (I) of any human activity such as agriculture is equal to the product of the human population size (P, mouths to feed), the affluence of the pop-ulation (A, income and related food prefer-ences), and the technology being employed in production (T).

Short-Run Adjustments

There are several ways in which food produc-tion systems adjust in the short run to equate demand and supply at different spatial scales. The most obvious adjustment mechanism is *the market* (Fig. 12.3). If demand exceeds supply in a given period, prices rise and provide an incen-tive for producers to increase supply in subse-quent periods. Alternatively, if supply exceeds demand, prices fall, causing consumption to rise, production to drop, and stocks in storage to be drawn down. Market equilibration does not occur instantaneously due to the length of crop and animal production cycles, marketing chains, and policy disincentives to change. As a result, market adjustments can have serious impacts on consumers when supplies are short, stocks are low, and prices are high. The consequences of price hikes are particularly serious for the world's poorest consumers who typically spend 50–75% of their incomes on food (Banerjee and Duflow 2007). Similarly, sharp price declines due to excess supplies can have devastating effects on producers, especially those who do not have diversified income sources, insurance, or savings.

Substitution plays a key role in producers' and consumers' responses to price—a concept that is often neglected in discussions of carrying capacity in food production systems. For exam-ple, when the price of rice rises in Asian mar-kets, consumers may switch to wheat or cassava as the staple in their diets. When the price of maize is relatively high, farmers in the Midwest USA may alter their crop rotation from soy to

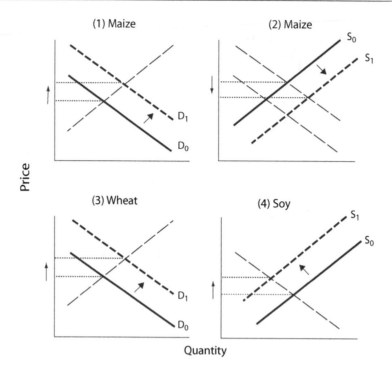

FIGURE 12.3. Market dynamics of agricultural supply and demand. D = demand curve; S = supply curve. Panel (1) – rising demand for maize leads to growth in supply along the curve that includes production at higher marginal costs. Panel (2) – longer run shift in supply due to technical change induced by higher prices. Panel (3) – higher maize prices increase demand for wheat in livestock markets, causing wheat prices to rise. Panel (4) – greater area sown to maize reduces area planted to soy, causing soy prices to rise. Redrawn from Naylor et al. (2007c).

maize. Substitution in production often occurs with a lag, particularly when major assets (e.g., large farm machinery) cannot be transferred across crops. It is worth noting that, although substitutions occur throughout the world in all income classes, extremely poor consumers and producers tend to be more limited in their options.

Four other important adjustment processes may also come into play in response to fluctuations in price. The first is through the drawdown of *stocks of grain*, which are typically carried as a reserve from one crop year to the next. Another adjustment mechanism is through *livestock*; when cereal and other feed prices rise, animals may be sold or butchered as a direct response. A third, and much starker, short-run adjustment to high prices is *human starvation*. Hunger results mainly from the lack of income to buy food, not from physical food shortages

alone. Finally, *natural ecosystems* may be the primary adjusters during high commodity price periods, as reflected in agricultural expansion into pristine rainforests or wetlands, excessive pumping of groundwater resources, and pollution to surrounding ecosystems.

Longer-Run Adjustments

Crop and animal production systems also adjust in fairly predictable ways over the medium to long run. There is a continuous process of **bottleneck breaking** in these systems (Evans 1998). For example, when the brown plant hopper became a major pest in intensive rice systems of Indonesia in the 1970s—as described in more detail in the Indonesian case study— the government scaled back pesticide subsidies (because the pests had become resistant to the pesticides), farmers adopted integrated pest-management practices (including crop rotations), and plant biologists continued to engineer cultivars that were more resistant to the pest. The process of bottleneck breaking can also occur on the demand side. For instance, when maize surpluses in the international market caused prices to fall and farm incomes to drop in the USA in the latter half of the twentieth century, food processors launched the maize-sweetener industry, and technology for the maize bio-ethanol began to be developed—both of which added new layers to overall maize demand. The federal government also implemented food aid programs to dispose of surplus grain, which have created disincentives in many recipient countries to invest in agriculture (Falcon 1991). Although bottleneck breaking increases the resilience of local food production systems, its net effect on regional or global resilience depends on market and policy responses.

The decline in maize prices noted above came as no surprise and reflects a second long-run adjustment in food production systems. Investments in infrastructure and technology, such as irrigation, genetic improvements, or planting and harvesting equipment, lead to increased supplies over time. Increase in agricultural productivity typically results in economy-wide income growth in agrarian societies. But as *Engel's Law* states above, income growth leads to a relative decline in expenditures on food over time. High growth in agricultural output with declining rates of growth in demand eventually results in excess supply in a closed economy setting. There are several ways to break free from this insidious feedback, including migration, reducing resource use in agriculture, promoting international trade (exports), and creating new forms of demand. The ongoing expansion of crop-based biofuels presents an interesting example, because energy demand tends to rise in lock-step with income growth and thus helps "fuel" the demand for agricultural commodities, even when relative expenditures on food are declining (Naylor et al. 2007c). Political support for biofuels in the USA and the EU is motivated to a large extent by the goal of revitalizing rural

economies, not just by the goals of expanding and diversifying fuel supplies.

A third adjustment process that occurs over the longer run is referred to as **induced innovation**—innovation that arises in response to price increases for scarce factors of production (Ruttan and Hyami 1984). For example, in Asia where arable land is scarce relative to labor, high-yielding seed varieties have been introduced as a land-saving technology. In the USA and Canada, where the opposite holds (agricultural labor is scarce relative to arable land), labor-saving innovations, such as planting and harvesting machinery and herbicides, have been introduced. The same principle can be applied to natural resource inputs for food production systems. The scarcity of water in relation to other factors of production has led to the design of water-saving innovations such as drip irrigation or desalinized water systems. In agricultural areas with high soil erosion, soil-saving technologies such as conservation tillage or cover crops have come into play. Like land- and labor-saving technologies, resource-saving innovations tend to be adopted when they are deemed economically profitable over a relevant time frame. Such calculations often involve present value accounting with discounting of future benefits—a process that can be controversial depending on the time horizon and which discount rate is used (Kolstad 1999, Portney and Weyant 1999; see Chapter 9). The basic point is that investments in new technologies or management practices require some sort of analysis—ranging from sophisticated calculations to more rudimentary weighting schemes—of alternative uses of limited natural and financial capital by current and future generations.

Role of Policy

Market forces shape the typical feedback mechanisms described above for food consumption and production. However, government policies are also important in influencing the dynamics of demand and supply and often override the market system. The food production sector is exceptional in terms of its heavy policy involvement (Naylor and Falcon In press). In most countries, governments have a hand in the agricultural sector via direct input subsidies, price or income supports, output taxes, credit programs, or land-reform measures. The agricultural sector may also be affected significantly by government policies implemented for nonagricultural purposes, such as controls on financial capital or exchange rate adjustments. Trade policies, such as tariffs and quotas on agricultural commodities or bio-ethanol, can similarly have direct and indirect impacts on food production and consumption. The list goes on. The main message is that the dynamics of crop and animal systems throughout the world are influenced by policies designed for a wide range of constituents and political purposes—including but not confined to the food and agricultural sector. The historical development of one policy overlaid by another, for reasons ranging from rural revitalization to food or energy self-sufficiency, is an underlying cause of social and environmental problems in food production systems in many countries.

Implications for Resilience-Based Management

Understanding the dynamics of food production systems raises some interesting dilemmas for resilience-based management. In many situations, market feedbacks help keep these systems on track, either by eliminating crops or production practices that are inferior or no longer valued by society, or by improving the efficiency of input use and resource allocation. Problems arise, however, when nonmarket consequences are involved—such as hunger, loss of cultural ties to the land, the destruction of pristine environments, and damages to human health and ecosystems from pollution—because the feedback mechanisms are less direct. In addition, policy often overrides market signals and biophysical responses that might otherwise enhance resilience. For example, government subsidies for irrigation lead to inefficient water use practices, excessive groundwater pumping, and salt-water incursion in many locations, obscuring obvious biophysical signals of scarcity and thus delaying appropriate action

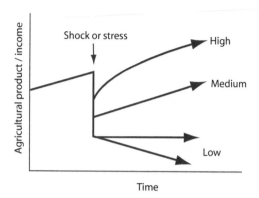

FIGURE 12.4. Resilience trajectory. Possible productivity or income trajectories of an agro-ecosystem after being subjected to a stress or shock. Redrawn from Conway (1997).

until large or irreversible changes may have occurred (see Chapter 9). Similarly, prolonged government support for a particular crop, such as irrigated wheat in the Sonoran desert or alfalfa in Southern California, leads to continued investments in that crop despite its inappropriate fit with natural and human resource base (Naylor and Falcon In press). Market and policy incentives might help a particular food production system cope with external shocks such as drought, pest infestations, war, and oil price spikes, as shown by the "high resilience" arrow in Fig. 12.4. But the underlying question remains: Is the original trajectory of that system designed for resilience, given the full scope of biophysical, economic, and cultural factors needed to ensure sustained production, consumption, and environmental quality over time?

Case Studies on the Resilience Challenge

Answering the question of whether food production systems are on a resilient trajectory requires a focus on the biophysical dynamics of the systems as well as the social and economic determinants of change. Many crop and animal systems are managed primarily for a single ecosystem service—the production of a consumable or marketable commodity (Robertson and Swinton 2005). Yet several other ecosystem services can also be provided by these systems, including soil stability, nutrient balance, pest and weed control, hydrological cycling, biodiversity protection, climate regulation, clean air and water, cultural value, and human nutrition (Daily 1997, Conway 1997). Unfortunately, the goal of managing crop and animal production for a variety of ecosystem services that support ecological resilience in the long run is often overridden by social and economic priorities in the short run—for example, to support political constituents, respond to global market opportunities, reduce poverty, or meet external demands for food, feed, and fuel. A decoupling of food production systems from the ecological systems on which they fundamentally depend diminishes their resilience (Robertson and Swinton 2005), as shown in the case studies that follow. With a continually rising global demand for food, feed, and fuel, it will become increasingly important to identify and promote policy and institutional mechanisms to couple or recouple agricultural and environmental systems.

Case I: Managing for Pro-poor Growth in Indonesia

Indonesia is the world's fourth most populous nation (~230 million people) and is extremely diverse culturally, with more Muslims than any other country but also with Catholics, Protestants, Hindus, Buddhists, and animists. The Indonesian archipelago has about 9000 inhabited islands, 400 language groups, and cropping systems that vary from Bali's manicured irrigated rice systems, to Nusa Tenggara Timor's (NTT's) rainfed corn and sorghum-based systems, the Maluka's rainfed cassava systems, Borneo's oil palm plantations, and Papua's sago palm and sweet potato systems. Rice is the main staple food in most, but not all, parts of the country. Monoculture is not typical except for flooded rice during the monsoon season on some islands. As Wallace's Line (a zoographic boundary between Asian and Australasian faunas) splits the country just east of Bali, the country contains wide geographic disparities

not only in soil, rainfall, flora, and fauna, but also in agriculture, rural incomes, and income growth.

Despite its inherent diversity, Indonesia is second only to China in terms of its success in poverty reduction during the past 50 years, advancing from about 70% below the poverty line in the late 1960s to 15% in the twenty-first century. Although early land reform and a variety of later fiscal transfers (such as school vouchers and health benefits) have been tested for poverty alleviation over the past half-century, agricultural development focused on pro-poor growth has demonstrated the only sustained success (Timmer 2005). This development process was tied primarily to the Green Revolution in rice production and illustrates a classic case of induced innovation. In most Asian countries including Indonesia—and particularly the island of Java, one of the world's most densely populated areas—land is scarce relative to labor. On Java, it is not uncommon to see a farmer working a quarter-hectare plot of land, or to see thirty unskilled workers show up to help harvest this farmer's plot in exchange for a share of their individual harvest output (Naylor 1994). Forty years ago, human pressure on this limited land area was so severe that Clifford Geertz (1963), a great anthropologist and agricultural development specialist, coined the term "agricultural involution" to depict a fragile balance of existence between humans and the rice ecosystems on which they survived.

The scarcity of land in relation to labor—and the extreme poverty it created in many parts of Asia—induced the **Green Revolution** in rice and wheat in the late 1960s and early 1970s. In simplest terms, the Green Revolution consisted of the design and dissemination of high-yielding seed varieties for the major cereal crops; because the seeds are scale-neutral, farmers on small and large plots alike could adopt the technology (Conway 1997). The new varieties contained dwarf genes, had shorter maturities, and an absence of photoperiod sensitivity, which allowed a shift in crop patterns from one or two rice crops per year to three crops (or five crops over 2 years) without rotation or fallow. Annual productivity of rice thus increased substantially. However, a reliable water source and added nutrients were also required to maximize the potential of the new crop varieties, and, as a result, the greatest early successes were seen in irrigated areas where farmers also had access to credit, affordable inputs (particularly synthetic fertilizers), and transportation infrastructure.

In Indonesia, the Green Revolution for rice was promoted as part of a pro-poor growth strategy. High-yielding seed varieties were adopted and disseminated in the late 1960s and throughout the 1970s, along with nitrogen fertilizers and improved irrigation infrastructure (mostly "run of the river" systems). Policy incentives to enhance adoption of the new technology were created, such as the "farmer's formula" in 1972 that linked very cheap fertilizers to the price of paddy to enhance profitability (Mears 1981, Timmer 1975). Improved microcredit programs came 15 years later. The rice sector served as the engine for rural economic growth throughout the 1970s and 1980s, and led to steady increases in real wages and incomes for unskilled labor (Naylor 1991, 1994). During this period, poverty reduction also resulted from expanded education (important especially for girls), better rural health and family-planning practices, and a macro policy that did not stifle agriculture or agricultural exports (Timmer 2005).

The introduction of the Green Revolution and the implementation of pro-poor growth strategies in Indonesia caused rice prices to fall for net consumers and lifted a large number of people out of poverty. But the relevant question for this chapter is: Did it lead to a resilient agricultural system—one that can withstand shocks and provide multiple ecosystem services for long-term benefits to the human population and the environment? The answer is "not completely." There has been a marked decline in the diversity of rice varieties used by farmers over the decades (Fox 1991). One of the most successful early varieties, IR36, was released in Indonesia in 1977. By 1982 it was planted on 40% of total rice acreage, and in East Java it was planted on 77% of the wet season and 85% of the dry season acreage. Between 1975 and 1985, rice output grew by almost 50% in Indonesia. But unfortunately IR36 was susceptible to some

major rice pests such as the brown plant hopper and grassy stunt, and by 1985 the variety began to fail. Substantial yield losses in IR36 led to the release of an improved cultivar, IR64, whose use grew even more rapidly than IR36. This classic case of bottleneck breaking led to productivity gains in the short- and medium-term. But the rapid dissemination of IR64 has caused the country's field- and landscape-level genetic base for rice to become even more slender (Fox 1991), potentially lowering the resilience of the rice sector in the long run.

To add to the loss of genetic diversity seen in farmers' fields, pest problems associated with IR36 led to the introduction of policies to subsidize pesticides (up to 80% of the retail cost) and encourage prophylactic spraying in the 1970s and early 1980s. This pesticide policy was a clear case of bad science, because it soon led to resistance within the brown planthopper population and greater pest infestations over large areas. Moreover, the policy was driven by corruption, as a senior agricultural official was the principal owner of the pesticide plant. Once the subsidy policy was dismantled, **integrated pest management** (IPM) became a more popular and successful practice—in conjunction with the development of new host plant resistance and limited spraying of chemicals—for stabilizing yields. The IPM practices reestablished crop rotations, although rice remains the dominant crop in the monsoon season in many areas, particularly on Java.

The agricultural sector now faces a new set of challenges, such as crop diversification, urbanization and rising labor prices, new demands, and marketing arrangements with the rise of supermarket chains, and global climate change (Timmer 2005, Naylor et al. 2007a). It will be important for farmers and policymakers to preserve genetic diversity, minimize the use of harmful chemicals (including particular herbicides as chemical weed control replaces hand weeding with increased labor costs), and balance incentives for productivity growth and rural poverty alleviation. The Indonesian rice sector has thus far been remarkably adaptable and resilient but will undoubtedly continue to be tested in biophysical, economic, and cultural terms.

Case II: Managing for Globalization in the Yaqui Valley, Mexico

The story of the Yaqui Valley in Sonora, Mexico, also pertains to the Green Revolution, but in a very different context. The Yaqui Valley, located in Northwest Mexico along the Gulf of California, was actually the home for the Green Revolution in wheat in the late 1960s. Because the region is agro-climatically representative of 40% of the developing world wheat-growing areas, it was selected as an ideal place for the early wheat-improvement program. It was here where Norman Borlaug, an agricultural scientist who was later awarded a Nobel Peace Prize, introduced the first high-yielding dwarf varieties of wheat (Matson et al. 2005). Using a combination of irrigation, high fertilizer rates, and modern cultivars, Yaqui farmers produce some of the highest wheat yields in the world (Matson et al. 1998). The Valley consists of 225,000 ha of irrigated wheat-based agriculture and is one the country's most productive breadbaskets (Naylor et al. 2000). However, its agricultural productivity has been threatened repeatedly in recent decades by drought, pests and pathogens, increased salinization, price shocks, and policy forces (Naylor and Falcon In press), making it an interesting case study for resilience.

Although the Yaqui Valley has some similarities with Indonesia in terms of the high-yielding seed technologies used, the story differs on at least four counts. First, the Valley is located in the Sonoran desert and depends fundamentally on irrigation from reservoirs, and to a lesser extent from groundwater pumping, for its intensive agricultural production. Second, while there are a large number of poor farmers in the region—mainly in the *ejido* (collective agriculture) sector (Lewis 2002)—the Valley is increasingly characterized by larger, more wealthy farmers who operate over 50 ha apiece and in some cases hundreds of hectares. Third, Yaqui farmers use high rates of chemical inputs; nitrogen applications for wheat, in particular, are among the highest in the world and result in major losses to the environment through various biogeochemical pathways (Matson et al. 1998). Finally, the main economic and policy determinants of change for Yaqui Valley farmers originate at national and international scales and are often exogenous to production practices in the region (Naylor and Falcon In press).

With a semiarid climate and variable precipitation rates, the Valley has relied on the development and maintenance of irrigation reservoirs for agricultural intensification. By 1963 three major dams had been constructed supplying irrigation water to 233,000 ha (Naylor et al. 2000). However, the construction of these reservoirs did not eliminate the region's sensitivity to climatic extremes. For example the prolonged drought during 1994–2002 led to dramatic declines in total reservoir volume, increases in well pumping, and reduced water allocations to farmers, resulting in less than 20% of the total area in production in 2003 (Matson et al. 2005). By this time, Valley farmers had completely drained the 64-km^2 reservoir from the Rio Yaqui. Increased dependence on groundwater from wells—water that tends to be more saline than the high-quality fresh water from reservoirs—in turn raised the risk of salinization, a problem that affects roughly one third of the soils in the Valley. Rains returned to the region by 2005 and provided water to the reservoir once again, but policy incentives supporting the production of crops that are not drought or salt-tolerant continue to weaken the resilience of this desert agro-ecosystem system.

Policy has played an enormous role in agricultural development in the Valley, but it has often worked at odds with environmental quality and ecosystem health—and on occasion even with farm profitability (Naylor and Falcon In press). Many of the policies affecting agriculture in the region have been macroeconomic, focused on trade, exchange rates, interest rates, and national financial portfolio balances. The agricultural policies at the microlevel have been implemented in Mexico City for the country as a whole and not necessarily for the benefit of Yaqui farmers.

During the 1980s, government involvement in almost all phases of the Mexican food system was pronounced. Significant price supports for agricultural products, large input subsidies on water, credit, and fertilizer, and major consumption subsidies on basic food products were justified primarily as poverty-alleviation policies. In the early 1990s the government under President Salinas began to withdraw government support in agriculture as part of a broader liberalization process that was occurring in other sectors of the economy. New international trading arrangements were implemented for agriculture, mainly via NAFTA, which reduced trade barriers, motivated large changes in prices of many agricultural inputs and outputs, and thus dramatically altered relative prices to producers. Producer incentives were also altered by replacing price supports for wheat and other agricultural products with income supports—thereby decoupling government payments from total production, which hurt farmers in this high-yield region of Mexico. During this period, the government reduced its institutional involvement in agriculture, e.g., by reducing consumer food subsidies, privatizing the nationalized Mexican Fertilizer Company (FERTIMEX), removing or reducing government credit subsidies, and largely eliminating public extension services. Finally, the operating authority and funding responsibilities for irrigation systems were decentralized from federal to local water-user groups via the Water Laws of 1992 and 1994, and a constitutional change in land rights (Article 27) in 1992 made possible the (legal) sale and rental of *ejido* land (Naylor et al. 2000, Naylor and Falcon In press).

The overall intentions of these numerous policy changes were to integrate Mexican agriculture into the global economy, improve efficiency of production, and increase private sector involvement. But Yaqui farmers were exposed to markets in unprecedented ways. They were also hit by a series of external shocks—pest attacks, drought, large fluctuations in world commodity markets, and a major devaluation of the exchange rate—that fundamentally altered the economic and biophysical environment in which they operated. In principle, the full suite of policies could have led to greater input use efficiency (e.g., water and nitrogen), resulting in "win–win" solutions for farm profitability and the environment. However, despite higher marginal costs, farmers did not reduce fertilizer and water applications until forced to do so with the drought (Manning

2002, Addams et al. In press). Moreover, Mexico's main agricultural trading partner—the USA—continues to subsidize its wheat farmers in many indirect ways, leaving Yaqui farmers at a competitive disadvantage unless they are also subsidized. Some farmers have attempted to diversify production into high-valued crops, livestock, or aquaculture, but the economic and biophysical risks of doing so remain high.

Economic and policy changes affecting agricultural decision-making in the Yaqui Valley in the 1990s were largely exogenous to the farming system. However, farmers in this highly commercialized region of Mexico are not just "policy-takers." They also *do* have a voice in the development of national agricultural policies. Their declining competitive position relative to the USA, coupled with the fall in international commodity prices in the latter part of the decade, induced an intensive lobbying effort that ended with the reenactment of commodity price supports in 2000 for the bulk crops such as wheat, maize, and cotton. Yaqui farm groups were not solely responsible for the renewed protection, but they had— and continue to have—a persuasive influence on other farm groups in Mexico and a history of political power in Mexico City. Three Mexican presidents came from the Valley's main city, Ciudad Obregon, and the connection between large farm organizations in the Yaqui Valley, and politicians in Hermasillo (the state capital) and Mexico City have traditionally been strong. When lobbying has not worked, farm groups have resorted to other tactics, such as threatening to close down the highway that runs through the state of Sonora to the US border.

The policy shift back toward protection at the turn of the twenty-first century raises important questions for the future sustainability and resilience of Yaqui Valley agriculture. Will farmers continue to grow wheat, corn, and cotton as a result of the economic security stemming from policy support—despite the fact that these crops demand high fertilizer, pesticide, and water inputs? If so, will improvements in input use efficiency be sufficient to reduce the cost-price squeeze that farmers have experienced in the past, and to lessen the impact on the environment? Will farmers continue to band together in the promotion of these crops in order to preserve their "safety in numbers"? Answers to these questions will depend on agricultural policy dynamics within Mexico—driven in large part by the tension between northern commercial interests versus southern social interests—and between Mexico, the USA, and Europe. Managing for resilience thus requires the participation of policymakers outside of the agricultural system in question. As long as the USA and the EU persist in subsidizing wheat, maize, and cotton through various policy instruments, Mexico will likely be forced politically to subsidize these crops as well. The future course of commodity protection in Mexico will depend importantly on progress related to agricultural trade negotiations within the World Trade Organization, the 2008 US Farm Bill, EU farm policy, and trends in international commodity prices. This progress will be influenced, in turn, by a set of emerging issues within the world food economy, including the industrial livestock revolution, the biofuels boom, and global climate change.

Emerging Issues for Resilience-Based Management

A key feature of resilience-based management is to anticipate change and adjust accordingly in order to preserve long-run ecological functioning and human welfare (see Chapters 2 and 3). Beyond the specific, anticipated sorts of changes in a system, resilience-based management needs to consider the possibility of complete surprises and uncertainties, such as the confluence of events that led to the world food crisis in 2008 (Walker and Salt 2006). There are at least three major transitions confronting global food production systems today that are worth considering in this context: growth in industrial livestock systems, the rising use of crops for fuel, and global climate change. The first two cases can be thought of as demand-driven changes, while the third is primarily supply-driven. In all cases, policies directed toward both the agriculture and the energy sectors will play a role in determining the resilience of food production systems.

Industrial Livestock

Domesticated livestock have played a role in human societies and evolution for the past 10,000 years, providing important sources of protein, fertilizer, fuel, traction, and transport (Diamond 1997, Smith 1998). Traditionally, livestock have been an integral part of agricultural systems, raised close to their food source, and used as an input (soil nutrients and traction) in crop production. In recent decades, however, livestock have become industrialized, often raised far from its feed source and traded internationally (Naylor et al. 2005, Galloway et al. 2007). At the heart of this transition is very rapid income-driven growth in meat demand, particularly in parts of the developing world such as China, Southeast Asia, and Latin America (Steinfeld et al. 2006). In addition, relatively inexpensive feeds, improved transportation, technological innovations in breeding and processing, concerns over food safety, and vertical integration of the industry have led to industrialization and spatial concentration of intensive livestock systems. Urbanization and development of large-scale retail chains further contribute to intensification of these systems (Steinfeld et al. 2006).

A dominant feature of the geographic concentration of livestock—and one that has major implications for the resilience of the sector—is the de-linking of animal production from the supporting natural resource base. Feed is sourced on a least-cost basis from international markets, and the composition of feed is changing from agricultural by-products to grain, oilmeal, and fishmeal products that have higher nutritional and commercial value (Naylor et al. 2005, Galloway et al. 2007). Synthetic fertilizers as opposed to animal manure are used to fertilize crops, and machines are used instead of animal traction to plow the land—both contributing to higher fossil fuel inputs and greater greenhouse gas emissions. The pattern of industrialization is particularly striking for monogastric animals (poultry and hogs), which utilize

concentrated feeds more efficiently than rumi-
nants (cattle, sheep, and goats) and which
have short life cycles that favor rapid genetic
improvements (Smil 2002). Pork and poultry products also tend to be less expensive
for developing-country consumers; as a result,
industrial livestock production is expected to
meet most of the income-driven doubling in
meat demand forecast for developing countries
in the coming decades (FAO 2004).

For meat and other livestock products the
income elasticity of demand is high—that is,
when incomes grow, expenditures on livestock
products grow rapidly (Steinfeld et al. 2006).
A classic relationship exists between incomes
and direct (food) versus indirect (feed) demand
for grains (Fig. 12.5). As incomes rise, more
grain and oilseed crops are needed to feed
the human population via the livestock sector, which in turn has detrimental effects on
global land and water resources. Through grazing and feed crop production, livestock is the
largest global user of land resources, occupying
almost one third of the ice-free terrestrial surface of the earth (Steinfeld et al. 2006, Galloway
et al. 2007; see Chapter 8). With industrialization, land use change associated with livestock
is being driven increasingly by feed crop production as opposed to grazing, although grazing
remains a major form of land use (see Chapter 8
and Fig. 12.2). A similar pattern holds for water

resources: agriculture dominates global water
use, and a growing share of crop production is
now devoted to animal feeds (see Chapter 9).

The de-linking of livestock production from
the land base, and the increasing intensity of
irrigation and synthetic fertilizer applications
for feed crop production, has internal and external impacts that are often obscured by the lack
of appropriate market valuation for agricultural
water use and pollution. Moreover, with international trade in livestock products now growing faster than production, the link between
consumers and producers has weakened—
effectively eliminating the accountability that
livestock consumers might feel in relation to the
products they eat (Galloway et al. 2007). Policies also play a role in promoting feed grain and
industrial livestock production in many countries, even when factor scarcity or resource constraints might otherwise lead to a decline in
the livestock sector. Finally, the emerging dominance of industrial livestock has curbed the
market access for livestock producers in small
scale or extensive pastoral systems, most of
whom live in poor rural communities (Steinfeld
et al. 2006). The intensification of livestock has
increased global protein consumption in both
developing and developed countries. However,
it is essential that these systems be managed
to strengthen economic and ecological feedbacks through improved nutrient cycling, efficient water use, and food safety measures. If
livestock production remains decoupled from
its supporting resource base, the resilience of
food production systems at local to global scales
remains in question.

The Biofuels Boom

The integration of the global agricultural and
energy sectors caused by recent and rapid
growth in the biofuels market raises even more
serious questions than industrial livestock in
terms of the resilience of food production systems. Investments in crop-based biofuels production have risen recently around the world
as countries seek substitutes for high-priced
petroleum products, greenhouse gas-emitting
fossil fuels, and energy supplies originating

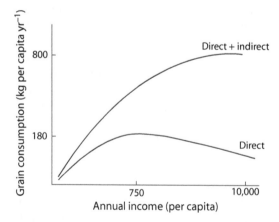

FIGURE 12.5. Direct and indirect (to support livestock production) grain consumption as a function of
income.

from politically unstable countries. Some countries such as the USA are also supporting crop-based biofuels production as a means of rural revitalization. Growth in the biofuels sector raises two important questions concerning the resilience of food production systems. First, can agro-ecological systems be sustained environmentally given the degree of intensification needed to meet biofuels production targets in various countries over time? And second, as greater demand pressure is placed on agricultural systems—and as agricultural commodity prices rise in response—can food security be maintained for the world's poorest populations? In answering these questions, a few trends seem clear. Total fuel energy use will continue to escalate as incomes rise in both industrial and developing countries, and biofuels will remain a critical energy development target in many parts of the world if petroleum prices remain high. Even if petroleum prices dip, policy support for biofuels as a means of boosting rural incomes in several key countries will likely generate continued expansion of biofuels production capacity over the next decade (Naylor et al. 2007a).

The rising use of food and feed crops for fuel is altering the fundamental economic dynamics that have shaped global agricultural markets for the past century. Although both energy and food demand rise with income growth, the rate of increase is much greater for energy, as shown in Fig. 12.6. Engel's Law, coupled with impressive increases in world food production, has led to a steady trend decline in real food prices in international markets for the past several decades. But this pattern is changing with the new linkages between the agriculture and the energy sectors. As energy markets increasingly determine the value of agricultural commodities (Cassman et al. 2006, Schmidhuber 2007), the long-term trend of declining real prices for most agricultural commodities is likely to be reversed and Engel's Law overridden (Naylor et al. 2007c).

Over the short term this reversal, while potentially helping net food producers in poor areas, could have large negative consequences for the world's food insecure, especially those who consume staple foods that are direct or indirect substitutes for biofuel feedstocks. Sugarcane, maize, cassava, palm oil, soy, and sorghum—currently the world's leading biofuel feedstocks—comprise about 30% of mean calorie consumption by people living in chronic hunger around the world (Naylor et al. 2007c). The use of these crops for global fuel consumption could thus increase the risk of global food insecurity, particularly if rural income growth is not rising in parallel. Rising commodity prices for feedstock crops and their substitutes (e.g., maize, soy, wheat, and cassava) also have a direct impact on the livestock industry, since these crops are an important feed ingredient, particularly for pork and poultry.

The risks of food insecurity associated with biofuels development were realized in 2008 with the sharp run-up of agricultural

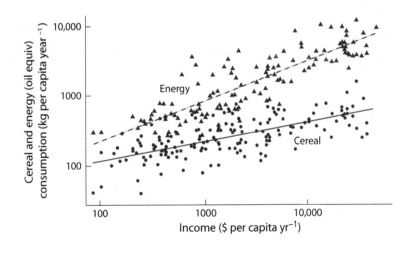

FIGURE 12.6. Per capita cereal consumption (circles) and energy consumption (triangles) by country as a function of income in 2003. Redrawn from Naylor et al. (2007c).

commodity prices worldwide. Increased demand for biofuel feedstocks (on top of the rising demand for meat in China and other emerging economies)—coupled with some regional drought- and disease-related supply shocks—led to heightened speculation in financial markets and panic among many governments facing food riots and political unrest. Although biofuels were not the sole cause of the world food crisis as it unfolded in 2008, the tight linkage between agricultural and energy markets raises serious questions for the future resilience of food production systems and the ability of poor people to afford adequate nutrition.

On the supply side, the growth in biofuels globally and regionally creates risks of environmental decay and resource exhaustion (Naylor et al. 2007c). For example, the USA—now one of the two largest global bio-ethanol producers along with Brazil—is fundamentally constrained in how much maize can be produced for fuel by both land area and yield potential. Although maize area has expanded at the cost of other crops such as soy in recent years, price feedbacks limit the amount of area substitution that will occur over time (Fig. 12.3). Some agricultural lands that were previously removed from production for programs like the Conservation Reserve Program (CRP) are now being brought back into maize cultivation (Imhoff 2007). Introducing monocultures into CRP lands will likely have adverse effects on biodiversity and wildlife habitat—two main CRP goals. The yield potential for US maize is also limited in terms of genetic gains (Cassman and Liska 2007); most yield growth is likely to come from additional inputs (fertilizers, water), creating additional environmental stress to supporting and surrounding ecosystems unless input use efficiency improves dramatically (Cassman et al. 2003, 2006).

There are several other examples where crop production for biofuels potentially lowers the ecological resilience of food production systems worldwide (Naylor et al. 2007c). For example, oil palm produced in the rainforests of Borneo for biodiesel and food can cause habitat destruction, a loss of biodiversity, degraded water and air quality, the displacement of local commu-

nities, and a change in regional climate (Curran et al. 2004). Land clearing through fire contributes to especially large environmental threats of regional air pollution, biodiversity loss, and net carbon emissions (Tacconi 2003). Crop production in low-productivity, hillside areas of China for bio-ethanol can cause soil erosion and flooding. The substitution of maize for soy in the USA can lead to increased soy production in the Amazonian rainforest, thus potentially causing biodiversity loss and a change in regional climate. The wide range of ecological, environmental, and food security effects of biofuels production are just starting to be measured and documented. Strategies for managing crop-based biofuels development for resilience need to focus on the implications for land use (soil erosion, biodiversity), water use (quantity and quality), air quality, and net energy and climate outcomes at local and global scales.

A key to enhancing the resilience of food production systems in an era of biofuels will be the emergence of economically and technologically feasible sources of cellulosic fuels that can be grown on degraded lands (Tilman et al. 2006). Current cellulosic biomass-to-fuel conversion systems are not yet cost-effective and require large amounts of water (Naylor et al. 2007c, Wright and Brown 2007). The technology for large-scale deployment of cellulosic biofuels production is probably at least 10 years away (Himmel et al. 2007), although smaller-scale biomass systems using more rudimentary technology have long been viable for local fuel production. The coupling of the agricultural and energy sectors at regional to global scales through the development of crop-based biofuels is thus likely to play an important role in the resilience of food production systems for decades to come.

Global Climate Change

A third—and arguably the dominant— emerging influence on resilience in food production systems worldwide is global climate change. The fourth report of the Intergovernmental Panel on Climate Change (IPCC) released in 2007 presented strong scientific data and consensus once again that the global climate is changing, and that humans are both causing and will be damaged by this change (IPCC 2007a,b). The agricultural sector is likely to be affected more directly than any other sector by rising temperatures throughout the world, sea level rise, changing precipitation patterns, declining water availability, new pest and pathogen pressures, and declining soil moisture in many regions. Vulnerability is created through biodiversity loss and simplification of landscapes in the face of climate change (see Chapter 2). The agricultural sector is also contributing to climate change through nitrous oxide emissions from fertilizer use (denitrification) and methane release from rice fields and ruminant livestock (e.g., cattle and sheep; Steinfeld and Wassenaar 2007).

Although predicting future climate conditions involves many uncertainties (see Chapter 14), there is broad scientific consensus on three points (IPCC 2007a). First, all regions will become warmer. The marginal change in temperature will be greater at higher latitudes, although tropical ecosystems are likely to be particularly sensitive to projected temperature changes due to the evolution of species under more limited seasonal temperature variation (Deutsch et al. 2007). Second, soil moisture is expected to decline with higher temperatures and evapotranspiration in many areas of the subtropics, leading to sustained drought conditions in some areas and flooding in other areas where rainfall intensity increases. In general, wet regions are expected to become wetter, and dry regions are expected to become drier, although changes in precipitation are much less certain than those of temperature. Third, sea level will rise globally with thermal expansion of the oceans and glacial melt.

How the anticipated changes in global climate will affect agricultural productivity around the world and the resilience of food production systems depends on regional patterns of change and the ability of countries within each region to adapt over time. Again, there is much uncertainty in the distribution of future impacts, but some regional predictions can be made with reasonable confidence—ignoring, for the moment, adaptation. For example, sea level rise

will be most devastating for small island states and for countries such as Bangladesh that are low-lying and highly populated. Large areas of Bangladesh already flood on an annual basis and are likely to be submerged completely in the future, leading to a substantial loss of agricultural land area, even for deep water rice. Moreover, the rapid melting of the Himalayan glaciers, which regulate the perennial flow in large rivers such as the Indus, Ganges, Brahmaputra, and Mekong, is expected to cause these river systems to experience shorter and more intense seasonal flow and more flooding—thus affecting large tracks of agricultural land. Moderate temperature changes are likely to be more positive for agricultural yields in high latitudes than in mid- to low-latitudes (IPCC 2007b). In addition, CO_2 fertilization will benefit some crops in the mid-latitudes (provided that temperature changes are not extreme and sufficient water is available for crop growth) until mid-century—at which time the deleterious effects of temperature and precipitation changes are expected to offset CO_2 fertilization benefits. Food insecure populations, particularly in southern Africa and South Asia, will likely face greater risks of low crop productivity and hunger due to increased temperature; yield losses could be as high as 30–50% for some crops in these regions if adaptation measures are not pursued (Lobell et al. 2008). Africa as a whole is particularly vulnerable to climate change since over half of the economic activity in most of the continent's poorest countries is derived from agriculture, and over 90% of the farming is on rainfed lands (Parry et al. 2004, Easterling et al. 2007, World Bank 2007).

Given these large potential impacts, adaptive capacity is critical to the resilience of any particular agricultural system (see Chapter 3). Adaptation measures range in size and expense, with activities such as the shifting of planting dates or substitution among existing crop cultivars at the low end, to the installation of new irrigation infrastructure or sea walls at the high end (Lobell et al. 2008). A common assumption is that agricultural systems will shift geographically over time to regions with suitable agro-climatic conditions (e.g., crops will move poleward with warmer temperatures)—resulting in little net impact on global food supplies in the future. However, the necessary extensive genetic manipulation through breeding will require continued collection, evaluation, deployment, and conservation of diverse crop genetic material (Naylor et al. 2007b). Because climate change will also affect wild relatives of plants and hence the *in situ* genetic stock on which the agricultural sector fundamentally depends (Jarvis et al. 2008), there is an urgent need to invest in crop genetic conservation for resilience, particularly in cases where *ex situ* genetic resources (genetic material stored in gene banks) are scarce (Fowler and Hodgkin 2004).

Overall, there are three important points to consider with respect to the resilience of food production systems under a changing climate. The first is that poor farmers tend to have fewer resources at their disposal than wealthy farmers—either at the household, community, or state levels—for adaptation. They typically do not have access to irrigation, a wide selection of seed varieties, knowledge of alternative cropping systems, credit, or personal savings (Dercon 2004, Burke et al. 2008). Second, because adaptation strategies that benefit the greatest numbers of producers involve large-scale investments (e.g., irrigation, breeding for new cultivars), there is a "public good" aspect to adaptation that cannot be ignored. Governments and the international donor community have a role to play in ensuring that adaptation can occur in both rich and poor countries to protect the resilience of food production systems. Finally, providing information to policymakers, farmers, and agribusiness throughout the world on climate change and its potential impacts is needed *now*—particularly in poor countries where such information is often lacking—in order to motivate private sector solutions to the problem in both the short and long run. Substantial investments in education, information, infrastructure, and new crop varieties suited to the projected climate will be required to sustain all aspects of food production systems: food security, rural communities, agro-ecosystems, crop genetic diversity, and agricultural yield growth. Without a forward-looking vision, the resilience of agricultural and livestock systems will undoubtedly erode.

Conclusion

The twenty-first century marks a new era for global food production systems along several axes. There has never been a time when the human population has approached 8 billion people, with growing demands for food and animal feeds. Nor has there been a time when the agricultural sector promises to be so tightly linked with the energy sector in terms of price level and volatility. Widespread urbanization is consuming fertile agricultural land and water and at the same time is creating greater demands for food from the arable land that remains in production (see Chapter 13). Globalization is leading to the integration of agricultural commodities, inputs, and financial markets over space and time that increasingly decouple consumption from production and leave farmers vulnerable to swings in factor and output prices and to volatility in the cost of financial capital (see Chapter 14). And perhaps most daunting, climate change over the course of the twenty-first century will alter modern food production systems in unprecedented ways and threaten food security in many poor regions of the world. Crop yields are predicted to decline dramatically in many areas as a result of climate change—just at a time when rising demands from population growth, income growth, biofuels expansion, and urbanization require significantly higher yield growth than in the past.

Designing adaptive strategies to ensure resilience in food production systems in this new biophysical and socioeconomic era will require focused attention by agricultural scientists, policymakers, international development agencies and donors, industry leaders, entrepreneurs, resource managers, crop seed collectors and conservation experts, NGOs, and food producers and consumers throughout the world. How should agricultural systems be designed and managed to meet the challenges ahead? What incentives should be provided to farmers to guarantee a wide range of ecosystem services from agriculture beyond "the global pile of grain"? How can food security, food quality, and environmental quality be ensured for the generations to come?

Answering these questions is not an easy task, particularly given the diversity of groups involved in the global agricultural enterprise and their differing priorities with respect to production, income, culture, and environmental goals now and in the long-run future. Despite such complexity, the transition toward resilient food production systems will require three main steps (Lambin 2007). The first step involves a widespread recognition that the biophysical and socioeconomic conditions of the twenty-first century are markedly different than the past and that many crop and animal production systems will need to be redesigned fundamentally (see Chapter 5). Some principles could help shape this process of redesign. For example, given the growing resource constraints on agriculture, it is time to focus more on adapting crops and animals to the local resource base and nutritional needs, rather than transforming the resource base to meet specific commodity production targets (see Chapter 2). Does it make sense to grow water-intensive crops in irrigated desert ecosystems or erosion-prone crops on hillsides? Or does it make sense to decouple crops from livestock and use large amounts of fertilizers to produce feed? Should native legume-based systems or intercropping be further encouraged in poor areas of Sub-Saharan Africa or South Asia where soil fertility is low, fertilizers expensive, and protein deficiencies high? The shift toward production systems that are more compatible with the human and natural resource base leads to another principle for redesign: the promotion of crop diversity on local to global scales (see Chapter 2). With the projected magnitude of climate changes to come, it will be necessary to encourage a wide range of cropping systems that are suitable for the future climate, not just to adapt existing crops—particularly the major crops of maize, rice, and wheat—to increased temperature or drought tolerance in existing areas. Creating a resilient path for food production systems worldwide begins, in the words of Wendell Berry (2005) "in the recognition and acceptance of limits." It also begins with a broad understanding of the economic, social, and climate forces that are changing the agricultural landscape (see Chapters 2 and 3).

In order to promote a vision of redesign operationally, the second step is that government policies influencing the trajectory of food production systems must be altered in order to re-couple agriculture with its environmental support systems (Robertson and Swinton 2005). For example, distortionary policies that encourage crop or animal production that is incompatible with resource constraints should be discarded, and incentives to improve the efficiency of input use and eliminate external impacts from agriculture should be introduced. The difficulty lies in the many policies created without agriculture or livestock in mind (such as macro policy), which affect the profitability and structure of food production systems. Moreover, a wide range of policies initiated by different agencies with little or no overlap typically shape food production systems within any country, a point made clearly by the current role of energy policy and agricultural policy in the USA. Given the enormous influence of policies on crop and animal systems throughout the world, aligning economic incentives for resilience is perhaps the most challenging step.

The third and final step toward the resilience transition is to promote the scientific and information tool kit that enables farmers to anticipate and respond to change within the local cultural context. The scientific tool kit is broad and includes emerging knowledge from a wide range of disciplines (e.g., genetics, agronomy, biology, hydrology, climate science, economics) to improve management practices, breeding efforts, cultivar development and deployment, local adaptation to climate change, water availability, and options for income and nutritional enhancement. Advanced genetics will play a key role in identifying desired crop traits and increasing productivity of both major and minor crops under evolving stresses; this field extends beyond the use of genetically modified organisms to include marker-assisted breeding and bioinformatics (Naylor et al. 2004). Integrated management practices that improve input use efficiency (water, nutrients) will be equally, if not more important for enhancing crop productivity, incomes, and environmental services in all areas of the world (Cassman et al. 2003). Finally, the use of new information technology, such as remote sensing images and geographic information systems (GIS) to identify yield gaps, input use efficiencies, soil and water constraints, and climate projections on regional scales, could help transform the ability of farmers in both rich and poor countries to respond to resource and climate changes over time (Cassman 1999, Lobell and Ortiz-Monasterio 2008).

The main point is that efforts are needed to develop the full tool kit as an integrated package throughout the world, thus helping to preserve options for future adaptation to change (Solow 1991). Such efforts will require the institutional commitment of international agriculture and development agencies and donors, as well as the commitment of national governments, to reach farmers in all regions (Falcon and Naylor 2005). Many farmers are already redesigning their agricultural systems to meet local needs and constraints. It is now time for the global community to embrace a vision of redesign of food production systems to meet human needs without compromising future options, cultural integrity, and environmental quality. The world food crisis that hit agricultural commodity markets in 2008 is a stark reminder that all countries need to build resilient food production systems.

Review Questions

1. What are the major differences between agricultural systems and other ecosystems like drylands or forests? What implications do these differences have for the types of ecosystem stewardship issues that arise in agriculture and the potential challenges and opportunities for solving them?
2. What are the relative costs and benefits of intensification and extensification of agriculture as a way to meet global food needs? How might the costs and benefits be optimized in a developing nation like Indonesia and in a developed country like Sweden or the USA?
3. How do supply and demand for food change with income in a developing nation? How is this affected by eating habits, fuel demand,

and the politics of globalized trade in developed nations?
4. What are important short-term and long-term adjustments to food shortage? What policies might a developing nation pursue to maximize long-term social–ecological sustainability within the context of meeting future needs for food?
5. How might food systems be redesigned to meet the challenges of a rapidly changing planet? What changes in policies might foster the stewardship of agricultural systems, and how might these policy changes be achieved?

Additional Readings

Cassman, K.G. 1999. Ecological intensification of cereal production systems: Yield potential, soil quality, and precision agriculture. *Proceedings of the National Academy of Sciences* 96:5952–5959.

Conway, G. 1997. *The Doubly Green Revolution: Food for All in the 21st Century*. Cornell University Press, Ithaca.

Evans, L.T. 1998. *Feeding the Ten Billion: Plants and Population Growth*. Cambridge University Press. Cambridge.

Matson, P.A., W.J. Parton, A.G. Power, and M. J. Swift. 1997. Agricultural intensification and ecosystem properties. *Science* 227:504–509.

Naylor, R.L. 2000. Agriculture and global change. Pages 462–475 *in* G. Ernst, editor. *Earth Systems: Processes and Issues*. Cambridge University Press, Cambridge.

Naylor, R.L., A. Liska, M. Burke, W. Falcon, J. Gaskell, et al. 2007. The ripple effect: Biofuels, food security and the environment. *Environment* 49(9):30–43.

Power, A.G. 1999. Linking ecological sustainability and world food needs. *Environment, Development, and Sustainability* 1:185–196.

Robertson, G.P., and S.M. Swinton. 2005. Reconciling agricultural productivity and environmental integrity: A grand challenge for agriculture. *Frontiers in Ecology and the Environment* 3:38–46.

Smil, V. 2000. *Feeding the World: A Challenge for the 21st Century*. MIT Press, Cambridge, MA.

Tilman, D., K. Cassman, P. Matson, R. Naylor, and S. Polasky. 2002. Agricultural sustainability and intensive production practices. *Nature* 418: 671–677.

13
Cities: Managing Densely Settled Social–Ecological Systems

J. Morgan Grove

Introduction

Why are Cities and Urbanization Important?

The transition from a rural to urban population represents a demographic, economic, cultural, and environmental tipping point. In 1800, about 3% of the world's human population lived in urban areas. By 1900, this proportion rose to approximately 14% and now exceeds 50% in 2008. Nearly every week 1.3 million additional people arrive in the world's cities (about 70 million a year), with increases due to migration being largest in developing countries (Brand 2006, Chan 2007). People in developing countries have relocated from the countryside to towns and cities of every size during the past 50 years. The urban population on a global basis is projected by the UN to climb to 61% by 2030 and eventually reach a dynamic equilibrium of approximately 80% urban to 20% rural dwellers that will persist for the

foreseeable future (Brand 2006, Johnson 2006). This change from 3% urban population to the projected 80% urban is a massive change in the social–ecological dynamics of the planet.

The spatial extent of urban areas is growing as well. In industrialized nations the conversion of land from wild and agricultural uses to urban and suburban settlement is growing at a faster rate than the growth in urban population. Cities are no longer compact (Pickett et al. 2001); they sprawl in fractal or spider-like configurations (Makse et al. 1995) and increasingly intermingle with wildlands. Even for many rapidly growing metropolitan areas, suburban zones are growing faster than other zones (Katz and Bradley 1999). The resulting new forms of urban development include edge cities (Garreau 1991) and a wildland–urban interface in which housing is interspersed in forests, shrublands, and desert habitats.

Accompanying this spatial change is a change in perspectives and constituencies. Although these habitats were formerly dominated by agriculturists, foresters, and conservationists, they are now increasingly dominated by people possessing resources from urban systems, drawing upon urban experiences, and expressing urban habits.

J.M. Grove (✉)
Northern Research Station, USDA Forest Service, Burlington, VT 05403, USA
e-mail: mgrove@fs.fed.us

F.S. Chapin et al. (eds.), *Principles of Ecosystem Stewardship*,
DOI 10.1007/978-0-387-73033-2_13, © Springer Science+Business Media, LLC 2009

An important consequence of these trends in urban growth is that cities have become *the* dominant global human habitat of this century in terms of geography, experience, constituency, and influence. This reality has important consequences for social and ecological systems at global, regional, and local scales, as well as for natural resource organizations attempting to integrate ecological function with human desires, behaviors, and quality of life.

Urbanization is a Dynamic, Social, and Ecological Phenomenon

Urbanization is having significant and unpredicted effects on the human global population. According to UN projections, the world's overall fertility rate will decline below replacement levels by 2045, due in large part to declining fertility in cities. In cities women tend to have both more economic opportunities and more reproductive control, and the economic benefit of children depends less on the quantity of children than on their quality, particularly in terms of education. This trend is amplified by the fact that the social and financial costs of childbearing and childrearing continue to rise in cities (Brand 2006).

Birthrates on a national basis have already dropped as a result of rapid urbanization in both developed and developing nations. In the case of the developing world, the fertility rate has declined from six children per woman in 1970 to 2.9 currently. In twenty emerging-economy countries—including China, Chile, Thailand, and Iran—the fertility rate has declined below the replacement rate of 2.1 children per woman (Brand 2006).

What this means in a global, long-term demographic context is that, although the world's population doubled in a single generation for the first time in human history, from 3.3 billion in 1962 to 6.5 billion now, this is unlikely to occur again. The "population momentum" of our current global population and our children will carry the world population to a peak of 7.5–9 billion around 2050 and then decline (Brand 2006).

Urbanization creates both ecological vulnerabilities and efficiencies. For instance, coastal areas, where many of the world's largest cities occur, are home to a wealth of natural resources and are rich with diverse species, habitat types, and productive potential. They are also vulnerable to land conversion, changes in hydrologic flows, outflows of waste, and sea level rise (see Chapter 12; Grimm et al. 2008). In the USA, 10 of the 15 most populous cities are located in coastal counties (NOAA 2004) and 23 of the 25 most densely populated US counties are in coastal areas. These areas have already experienced ecological disruptions (Couzin 2008).

The link between urbanization and coastal areas is evident on a global basis as well. Because of the coastal locations of many major cities, urban migration also brings people to coastlines around the world in one of the greatest human migrations of modern times. The most dramatic population growth has occurred in giant coastal cities, particularly those in Asia and Africa. Many experts expect that cities will have to cope with almost all of the population growth to come in the next two decades, and much of this increase will occur in coastal urban centers (Brand 2006, Johnson 2006).

While ecological vulnerabilities are significantly associated with urban areas, urbanization also fosters ecological efficiencies. The ecological footprint of a city, i.e., the land area required to support it, is quite large (Folke et al. 1997, Johnson 2006, Grimm et al. 2008). Cities consume enormous amounts of natural resources, while the assimilation of their wastes—from sewage to the gases that cause global warming—also are distributed over large areas. For example, London occupies 170,000 ha and has an ecological footprint of 21 million hectares—125 times its size (Toepfer 2005). In Baltic cities, the area needed from forest, agriculture, and marine ecosystems corresponds to approximately 200 times the area of the cities themselves (Folke et al. 1997).

Ecological footprint analysis can be misleading, however, for numerous reasons (Deutsch et al. 2000). It ignores the more important question of efficiency, defined here as persons-to-area: how much land area (occupied area and footprint area) is needed to support a certain number of persons? From this perspective, it becomes clear that urbanization is critical to delivering a more ecologically sustainable and resource-efficient world because the per-person environmental impact of city dwellers is generally lower than people in the countryside, and it can be reduced still further (Brand 2006, Johnson 2006, Grimm et al. 2008). For instance, the average New York City resident generates about 29% of the carbon dioxide emissions of the average American. By attracting 900,000 more residents to New York City by 2030, New York City can actually save 15.6 million metric tons of carbon dioxide a year relative to the emissions of a more dispersed population (Chan 2007).

The combined effects of urbanization on migration, fertility, and ecological efficiency may mean that social–ecological pressures on natural systems can be dramatically reduced in terms of resources used, wastes produced, and land occupied. This may mean that cities can provide essential solutions to the long-term social–ecological viability of the planet given current population trends for this century.

Are Urban Areas Ecosystems?

Earlier chapters about low-density, social–ecological systems such as drylands, forests, and oceans took pains to point out the pervasive importance of social processes in governing social–ecological dynamics. Conversely, the fundamental importance of *ecological* processes is sometimes overlooked in cities. An **ecosystem** is an assemblage of organisms interacting with the physical environment within a specified area (see Chapter 1; Tansley 1935, Bormann and Likens 1979). When Tansley (1935) originated the term ecosystem, he carefully noted that "... *ecology must be applied to conditions brought about by human activity. The 'natural' entities and the anthropogenic derivatives alike must be analyzed in terms of the most appropriate concepts we can find.*" Since the 1950s, social scientists have contributed to an expanded view of ecosystems inclusive of humans along a continuum from wilderness to urban areas (Hawley 1950, Schnore 1958,

Duncan 1961, 1964, Burch and DeLuca 1984, Machlis et al. 1997). Public health researchers and practitioners have provided supplemental perspectives (Northridge et al. 2003), and transdisciplinary approaches have been proposed to implement a social–ecological framework (Chapters in this book; Elmqvist et al. 2004, Collins et al. 2007).

An urban ecosystem perspective retains a concern with ecological structure and function, including biophysical fluxes (Stearns and Montag 1974, Boyden et al. 1981, Burch and DeLuca 1984a, Warren-Rhodes and Koenig 2001) and ecological regulation of system dynamics (Groffman et al. 2003, Pickett et al. 2008). At the same time, demographic, social, and economic structures and fluxes clearly exert important controls over these dynamics as well (Burch and DeLuca 1984a, Grove and Burch 1997, Machlis et al. 1997). Integrating social–ecological structure, function, and regulation of urban ecosystems is therefore essential to an understanding of the ecology of cities (Grimm et al. 2000, Pickett et al. 2001) as open complex adaptive systems that can be characterized in terms of vulnerability and resilience. In contrast to rural areas, urban social–ecological systems are distinguished by a high population density, the built environment, and livelihoods that do not directly depend on the harvest or extraction of natural resources. Finally, ecosystem service concerns are likely to differ between cities and many rural areas, particularly cultural services such as social identity, knowledge, spirituality, recreation, and aesthetics.

Why Use the Approaches Described in This Book?

In very broad historical terms we have begun a new paradigm for cities. Since the 1880s, a great deal of focus has centered on the "Sanitary City," with concern for policies, plans, and practices that promoted public health (Melosi 2000). While retaining the fundamental concern for the Sanitary City, we have begun to envelope the Sanitary City paradigm with a concern for the "Sustainable City," which places urbanization in a social–ecological context on a local, regional, and global basis.

Urban ecology has a significant role to play in this context. Already, urban ecology has an important applied dimension as an approach used in urban planning, especially in Europe. Carried out in city and regional agencies, the approach combines ecological information with planning methodologies (Hough 1984, Spirn 1984, Schaaf et al. 1995, Thompson and Steiner 1997, Pickett et al. 2004, Pickett and Cadenasso 2007).

Cities face challenges that are increasingly complex and uncertain. Many of these complexities are associated with changes in climate, demographics, economy, and energy at multiple scales. Because of these complex, interrelated changes, concepts such as resilience, vulnerability, and ecosystem services may be particularly useful for addressing current issues and opportunities as well as preparing for potential future scenarios requiring long-term, and frequently capital-intensive, change.

Cities have already begun to address these challenges and opportunities in terms of policies, plans, and management. For example, on June 5, 2005, mayors from around the globe took the historic step of signing the Urban Environmental Accords—Green City Declaration with the intent of building ecologically sustainable, economically dynamic, and socially equitable futures for its urban citizens. The Accord covered seven environmental categories to enable sustainable urban living and improve the quality of life for urban dwellers: (1) energy, (2) waste reduction, (3) urban design, (4) urban nature, (5) transportation, (6) environmental health, and (7) water (www.urbanaccords.org). International associations such as *ICLEI-Local Governments for Sustainability* (http://www.iclei.org/) are developing and sharing resources to address these issues.

The ability to address these seven categories will require numerous, interrelated strategies. New York City's plaNYC for *A Greener, Greater New York* (http://www.nyc.gov/html/planyc2030/downloads/pdf/full_report.pdf), for example, includes 127 different but interrelated strategies for making the city more sustainable, dynamic, and equitable. However, many cities are managed in disciplinary and fragmented

ways. In some sense, city agencies and non-governmental organizations (NGOs) too often resemble traditional university departments separated by academic disciplines. The essence of this situation readily maps to Yaffee's (1997) "recurring nightmares" (Chapter 4): (1) a process in which short-term interests out-compete long-term visions and concerns; (2) conditions in which competition supplants cooperation because of the conflicts that emerge in management issues: (3) the fragmentation of interest and values; (4) the fragmentation of responsibilities and authorities (sometimes called "functional silos" or "stove pipes"); and (5) the fragmentation of information and knowledge, which leads to inferior solutions.

To address these "recurring nightmares," universities and cities alike have begun to reorganize themselves in part by creating Offices of Sustainability (Deutsch 2007). A fundamental challenge to these types of offices and the polycentric networks in which they exist will be to understand urban ecosystems as complex adaptive systems in order to build resilient urban futures that are ecologically sustainable, economically dynamic, and socially equitable.

The following section applies several of the resilience principles described earlier in this book for understanding and building more resilient urban futures: (1) cities are open, and multiscale systems, (2) cities are heterogeneous and ecosystem composites, and (3) cities are complex adaptive systems.

Principles

Cities are Open, and Multiscale Systems

The recognition that cities are open multiscale systems has only recently become evident in the ecological study of urban areas (Pickett et al. 1997a, Grimm et al. 2000). Urban ecology began as the study of "ecology *in* cities," which focused historically on ecologically familiar places and compared urban and nonurban areas: parks as analogs of rural forests (e.g. Attorre et al. 1997, Kent et al. 1999) and vacant lots as analogs of fields or prairies (Vincent and Bergeron 1985, Cilliers and Bredenkamp 1999). Urban streams, rock outcrops, and remnant wetlands were the object of ecological studies similar in scope and method to those conducted in nonurban landscapes. There is a long European tradition of these types of ecological studies *in* cities (Sukopp et al. 1990, Berkowitz et al. 2003).

The study of "ecology *of* cities" builds upon the focused efforts of "ecology *in* cities," while incorporating a more expansive approach to cities that is consistent with the social–ecological approaches described in this book (see Chapters 1–5). In particular, the ecology-*of*-cities approach developed in response to the recognition of the open and multiscale nature of cities. Input–output budgets of a city were the first type of ecology-*of*-cities approach addressing the open nature of cities. This budgetary approach relies on a "closed box" approach to ecosystems. Inputs and outputs are measured and the processes within the system are implicitly assumed to be homogeneous. This approach is similar to the ecosystem ecology of the 1960s and 1970s and has been used by ecologists (Bormann and Likens 1967), environmental historians (Cronon 1991), and social scientists (Stearns and Montag 1974). The material and energy budget of Hong Kong (Boyden et al. 1981) and the nitrogen budget of New Haven, Connecticut (Burch and DeLuca 1984) are examples. The lack of interdisciplinary experts noted by Boyden et al. (1981) and the apparent lack of interest by mainstream ecology constrained this approach to cities. However, the two urban projects of the Long-Term Ecological Research (LTER) Network, the Baltimore Ecosystem Study (BES) and Central Arizona Phoenix (CAP) program, have developed nitrogen budgets in terms of both internal dynamics and inputs–outputs for their urban ecosystems (Baker et al. 2001, Groffman et al. 2004).

Ecology *of* cities in its contemporary form incorporates new approaches from ecology in general and from ecosystem ecology in particular. It also benefits from relatively new specialties such as landscape ecology, which focuses on the functional consequences of spatial heterogeneity. It further benefits from increasing interdisciplinary work and training. Together, these developments make the inclusive approach to ecology *of* cities very differ-

ent from the examples from the 1970s and early 1980s. There are several reasons for this difference. First, the ecology *of* cities addresses the whole range of habitats in metropolitan systems, not just the green spaces that are the focus of ecology *in* cities. Second, spatial heterogeneity, expressed as gradients or mosaics, is critical for explaining interactions and changes in the city. Third, the role of humans at multiple scales of social organization, from individuals through households and ephemeral associations, to complex and persistent agencies, is linked to the biophysical scales of the metropolis. Finally, humans and their institutions are a part of the ecosystem, not simply external, negative influences. This opens the way toward understanding feedbacks among the biophysical and human components of the system, toward placing them in their spatial and temporal contexts, and toward examining their effects on ecosystem inputs and outputs at various social scales, including individuals, households, neighborhoods, municipalities, and regions (Grove and Burch 1997).

Cities are Heterogeneous and Ecosystem Composites

Urban ecosystems are notoriously heterogeneous or patchy (Jacobs 1961, Clay 1973). Biophysical patches are a conspicuous layer of heterogeneity in cities. The basic topography, although sometimes highly modified, continues to govern important processes in the city (Spirn 1984). The watershed approach to urban areas has highlighted the importance of slopes, and of patchiness along slopes, in water flow and quality (Band et al. 2006). Steep areas are often the sites of remnant or successional forest and grassland in and around cities. Soil and drainage differ with the underlying topography. Vegetation, both volunteer and planted, is an important aspect of biophysical patchiness. The contrast in microclimate between leafy, green neighborhoods versus those lacking a tree canopy is a striking example of biotic heterogeneity (Nowak 1994). Additional functions that may be influenced by such patchiness include carbon storage (Jenkins and Riemann 2003), animal biodiversity (Adams 1994,

Hostetler 1999, Niemela 1999), social cohesion (Grove 1995, Colding et al. 2006), and crime (Dow 2000, Troy and Grove 2008).

Social and economic heterogeneity is also pronounced in and around cities. Patchiness can exist in such social phenomena as economic activity and livelihoods, family structure and size, age distribution of the human population, wealth, educational level, social status, and lifestyle preferences (Burch and DeLuca 1984, Field et al. 2003).

Temporal dynamics are just as important as spatial pattern, since none of these social patterns are fixed in time. This insight is a key feature of the socio-spatial (Gottdiener and Hutchinson 2001) and patch dynamics approaches to urban ecosystems (Pickett et al. 1997b, Grimm et al. 2000, Pickett et al. 2001). It is a critical feature for including the built nature of cities as well. Most people, and indeed most architects and designers, assume that the built environment is a permanent fixture. However, buildings and infrastructure change, as do their built and biophysical context. This elasticity in the urban system suggests a powerful way to reconceptualize urban design as an adaptive, contextualized pursuit (Pickett et al. 2004, Shane 2005, Colding 2007, McGrath et al. 2008). Such dynamism combines with the growing recognition of the role of urban design in improving the ecological efficiencies and processes in cities. Although this application of patch dynamics is quite new, it has great promise to promote the interdisciplinary melding of ecology and design and to generate novel designs with enhanced environmental benefit (McGrath et al. 2008). Thus, patches in urban systems can be characterized by biophysical structures, social structures, built structures, or a combination of the three at multiple scales (Cadenasso et al. 2006).

Not only are urban areas heterogeneous, they are ecological composites, constituted by many of the ecosystem types described in this book: forests, drylands, freshwaters, estuaries, coastal areas, and urban gardens (see Chapters 8–12), combined with the social attributes described in Chapters 3–4. Because cities are ecological composites, they often include the ecological and social characteristics associated with these individual ecosystem types. In the case of freshwaters, for example, cities bordering lakes and rivers are affected by the spatial heterogeneity of flows; the interaction of fast and slow variables; nonlinearities and thresholds in system change; the need to account for blue and green water; and management systems that are fragmented and require continuous adaptation. Likewise, in the case of estuaries, social sources of resilience depend upon monitoring programs, spatial complexity, some degree of localized management control, and a willingness to entertain and implement actively adaptive, experimental management policies (Felson and Pickett 2005). Because of this urban ecological composition, there is a great deal to learn and adapt from the experiences and knowledge of these particular ecosystem types to urban settings.

Cities are Complex Adaptive Systems

The fact that cities are complex adaptive systems is manifest in the definition provided in Chapter 1: *Systems whose components interact in ways that cause the system to adjust or "adapt" in response to changes in conditions. This is a simple consequence of interactions and feedbacks.* These interactions and feedbacks are expressed in urban areas in several ways.

Cities, like all social–ecological systems, exist in a state of nonequilibrium (Pickett and Cadenasso 2008) as a result of both major disruptions (pulses) and chronic stresses (presses). Pulses include disease epidemics, droughts, famines, floods, earthquakes, fires, and warfare. Long-term presses result from demographic changes caused by immigration, emigration, and/or aging; changes in economy through transitions from agriculture, manufacturing, shipping, and service economies (Fig. 13.1); and changes in transportation systems including water, rail, auto, and air. The dynamic results of these long-term press and pulse forces are manifest in Batty's (2006) long-term rank clocks for urban areas in the USA (1790–2000) and the planet (430 BC–AD 2000), which illustrate the

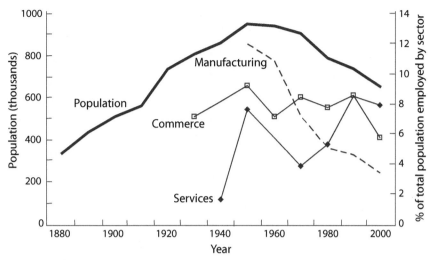

FIGURE 13.1. Long-term trends in population and economy for the city of Baltimore, USA, 1880–2000. Data from the LTER Ecotrends Project: Socioe-conomic Catalogue (http://coweeta.ecology.uga.edu/trends/catalog trends base2.php).

long-term dynamics of cities in terms of population size over time (Figs. 13.2, 13.3).

Social institutions (the rules of the game) play a key role in the adjustments or adap-tations of cities to changing conditions (see Chapter 4, Burch and DeLuca 1984, Machlis et al. 1997). Institutions, such as property rights, direct the allocation of resources to individuals

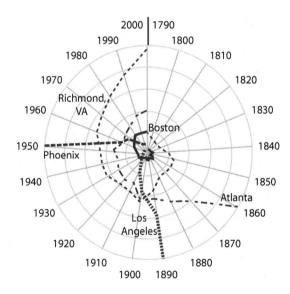

FIGURE 13.2. The trajectory of relative population rank of four of the 100 most populous US cities from 1790 to 2000. The most populous city is at the center of the rank clock, and the hundredth city is at the periphery. Boston, an important colonial city in the northeastern USA, has always been one of the largest US cities. Richmond, another impor-tant colonial city, declined in importance during the early twentieth century, as new cities like Los Ange-les in the western USA became important. Phoenix, a desert city attractive to retired persons, became important only in the last 50 years. Very few cities have remained among the most populous US cities throughout their history. Modified from Batty (2006).

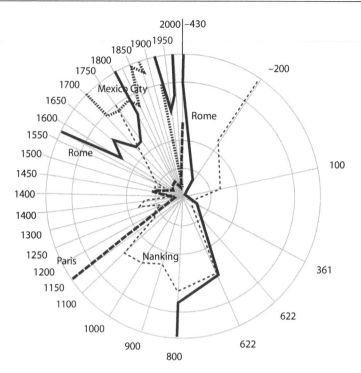

FIGURE 13.3. The trajectory of relative population rank of 4 of the 50 most populous world cities from 430 BC to 2000 AD. The most populous city is at the center of the rank clock, and the fiftieth city is at the periphery. Rome (Italy) and Nanking (China) remained important throughout most of this 2500-year history. Most cities, however, have had a highly volatile population history, with many cities of the developed world (e.g., Paris, France) declining in relative rank during the industrial revolution and cities from the developing world (e.g., Mexico City) becoming more populous, a trend that is likely to continue in the future. Modified from Batty (2006).

and organizations and affect human interactions. Social institutions can be thought of as dynamic solutions to universal needs, including health, justice, faith, commerce, education, leisure, government, and sustenance (Machlis et al. 1997). While the structure of these institutions is important, structure should not be mistaken for function: the health of individuals and populations, the exchange of goods and services, the provisioning of food, water, energy, and shelter (Machlis et al. 1997). In this context, different forms of social institutions might yield identical functions. The ability of social institutions to change in form and yet continue to yield comparable institutional functions is a key element to the adaptive capacity of urban social–ecological systems.

Social institutions are interrelated and often depend upon polycentric governance and social capital within and among cities (see Chapters 4 and 5). For instance, social capital is a crucial factor differentiating "slums of hope" from "slums of despair" (Box 13.1; Brand 2006). This is where community-based organizations (CBOs) and NGOs that support local empowerment play critical roles (Colding et al. 2006, Lee and Webster 2006, Andersson et al. 2007). Typical CBOs include, according to a 2003 UN report, "community theater and leisure groups; sports groups; residents associations or societies; savings and credit groups; child care groups; minority support groups; clubs; advocacy groups; and more.... CBOs as interest associations have filled an institutional vacuum, providing basic services such as communal kitchens, milk for children, income-earning schemes and cooperatives" (Brand 2006). CBOs and NGOs can be diverse, not necessarily focused on "environmental" issues,

---

Box 13.1. Slums: Past and Present

It can be argued that cities have rarely been the result of grand master plans and that all of the world's current major cities began as disreputable shantytowns (Brand 2006, Neuwirth 2006). Slums remain a prominent issue of concern today and, in many ways, the megacities of the developing world are struggling with the same issues of uncharted and potentially unsustainable growth that industrial cities in Europe and the USA faced in the late 1800s and early 1900s (Johnson 2006). By 2015 for instance, the five largest cities on the planet will be Tokyo, Mumbai, Dhaka, Sao Paulo, and New Delhi, with populations greater than 20 million each. Most of this growth will occur in shantytowns: built on illegally occupied land without the guidance of civic planning professionals or traditional infrastructure to support its growth. By some estimates, 25% of the world's population will live in shantytowns by 2030 (Johnson 2006).

Although shantytowns lack the formal plans and infrastructure of urban areas in developed countries, they are dynamic places of social innovation and creativity with economic activities of ordinary life: shops, banks, and restaurants. All of this has been accomplished without urban planners, without government-created infrastructure, and without formal property deeds. While these shantytowns might offend some persons' sense of order, these shantytowns are not exclusively places of poverty and crime. In fact, they are where the developing world goes to get out of poverty (Brand 2006). They are a reminder that different social forms might yield identical functions; that the ability of social institutions to change in form yet continues to yield comparable institutional functions is a key element to the adaptive capacity of urban social–ecological systems.

It might be negligent to claim shantytown residents do not need nor want formal civic resources in terms of expertise, investments, and opportunities in areas such as epidemiology, public infrastructure, education, engineering, waste management, and recycling (Brand 2006, Johnson 2006). But it would also be negligent to not recognize the enormous variety of shantytown experiences among the thousands of "emerging cities with different cultures, nations, metropolitan areas, and neighborhoods. From this variety is emerging an understanding of best and worst governmental practices—best, for example, in Turkey, which offers a standard method for new squatter cities to form; worst, for example, in Kenya, which actively prevents squatters from improving their homes. Every country provides a different example. Consider the extraordinary accomplishment of China, which has admitted 300 million people to its cities in the last 50 years without shantytowns forming, and expects another 300 million to come" (Brand 2006:13). And in this process it might be important to examine the ecological footprint of shantytowns, with their extreme density, low-energy use, and ingenious practices of recycling everything. Maybe there are ideas there that could be generalized on a global basis (Brand 2006).

Religious groups play significant roles as well. According to Davis (2006) "Populist Islam and Pentecostal Christianity (and in Bombay, the cult of Shivaji) occupy a social space analogous to that of early twentieth-century socialism and anarchism. In Morocco, for instance, where half a million rural emigrants are absorbed into the teeming cities every year, and where half the population is under 25, Islamicist movements like 'Justice and Welfare,' founded by Sheik Abdessalam Yassin, have become the real governments of the slums: organizing night schools, providing legal aid to victims of state abuse, buying medicine for the sick, subsidizing pilgrimages and paying for funerals." He adds that "Pentecostalism is...the first major world religion to have grown up almost entirely in the soil of the modern urban slum" and "since 1970, and largely because of its appeal to slum women and its reputation for being colour-blind, [Pentecostalism] has been growing into what is arguably the largest self-organized movement of urban poor people on the planet" (Brand 2006, Johnson 2006).

The networks among cities are important too. As New York City prepared its *Greener, Greater New York* plan, David Doctoroff, deputy mayor for Economic Development, noted that, "We shamelessly stole congestion pricing from London and Singapore, renewable energy from Berlin, new transit policies from Hong Kong, pedestrianization and cycling from Copenhagen, bus rapid transit from Bogota, and water-cleaning mollusks from Stockholm" (Chan 2007).

Many of the social and ecological interactions and feedbacks in urban ecosystems confer resilience: the capacity of a social–ecological system to absorb a spectrum of shocks or perturbations without fundamentally altering its structure, functioning, and feedbacks. Resilience depends on (1) adaptive capacity; (2) biophysical and social legacies that contribute to diversity and provide proven pathways for rebuilding; (3) the capacity of people to plan for the long term within the context of uncertainty and change; (4) a balance between stabilizing feedbacks that buffer the system against stresses and disturbance and innovation that creates opportunities for change; and (5) the capacity to adjust governance structures to meet changing needs (see Chapter 1; Gunderson and Holling 2002, Folke 2006, Walker and Salt 2006).

Biophysical and social legacies can significantly affect the resilience of urban systems. These legacies or dependencies can be temporal or spatial. For instance, historic residential segregation and industrial development in Baltimore, Maryland, created a situation in which predominantly white neighborhoods are in proximity to TRI sites (toxic release inventory sites; Boone 2002). Legacies [dependencies] can be spatial too. For example, as police departments decide how to allocate scarce resources, part of their decision-making process is to label neighborhoods as green (no crime problems), yellow (some crime problems), and red (severe crime problems). Whether the police department expends resources in a neighborhood depends upon their assessment of that neighborhood as well as adjacent neighborhoods. Thus, if a neighborhood is coded yellow and is bordering green neighborhoods, police resources are dedicated to the yellow neighborhood to reduce the risk of spillover or contagion into green neighborhoods. If a neighborhood is coded yellow and is bordering red neighborhoods, police resources may not be invested because the likelihood of decline from yellow to red is too great. Finally, temporal and spatial legacies can be **prospective**. In other words, people take specific actions in specific places today because they believe those actions will influence the capacity of future residents and governance networks to meet short-term and long-term challenges and opportunities.

Working prospectively to create social-ecological legacies is a profound challenge. As Doctoroff notes (Chan 2007), "sustainability is an almost sacred obligation to leave this city better off for future generations than we who are here today have found it." The greatest challenges are not matters of technology, but rather issues of "will and leadership." Short-term sacrifices for long-term gains "are not things that, by its very nature, the political system is equipped to decide."

While Doctoroff's observation is well-founded, there is already evidence for urban resilience. Cities are the most long-lived of all human organizations. The oldest surviving corporations, the Sumitomo Group in Japan, and Stora Enso in Sweden, are about 400 and 700 years old, respectively. The oldest universities in Bologna and Paris have been in place more than 1,000 years. The oldest living religions, Hinduism and Judaism, have existed more than 3,500 years. In contrast to these corporations and religions, the town of Jericho has been continuously occupied for 10,500 years and its neighbor, Jerusalem, has been an important city for 5,000 years, though it has been conquered or destroyed 36 times and experienced 11 religious conversions (Brand 2006). Many cities die or decline to irrelevance, but some thrive for millennia (Batty 2006, Brand 2006, Johnson 2006; Fig. 13.3).

Part of a city's durability is associated with the fact that it is constantly changing. In Europe, cities replace 2–3% per year of their material fabric (buildings, roads, and other construction) by demolishing and rebuilding it. In the USA and the developing world, that turnover occurs even faster. Yet within all of that turnover something about a city remains deeply constant and self-inspiring (Brand 2006). Some combination of geography, economics, and cultural identity ensures that even a city destroyed by war (Warsaw, Dresden, Tokyo) or fire (London, San Francisco) will often be rebuilt (Johnson 2001, Brand 2006, Johnson 2006). Thus, the resilience of urban ecosystems rarely depends upon a single factor, but a diversity of interacting social–ecological feedbacks.

Cities as Functional Social–Ecological Systems

Ecosystem Services and the Dynamics of Cities

The existence, significance, and dynamics of ecosystem services—supporting, provisioning, regulating, and cultural services—have been only partially characterized in urban areas (Bolund and Hunhammar 1999, Colding et al. 2006, Farber et al. 2006, Andersson et al. 2007). Their existence in urban areas is increasingly well-documented, although frequently not identified as ecosystem services per se. For instance, there is growing knowledge and data about urban ecosystems describing soil dynamics and nutrient flux regulation (supporting); production of freshwater, food, and biodiversity (provisioning); modification of climate, hydrology, and pollination (regulation); and links to social identity and spirituality and importance to recreation and aesthetics (cultural).

The significance of these ecosystem services is often poorly understood. One approach is to estimate the monetary value of ecosystem services, for example, the capacity of ecosystems to purify water for New York City (see Chapter 9).

Although this approach is valuable in addressing certain issues, it may miss important aspects of the dynamics of urban ecosystems: "Do variations in ecosystem services affect the desirability of cities?" In other words, do ecosystem services affect where households (families) and firms (businesses) choose to locate in order to avoid some places (push) and seek other places (pull)? Clearly, this may be the case as households and firms seek places that afford, for instance, clean air and water, recreation and communal opportunities, and efficiencies in energy and transportation. In contrast to other types of ecosystems, provision of and access to these ecosystem services is regulated by a combination of ecological, social, and built systems.

The dynamics or interactions among ecosystem services in urban areas are poorly understood. Preferences for ecosystem services may vary over time. For instance, in the early 1900s in Baltimore, Maryland, households preferred to live close to industrial factories where they worked and were not concerned with air quality. With growing knowledge about the relation between air quality and health, households now value clean air more than proximity to their work place, and desirability of these neighborhoods has declined (Boone 2002).

Preferences for ecosystem services may vary among social groups. Certain ethnic groups may prefer different recreation or communal opportunities. Demographics or life-stage is important too. Young families with children may prefer houses with large yards and nearby play-parks for their children, while retired couples may prefer to live in apartments close to greenways and waterfront promenades for their daily walks.

Preferences for ecosystem services may be conditioned by interacting factors. For instance, living close to a park in Baltimore is generally highly desirable. Proximity to a park increases the value of a home in neighborhoods with low levels of crime. However, in neighborhoods with high levels of crime, living close to a park actually depresses the value of a home (Troy and Grove 2008).

Finally, preferences for ecosystem services may be nonlinear and characterized by thresholds. In other words, more is not always better. For example, increases of tree canopy cover in Baltimore led to increased environmental satisfaction up to a threshold of about ˜60%, above which environmental satisfaction no longer increased.

In sum, we increasingly understand and appreciate the existence and importance of ecosystem services in urban areas. However, we are only beginning to understand the dynamics of human responses to variation in these services. Understanding these dynamics is crucial for understanding the push–pull drivers of urban ecosystems and their resilience over time (Borgstrom et al. 2006).

Management of Complex Adaptive Systems

Cities rarely are the result of grand master plans. Rather, they often exhibit emergent properties that are the result of bottom-up processes driven by diverse interests, agencies, and events (Jacobs 1961, Johnson 2001, Shane 2005, Johnson 2006, Batty 2008, Grimm et al. 2008). Given the bottom-up nature of these processes and emergent properties, it is remarkable how similar cities tend to be in their functions (Batty 2008, Grimm et al. 2008).

Polycentric governance (see Chapter 4) among different types of organizations—government agencies, NGOs, and CBOs—and across scales often exist. These networks tend to focus on a specific issue or interest: the stovepipes that are part of Yaffee's recurring nightmares. The ability of polycentric governance networks to contribute to urban resilience depends upon their capacity to (1) address the essential interdependence of demographic, economic, social, and ecological challenges and solutions that cities face; (2) plan for the long term within the context of uncertainty and change; and (3) adjust governance structures to meet changing needs (see Chapter 5). This requires the ability to sense and interpret patterns and processes at multiple scales. At a local scale, there is a growing trend among city governments to develop GIS-based systems that monitor a wide range of indicators in nearly real-time. For

instance, a number of US cities have developed **CitiStat-type programs** that provide accurate and timely intelligence, develop effective tactics and strategies, rapidly deploy resources, and facilitate follow-up and assessments (http://www.baltimorecity.gov/news/citistat/).

New York City has taken the CitiStat approach to a new level with its 311 system. The 311 system functions in three ways, citizens reporting problems to the city, citizens requesting information about city services, and as a mutually learning system that builds upon the first two functions. The novel idea behind the service is that this information exchange is genuinely two-way. The government learns as much about the city as the 311 callers do. In a sense, the City's 311 system functions as an immense extension of the city's perceptual systems, harnessing millions of ordinary "eyes on the street" to detect emerging problems or report unmet needs (Johnson 2006).

New York City's 311 approach makes manifest two essential principles for how cities can generate and transmit good ideas. First, the elegance of technologies like 311 is that they amplify the voices of local amateurs and "unofficial" experts and, in doing so, they make it easier for "official" authorities to learn from them. The second principle is the need for lateral, cross-disciplinary flow of ideas that can challenge the disciplinary stovepipes of knowledge, data, interests, and advocacy associated with many government agencies, NGOs, and the training of professionals (see Chapters 4 and 5). This second principle is increasingly realized by the polycentric and interdisciplinary nature of sensing and interpretation facilitated by the Web and new forms of amateur cartography built upon services like Google Earth and Yahoo! Maps. Local knowledge that had so often remained in the minds of neighborhood residents can now be translated into digital form and shared with the rest of the world. These new tools have begun to unleash a revolution in the exchange and interpretation of data because maps no longer need to be created by distant professionals. They are maps of local knowledge created by local residents. And these maps are street-smart and multimedia. They can map blocks that are not safe after dark, playgrounds that need to be renovated, community gardens with available plots, or local restaurants that have room for strollers. They can map location of trees, leaking sewers, and stream bank erosion. These tools enable locals to map their local history as well—where things were or what things were like—creating and sharing long-term, local knowledge (Johnson 2006).

The scale of these observations can broaden from a neighborhood to an entire planet as these local data and knowledges become networked. Formal examples of these types of systems already exist. Public health officials increasingly have global networks of health providers and government officials reporting outbreaks to centralized databases, where they are automatically mapped and published online. A service called GeoSentinel tracks infectious diseases among global travelers and the popular ProMED-mail email list provides daily updates on all known disease outbreaks around the world (Johnson 2006). Although these types of systems are intended to be early warning systems on specific topics, they are likely to become interdisciplinary as the interdependence of challenges and solutions are recognized and facilitate comparisons and learning among both official and unofficial experts.

Summary

Cities continue to be tremendous engines of wealth, innovation, and creativity. They are becoming something else as well: engines of health that are both public and environmental, Sanitary and Sustainable. Two great threats loom over this new millennium: global warming and finite supplies of fossil fuel. These two threats may have massively disruptive effects on existing cities in the coming decades. They are not likely to disrupt the macro-pattern of global urbanization over the long term, however (Johnson 2006). The energy efficiencies of cities can be part of the solution for both global warming and energy demands.

The long-term challenges that cities face are social, ecological, and interrelated, including the growth and aging of urban populations; aging infrastructure and incorporation of adaptive technologies; environmental changes and resource limitations; and governance problems, particularly inequality. These challenges *and* their solutions are completely interdependent: sustainability and economic growth can be complementary goals (Chan 2007).

However profound the threats are that confront us today and for the near future, they are solvable (Johnson 2006). It will require approaches that perceive cities as complex, dynamic, and adaptive systems that depend upon interrelated ecosystem services at local, regional, and global scales. Polycentric governance networks will contribute to urban resilience depending upon their adaptive capacity to (1) address the essential interdependence of demographic, economic, social, built, and ecological challenges and solutions that cities face; (2) plan for the long term within the context of uncertainty and change; and (3) adjust governance structures to meet changing needs. This will increasingly involve the ability to sense and interpret patterns and processes at multiple scales. Tools that harness diverse local knowledges and "eyeballs on the street" to engage in genuine exchanges and evaluations of information will become less novel and more routine. The exchange of knowledge among cities will be important on a global basis, as we learn that there are multiple pathways to similar solutions for resilient cities that are ecologically sustainable, economically dynamic, and socially equitable.

Review Questions

1. Are the populations of cities growing more rapidly than the global population? Is this likely to continue indefinitely? Why or why not?
2. In what ways are cities similar to or different from social–ecological systems that have lower population density?
3. Are urbanization and the shift from the Sanitary to the Sustainable City directional changes? Why or why not?

4. How might demands for ecosystem services be similar and different in urban versus less densely settled areas?
5. In what ways is the transition from an "ecology *in* cities" to an "ecology *of* cities" important to understanding cities as open and multiscalar?
6. In what ways does expanding urbanization influence ecological vulnerability, resilience, and efficiency of resource use on a local, regional, and global basis?
7. Evaluate the claim that cities can provide essential solutions to the long-term social–ecological viability of the planet given population trends for this century. How is this true or not true?
8. What are some of the social challenges and opportunities to developing resilient cities?

Additional Readings

Batty, M. 2008. The size, scale, and shape of cities. *Science* 319:769–771.

Grimm, N., J.M. Grove, S.T.A. Pickett, and C.L. Redman. 2000. Integrated approaches to long-term studies of urban ecological systems. *BioScience* 50:571–584.

Grimm, N.B., S.H. Faeth, N.E. Golubiewski, C.L. Redman, J. Wu, et al. 2008. Global change and the ecology of cities. *Science* 319:756–760.

Johnson, S. 2001. *Emergence: The Connected Lives of Ants, Brains, Cities, and Software*. Scribner, New York.

Northridge, M.E., E.D. Sclar, and P. Biswas. 2003. Sorting out the connection between the built environment and health: A conceptual framework for navigating pathways and planning healthy cities. *Journal of Urban Health: Bulletin of the New York Academy of Medicine* 80:556–568.

Pickett, S.T.A., M.L. Cadenasso, and J.M. Grove. 2004. Resilient cities: Meaning, models and metaphor for integrating the ecological, socioeconomic, and planning realms. *Landscape and Urban Planning* 69:369–384.

Pickett, S.T.A., P. Groffman, M.L. Cadenasso, J.M. Grove, L.E. Band, et al. 2008. Beyond urban legends: An emerging framework of urban ecology as illustrated by the Baltimore Ecosystem Study. *BioScience* 58:139–150.

Spirn, A.W. 1984. *The Granite Garden: Urban Nature and Human Design*. Basic Books, Inc., New York.

14
The Earth System: Sustaining Planetary Life-Support Systems

Oran R. Young and Will Steffen

What is the Issue?

The Earth as a whole can be viewed as a social–ecological system; in fact, the largest such system that can exist. The increasing evidence that human activities are now interacting with the natural environment of the Earth at the scale of the planet lends credence to this perspective. The scientific understanding of the human imprint on the planet is well recognized throughout the policy and management sectors and is raising severe challenges to governance structures. Never before has humanity had to devise and implement governance structures at the planetary scale, crossing national boundaries, continents and large biogeographic regions. Responsible stewardship of the global social–ecological system is the ultimate challenge facing humanity, as it entails safeguarding our own life-support system. The Earth as a social–ecological system is a very recent phenomenon. For nearly all of its existence, Earth

has operated as a biophysical system, without the social component, as fully modern *Homo sapiens* arose only about 200,000–250,000 years ago. This long evolution of Earth as a biophysical system provides the canvas on which the human enterprise has exploded with exponentially growing impact in the last micro-instant of Earth's existence. A full understanding of the implications of this phenomenon requires an understanding of Earth as a system, and particularly the natural envelope of environmental variability that provides the conditions for human life on the planet.

Our planet is about 4.6 billion years old, and life on Earth originated about 4 billion years ago. The earlier notion that life has been maintained over this long period because the geophysical conditions of the planet happened to be amenable to life has been modified to acknowledge the role of life itself in maintaining its own environment. Biological processes interact with physical and chemical processes to create and maintain the planetary environment, with life playing a much stronger role than previously thought in modifying its own environment (Lovelock 1979, Levin 1999, Steffen et al. 2004).

The term **Earth System** is increasingly used to describe the complex set of interacting

O.R. Young (✉)
Bren School of Environmental Science and
Management, University of California, Santa Barbara,
CA 93106-5131, USA
e-mail: young@bren.ucsb.edu

F.S. Chapin et al. (eds.), *Principles of Ecosystem Stewardship*,
DOI 10.1007/978-0-387-73033-2_14, © Springer Science+Business Media, LLC 2009

physical, chemical, biological, and anthropogenic processes that define the planet's environment (Oldfield and Steffen 2004). The Earth System has the following major characteristics:

- It is a "materially closed system" – i.e., it does not exchange significant amounts of matter with space – and it has a single primary energy source, the Sun.
- The major dynamical features of the Earth System are (1) the transport and transformation of materials and energy internally throughout the system and (2) a large array of complex feedback processes that give the planetary environment its characteristic systemic nature and, importantly, limit variability within well-defined bounds.
- Human societies are an integral part of the Earth System, not an outside driver perturbing an otherwise natural system. The rapid growth of the human enterprise over the last century or two to become a global geophysical force means that the processes of human societies are now influencing the large feedback loops that modulate the global environment. The Earth System is now truly a social–ecological system and is not just a biophysical system.

The most dramatic evidence for the systemic nature of the planetary environment during the period of human existence comes from the Antarctic ice cores, such as the Vostok record (Fig. 14.1a; Petit et al. 1999). The Vostok ice core shows the variation in temperature and atmospheric gas concentrations (carbon dioxide (CO_2) and methane (CH_4)) over the entire period of human existence on Earth and beyond. Three features of the record are striking. First, the Earth's environment cycles through cold periods ("ice ages") and shorter, intervening warm periods at regular – ca. 100,000 years – intervals, ultimately paced by the Earth's orbit around the Sun. Second, the concentration of CO_2 and CH_4, both greenhouse gases, in the atmosphere and the temperature are closely coupled. This is, however, not a simple cause–effect relationship, but rather a complex coupling involving several global-scale feedback loops. Third, there are well-defined upper and lower limits to the gas concentrations

and to the temperature as Earth cycles through glacial and interglacial states. All of these features are typical of a system with a high degree of internal self-regulation.

Feedback loops involving physical, chemical, and biological processes link physical climate and the carbon cycle (Fig. 14.2; Ridgwell and Watson 2002). Feedback loops consist of processes that either amplify or dampen an initial perturbation (see Chapter 1). Combinations of such processes form linked loops that feed back on themselves, leading to emergent phenomena that can either stabilize the global environment or propel it from one alternative state to another. Understanding the nature of these planetary feedback loops is essential for effective governance of Earth as a social–ecological system.

For nearly all of human existence, but especially during the last 10,000 years, the dynamics of the Earth System have provided an exceptionally accommodating and resilient environment that has allowed our species to grow in number and with ability to modify the environment around us (Fig. 14.1b). This has benefited human society, as we developed agriculture and then villages, cities, and civilizations. This evolution of the human enterprise, from hunter-gatherers to a global geophysical force, is also mirrored in the increase in scale of social–ecological interactions, from local to global.

Humans have been hunter-gatherers living in small groups and, for the most part, in nomadic conditions throughout most of their existence as a distinct species. During this period, however, we slowly learned to manipulate the environment, most notably through the use of fire (Pyne 1997). Other important modifications of the environment include the wave of late Pleistocene extinctions of megafauna, due at least in part to human hunting pressures (Martin and Klein 1984), and the domestication of animals, beginning with the dog about 100,000 years ago. These modifications of the environment were largely carried out at local scales, and, for the most part, were modifications of the natural dynamics of ecosystems rather than wholesale conversions of ecosystems. The imprint of these human activities therefore did not reach

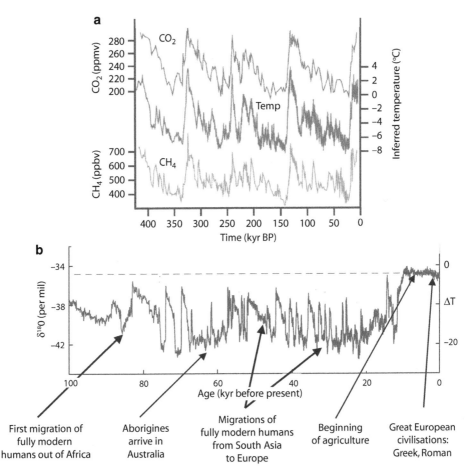

FIGURE 14.1. (a) The 420,000-year Vostok (Antarctica) ice core record, showing the regular pattern of atmospheric CO_2 and CH_4 concentration and inferred temperature through four glacial-interglacial cycles (Petit et al. 1999). Redrawn from Steffen et al. (2004). (b) The last glacial cycle of ^{18}O (an indicator of temperature) and selected events in human history. Information from Oppenheimer (2004).

the global scale, nor did it modify the natural dynamics of the Earth System.

The advent of sedentary agriculture about 10,000 years ago was a major turning point in the development of the human enterprise, with significantly more extensive modifications of the environment at larger scales. The two most important in terms of biophysical impacts were the clearing of forests and the tilling of grasslands for agricultural crops and the irrigation of rice. In some cases, these practices dramatically altered landscapes at a regional level. However, the rate and geographical scale of the spread of agriculture were still modest enough to produce no significant impacts on the dynamics of the Earth System. There is no strong evidence from global environmental records of human impacts associated with early agriculture. Social–ecological systems operated at the local and regional scales only.

All of these changed with the beginning and spread of industrialization around 1800 (Turner et al. 1990). The critical feature of industrialization in terms of the Earth System was the use of fossil fuels as an energy source. Coal, and later oil and gas were accessed from beneath the Earth's surface. These carbon-based fuels had been locked away

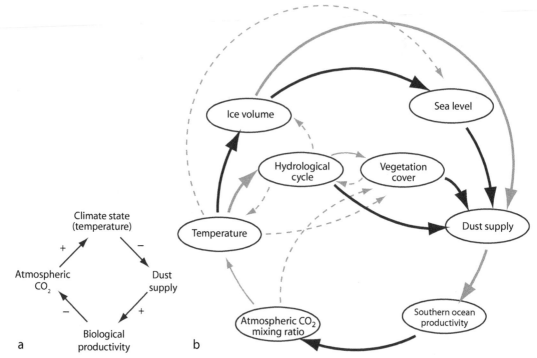

FIGURE 14.2. Feedbacks in the climate system. Different components of the Earth system can directly interact in three possible ways; a positive influence (i.e., an increase in one component directly results in an increase in a second), a negative influence (i.e., an increase in one component directly results in a decrease in a second), or no influence at all. An even number (including zero) of negative influences occurring within any given closed loop gives rise to an amplifying (positive) feedback, the operation of which will act to amplify an initial perturbation. Conversely, an odd number of negative influences gives rise to a stabilizing (negative) feedback, which will tend to dampen any perturbation. In the two schematics, positive influences are shown in grey and negative in black (Ridgwell and Watson 2002). Redrawn from Steffen et al. (2004). (a) Schematic of a simplified feedback system, involving dust, the strength of the biological pump, CO_2, and climatic state (represented by mean global surface temperature). Because there is an even number of negative influences (=2), this represents an amplifying feedback, with the potential to amplify an initial perturbation (in either direction). (b) Schematic of the hypothetical glacial dust-CO_2-climate feedback system, with explicit representation of the various dust mechanisms that we have identified. Primary interactions in the dust-CO_2-climate subcycle are indicated by thick solid lines, while additional interactions (peripheral to the argument) are shown dotted for clarity. For instance, the two-way interaction between temperature and ice volume is the ice-albedo feedback. Four main (amplifying) feedback loops exist in the system: (1) dust supply → productivity → CO_2 → temperature → ice volume → sea level → dust supply (four negative interactions), (2) dust supply → productivity → CO_2 → temperature → hydrological cycle → vegetation → dust supply (two negative interactions), (3) dust supply → productivity → CO_2 → temperature → hydrological cycle → dust volume → dust supply (two negative interactions), (4) dust supply → productivity → xCO_2 → temperature → ice volume → dust supply (two negative interactions).

from the active biogeochemical cycling of carbon between land, ocean, and atmosphere for millions of years. The usage of these fuels began to noticeably perturb the carbon cycle at the global scale. The flow-on effects from the use of fossil-fuel energy systems were even more

dramatic. The ability to chemically "fix" nitrogen from the atmosphere to produce fertilizers led to even more profound changes to the nitrogen cycle than to the carbon cycle. The ability to clear land and to extract resources from coastal and marine ecosystems accelerated with the use of fossil-fuel-driven machinery and the transformation of the environment became global in scope.

The advent of industrialization as the beginning of a new planetary epoch, now called the **Anthropocene** (Crutzen 2002), marks the beginning of Earth itself as a social–ecological system. Even more dramatic has been the explosion of the human enterprise in terms of both population and economic activities after World War II, now labeled the **Great Acceleration** (Fig. 14.3; Hibbard et al. 2006, Steffen et al. 2004). Human population doubled to over 6 billion people in just 50 years, while the global economy expanded fifteen-fold. Resource use has expanded, while mobility and communication have accelerated to levels that could scarcely have been imagined in the first half of the twentieth century (McNeill 2000). The human imprint on the global environment is now unmistakable (Fig. 14.3; Steffen et al. 2004). Climate change is the most well-known aspect of human-influenced change at the global scale, but it is now certain that human activities have also modified the nitrogen, phosphorus, and hydrological cycles and have significantly altered the structure and composition of the atmosphere, oceans (particularly coastal seas), and land.

Directional Changes at the Planetary Scale

Taken together, the recent changes in human actions and Earth System responses (Fig. 14.3) demonstrate without doubt that the Earth has rapidly evolved from its earlier status as a biophysical system to its current status as a social–ecological system. To understand the nature of this planetary social–ecological system, we need to examine in more detail the role of human actions in large-scale biophysical changes, often called global environmental changes, the parallel changes in large-scale social processes, properly interpreted as global social changes, and the interdependencies and interactions between the two. The basic message of this brief account is that the Earth System is a complex and highly dynamic system whose future trajectory is difficult to forecast.

Global Environmental Changes (GECs)

It is helpful to divide the human imprint into two types – *systemic* and *cumulative*. **Systemic processes** are those for which changes anywhere on the planet rapidly affect the Earth System at the global scale. The classic example is emissions of greenhouse gases, which, because of the rapid mixing of the atmosphere, affect the climate at the global scale. **Cumulative processes** are those whose initial impact is local but whose effects occurring in different parts of the world aggregate to produce global consequences. For example, the loss of biodiversity is largely a local process with local consequences, but over the past two centuries, incremental losses of biodiversity on every continent have aggregated to become a global crisis that challenges human well-being (MEA 2005d).

The best-known systemic human imprint on the Earth System is a suite of changes in the composition of the atmosphere. CO_2 concentration now stands at 385 ppm (parts per million), 100 ppm higher than the preindustrial value (the maximum value observed during previous interglacial periods was 280–300 ppm). The current rate of increase is 2–3 ppm per year. Concentrations of other important greenhouse gases, such as methane and nitrous oxide, have also risen significantly over the past two centuries because of human activities.

The atmosphere has also changed as a result of human-driven emissions of aerosols, small particles of various sizes and composition. Unlike greenhouse gases, which have long lifetimes in the atmosphere and are thus well mixed at the global scale, aerosols have lifetimes on the order of days and sometimes a week. They are therefore largely local and regional phenomena and give rise to consequences at that scale, primarily involving human health and the functioning of the terrestrial biosphere.

Climate change is the most obvious consequence of human-driven changes to

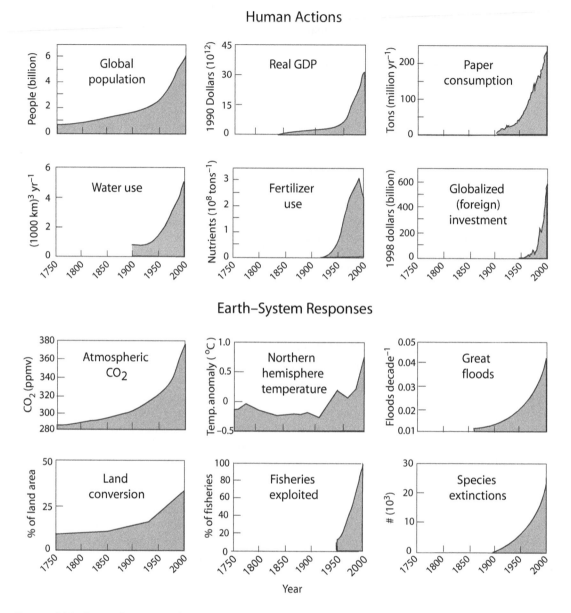

FIGURE 14.3. Increasing rates of change in human actions and Earth System responses since the beginning of the Industrial Revolution. Significant increases in rates of change occur around the 1950s in each case and illustrate how the past 50 years have been a period of dramatic and unprecedented change in human history. Adapted from Steffen et al. (2004).

atmospheric composition. The most recent assessment of the science of climate change (IPCC 2007a) concluded that the warming of the global climate system is unequivocal, as shown by the rise in global mean temperature, the widespread melting of snow and ice, rising sea levels, and other changes in the Earth System. In addition to rising temperature, changes in other aspects of climate have been observed at the scale of continents and ocean basins – wind patterns, precipitation, ocean salinity, sea ice, ice sheets, and aspects of extreme weather.

The Intergovernmental Panel on Climate Change (IPCC) has concluded that increases in anthropogenic greenhouse gas emissions have caused most of the observed increase in global mean temperature since the mid-twentieth century. Although not a focus of attention in the cautious and consensus-based findings of the IPCC reports, many experts now believe that abrupt climate change – taking such forms as the disintegration of ice sheets or the shutting down of the thermohaline circulation and occurring in periods of a few years or at most a decade or two – is a distinct risk.

The projections for climate change for the twenty-first century present a daunting challenge to the stewardship of Earth as a social–ecological system. The most recent estimates suggest a rise of global mean temperature by 2100, compared to preindustrial temperatures, of at least 1.5°C and possibly up to 6°C or slightly more. A global debate has begun on what constitutes "dangerous climate change," with some suggesting that 2.0°C should be an upper limit, beyond which damage to natural, social, and economic systems will be unacceptably high (Stern 2007). Given that the current rate of increase of greenhouse gas emissions is tracking at the upper limit of the projected range (Raupach et al. 2007) and that both global mean temperature and sea-level rise are also tracking at the upper limit of their projected ranges (Rahmstorf et al. 2007), the goal of limiting the temperature rise to 2°C or less seems unattainable without very large and rapid reductions in greenhouse gas emissions. The ways in which societies around the world tackle the climate-change challenge over the next decade will be a turning point for the evolution of the Earth System over the next century and beyond.

Cumulative changes are just as important and widespread as systemic ones, but perhaps not as widely publicized. For example, the human imprint on Earth's land surface is staggering, with extensive human domination of large areas of the planet. It is estimated that about 50% of the ice-free land surface has been converted or extensively modified for human use. As for the remaining 50%, most has been modified to some extent for human use; virtu-ally no land area outside of the most extreme environments has been left untouched (Ellis and Ramankutty 2008). Using another measure, between 5 and 50% of the net primary production of the terrestrial biosphere – the net amount of carbon assimilated by the biosphere from the atmosphere – is used, co-opted, or diverted from its natural metabolic pathway by human activities (Vitousek et al. 1997).

The oceans are now also significantly modified by humans, although not as extensively as the land surface. The human imprint is most clearly seen on fisheries, with 47–50% of fisheries for which sufficient information is available being fully exploited, 15–18% overexploited, and 9–10% depleted (see Chapter 10). In a way analogous to human appropriation of net primary production on land, it has been estimated that humans harvest about 8% of the total primary production of the oceans, with much higher percentages for the continental shelves and regions of major upwelling.

The most rapidly changing component of the biophysical Earth System is probably the coastal zone, with about 50% of the human population now living within 100 km of a coastline, a percentage that will likely continue to rise through this century (see Chapters 11 and 13). The human imprint on the coastal environment has traditionally been focused on urban areas and port facilities, where the geomorphology of the coast has been extensively modified. The last several decades, however, have seen much more extensive modification of the coastline, especially in the tropics, driven by the demands of a globalizing food system. About 50% of mangrove forests globally have now been converted to other uses of direct benefit to humans (e.g., for the production of prawns; see Chapter 2).

Although climate change has focused much of the public's attention on global environmental changes, the accelerating loss of biological diversity may be just as serious, or perhaps even more so, in terms of the long-term functioning of the Earth System. The current rate of extinctions is at least 100 times the background rate, or perhaps even 1000 times greater (MEA 2005d). This suggests that the Earth is in the midst of its sixth great extinction event, the first caused by a biotic force – *Homo sapiens*. The estimated future decline in the species richness of mammals, fish, birds, amphibians, and reptiles – 20% of birds are threatened with extinction, 39% of mammals and fish, 26% of reptiles, and 30% of amphibians – is serious enough, but masks the fact that many species that do not become extinct will have greatly reduced ranges or will have become ecologically extinct, that is, be so few in number that they cannot carry out the ecological functions that they previously performed and that are essential in the provision of ecosystem services for human well-being.

Water resources are central to well-being, and their management has been the defining characteristic of many past and present civilizations. It is therefore not surprising that the hydrological cycle is one of the most pervasively modified of the great material cycles of the planet (see Chapter 9; Gleick et al. 2006). At present, about 40% of the total global runoff to the ocean is intercepted by large dams and diverted to human uses. In the northern hemisphere, only about 23% of the flow in 139 of the largest rivers is unaffected by reservoirs. As the availability of surface water becomes increasingly limited in many parts of the world, societies are turning toward groundwater as a resource. Although much less is known about the rate of extraction of groundwater, it has already become clear in many areas, particularly arid and semiarid areas, that groundwater is being extracted much faster than reservoirs can recharge, a situation equivalent to the mining of water resources rather than their sustainable use.

In summary, the advent of humans as a global geophysical force has led to dramatic changes in many features of the Earth System's functioning in a remarkably short period of time as compared with the natural rhythms of this system. Changes in the physical part of the Earth System include the rapid warming of the climate system, the diminution of snow and ice cover, and rising of sea levels. Changes in the chemical part of the Earth System from the impact of the human enterprise include the acidification of the oceans due to increasing dissolution of atmospheric CO_2 and the increasing acidifi-

cation of soils and landscapes due to the application and deposition of reactive nitrogen compounds. Biological aspects of the Earth System have also been extensively modified by human action, in general trending toward a homogenized and simplified biosphere with as yet unknown implications for the functioning of the Earth System.

Global Social Changes (GSCs)

The preceding paragraphs focus mainly on the biophysical aspects of the Earth System, with the human components of the system appearing only as drivers and targets of change. For nearly all of the Anthropocene, there has been a large conceptual disconnect between the burgeoning human enterprise and its imprint on the planet. We have typically assumed that the Earth System is large and robust enough to continue indefinitely to provide an accommodating and pleasant life-support system for humans, no matter what the nature and size of the human enterprise. We have also assumed, at least tacitly, that environmental impacts of human actions arise only at local and regional scales and can be ameliorated as societies become wealthier.

The realization that human activities can impact significantly the functioning of the Earth System as a whole has come only in the last decade or two. The Antarctic ozone hole and climate change have been the main triggers for this growing realization, with the 1987 Montreal Protocol and the 1997 Kyoto Protocol constituting preliminary attempts at creating global institutions to deal with planetary environmental problems. Thus, the last decade or so has seen the first real *recognition* of the Earth as a complex social–ecological system, although the development of this system began with the advent of industrialization about two centuries ago.

Much like GECs, global social changes occur in both systemic and cumulative forms. Under most circumstances, the changes that occupy human attention are social rather than environmental in nature. A number of these social changes have far-reaching consequences for the

Earth System. But for the most part, they are unintended byproducts of social changes; they are seldom taken into account in efforts to measure social welfare at the national level, much less at the global level.

A variety of systemic social changes are commonly lumped together under the heading of globalization. But this concept encompasses a host of distinct processes that are worth differentiating in this discussion (Held et al. 1999). Global social change involves the rise of trade and monetary systems that are planetary in scope. Exports as a percent of gross domestic product (GDP) in developed countries (in constant prices) rose from 8.3% in 1950 to 23.1% in 1985 (Held et al. 1999). Today, this figure for all the countries of the world has risen to over 40% (World Bank 2004). Among other things, this has given rise to multinational corporations that now rival or exceed many nation states in terms of their global influence and economic size. The growth in foreign direct investment (FDI) has been equally dramatic during this period. FDI now stands at 15–20% of GDP (World Bank 2004). Private investment that crosses national boundaries has now eclipsed official development assistance to become a major economic force at the global level.

Two consequences of these economic developments stand out. These elements of globalization affect everyone, even those living in small communities that are now tied to global markets for energy, wood products, agricultural commodities, and so on. At the local level, this leads to asymmetrical dependence in which global economic developments can have dramatic impacts on communities that are powerless to influence global trends. In addition, monetary crises, such as the panic of 1997, can ripple through the entire system overnight. In the absence of well-developed mechanisms to counter such phenomena, monetary crises can easily lead to self-reinforcing processes.

The scope of globalization extends to technological developments as well as matters of trade and money. Information systems, based initially on the fax machine but now mainly on the computer and its links to the Internet, have revolutionized our ability to communicate rapidly and extensively with others located anywhere on the planet. By 2005, over half of the people living in developed regions were users of the Internet; the number of users is growing rapidly in developing countries as well. New technologies emerge rapidly and spread immediately at the global level, despite the efforts of individual inventers and companies to control this process through the use of patents enforceable by law (Stiglitz 2006).

Along side these economic and technological changes are several social and institutional changes that are likely to have equally large impacts on human–environment interactions at a planetary scale. Although the accuracy of treating international society as a society of states has always been subject to challenge, major changes at this level are now well underway (Keene 2002). It is not that the state is likely to fade away as a major player at the planetary level any time soon. But nonstate actors of various kinds, including NGOs and multinational corporations, are increasingly powerful players at the global level. A suite of developments that deserves the label "global civil society" has altered the organizational landscape in ways that have major implications for Earth System governance (Wapner 1997, Kaldor 2003, Keane 2003). It is no longer sufficient to focus on the actions of states in thinking both about the causes of stress on planetary life-support systems and about the range of strategies available for coming to terms with these stresses.

Cumulative social changes are also striking. They include population pressures, industrialization, changing consumer preferences, urbanization, and political decentralization. Economic growth and industrialization have been occurring in China since the late 1970s at a sustained pace never before seen in the industrial age. India is not far behind. Between them, China and India contain more that a third of the world's human population. People living in these countries aspire to lead more affluent lives and, so far at least, this is taking the form of a rising demand for individual residences, private automobiles, and the full range of consumer durables that constitute a prominent feature of contemporary life in the Organization for Economic Cooperation and

Development (OECD) countries. It is apparent that the potential environmental impacts of these cumulative social changes are enormous. With respect to natural resources and ecosystem services, it would take two to three planets to bring everyone on Earth up to current living standards in Europe and five planets to emulate living standards in North America (www.footprintnetwork.org).

These developments have fueled both social processes like urbanization and political processes like decentralization intended to provide enhanced authority to those living outside dominant urban centers. Lured by the prospect of jobs, people continue to stream into cities, despite urban living conditions that are often squalid. City dwellers now constitute approximately 50% of the Earth's human population, and this figure is destined to continue to rise during the course of this century (see Chapter 13). Chinese planners expect that 75% of China's population will reside in urban areas by 2050.

Legal and political decentralization is driven partly by a legitimate interest in putting authority into the hands of those who understand biophysical systems best but also in part by a desire on the part of central government to shed responsibility for environmental problems without providing regional or local governments with the resources needed to handle such tasks effectively. The result is a situation in which it is often easy for unscrupulous individuals to exploit social–ecological systems to their own advantage. The remarkable size of the informal economy in many parts of the world is a product, in part, of the actions of corrupt officials and private citizens who engage in illegal trade in such products as drugs in Columbia, animal parts in Africa, wood products in Indonesia, and so on. (see Chapter 7; Bardhan and Mookerjee 2006).

Environmental–Social Interactions

Interactions between global environmental changes and global social changes are rising rapidly and already leading to profound impacts on major ecosystems. There is every

reason to believe that this trend will continue in the absence of a concerted effort at the global level to regulate it. A few examples will suffice to make this proposition more concrete.

In response to the emergence of a global market together with pressures to embrace export-led growth, many communities in Central America have cleared sizable areas of forest and established coffee plantations. But the market for coffee is both global and notoriously volatile (Bates 1997). Changes in world market prices, over which local communities have little control, can affect their welfare dramatically. While it is illogical at the macro-level, it is not hard to understand why individual communities tend to respond by clearing more forest and replacing it with coffee plantations in an effort to maintain an adequate flow of income and foreign exchange on which they now depend.

The government of Brazil, motivated largely by a desire to enhance security along with the central government's control over outlying regions, built the Trans-Amazonian Highway during the 1960s and 1970s. Whatever its success in terms of security, the construction of the highway opened large areas of the Amazon Basin to individual settlers and to corporations desiring to clear the forest in order to establish ranches to supply meat to fast-food restaurants in North America and, to a lesser extent, Europe. The environmental impacts have been predictable (see Chapter 7). Settlers gradually clear larger areas of forest; the ranchers move on to clear new areas in the forest whenever the land initially cleared loses its fertility. The destruction of the forest has severe environmental impacts both in terms of the release of carbon stored in the trees of tropical forest and in terms of the loss of habitat for endemic species (see Chapters 6 and 7).

The growth of trade on a global basis can be beneficial, especially to those seeking to promote export-led economic growth. But it can prove costly in terms of damage to the environment and threats to the human health. Invasive species, spread around the world as an unintended byproduct of the growth of trade, have now emerged as one of the two or three most important causes of the loss of biological diversity (see Chapter 2). The influx of Eurasian milfoil, for instance, has brought about dramatic changes in many North American lakes and ponds. Trade coupled with the rapid growth in long-distance travel on the part of humans has opened up the prospect of disease vectors that are global in scope. Fears regarding the prospect of a rapid and uncontrollable global spread of AIDS, SARS, and bird flu may be somewhat exaggerated. Yet it is hard to deny that globalization has brought with it the potential for the development of disease vectors whose impacts could dwarf those of the 1918–1919 epidemic of influenza (Kolata 2000).

Now that the Earth as a whole has emerged as a social–ecological system, we must direct attention to global environmental changes, global social changes, and their interactions or, to put it in simpler terms, adjust our conceptual lens to view the Earth as a coupled social–ecological system (Walker and Salt 2006). This is especially important to the extent that we are concerned about sustainable development as a component of environmental protection (WCED 1987). Large-scale biophysical changes, such as substantial shifts in Earth's climate system, raise profound questions about the future viability of contemporary human lifestyles. Yet many human beings are still too preoccupied with efforts to achieve food security or to cope with chronic shortages of freshwater to focus on issues like climate change (see Chapter 3). Others who are beginning to experience the benefits of industrialization wish to emulate the lifestyles of affluent westerners, regardless of the consequences of their actions for the Earth System.

Common Threads

The systemic and cumulative directional changes we have noted seem quite disparate, at least on the surface. What is the link between forces leading to the destruction of forests in Amazonia and the potential spread of SARS or bird flu from China to the rest of the world? How should we think about the links between industrialization and the loss of biological diversity?

Even in biophysical terms, links between and among these directional changes are often substantial and sometimes surprising. There are obvious connections between the climate change and the progressive loss of biological diversity; both are tightly coupled with the forces leading settlers and ranchers to clear forest lands in the Amazon Basin. Similar remarks apply to the environmental impacts of the creation of coffee plantations in Central America.

Here, we identify and explore briefly several key features of the directional changes identified in the preceding section that pose problems for finding ways to govern these processes through national and international measures. Directional changes at the planetary level pose challenges for governance that are unprecedented in a number of respects and are unlikely to yield to familiar tools in our governance toolkit.

To begin with, the impacts of human actions at a planetary scale now extend far beyond anything we have experienced before (Vitousek et al. 1997). As a result, it is essential to think about coupled social–ecological dynamics rather than treating planetary processes as biophysical systems that are occasionally perturbed by human actions. In this context, the one thing we can hope to change directly is the behavior of human beings. We may desire to lower atmospheric CO_2 concentrations or to improve efficiency in the use of water for irrigation. But all such efforts require changes in human behavior.

Uncertainty looms large in any effort to understand, much less to forecast, the dynamics of the Earth System (Wilson 2002). The complex and dynamic nature of the planet as a social–ecological system can lead to changes that are nonlinear, abrupt, irreversible, and sometimes nasty, at least from a human perspective. Simulations with Earth System models show that slight changes in initial conditions can lead to profound differences in outcomes over time. It is worth differentiating first-order and second-order uncertainty in this connection (see Chapter 5). We do not know what atmospheric concentrations of greenhouse gases are compatible with limiting temperature increases to 2°C. With regard to the loss of biological diversity, on the other hand, we are uncertain not only about current extinction rates but also about the total number of species existent on the planet. Answering questions about the proportion of species likely to go extinct during a given period of time is therefore an order of magnitude more difficult than addressing the effects of increases in surface temperatures. It follows that we cannot wait to take action regarding large-scale changes in the Earth System until we are certain about both the changes and the probable consequences of our deliberate interventions. This would only become a recipe for paralysis in the face of profound challenges to human welfare. Yet our skills at decision making under uncertainty remain limited (Tversky and Kahneman 1974, Kahneman and Tversky 1979).

A third common thread has to do with asymmetries in rates of change in social–ecological terms and in our capacity to develop effective governance systems, especially at a global scale. Even at the national level, governance systems are hard to adapt or adjust to major changes in the demand for governance. We generally assume that problems will evolve slowly enough to give us time to overcome problems arising from governance systems that are dominated by special interests. But we now face a dilemma in these terms. Global environmental changes and global social changes are occurring at an accelerated pace. As a result, fish stocks collapse before we can agree on ways to reduce harvest levels; habitat destruction occurs before we are able to muster agreement on the need for protection; uses of the atmosphere as a repository for various types of waste rise faster than we can address the sources of such developments at the level of policy. What is needed, under the circumstances, are governance systems that can produce substantial changes in a timely manner, without falling prey to various types of corruption.

We are also poorly prepared for situations in which large systems change at a pace that equals or exceeds rates of change in small systems (Young et al. 2006). We understand that local weather patterns will change rapidly and even dramatically, but we expect the Earth's climate system to remain unchanged. We are

not surprised to experience large swings in the economies of local communities, but we expect the economy of the Earth System to remain stable. We anticipate fluctuations in human population at the local level, but we are poorly prepared to address rapid demographic changes at the national level, much less at the global level. Global environmental and social changes have now called these expectations into question. Financial crises that spread like wildfire have already occurred. Demographic transitions that will pose profound questions for public policy are already underway. Abrupt climate change is a distinct possibility. As a result, we can no longer assume that large-scale systems are stable enough to justify focusing attention exclusively on small-scale reform.

Taken together, these common threads demonstrate a need for new systems of governance for sustainable development. Once we shift our paradigmatic perspective to recognize that the Earth is a complex and dynamic social–ecological system, it is clear that we need to think in terms of coupled systems in which change is large-scale, often nonlinear, frequently fast, and sometimes irreversible. From a human perspective, the consequences of this sort of change may prove not only unfamiliar but also nasty. Still, introducing new and effective governance systems at a planetary scale is a daunting task. It will certainly loom large as one of the great issues of the twenty-first century (see Chapter 15; Young 2002b, Biermann 2007, Young et al. 2008).

The Challenge of Earth System Governance

Governance, in every setting, centers on the development and operation of mechanisms designed to steer societies away from bad outcomes (e.g., the tragedy of the commons) and toward good outcomes (e.g., sustainable use of ecosystem services). As we move deeper into the Anthropocene, an era in which human actions are major determinants of the condition of biophysical systems on a planetary scale, environmental governance becomes increasingly a matter of limiting or constraining disruptive impacts of anthropogenic forces on planetary life-support systems rather than simply a matter of achieving efficiency and equity in human uses of natural resources.

Successful governance in a world featuring both global environmental changes and global social changes requires the development of arrangements that are capable of dealing with cross-level interactions and finding effective means of coping with uncertainty.

Local developments often generate effects that are significant at a global scale (e.g., the contributions of deforestation in specific places to concentrations of carbon dioxide in the Earth's atmosphere). Conversely, global developments (e.g., climate change) can produce severe impacts on social welfare at a local level. It follows that we need various forms of multilevel governance or, in other words, distinct arrangements operating at different levels of social organization that interact in a mutually reinforcing or synergistic manner to provide effective Earth System governance. The idea of subsidiarity, based on the assumption that there is a proper level at which to address specific problems, will not suffice to solve Earth System problems.

Because the problems at stake are highly complex – featuring nonlinear and often abrupt changes – the governance systems we create to address them must be able simultaneously to arrive at decisions in situations under conditions of severe uncertainty and to respond in an adaptive manner as new information relating to problems like climate change becomes available. This puts a premium not only on the development of early warning systems but also on the capacity to assimilate new information and to overcome rigidities giving rise to path dependence when it becomes clear that prior decisions or policies are outmoded.

How can governance systems address specific problems (e.g., climate change, loss of biological diversity) in settings of this sort? Three approaches can prove helpful in turning general objectives like sustainable development into well-defined and operational goals in specific cases. One approach centers on translating general goals (e.g., avoiding dangerous inter-ference with the Earth's climate system) into clear-cut measures to be used at the operational level. Efforts to set specific goals relating to climate change (e.g., no more than 450 ppm of CO_2 in the Earth's atmosphere or no more than a temperature increase of $2°C$) exemplify this approach. So also does the setting of goals calling for measurable changes in some dependent variable (e.g., the effort to halve the number of people without safe drinking water by 2015 as spelled out in the UN Millennium Development Goals). As these examples suggest, however, this approach has weaknesses as well as strengths. What reason do we have for setting 450 ppm as a cap on the concentration of CO_2 in the Earth's atmosphere? Is the number of people lacking safe drinking water a decision variable that we can hope to pursue actively and effectively as a matter of public policy?

A second way of articulating the goals of governance in operational terms focuses on the development of safeguards to prevent or control runaway processes like abrupt disintegration of ice sheets or financial panics. This approach calls for an enhanced interest in strengthening adaptive capacity with regard to climate change and devising countercyclical mechanisms to prevent various kinds of destructive spirals made possible by globalization. Some may regard this as an overly pessimistic strategy. But it is worth noting that we adopt approaches of this sort all the time at the domestic level by creating mechanisms to prevent escalating financial crises or by emphasizing prevention and preparedness as well as response with regard to extreme events like hurricanes or tsunamis.

Going a step further, there may be a convincing case in some situations for adopting a form of worst-case analysis in thinking about the Earth as a social–ecological system. We use this type of thinking in the realm of national security as a matter of course. We focus on the capabilities of potential opponents without considering their goals or motives, and we tolerate the expenditure of hundreds of billions of dollars per year to increase our sense of security, without any serious debate about the wisdom of doing so. Why not take a similar approach to an issue like climate change where most experts believe an investment of something like $100–200 billion in the near future could offset much larger costs brought on over a period of years or decades by the intensification of climate change (Stern 2007)? It seems clear that this is more a matter of mindsets or discourses than a lack of the capability needed to act vigorously to reduce emissions of greenhouse gases, at least among the advanced industrial states in the world today.

Applications – Climate and the MDGs

Because we are dealing with large, complex systems that are subject to nonlinear and sometimes abrupt changes and that are typically sensitive to initial conditions, there is little prospect of devising simple recipes for addressing the demand for governance across the full range of Earth System issues. What we can say is that successful governance systems require careful attention to matching the properties of the social–ecological systems in question with the attributes of the governance systems created to manage them (Young 2002b, Galaz et al. 2008).

We do know that governance systems operating on a global scale must be able to monitor the dynamics of planetary processes closely, provide early warning regarding the approach of tipping points or thresholds, and respond to such developments with appropriate adjustments in a timely manner. Fulfilling these requirements with respect to specific problems will require close attention to the special features of individual issues. Addressing climate change, for instance, requires actions that differ from those required to stem the loss of biological diversity. To explore the implications of these observations concretely, we turn in this section to brief case studies of efforts (1) to address the problem of climate change and (2) to meet the Millennium Development Goals. Climate change constitutes a systemic problem that requires an unprecedented response at a global scale. Efforts to address this problem have produced meager results so far. Fulfilling the Millennium Development Goals, by contrast, is a matter of tackling a cumulative problem that requires action in spatially-defined settings but that has planetary-scale consequences in the aggregate. At this stage, the results of efforts to address this problem are mixed.

Controlling Climate Change

Climate change is arguably the most complex governance issue that humanity has faced. It affects all countries (but not in the same way); it has a very long timeframe, and it has the potential to threaten the viability of contemporary civilization. The first serious attempt at controlling climate change was the 1992 United Nations Framework Convention on Climate Change (UNFCCC), which entered into force in 1994 and which nearly all countries have ratified. The critical component of the UNFCCC is Article 2, which states the ultimate objective of the convention as stabilizing "... greenhouse gas concentrations in the atmosphere at a level that prevents dangerous anthropogenic interference with the climate system." The 1997 Kyoto Protocol, the first attempt to translate this general goal into an explicit approach to reducing greenhouse gas emissions by assigning targets and timetables, became embroiled in international politics and has not achieved universal coverage. Although the protocol finally entered into force in 2005, the USA has refused to ratify it. The Kyoto Protocol and other efforts outside it to reduce greenhouse gas emissions have had no measurable effect, as can be seen in the observed record of greenhouse gas emissions through 2006 (Fig. 14.4; Raupach et al. 2007). The emissions trajectory is tracking on a business-as-usual line, well above all concentration stabilization trajectories.

There are several severe challenges that make it difficult to come to terms with climate change from the perspective of Earth System governance and that global society has failed so far to meet:

- *Equity issues.* The climate change challenge is bedeviled by equity issues of several types. First, most of the increased greenhouse gases in the atmosphere have been emitted by the industrialized world, yet the consequences

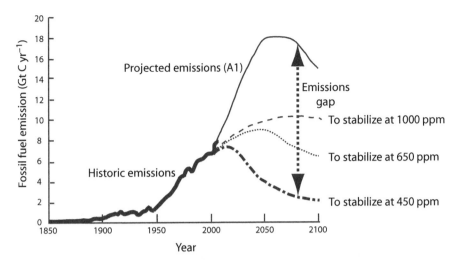

FIGURE 14.4. Observed global CO_2 emissions compared with emission scenarios and stabilization trajectories. Redrawn from Raupach et al. (2007).

of climate change will be borne disproportionately by the developing world. This is exacerbated by the generally lower adaptive capacity in developing countries. Second, although there is much focus now on China and India as future larger emitters, past emissions have been dominated by the USA and Europe. There is a tendency to focus so much on the rapidly developing Asian giants that the historical legacy is overlooked. Finally, countries tend to quantify emission rates in ways that favor their own positions. For example, Australia, with only 20 million inhabitants, argues that its emissions are a very small fraction of the global total and are already dwarfed by China's emissions. But Australia's per capita emissions vie with those of the US as the highest on the planet, much higher than China's. This raises a fundamental question. Should emissions permits in any global scheme be allocated to countries or to individuals on the premise that every human being has an equal right to use the "atmospheric sink"? This question is one of the most controversial in the climate change debate. Until it is resolved, little progress will be made in achieving a globally acceptable emissions reduction scheme.

- *Long lag times.* The biophysical component of the climate system has exceptionally long lag times that challenge our ability to devise appropriately "fitted" governance systems. The momentum already built into the climate system, due largely to the thermal inertia in the oceans, implies that we are committed to another 0.5°C or 0.6°C of global mean temperature rise, regardless of the success of efforts to reduce future emissions. Thus, the nature of the Earth's climate in 2030 can already be predicted with a high degree of certainty (in terms of temperature; Plate 8; IPCC 2007a); no matter how stringent measures we take in the next decade or two, it will make little difference by 2030. What government can convince its electorate to make large sacrifices now with no perceptible effect for 20 years? There are even longer lags to deal with. If temperature rises to 2.5 or 3.0°C above the preindustrial level, it is highly likely that most of the Greenland ice sheet and parts of the West Antarctic ice sheet will be committed to melting, although it may take a few centuries for this to play out. This would lead to a sea-level rise of 6–7 m. Once the "tipping point" is crossed, there is no going back even if we could reduce greenhouse gas emissions to zero overnight.

- *Dangerous climate change.* The UNFCCC makes reference to dangerous interference in the climate system but gives little guidance as to what constitutes "dangerous climate change." Although scientific research on the nature of climate impacts can inform the debate, deciding what is really "dangerous" in terms of climate change is an individual and societal value judgment that could vary markedly around the world. The inhabitants of low-lying Pacific Island states, for instance, have already decided that the current level of climate change (a global mean temperature rise of about 0.7°C) is dangerous; they are preparing to leave their countries in large numbers for New Zealand. On the other hand, Russia may benefit from further warming. Agriculture could move northwards; the country's vast boreal forests could expand into the present tundra regions, and large deposits of oil and gas in the Arctic Ocean could become accessible as the sea ice retreats. A large number of interpretations of "dangerous climate change" lie between these two extremes. But the point is that societal judgments of what is dangerous and what is not will vary widely around the world, adding another layer of complexity to the task of building a global governance system to control climate change.

- *Nonlinearities in the climate system.* The dynamics of the climate system are highly nonlinear in many aspects. In that regard, CO_2 behaves as a "threshold gas," a small incremental change (an increase of a few parts per million) could tip the climate system into a cascade of feedbacks that could propel the Earth into another climatic and environmental state (see Chapter 5). Such a rapid and irreversible (in human timeframes) transformation could be triggered, for example, by large-scale feedbacks in the carbon cycle – large outgassing of CO_2 from the soil, increases in wildfires in the boreal and tropical zones, activation of methane clathrates buried under the coastal seas, and the rapid loss of methane from tundra ecosystems as they warm. This raises the issue of overshoot. The pathway to a given stabilization target may be as important as the ultimate target itself. If atmospheric concentration rises too high during a particular stabilization trajectory, this carbon cycle threshold may be crossed and the natural feedbacks described above may be activated, rendering the original target unattainable. In a worst-case (but by no means impossible) scenario, such a "runaway" effect could lead to the collapse of modern civilization (i.e., an uncontrollable decline in economic activity, population, social cohesion, and individual and societal well-being). How can current decision-making structures deal with such a momentous issue? What type of governance/institutional system can be devised to "manage" our own life-support system for the benefit of humanity as a whole?

There is a growing urgency to find workable solutions to these institutional challenges. Many societies have faced momentous environmental challenges of various types. Some could not change in innovative and adaptive ways and collapsed; others found new approaches to deal with the challenge and transformed themselves into flourishing and resilient societies (Tainter 1988, Diamond 2005). What does this imply for contemporary, globalized society and the climate change challenge? The complexity of this particular environmental challenge dwarfs those of the past and, ultimately, implies that we must become effective and successful stewards of our own life-support system, which we have now transformed into the largest and most complex social–ecological system possible. To fail at this challenge has implications for an increasingly connected global society that are hard to imagine.

Because climate change is a systemic issue, avoiding dangerous interference in the climate system will require coordinated human actions on a planetary scale. But the fact that the maintenance of a viable climate system also exhibits the basic features of a public good (see Chapter 4) means that many actors in the system will experience strong incentives to become free riders, leading to suboptimality or even outright failure in efforts to supply this good.

Experience in other realms suggests that the best prospect for overcoming this problem may be to start with a coalition of key players and to build out from there to draw in free riders step-by-step (Schelling 1978). This is what happened in the case of ozone depletion. Of course, the two problems differ in a number of significant ways, but it seems likely that addressing climate change will require a similar process.

Finally, climate change, although it now dominates the media and the political arena, is not the only global environmental change that threatens the well-being of humanity and the stability of the Earth System. The Millennium Ecosystem Assessment (2005a) highlighted many other global-scale changes that affect the ecosystem services on which humanity ultimately depends. Most of these issues, if not all of them, are not as well developed in terms of ongoing scientific assessments, international negotiations, and global organizations as is climate change. Nevertheless, ensuring that the Earth System continues to provide the suite of essential ecosystem services on which we depend is just as important as stabilizing the climate system in a state within which we can survive and prosper.

Fulfilling the Millennium Development Goals

The millennium development goals (MDGs) are a set of eight broad policy targets intended to improve the lives of people in developing countries (Table 14.1). The problem here is cumulative in the sense that it requires action in many specific locations but at the same time global because overall success or failure in fulfilling the MDGs will have planetary consequences. Adopted initially by all the member states of the UN in the Millennium Declaration of 2000, the commitment to the MDGs was reinforced at both the 2002 Monterrey Conference on Financing for Development and the 2002 World Summit on Sustainable Development in Johannesburg. Individual goals cover a variety of issues ranging from the alleviation of poverty to the improvement of health and the

adjustment of trade and financial arrangements to facilitate efforts on the part of developing countries to improve their lot.

Most of the MDGs call for specific reductions in well-defined conditions (e.g., the number of people living on less than a dollar a day, the incidence of HIV/AIDS, malaria, and other diseases) with a target date of 2015. It is possible in a number of cases to track progress toward meeting these goals in a quantitative manner. Goal #1, for instance, calls for halving "... between 1990 and 2015, the proportion of people whose income is less than $1 a day." Goal #7 envisions halving "... by 2015, the proportion of the population without sustainable access to safe drinking water and basic sanitation." Goal #8 speaks to the need to "... develop further an open, rule-based, predictable, nondiscriminatory trading, and financial system."

Have we made progress toward meeting these goals and, in the process, addressing these cumulative social–ecological concerns through initiatives launched at the global level? A progress report released by the UN in July 2007 concludes that "[h]alf way to a 2015 deadline, there has been clear progress toward implementing the Millennium Development Goals ... But their overall success is still far from assured ..." (UN 2007b). Let us take a closer look at progress relating to Goals #1, 7, and 8 and delve into the issues of governance on a planetary scale that they pose. The results are illuminating in terms of the insights they provide regarding issues of governance on a large scale.

There is relatively good news regarding poverty. The number of people in developing

TABLE 14.1. UN Millennium Development Goals. Information from United Nations (2007a).

1. Eradicate extreme poverty and hunger
2. Achieve universal primary education
3. Promote gender equality and empower women
4. Reduce child mortality
5. Improve maternal health
6. Combat HIV/AIDS, malaria, and other diseases
7. Ensure environmental sustainability
8. Develop a global partnership for development

countries living on less than a $1 a day fell from 1.25 billion in 1990 to 980 million in 2004 or in other words from nearly a third living in extreme poverty "... to 19 percent over this period" (UN 2007a: 6–7). As the UN report points out, however, progress is unequally shared. Most of the progress occurred in East and South Asia, while "...poverty rates in Western Asia more than doubled between 1990 and 2005" (UN 2007a: 7). With respect to safe drinking water and basic sanitation, progress has been made but much more must be done to meet Goal #7. Thus, "... if trends since 1990 continue, the world is likely to miss the target by almost 600 million people" (UN 2007a: 25). Progress toward meeting Goal #8 (fair trade) is harder to track in any quantitative way. Yet there are reasons for concern in this area. By the middle of 2007, the Doha Round of trade negotiations had failed to reach agreement "... on the overall programme of measures to be adopted" (UN 2007a: 20). Most donor countries have failed to meet the targets for official development assistance (ODA) set at the 2002 Monterrey Conference. ODA actually fell from 2005 to 2006, and "... aid to the least developed countries (LDCs) has essentially stalled since 2003" (UN 2007a: 29).

What should we make of this mixed record regarding fulfillment of the MDGs and especially of the lackluster performance in several key areas. The evidence suggests that the main drivers in this domain include both internal and external factors (Collier 2007). Internally, the existence of good governance appears to be a critical condition for success. This is not, at least in the first instance, a matter of conforming to some ideal standard of democracy. More important with respect to issues like sanitation and poverty are conditions like the absence of civil war and limitations on the ability of exploitative individuals to find ways to feather their own nests and, more often than not, manage to hold their assets in offshore accounts.

Externally, key factors are the rules of the game and the willingness of wealthy countries to provide properly targeted and well-managed ODA. It is generally acknowledged, for instance, that the agricultural polices of the EU and the US make it difficult for farm-ers in developing countries to compete in the markets of these advanced industrial countries. And there is no reason to expect that these policies will change in any fundamental way, even if the Doha Round of trade negotiations produces agreement on some significant issues.

The story regarding ODA is equally troubling. Using development assistance to make progress toward broader economic, political, and social goals is always tricky; things can and often do go wrong in the provision of development assistance. Still, we cannot help being concerned by trends regarding the availability and use of ODA. The fact that most wealthy countries have failed to make good on the commitments they agreed to in 2002 at the Monterrey Conference is telling. Overall, ODA today is less than half of the target of 0.70% of GDP for the wealthy countries as a group. In some of these countries, willingness to provide development assistance is actually waning. A particularly serious concern is the failure to make good on the pledge to "... double aid to Africa by 2010 at the summit of the Group of eight industrialized nations in Gleneagles in 2005" (UN 2007: 29).

What lessons can we draw from this case study that are relevant to the challenge of Earth System governance? First, and in some respects foremost, is the limited capacity of a large number of developing countries to participate in a meaningful fashion in addressing planetary issues like climate change. Leaders of countries preoccupied with problems of poverty, disease, and a lack of safe drinking water have little time and energy to consider systemic problems like climate change. It is not that they do not care about climate and fail to grasp the seriousness of this problem for social welfare in their own countries. The sad fact is that they do not have the resources needed to deal with all their problems at the same time and that issues like climate change seem like concerns that can be put off, at least in the short run (see Chapter 3).

What is to be done under these conditions? Several strategies, implemented simultaneously seem promising as answers to this question. The first objective is to provide people right down to the local level with incentives to take large-scale problems seriously. A striking example involves the rise of tourism in Southern Africa and the incentives associated with this development to treat wild elephants as a source of income rather than as pests. The fact that this example relates to biological diversity rather than to climate change is not accidental. It may well be easier to generate interest at the local level in taking steps that pertain to issues, like managing elephant populations, that seem tangible and close to home.

Beyond this, it obviously helps to have external sources of funding available on favorable terms. The evidence is strong that the Montreal Protocol Multilateral Fund played a significant role in drawing developing countries into the effort to protect stratospheric ozone. Even more to the point are the efforts of the Global Environment Facility (GEF), a multilateral funding mechanism administered jointly by UNEP, UNDP, and the World Bank and dedicated to providing funds needed to help developing countries address large-scale environmental problems like climate change. Targeted on issues like climate change, biological diversity, and freshwater, the GEF has emerged as a particularly important source of funds in light of the failure of most developed countries to live up to the financial commitments they made in Monterrey in 2002 and later that same year in Johannesburg (Stiglitz 2002).

Addressing cumulative issues, like those spelled out in the MDGs, differs significantly from efforts relating to systemic issues. It is to be expected that initiatives aimed at reducing poverty or providing safe drinking water will be far more effective in some countries or regions than in others (Collier 2007). Changes in the rules of the game, like adjustments designed to help developing countries to export agricultural products, will be more important to some developing countries than others. The capacity of individual developing countries to absorb and make good use of development assistance varies dramatically. As a result, any search for general or universal prescriptions in this issue area is bound to fail (see Chapter 4). This means that we are unlikely to find ourselves seeking to build comprehensive regimes, like the global regime dealing with climate change, when it comes to battling poverty or improving health care in the developing countries. Even so, fulfilling the MDGs must be treated as a high priority in any effort to protect the planet's life-support systems.

Synthesis and Conclusions

There is now an inescapable demand for effective governance at the planetary level or, in other words, Earth System governance. Both the biophysical and the social parts of the Earth System are capable of rapid and irreversible changes that could be detrimental to the well-being of humans and the viability of modern societies. Interactions between global environmental changes and global social changes may trigger events that are nonlinear, abrupt, irreversible, and nasty, at least from a human perspective. To minimize the risk that human actions will trigger such events, we need to develop a capacity to manage major components of the Earth System in such a way as to keep close to the bounds of natural variability. This presents unprecedented challenges to humanity to conceptualize the Earth as a complex social–ecological system and to move from being clumsy exploiters to sensitive stewards of our own life-support system. This is the fundamental challenge of the twenty-first century.

How can we prepare to meet this challenge in an effective manner? We cannot succeed in efforts to address these problems by focusing on biophysical systems and treating human actions as perturbations or minor disturbances to be set aside or ignored in most cases. Understanding the resultant social–ecological systems will require the development of a new generation of models and methods capable of capturing the dynamics of coupled systems in which feedback loops linking the biophysical components and the human components are central concerns. Governance systems treated as steering mechanisms can play an important role in maintaining the resilience of these social–ecological systems. But they interact with numerous other drivers and must be designed to fit the distinctive properties of the social–ecological systems in question in order to prove effective.

What recommendations can we offer on the basis of this analysis that may prove helpful to those responsible for addressing specific problems of Earth System governance? Here we list and briefly clarify six lessons for policymakers who are responsible for addressing both systemic and cumulative issues on a planetary scale:

- *Draw on multiple types and sources of knowledge.* It is important to avoid the pretense that mainstream, western scientific analyses are superior to those arising from other approaches to the production of knowledge. The ability to integrate a number of types of knowledge will emerge as an essential skill (see Chapter 4).
- *Pay attention to long-term consequences.* In addressing issues of the type we have discussed, it is important to avoid the dismissal of long-term consequences arising both from the idea of discounting future benefits and costs and from the dominance of the electoral cycle in policymaking. The importance of considering future needs as well as current needs is the basic message arising from the idea of sustainable development (see Chapters 7 and 9).
- *Learn how to cope with uncertainty.* In dealing with the planetary issues discussed in this chapter, policymakers will always be faced with the need to make decisions under conditions of uncertainty. This suggests the importance of thinking in terms of satisficing (i.e., selecting satisfactory options rather than holding out indefinitely to find the perfect option) and making use of heuristics or rules of thumb. It also provides a strong rationale for erring on the side of caution in estimating subjective probabilities regarding events that could push the Earth System across thresholds and into irreversible changes (see Chapter 1).
- *Create sensitive monitoring systems.* In dealing with large-scale social–ecological systems, there is a need for sophisticated monitoring systems capable of picking up early signs of changes and providing information needed to respond to changes in a timely manner. The more dynamic the system and the higher the level of uncertainty, the more important it is to allocate scarce resources to the creation and operation of early warning systems.
- *Emphasize social learning as well as adaptation management.* The idea of adaptive management with its emphasis on approaching policymaking with an experimental frame of mind is a step in the right direction. But in dealing with complex and dynamic social–ecological systems, it is important to go a step further to focus on social learning or, in other words, to consider adjusting goals or reframing cognitive models as well as attending to instrumental matters like the pros and cons of alternative means to pursue existing goals (see Chapters 4 and 5).
- *Prepare for crises as periods of opportunity.* Crises are, of course, dangerous situations in which large systems can cross thresholds and move into different – and often less desirable – basins of attraction. But they are also periods of opportunity during which it is possible to make important changes in prevailing governance systems. Because there is never time to engage in extensive analysis during a crisis, it is highly desirable to think about the relative merits of different approaches to institutional reform before a crisis erupts (see Chapter 5).

Review Questions

1. What developments during the second half of the twentieth century and the first years of the present century have made it appropriate to treat the Earth System as a whole as a social–ecological system?
2. What is the significance of introducing the concept of the Anthropocene to describe the current era in the earth's history?
3. In what ways do global environmental changes and global social changes interact with one another to form a coupled planetary system?
4. How seriously should we take the prospect of nonlinear, abrupt, and irreversible changes in the Earth System?

5. What is the difference between systemic changes and cumulative changes, and why does this distinction matter?
6. How should we think about sustainable development as a criterion for evaluating the results of Earth System governance?
7. How can we overcome the obstacles that impede our efforts to strengthen the climate regime?
8. Why are we making more progress in meeting MDG #1 than MDG #7?
9. How do we need to adjust our approach to governance to address problems arising in very large and highly dynamic systems like the Earth System?

Additional Readings

Biermann, F. 2007. Earth system governance as a crosscutting theme of global change research. *Global Environmental Change* 17:326–337.

Costanza, R., L.J. Graumlich, and W. Steffen, editors. 2007. *Sustainability or Collapse? An Integrated History and Future of People on Earth*. MIT Press, Cambridge, MA.

Diamond, J. 2005. *Collapse: How Societies Choose to Fail or Succeed*. Viking, New York.

Gunderson, L.H. and C.S. Holling, editors. 2002. *Panarchy: Understanding Transformations in Human and Natural Systems*. Island Press, Washington.

McNeill, J. 2000. *Something New under the Sun: An Environmental History of the Twentieth Century*. Penguin Press, London.

Schellnhuber, H.-J., P.J. Crutzen, W.C. Clark, and M. Claussen, editors. 2004. *Earth System Analysis for Sustainability*. MIT Press, Cambridge, MA.

Steffen, W.L., A. Sanderson, P.D. Tyson, J. Jäger, P.A. Matson, et al. 2004. *Global Change and the Earth System: A Planet under Pressure*. Springer-Verlag, New York.

Tainter, J. 1988. *The Collapse of Complex Societies*. Cambridge University Press, Cambridge.

Walker, B.H., and D. Salt. 2006. *Resilience Thinking: Sustaining Ecosystems and People in a Changing World*. Island Press, Washington.

Young, O.R. 2002. *The Institutional Dimensions of Environmental Change: Fit, Interplay, and Scale*. MIT Press, Cambridge, MA.

Young, O.R., L.A. King, and H. Schroeder, editors. 2009. *Institutions and Environmental Change: Principal Findings, Applications, and Research Frontiers*. MIT Press, Cambridge.

Part III
Integration and Synthesis

15
Resilience-Based Stewardship: Strategies for Navigating Sustainable Pathways in a Changing World

F. Stuart Chapin, III, Gary P. Kofinas, Carl Folke, Stephen R. Carpenter, Per Olsson, Nick Abel, Reinette Biggs, Rosamond L. Naylor, Evelyn Pinkerton, D. Mark Stafford Smith, Will Steffen, Brian Walker, and Oran R. Young

Introduction

Accelerated global changes in climate, environment, and social–ecological systems demand a transformation in human perceptions of our place in nature and patterns of resource use. The biology and culture of *Homo sapiens* evolved for about 95% of our species' history in hunting-and-gathering societies before the emergence of settled agriculture. We have lived in complex societies for about 3%, and in industrial societies using fossil fuels for about 0.1% of our history. The pace of cultural evolution, including governance arrangements and resource-use patterns, appears insufficient to adjust to the rate and magnitude of technological innovations, human population increases, and environmental impacts that have occurred. Many of these changes are accelerating, causing unsustainable exploitation of ecosystems, including many boreal and tropical forests, drylands, and marine fisheries. The net effect has been serious degradation of the planet's life-support system on which societal development ultimately depends (see Chapters 2 and 14).

Efforts to redirect exploitation and foster sustainability have led to a gradual shift in resource management paradigms that often follow a transition from an intensive **resource exploitation phase** to a **steady-state resource management** paradigm aimed at maximum or optimum sustained yield (MSY or OSY) of a single resource, such as fish or trees, and subsequently to **ecosystem management** to sustain a broader suite of interdependent ecosystem services in their historic condition (Fig. 15.1). Despite its sustainability goal, management for OSY often results in overexploitation because of overly optimistic assumptions about the capacity of resource managers to sustain productivity, avoid disturbance and pest outbreaks, regulate harvesters' actions, and anticipate surprises such as extreme economic or environmental events (see Chapters 1, 4, and 5). Actions to address emerging problems are often delayed until research can provide a more complete understanding of likely system response. Even ecosystem management, which is widely viewed as the state-of-the-art management paradigm for sustainability, is constrained by its focus on historic conditions as a reference point for management and conservation planning.

F.S. Chapin (✉)
Institute of Arctic Biology, University of Alaska Fairbanks, Fairbanks, AK 99775, USA
e-mail: terry.chapin@uaf.edu

F.S. Chapin et al. (eds.), *Principles of Ecosystem Stewardship*,
DOI 10.1007/978-0-387-73033-2_15, © Springer Science+Business Media, LLC 2009

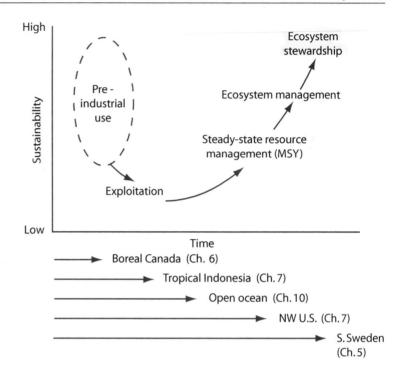

FIGURE 15.1. Evolution of resource-management paradigms and their focus on the balance between exploitation and sustainability. Arrows at the bottom show the history (up to the present) in representative locations and the chapter (Ch) in which each is described. A major opportunity for developing nations is to "leap-frog" from pre-industrial or exploitative phases directly to ecosystem stewardship.

Given the challenges of sustaining social–ecological systems in a rapidly changing world, we advocate a shift to *ecosystem stewardship* (Table 15.1). The central goal of ecosystem stewardship is to sustain the capacity of ecosystems to provide services that benefit society by sustaining or enhancing the integrity and diversity of ecosystems as well as the adaptive capacity and well-being of society. This requires adaptive governance of coupled social–ecological systems to provide flexibility to respond to extreme events and unexpected changes. Rather than managing resource stocks and condition, ecosystem stewardship emphasizes adaptively managing critical slow variables and feedbacks that determine future trajectories of ecosystem dynamics (see Chapters 2, 4, and 5). Actions that foster a diversity of future options rather than a single presumed optimum provide resilience in the face of an unknown, but rapidly changing future. In this perspective, uncertainty and change become expected features of ecosystem stewardship rather than impediments to management actions. This shift

TABLE 15.1. Differences between steady-state resource management and ecosystem stewardship.

Characteristic	Steady-state resource management	Ecosystem stewardship
Reference point	Historic condition	Trajectory of change
Central goal	Ecological integrity	Social–ecological sustainability benefits
Predominant approach	Manage resource stocks and condition	Manage stabilizing and amplifying feedbacks
Role of uncertainty	Research reduces uncertainty before taking action	Actions maximize flexibility to adapt to an uncertain future
Role of resource manager	Decision maker who sets course for sustainable management	Facilitator who engages stakeholder groups to respond to and shape social-ecological change and nurture resilience
Response to disturbance	Minimize disturbance probability and impacts	Adapt to changes and sustain options
Resources of primary concern	Species composition and ecosystem structure	Biodiversity, well-being, and adaptive capacity

in paradigm builds on a wealth of knowledge and experience of many professionals, including resource managers and users, policy makers, and business leaders and integrates concepts from natural and social sciences and the humanities.

In Chapters 1–5, we presented a framework linking studies in vulnerability analysis, resilience theory, and social transformation to address ecosystem stewardship in a social–ecological context. We emphasized the key roles of adaptive co-management and adaptive governance in fostering both ecosystem sustainability and human well-being. We then illustrated in Chapters 6–14 the application of these ideas in specific social–ecological systems. In this chapter, we summarize practical approaches to implementing ecosystem stewardship through the integration of three broad sustainability strategies: (1) reducing vulnerability; (2) fostering adaptive capacity and resilience; and (3) navigating transformations to avoid, or allow escape from, undesirable social–ecological states. These three strategies are overlapping and complementary. They require more proactive and flexible approaches to defining the future state of the planet than have characterized either exploitative or equilibrial resource management paradigms of the past.

Reducing Vulnerability

Reduce Exposure to Hazards and Stresses

Vulnerability analysis entails assessing and minimizing hazards, stresses, and risks and reducing the sensitivity of social–ecological systems to those threats that cannot be adequately mitigated (Table 15.2). It is a logical starting point for implementing ecosystem stewardship, because it involves planning in the context of known current conditions and expected changes. Although vulnerability analysis traditionally emphasizes human vulnerability, we focus on the vulnerability of social–ecological systems in addressing ecosystem stewardship.

Assessment and mitigation of *currently recognized* hazards, stresses, and risks have always

TABLE 15.2. Examples of stewardship strategies to reduce vulnerability.

Reduce exposure to hazards and stresses
- Minimize known stresses and avoid or minimize novel hazards and stresses
- Develop institutions that minimize stresses originating beyond the managed unit
- Manage in the context of *projected* changes rather than historical range of variability

Reduce social–ecological sensitivities to adverse impacts
- Sustain the capacity of natural capital to provide multiple ecosystem services
- Sustain and enhance critical components of well-being, particularly of vulnerable segments of society
- Engage stakeholders in decision-making to account for variation in norms and values in assessing trade-offs
- Plan sustainable development to address the trade-offs among costs and benefits for ecosystems and multiple segments of society

been a fundamental goal of sound resource management and are routinely applied to managing individual stresses such as drought, overgrazing, pollution of freshwaters, and pest outbreaks (see Chapters 8–10). By monitoring trends in indicators of stresses and their ecological impacts, resource users can gauge changes from some historic reference point and take appropriate actions to reduce the stress. For example, overgrazing in drylands reduces the abundance of palatable grasses relative to unpalatable grasses or shrubs, indicating the need to reduce grazing pressure (see Chapter 8). Similarly, a decline in the size or trophic level of marine fish can be a sensitive indicator of overfishing (see Chapter 10). Social–ecological impacts of stresses are sometimes masked by ecosystem or social feedbacks, such as phosphorus sequestration in lake sediments (minimizing changes in water column phosphorus concentration; see Chapter 9) or perverse subsidies that motivate fishermen to increase fishing effort despite stock declines (minimizing declines in total catch; see Chapter 10). Such feedbacks make it valuable to identify multiple ecosystem and social indicators that are sensitive to initial phases of degradation (see Chapters 2 and 3), requiring knowledge and understanding of ecosystem dynamics and social–ecological interactions.

Mitigating *novel* stresses is an increasing challenge, because many of these stresses, such as climate change, acid rain, international fishing pressure, and global demand for biofuels, reflect global-scale economic, informational, and cultural linkages (see Chapters 4, 12, and 14). Mitigating global-scale stresses requires concerted global action, for which current governance mechanisms are inadequate (see Chapters 4 and 14). Nonetheless, effective collaboration among even a few key nations can accomplish a large proportion of the desired global mitigation, yielding disproportionately large global benefits (see Chapter 14). Global consensus is *not* required to make huge advances in the stewardship of our planet.

Given that many stresses are likely to continue or intensify, *trajectories* of expected change often provide more realistic management targets than do historical ranges of variability. Projections of climatic and demographic changes, while uncertain, provide an increasingly fine-scale framework for future planning (see Chapters 1 and 14). In cities, for example, planting trees and protecting green spaces ameliorate local impacts of increasingly frequent heat waves in urban heat islands, while providing other health and social benefits (see Chapter 13). Similarly, countries with trajectories of rapid population growth and rural poverty benefit from aid that empowers people to enhance the capacity for local food production (see Chapters 3 and 12) and at the same time avoid poverty traps (see Chapter 5). This contrasts with many current international aid programs that address immediate food needs in ways that undermine the capacity of local farmers to sell their crops or otherwise develop livelihoods that are sustainable within the constraints of locally available ecosystem services (see Chapter 12).

Reduce Sensitivity to Adverse Impacts of Hazards and Stresses

The current dynamics of most social–ecological systems result in part from a long coevolutionary history that has shaped governance and the capacity of people and other organisms to cope

with and adapt to a wide range of hazards and stresses. An important starting point for minimizing social–ecological sensitivity is therefore to sustain natural and social capital and livelihood opportunities and to understand adaptations to both the historical range of stresses and plausible future changes. In this context, resource management has traditionally emphasized the importance of protecting and sustaining the resources (e.g., water and soils) and organisms (e.g., population sizes of major functional groups) to provide multiple ecosystem services (Table 15.2; see Chapter 2). Similarly, cultural heritage and traditional economies are often important in sustaining livelihoods in human communities, although these economies may also reflect lack of access to more favorable opportunities, as in the former Apartheid system of South Africa (see Chapter 3). These general strategies of sustaining natural and social capital are at the core of steady-state ecosystem management and policy formulation. Nonetheless, achieving workable solutions is often difficult because of trade-offs among ecosystem services or societal outcomes and differences of opinion among stakeholders in evaluating these trade-offs (see Chapter 4). Because issues and trade-offs vary temporally and spatially, the application of general principles must be sensitive to local context. Thus, although the general guidelines are clear, their implementation commonly sparks conflicts that can only be addressed through stakeholder engagement in the decision-making process (see Chapter 4).

When social–ecological conditions change or new hazards or stresses are imposed, the mechanisms by which these systems coped with historical hazards and stresses may no longer suffice, for example, when confronted with new weapons, agricultural technologies, and globalized markets. In some places, the breakdown of traditional governance and resource-use systems combined with ready access to weapons cause violent local and regional conflicts that require fundamental solutions (see Chapter 8). These commonly involve renegotiation of resource access rights and access to new livelihood opportunities. In other cases, innovation may provide opportunities, as when chronic food shortages are addressed through more productive crop varieties or new agricultural practices (see Chapters 8 and 12). Innovations that are intended to reduce sensitivity to stress must be implemented cautiously, however, because they can create new vulnerabilities, such as susceptibility to crop diseases or increased dependence on a cash economy to buy expensive fertilizers or imported food at variable prices.

Policies that are intended to reduce sensitivity to stresses sometimes backfire because of an associated loss of resilience. For example, ecosystem management that is intended to reduce the sensitivity of timber economies to pests or fire can reduce the variability in these stresses—often leading to conditions to which many local organisms or social processes, particularly those that are important in postdisturbance reorganization and recovery, are less well adapted. When patterns of variability are no longer compatible with current (often changing) conditions, the whole system becomes vulnerable to large-scale change (see Chapters 1 and 5).

Within any social–ecological system, certain segments of society are particularly vulnerable to specific stresses and hazards. Targeted interventions that reduce the sensitivity of vulnerable segments of society are likely to be particularly effective in reducing *net* social impacts of shocks and stresses. For example, protection of the access rights of coastal fishermen can reduce vulnerability of both the fishermen and the fish stocks on which they depend (see Chapter 11). Similarly, effective systems of information exchange via the internet or telephone hotlines between urban residents, social networks, and city government can reduce the risks and sensitivity of urban slum dwellers to emerging environmental hazards, thereby reducing costs to society when a shock to the system occurs (see Chapter 13).

Careful analysis of current and projected hazards, sensitivities, and their interactions for multiple segments of society provides an excellent starting point for developing targets and pathways for sustainable development. Vulnerability analysis allows an assessment of trade-offs among costs and benefits for ecosystems

and various segments of society. The 1998 South African water policy, for example, established water reserves that guarantee access to domestic water for all segments of society and provided flows needed to sustain ecosystem integrity at times of drought. Thus domestic and ecosystem water rights took precedence over irrigated agriculture and industry (see Chapters 8 and 9). Such policies begin to address issues of long-term social–ecological resilience, discussed in the next section, as well as current vulnerabilities.

Enhancing Adaptive Capacity for Social-Ecological Resilience

At times of rapid directional social–ecological change, the historical state of the system may be poorly suited for adaptation to new conditions. Traditional informal property rights, for example, that may have been locally recognized and effective for centuries may be transferred from local users to newcomers who lack the knowledge or the ethical foundations to use ecosystem services sustainably. Key organisms may decline in response to novel disturbances, landscape structure, and pollutant levels. People may be disadvantaged in a more globalized economy or fail to cope with rising population pressures and shrinking resource supply. People and other organisms have always adapted to new opportunities through changes in their activities, ways of life, and locations (see Chapter 3). However, the novel character and rapid rate of recent changes challenge the capacity of social–ecological systems to adapt. The Earth System is moving into novel terrain.

Resilience-based ecosystem stewardship shifts the philosophy of resource management from *reactions* to observed changes to *proactive* policies that shape change for sustainability, while preparing for the unexpected. This paradigm shift is essential in a world undergoing rapid and directional change. A central approach to both reducing vulnerability and enhancing resilience is to enhance the adaptive capacity of social–ecological systems to both expected changes and unanticipated surprises. We describe four approaches to enhancing resilience by fostering (1) a diversity of options; (2) a balance between stabilizing feedbacks and creative renewal; (3) social learning and innovation; and (4) the capacity to adapt, communicate, and implement solutions (Table 15.3). These approaches provide ecosystem stewards with opportunities to think creatively about ways to sustain system attributes that society deems important and potential future pathways to sustain and enhance these attributes.

Foster Biological, Economic, and Cultural Diversity

Diversity, whether it is cultural, biological, economic, or institutional, is important because it increases the number of building blocks available to respond to and shape change. Diversity also broadens the range of conditions under which the system can function effectively (see

Chapter 1). A region whose economy depends entirely on one extractive industry, for example, is poorly buffered against market fluctuations or technological innovations that might reduce the value of that product. Similarly, low biological diversity constrains the capacity of ecosystems to adjust to large or rapid changes with subsequent losses of ecosystem services (see Chapter 2).

Although biodiversity warrants protection everywhere because of its functional importance and current high rates of loss (see Chapter 2), certain areas could contribute uniquely to planetary stewardship. For example, global hotspots of biodiversity or species of high cultural or iconic value (e.g., elephants, pandas, and polar bears) are likely to engender strong public support; areas with high topographic diversity and intact migration corridors enable species to migrate readily in response to rapid environmental change (see Chapter 2). Similarly, retention of corridors such as urban green spaces, hedgerows, and riparian corridors

TABLE 15.3. Examples of stewardship strategies to enhance social-ecological resilience.

Foster biological, economic, and cultural diversity
- Prioritize conservation of biodiversity hotspots and locations and pathways that enable species to adjust to rapid environmental change
- Retain genetic and species diversity that are underrepresented in today's landscapes
- Exercise extreme caution when considering assisted migration
- Renew the functional diversity of degraded systems
- Foster conditions that sustain cultural connections to the land and sea
- Foster retention of stories that illustrate past patterns of adaptation to change
- Subsidize innovations that foster economic novelty and diversity

Foster a mix of stabilizing feedbacks and creative renewal
- Foster stabilizing feedbacks that sustain natural and social capital
- Allow disturbances that permit the system to adjust to changes in underlying controls
- Exercise extreme caution in experiments that perturb a system larger than the jurisdiction of management

Foster social learning through experimentation and innovation
- Broaden the problem definition by learning from multiple cultural and disciplinary perspectives and facilitating dialogue and knowledge co-production by multiple groups of stakeholders
- Use scenarios and simulations to explore consequences of alternative policy options
- Test understanding through experimentation and adaptive co-management
- Explore system dynamics through synthesis of broad comparisons of multiple management regimes applied in different environmental and cultural contexts

Adapt governance to changing conditions
- Provide an environment for leadership to emerge and trust to develop
- Specify rights through formal and informal institutions that recognize needs for communities to pursue livelihoods and well-being
- Foster social networking that bridges communication and accountability among existing organizations
- Permit sufficient overlap in responsibility among organizations to allow redundancy in policy implementation

and retention of mobile species that link these habitats, such as pollinators, birds, and concerned gardeners, sustain diversity in human-dominated landscapes where people are most likely to lose their sense of connection to nature (see Chapter 2). Involvement of local residents in planning and implementing conservation efforts in lands that support their own livelihoods increases the likelihood that policies will be respected (see Chapter 6).

Areas that have already lost most of their biodiversity (e.g., production forests; mining, stream, or wetland restoration projects; or over-fished coastal systems; see Chapters 2, 7, 9, and 11) are unlikely to return to their former state, particularly in the context of rapid environmental change. Renewal of ecosystem services in highly modified ecosystems may occur more readily by developing functional diversity that is consistent with likely future conditions and societal goals than by trying to rebuild the historical predisturbance species composition. This redefines strategies of restoration ecology in a context of change.

The last several centuries have substantially reduced the cultural diversity of the planet as nation states, both internally and as colonial powers, sought to eliminate alternate cultures, ideologies, and modes of governance, especially where these involved claims to ownership and use of resources that differed from the claims of the state. The rise of nation states in Europe during the 17th and 18th centuries, for example, involved a deliberate effort to reduce cultural and political complexity, by removing people from historical relationships to land and water. The modern nation state standardized and thus simplified the formerly diverse and complex property rights and boundaries so that landscapes and seascapes could be mapped, administered, and taxed efficiently. Administration was concerned with extracting the maximum value from financially valuable resources and removing traditional uses of these or related culturally important resources (see Chapter 3). Similarly, in many colonial situations, indigenous people were forcibly removed from lands and waters they had previously controlled, so that the colonizers could lay claim to the vast majority of resources on these lands. In these cases, the elimination of the institutions and human relationships that were intrinsic to the cultural diversity was a logical part of exploiting resources.

More recently, during the mid-20th century, cultural assimilation was advocated globally as a way to rapidly improve opportunities for disadvantaged groups, for example, through education using a standardized curriculum in a single national language. These policies often further undermined cultural integrity through loss of language, local institutions, and cultural ties to the land and sea (see Chapter 3). Similarly, conservation plans that exclude people from their traditional homelands or disrupt rural economies, such as ranching, can stimulate poaching or land sales to developers that undermine the original policy intent (see Chapters 6–8, 11). Cultural diversity is valuable because it provides a diversity of knowledge systems, perspectives, and experience on ways to meet mutually agreed-upon goals in the face of large uncertain changes (see Chapter 4). On the other hand, cultural differences in preferences and belief systems can be sources of friction that are often essential to an equitable evaluation of trade-offs.

Under conditions of rapid change, some components of cultural, biological, and economic diversity are likely to be altered or lost. It is therefore important to identify, protect, and legitimize latent sources of diversity that may be underrepresented in current system dynamics. This can serve as insurance against excessive loss of current diversity components. For example, traditional crop varieties that have been replaced by higher-yielding hybrid or genetically engineered varieties assume a new importance as sources of genetic diversity to address the challenges of novel conditions and rapid spread and evolution of crop pests (see Chapter 12). Elders in most societies remember ways of doing things under a range of circumstances that enrich the options that are potentially available to address current and future changes. Many of these management practices reflect ecological understanding, such as indigenous burning in Australia and the USA, which reduced fuel loads, increased the proportion of food plants for both people and grazers, and reduced risk of large fires (see Chapter 6). As climate warming and a century of fire suppression increase risks of large fires, traditional indigenous management strategies provide alternative options for fire managers to consider. Other traditions such as gender inequality in education and voting or rigid property rights can reduce opportunities, indicating that some traditional institutions may reduce household and community resilience and ecosystem stewardship as social context changes.

Rapid climate change raises problematic decisions of whether and how to assist in the migration of long-lived immobile organisms like trees that are unlikely to keep pace with rapid environmental change. Should foresters plant a geographically diverse range of genotypes in forest regeneration projects rather than just the locally adapted genotype? Should organisms be transplanted to new, climatically favorable locations (assisted migration) when they are likely to go extinct in their current habitat (see Chapters 2 and 7)? Given the checkered history of species introductions and biocontrol programs, decisions that modify migration corridors or move species to new locations must be made cautiously and reflect a clear understanding of the trade-offs between risks and opportunities (see Chapter 9).

Like cultural and biological diversity, economic diversity increases the range of options for adjustment to change. Policies that provide short-term subsidies and incentives for innovation increase the opportunities to adjust to change (see Chapter 12). Conversely, perverse subsidies that sustain uneconomical practices, such as overfishing of stocks that would otherwise be uneconomical to fish, reduce the capacity of social–ecological systems to adjust to change (see Chapter 10). Economic subsidies frequently involve social–ecological trade-offs (e.g., short-term economic hardship to fishermen vs. long-term viability of a fish stock; viability of Scandinavian agriculture in an unfavorable economic climate vs. long-term food security; see Chapters 10–12). Decisions about subsidies often reflect power dynamics among affected stakeholders more than their value in ecosystem stewardship.

Foster a Mix of Stabilizing Feedbacks and Creative Renewal

In the short term, stabilizing feedbacks tend to sustain the properties of the system and keep it within its current state. Under circumstances of rapid directional change, however, these feedbacks can create a system that is increasingly out of balance with underlying driving variables. Therefore, a dynamic mix of stabilizing feedbacks and occasional disturbances that allow adjustments to changing conditions are most likely to sustain the fundamental properties and dynamics of the system—that is, to confer resilience. Managing stabilizing feedbacks has always characterized sound resource management. These include, for example, policies that minimize soil erosion and foster vegetation recovery after disturbance (see Chapters 7 and 12), market mechanisms that balance supply and demand (see Chapter 12), rangeland property rights that encourage pastoralists to care for the land while reducing resource use conflicts (Chapter 8), and fisheries quota systems that allocate access rights among fishermen (see Chapters 10 and 11).

Management that allows or fosters disturbance and renewal is often more controversial and may require greater search for creative solutions. Purchases of conservation easements that prevent residential development in rural scenic areas, for example, allow retention of fire as a natural ecological process in forests (see Chapter 7). Policies that allow rural residents to share revenues from big game hunting in African wildlife parks increase local support for wildlife conservation and reduce poaching (see Chapter 6). Policies that recognize the rights to public protest may create windows for policy adjustments to changing conditions (see Chapter 13).

Disturbance that creates the opportunity for renewal along either previous or new trajectories entails both opportunities and risks associated with new outcomes. Experimentation at small scales allows learning to occur so policies can be adjusted to either favor or reduce the likelihood of the changes that were observed. Experiments at larger scales than the system being managed are more dangerous because they can release changes that are beyond the control of individual resource managers to prevent undesirable outcomes (see Chapter 1). Current "experiments" with the climate, biodiversity, and cultural diversity of the planet are therefore of grave concern, including proposed geoengineering approaches to deal with climate change, which could lead to unintended feedbacks that are equally damaging to the planetary environment (see Chapter 14).

Foster Innovation and Social Learning

Although diversity provides the raw materials for adaptation, innovation and social learning are the core processes that build the adaptive capacity and resilience of a social–ecological system. The central roles of innovation and social learning in adaptation to changing social and economic conditions are universally recognized by business. However, they receive surprisingly little attention by resource managers and the public at large. Instead, there is often an emphasis on reducing variability, preventing change, and maintaining the *status quo*. This is no longer a viable management or policy framework under conditions of rapid directional social–ecological change. How can society shift from a mindset of fearing change to assessing its value as a way to cope with and realize new opportunities in a rapidly changing world? Not all changes are constructive. As discussed earlier, changes occurring at scales larger than the scale of management (e.g., the planet) should be approached cautiously.

An obvious starting point is to broaden the framework of problem definition by integrating a broader range of disciplines, knowledge systems, and approaches. Resource managers, for example, moved from the management of single species such as tigers, pines, or tuna to ecosystem management by acknowledging the importance of a broader range of ecosystem services and the key linkages between biophysical and social processes (see Chapters 1–4). Similarly, adaptive co-management of resources by agency managers and resource users provides opportunities to integrate a wealth of scientific

and local understanding to address challenging problems of social–ecological change (see Chapters 4–6).

In a rapidly changing world, however, knowledge of how to cope with previous conditions is often insufficient. Neither is it feasible to postpone actions until we can observe the performance of the system in equilibrium with new conditions. Instead, we must learn by doing without destroying future options—adaptively managing the global life-support system in which society is embedded (see Chapters 4 and 8). This requires educational transformations in both the school system and the workplace, including, as emphasized here, in resource management.

Adaptive management is a critical component of social learning because it embraces uncertainty and builds social learning into the management process. However, it is insufficient in a rapidly changing world because social–ecological systems, like other complex adaptive systems, are path-dependent, so management interventions that proved valuable in one circumstance may have different effects at other times and places (see Chapters 1 and 4).

Scenario modeling and analysis provides opportunities to explore those potential future conditions that cannot be readily predicted, for example, the consequences of new technologies or alternative policy strategies. This can help policy makers, researchers, resource managers, resource users, and the public envision potential futures, assess their fit to societal goals, and explore potential strategies and pathways to achieve desired ends (see Chapter 5). For example, scenarios of alternative water and land management policies in arid areas can broaden the discourse beyond fulfilling current needs to addressing long-term strategies that avoid unsustainable development (see Chapter 8).

Much can be learned about social–ecological dynamics and feedbacks from comparisons of similar management systems that have been applied in different social–ecological settings. Examples include marine reserves established in different marine ecosystems, democratic institutions or conservation strategies employed in different cultural settings, and integrated pest management applied to different agricultural systems. A great deal has been learned, for example, through comparative studies about the institutional arrangements and circumstances that foster sustainable use of common pool resources such as water, forests, fish, and rangelands (see Chapters 4 and 6–10). Co-management is not always conducive to flexibility and change. If co-management arrangements become overly codified and rigid, they constrain opportunities to adjust to change (see Chapter 6). Also, if responsibility is dispersed and unfocused, the hard choices may never be made. Leadership is essential in negotiating differences, providing vision, and building links between groups and their social networks at different levels of management and governance.

At the global scale, adaptive management, especially "management experiments," should be applied with extreme caution. There are thresholds or boundaries in the Earth System that would be dangerous to cross or to approach too closely. The best-known example is the concentration of carbon dioxide in the atmosphere. There is a threshold above which reinforcing feedbacks, such as weakening and then reversal of oceanic and terrestrial carbon sinks, could push the Earth System into another, much warmer state, state that is much less amenable for human life. At present, the precise location of this threshold is not well known; estimates vary between 350 and 600 ppm CO_2. Thresholds also exist in the chemical and biological components of the Earth System, but these are generally less studied and much less well known than the greenhouse-gas example. Given the existence of such thresholds and the serious consequences for humanity of crossing them, responsible stewardship of the Earth System as a whole requires careful attention to the precautionary principle (see Chapter 14).

Adapt Governance to Changing Conditions

Flexibility in governance structures that can deal with change is critical to long-term social–ecological resilience. Grazing systems in drylands (Chapter 8) offer models of how resource

access rules can change according to circumstances, contrasting with the rigidity of rules and processes of governments. The concept of "rules for changing the rules," as written into the constitutions of many countries, is a useful way to bring more flexibility to resource-use rights during this time of heightening uncertainty. Devolving the powers and resources of government to local scales can also enhance the responsiveness and adaptability to change in ways that sustain opportunities instead of constraining options, as long as it comes with the resources needed to navigate change and good systems of accountability. Bridging individuals and organizations such as NGOs or temporary public advocacy groups provide informal communication pathways that allow dialogue and negotiation to occur outside the rules and policies of formal institutions. Emergence of Ecuador's Watershed Trust Fund from a constellation of local groups, NGOs, and international aid agencies is an example (Chapter 9).

Decentralized, polycentric governance results in some overlap of responsibilities. This is analogous to biodiversity in providing redundancy in social–ecological functioning. State agencies, neighborhood groups, and national NGOs, for example, may all support actions that protect a certain species or valued habitat. When one of these groups "drops the ball" and fails to provide this governance function because of budget shortfalls or shifting priorities, the overlapping activities of other groups can sustain the basic need (see Chapter 9). Conversely, when national policies usurp the power of local institutions to protect valued local resources, the polycentric nature of governance is eroded, and ecosystem stewardship objectives are more likely to be threatened (see Chapter 6).

Navigating Transformations

Transformations are fundamental changes in social-ecological systems that result in different control variables defining the state of the system, new ways of making a living, and often changes in scales of critical feedbacks. In the context of ecosystem steward-ship, transformations involve forward-looking decisions to convert the system to a funda-mentally different, potentially more beneficial system (see Chapter 5). Unintended transfor-mations can also occur in situations where man-agement actions have prevented adjustment of the system to changing conditions. Rapid direc-tional changes in the factors that control sys-tem dynamics increase the likelihood that some critical threshold will be exceeded, so under-standing the actions that increase or decrease the likelihood of social–ecological transforma-tions is a critical component of resilience-based ecosystem stewardship (Table 15.4).

Preparing for Transformation

The first step in addressing potential transfor-mations (either desirable or not) is to iden-tify plausible alternative states and consider whether they are more or less desirable than the current state (Table 15.4). Because most trans-formations create both winners and losers, and the magnitudes of potential gains and losses are uncertain, stakeholder groups often dis-agree about how serious the problems are and whether or how to fix them. Important ini-tial steps in preparing for purposeful trans-

formation are mobilizing support for change; engaging stakeholders in identifying and rais-ing awareness of problems (e.g., rigidity traps or lack of institutional fit); and defining a col-lective vision for the future. Once the vision is defined, people are more willing to explore and agree on potential pathways to improved situa-tions. This includes identifying knowledge gaps, developing new governance and management approaches, identifying barriers to change, and developing strategies to overcome these barriers.

Shadow networks often play an impor-tant role in seizing windows of opportunity to make use of abrupt change. They can explore new approaches and experiment with social responses to uncertainty and change and thereby generate innovations that could trigger the emergence of new forms of governance and management of social–ecological systems (see Chapter 5). An important challenge is to pro-vide space for these networks to form through enabling legislation and financial, political, and moral support (see Chapter 5). Such learn-ing platforms can generate a diversity of ideas and solutions that can be drawn upon at crit-ical times. The challenge here is to establish structures like bridging organizations that allow for and support ecosystem stewardship for development.

TABLE 15.4. Strategies for purposeful navigation of transformations (see Chapter 5 for details).

Preparing for transformation
- Engage stakeholders to recognize dysfunctional states and raise awareness of the problem
- Identify, recruit, and support potential change agents
- Connect nodes of expertise and develop shadow networks of motivated actors
- Identify plausible alternative states and pathways
- Identify thresholds, potential crises, and windows of opportunity
- Identify the barriers to change and prepare strategies to overcome these

Navigating the transition
- Use crises or opportunities to initiate change
- Maintain flexible strategies for transition
- Negotiate the transformation with transparency and active stakeholder participation
- Foster structures that facilitate cross-scale and cross-organizational interactions

Building resilience of the new regime
- Create incentives and foster values values for stewardship in the new context
- Initiate and mobilize social networks of key individuals for problem-solving
- Mobilize new knowledge and external funding, when needed
- Foster the support of decision makers at other scales
- Shape the local context through adaptive co-management

Actively navigated transformations to alternate potentially more desirable states are frequently observed. These include the establishment of water management boards that guarantee domestic and green water flows in arid South Africa (see Chapter 9) and a shift from intensive logging to ecosystem management for multiple ecosystem services in the northwestern USA (see Chapter 7). Rigidity traps that still exist and require transformational rather than incremental solutions include persistent poverty in sub-Saharan Africa (see Chapter 12), failure of global governance to effectively address climate change (see Chapter 14), and repeated depletion of the planet's marine fish stocks by overfishing (see Chapters 10 and 11). **Every ecosystem type addressed in this book showed the potential to undergo purposeful transformations, although this did not always occur or sometimes reverted to the original system** (see Chapter 6). These observations suggest that transformations from degraded or dysfunctional states warrant consideration in most social–ecological systems. Indeed, this is *the* primary motivation of sustainable development projects undertaken in developing nations and warrants consideration in any ecosystem that confronts persistent social–ecological challenges.

Unintended transformations that have occurred include shift from slash-and-burn agriculture to intensive agriculture triggered by population growth and shortened cycles of forest recovery (see Chapter 12) and shift from production forests to residential housing associated with rising property values (see Chapter 7). **Every ecosystem type addressed in this book was observed to undergo unintended transformations in response to rapid directional changes in one or more environmental or social drivers** (see Chapters 6–14). The nature of many of these transformations was highly predictable from the known sensitivity of the system to critical controls, and the movement toward these state changes could be recognized from observed changes in drivers and known indicators of the transformation pathway. What is often unknown is the time course, abruptness, and sometimes the degree of irreversibility of the changes—hence the importance of fostering general resilience and the adaptive capacity to adjust to change. In other cases, the causes and nature of the transformation are unexpected. These observations suggest that unintended transformation should be taken as a serious possibility in all social–ecological systems and that resource managers should identify and respond to likely causes and indicators of movement toward transformation.

The properties of thresholds between alternative social–ecological regimes are poorly known and are currently an active area of research. Many thresholds are related to cascades through chains of positive feedbacks, such as trophic cascades in aquatic food chains (Chapter 9), spatial cascades of dryland degradation (see Chapter 8), fishing down marine food chains, as upper trophic levels are depleted (see Chapter 10), or agricultural transitions triggered by population increases (see Chapter 12). As ecosystems approach important thresholds, they respond more sluggishly to intervention, so the ecosystem becomes more difficult to control even as control is more desperately necessary. It has lost resilience. Ecosystems approaching thresholds may also become variable, flickering among alternate states in localized patches. Such flickering has been described for drought persistence in drylands. Conceptually, thresholds are expected in systems where human action is causing gradual change in slowly moving spatially extensive variables that contribute to strong feedback cycles with fast-moving variables. For example, gradual loss of habitat may eventually drive large predator populations below a threshold where they cannot persist, triggering cascades of change in lower-level consumers and plants. Preparing a social–ecological system for transformation therefore entails the capacity to detect early warnings and recognize potential thresholds.

Navigating the Transition

Preparing for and navigating transformational change is challenging because it depends on circumstances that are specific to each time and place. Transformational change is most likely to occur at times of crisis, when sufficient stakeholders agree that the current system is, by definition, dysfunctional. In such situations, the transformation may take many different alternative directions, some desirable, others highly undesirable. It is important that discussions about potential transformations be transparent, objective, and open to all stakeholders so that the process is not co-opted by a particular stakeholder group or agenda. Identifying potential crises that might provide windows of opportunity for transformations to promote ecosystem stewardship is a critical step in navigating transformations (see Chapter 5). For example, there is increasing recognition that the planet is approaching a point of "dangerous climate change"; this crisis represents an opportunity to implement global governance structures requiring more aggressive actions to reduce human impacts on the climate system (see Chapter 14). Prolonged droughts or catastrophic wildfires that burn the wildland–urban interface might trigger transformation in fire management and urban development, respectively (see Chapter 7). City infrastructure frequently has a lifetime of 50 years or less, and specific sections of the city may reinvent themselves even more frequently, providing opportunities to infuse new elements such as parks, green spaces, or efficient public transportation in ways that reshape urban dynamics (see Chapter 13). An important lesson from these examples is that the nature of crises that are likely to trigger transformation is often obvious ahead of time, providing opportunities to strategize and prepare for windows of opportunity.

Do we have to experience a crisis before we can change? How can we avoid or steer away from cascading ecological crises, unsustainable trajectories, and traps before they happen? There are at least four ways that crisis can lead to opportunities: (1) Resource stewards may actively prepare for change, so transitions happen smoothly; (2) a system may collapse locally, which raises awareness of the need for change; (3) actors may learn from crises happening in a similar system at other times or places; and (4) a crisis may happen in other sectors or at other scales but be seized as an opportunity to make changes. In Kristianstads Vattenrike in Sweden, for example, a local economic crisis coincided with rising national environmental concern because of pollution-related seal deaths along the Swedish coast (a national crisis) to initiate change.

Building Resilience of the New Regime

A successfully navigated transformation is best stabilized by building adaptive capacity and resilience through the processes described earlier (Tables 15.3 and 15.4). Because transformations create winners and losers, the resilience of the new system may initially be relatively fragile. Building resilience in new conditions can be strengthened by actions that build trust, identify social values among players of the new regime, and empower key stakeholders to participate in decisions that legitimize relationships and interactions of the new regime. Transformations often alter the nature of cross-scale interactions, providing both opportunities and challenges. Early attention to these cross-scale interactions that ensure good information flows, systems of accountability, and sensitivity to differing perspectives reduces the likelihood of reversion to earlier states or other unfavorable transformations. For example, the rapid transformation from rural to urban systems occurring in many parts of the world may require new systems of governance and patterns of social–ecological stewardship. While these are shaped by historical legacies, the resilience of the new system can be enhanced by eliminating barriers to cooperation among agencies and communicating a vision of opportunities provided by the transformed state throughout the public and private sectors. This can motivate and entrain actors and social networks to address new needs and opportunities that inevitably arise with transformation. Continuous evaluation and open discussion of the associated economic and noneconomic benefits and costs of change provide a basis for assessing progress in navigating relatively undefined structures and relationships that arise in novel social–ecological situations.

Comparisons of Vulnerability, Adaptive Capacity, and Resilience among Social-Ecological Systems

Contrasts among Types of Social–Ecological Systems

In the following paragraphs, we describe key vulnerabilities and sources of resilience that broadly characterize the types of social-ecological systems described in this book, recognizing that regional variation is substantial and that surprises will inevitably occur. Since no single type of institutional arrangement is the best for all situations, this broad social-ecological comparison illustrates the range of issues faced by ecosystem stewards and approaches that have proven useful to address the sustainability challenges in particular environments. Viewed together, the challenges and opportunities across this range of social-ecological systems provide a broader understanding of ecological stewardship.

Hinterlands (see Chapter 6) are typically occupied by small communities with strong cultural ties and informal institutions that link people to the ecosystems in which they live (Table 15.5). Nonresidents, however, often value these areas for conservation, recreation, or sources of resources (e.g., oil and gas) without considering the implications of these uses for hinterland residents. The integration of livelihoods into a biodiversity conservation ethic has been an integral part of many traditional cultures and can be a component of innovative solutions to current conservation challenges. Conversely, efforts to isolate parks from local livelihoods typically create highly artificial, fragile systems that have little historical precedent and are unlikely to be resilient to future environmental, economic, and social changes. Consequently, ecosystem stewardship of hinterlands requires adaptive governance at the local level that incorporates a range of perspectives on knowledge and the implications of change. Efforts to sustain traditional practices, as well as to facilitate smooth transformations to other forms of livelihood, must include institutions that pro-

vide strong linkages to regional and global processes that now shape opportunities in even the most remote locations. Resilience of hinterlands may also benefit from transformations that sustain and enhance ecosystem services for urban dwellers, including provisioning services like sustainable foods and cultural services such as recreational and existence values of wild landscapes and seascapes.

Forests (see Chapter 7) are dominated by plants that often have life spans lasting multiple human generations. Longevity generates stability and predictability for human residents of forests in terms of environment and ecosystem services (Table 15.5). However, large, rapid changes in environment, disturbance regime, and management goals can cause changes that are irreversible on the timescale of multiple human generations. Transformations of particular concern include high rates of tropical deforestation and temperate suburbanization, which may lead to large-scale regional and even Earth System feedbacks. These transformations are driven by global- or regional-scale processes and require intervention at these scales. The development and strengthening of institutions with a long-term view of the ecological and social conditions necessary to support forests and forest-dependent peoples are a foundation for sustainable forestry.

In contrast to forests, the resilience of **drylands** (see Chapter 8) derives primarily from effective adaptation of organisms, traditional societies, and local communities to high environmental variability. Local knowledge and governance arrangements are vitally important to this resilience at times of rapid environmental and social changes (Table 15.5). Cross-scale linkages and capacity to engage in diverse markets provide resources at times of crisis. These same characteristics render drylands vulnerable to large changes in slow variables, including prolonged drought and erosion of institutions (e.g., access rights) that allow society to cope with variability, particularly in environments of intermediate aridity where both population pressures and vulnerability to desertification are high. Natural variability slows down experiential learning, making the drylands vulnerable to rapid change. Cross-scale interactions can

TABLE 15.5. Representative sources of vulnerability and resilience and frequently observed transformations of Earth's major types of social–ecological systems. We consider only attributes that characterize each system type as a whole and ignore regional variations within systems.

Type of social–ecological system	Characteristic sources of vulnerability	Characteristic sources of resilience	Plausible undesirable transformations	Plausible desirable transformations
Hinterlands (Chapter 6)	Weak political voice; limited property rights to protect livelihoods	Diversity of knowledge systems and ecosystems; strong social–ecological coupling	Cultural assimilation; loss of biodiversity hotspots	Emergence of globally connected local governance
Forests (Chapter 7)	Rapid shifts in forest structure; rapid climate change	Ecological stabilizing feedbacks that sustain ecosystem services	Deforestation; spread of homes into wildland–urban interface	Reforestation; focused home developments that conserve forests
Drylands (Chapter 8)	Extended drought; overgrazing; distant voice	Adaptation to temporal and spatial variability; seasonal movements; mixed economies	Desert; fragmentation of pastoral lands that prevent seasonal movements	Sustainable water management; novel institutions that spread risk in time and space
Freshwater (Chapter 9)	Over-extraction; pollution with nutrients or toxins; invasive species	Riparian vegetation and wetlands; complex aquatic food webs; effective common-property institutions	No water; heavily polluted systems; extirpation of long-lived predatory fish;	Novel institutions for water commons that ensure sustainability
Open ocean (Chapter 10)	Overfishing; perverse subsidies; open access	Complex food webs; extensive spatial linkages	Fisheries collapse; novel food webs	Elimination of perverse subsidies for industrial fleets
Coastal (Chapter 11)	Loss of access rights; pollution; declining stocks from overfishing	Access to both marine and terrestrial resources; strong traditional knowledge and ties to coast	Aquaculture at the expense of other ecosystem services; extraction of nonrenewable resources (e.g., oil and gas) at the expense of fishing livelihoods;	Cultivation of new markets to sustain traditional livelihoods; strong co-management systems that sustain, and where needed transform, local communities
Agriculture (Chapter 12)	Loss of crop diversity; dependence on energy sector; economic globalization	High crop and cultural diversity; integrated pest management; flexible food production systems	Collapse of local agriculture in developing nations as food aid eliminates local markets	Redesign of food production systems to enhance other ecosystem services
Cities (Chapter 13)	Separation of people from food production	High cultural and structural diversity	Collapse of economies that support urban livelihoods and infrastructure; widespread disease	Renewal of cities as models of modern sustainable living; urban–rural function as complementary systems
Planet Earth (Chapter 14)	Sensitivity of climate system to surface properties; inadequate governance of the global commons	Strong biophysical feedbacks; diversity of knowledge systems	Large scale collapse of Earth as a life-support system; global-scale conflict due to geopolitical issues of climate change and energy demands	Emergence of coordinated systems of global governance

have massive effects on drylands that are usually poorly equipped to deal with these external forces because of their distant voice. Frequent transformations in drylands include desertification due to prolonged drought and/or overgrazing; conversion of mesic drylands to agriculture, which constrains access rights of pastoralists; and conversion to ranchettes or game ranches. Policies that mitigate these external forces for change sustain the natural resilience of drylands.

Most of the vulnerabilities of **freshwater ecosystems** are associated with intense biotic interactions that can drive rapid extirpation or evolution (see Chapter 9; Table 15.5). Freshwaters and their living resources are vulnerable to overextraction, pollution with nutrients or toxins, species invasions, and trophic cascades due to overexploitation of top predators. Resilience of freshwater ecosystems is supported by both ecological and social factors. On the ecological side, riparian vegetation, variable flow regimes, and complex food webs with long-lived top predators are keys to resilience. Over thousands of years, people have developed many effective institutions for dealing with water shortages. As these systems change, adaptive governance that is grounded in strong linkages between monitoring and decision-making provides an important source of resilience. There are both bad and good examples of transformation for freshwaters. In recent decades, many freshwater resources have been impaired and some have disappeared altogether. In some cases, long-standing systems have been effective and in others emerging problems have prompted people to develop new institutional arrangements for threatened freshwater commons.

In **open oceans** (see Chapter 10), the lack of tight coevolutionary interactions between people and fisheries has caused people to treat many of these fish stocks as open-access resources, making them highly vulnerable to overfishing. Past and current levels of resource use in open oceans are the primary cause of recent collapses in marine fisheries (Table 15.5). This problem is social in nature, resulting from both increased demand for fish and increased capacity to catch fish via fossil-fuel-dependent technologies. Perverse subsidies to industrial fleets drive most of the overfishing that has occurred. Subsidies prevent the fishery from reaching a bioeconomic equilibrium in which fishing pressure declines in response to declining stocks. The essential feedbacks of the social–ecological fisheries systems have been masked. Fisheries management is currently moving into a new era of ecosystem management involving multispecies approaches; establishment of marine protected areas; regulation of bycatch and habitat disruption (e.g., by trawling); and enhancement of fishery production (through hatchery and habitat enhancement programs). Given the emergent conditions, adaptively managed fisheries programs of experimentation are important in evaluating the currently unknown potential of these programs to enhance fishery sustainability and resilience.

In contrast to the open ocean, **coastal oceans** (see Chapter 11) are tightly coupled social–ecological systems (Table 15.5). The diversity of species and the sustainable patterns of livelihood that developed in coastal zones in both pre- and postindustrial times are products of a multiplicity of possible combinations of favorable conditions that permit coastal communities to achieve conservation of adjacent marine resources and to remain resilient to change. Sources of resilience are tied to the design and performance of governance that ensures the sustainability of local communities, such as community rights to access and to participate in management decisions, especially through adaptive co-management arrangements and to access both marine and terrestrial resources at times of scarcity. Depletion of coastal fisheries and economic globalization have led to the spread of one-species aquaculture, which ties local economies to global markets and marine systems worldwide.

The agricultural production of crops and animals for human consumption epitomizes the tight social–ecological linkages that are less obvious in some other systems. **Agricultural systems** (see Chapter 12) have adapted to a wide variety of environmental and social conditions, giving rise to a broad spectrum of genetic, crop, and cultural diversity (Table 15.5). The resulting resilience can be enhanced by selecting food production systems that are suitable to local social–ecological conditions, rather than modifying the environment to suit particular nonnative crops. When this does not occur, agricultural systems become vulnerable due to loss of local crop diversity and local knowledge, which could seriously constrain the capacity to adapt to directional climatic changes. Globalization of the economy creates additional vulnerabilities by decoupling food and environment spatially and temporally and distorting agricultural markets. Plausible transformations to address these vulnerabilities could occur by redesigning food production systems in three ways: selection of crops suitable to the environment; recoupling food production with its ecosystem base by removing distortionary policy incentives; and investment in a broad tool kit for agricultural improvement, including breeding, sustainable management practices and soil improvement techniques, multiple cropping systems, enhanced food quality, advanced genetics, information systems, and consumer labeling.

Cities (see Chapter 13) are the most rapidly expanding social–ecological systems on Earth and therefore provide huge opportunities for innovation in social–ecological stewardship and purposeful transformation to systems that maximize efficiencies of energy and resource use and reduce pressures on the remainder of the planet. They also represent huge concentrations of power, with potential to ignore rural issues and needs for sustainable development. Similarly, the tremendous human capital of urban systems, their mixing of diverse global cultures, and rapid turnover of infrastructure provide both opportunities for innovation and potentials for conflict and social–ecological degradation. Local social networks and bridging organizations can contribute to urban social–ecological resilience to (1) address new urban challenges and opportunities (e.g., rapid rural-to-urban migration, education and job opportunities for women that reduce population growth rates, rapid infrastructure turnover); (2) plan for the long-term in ways that reduce resource extraction from, and impacts on, nonurban regions; and (3) increase flexibility of governance structures to meet changing needs. In this context, it becomes essential that city inhabitants recognize the significance of life-support ecosystems in rural areas worldwide for their own well-being and that of the planet.

Given the scale and rate of changes, there is now an inescapable need for effective governance at the level of the **Earth System** (see Chapter 14). Both the biophysical and the social components of the Earth System are vulnerable to rapid and irreversible changes that could be detrimental to human well-being and the viability of modern societies (Table 15.5). Agreements are needed on issues like trade, security, migration, disease, climate, and other large-scale challenges that take into account the significance of resilience-based stewardship. Global social–ecological collaborations and the emergence of multilevel governance systems are needed to cope with, adapt to, and make use of rapid and directional change to shape transformations toward sustainability. Given that global-scale experiments (e.g., climate change and widespread biodiversity loss) result from decisions made at multiple scales, effective global governance must be linked to actions at many scales. This is *the* fundamental challenge of the 21st century.

General Patterns

From our comparative analysis of social–ecological systems of the world, several clear messages emerge:

- Resilience-based stewardship requires actions that recognize the coupled, interdependent nature of social–ecological systems.
- Every system exhibits critical vulnerabilities that tend to become exacerbated as directional environmental and social changes push these systems beyond the range of conditions to which organisms and cultures are adapted. However, the nature of these vulnerabilities differs among social–ecological systems.
- Every system has sources of biological, cultural, and institutional diversity and a substantial capacity to adapt to change.

This adaptive capacity can be enhanced through appropriate ecosystem stewardship supported by social–ecological governance from local-to-global scales.

- Every system has sources of resilience that provide opportunities for transformation to alternative, potentially more desirable social–ecological states. Human actions can enhance or degrade this resilience.

These broad conclusions suggest that ecosystem stewardship has important contributions to make in all systems. There is no region so resilient that policy makers and resource managers can ignore potential threshold changes, and we doubt that any region is beyond hope of substantial enhancement of well-being, adaptive capacity, and resilience. Meeting the goals of sustaining the important properties affecting ecosystem services will, however, require reconnecting people's perceptions, values, institutions, actions, and governance systems to the processes and dynamics of the biosphere and an active stewardship of ecosystems.

Critical Opportunities and Challenges for the Future

Resilience-based ecosystem stewardship provides a framework and guidelines that can ameliorate many problems and increase social–ecological sustainability. Nonetheless, serious challenges remain that constitute both major risks and opportunities for society. We highlight four social–ecological challenges that, if addressed effectively, would greatly enhance the sustainability of our planet.

- *Linking global, national, regional, and local governance*: The planet appears to be rapidly approaching a tipping point of dangerous climate change requiring new governance systems to prevent this from occurring. Scattered local and national initiatives are beginning to address this issue. How can governance across the full range of scales be linked to respond appropriately to climate change and avoid dangerous conditions? How can governance systems be designed to

facilitate local adaptation to ecological and social changes that are already occurring and reduce the risk of major geopolitical and social problems?

- *Sustaining cultural and biological diversity in a globalized world*: Irreversible losses of cultural and biological diversity are reducing resilience at all scales and the options for addressing an uncertain and rapidly changing future—often as an unintended consequence of a globalized market economy that responds to increasing demand for renewable and nonrenewable resources. How can the costs of globalization be taken into account in ways that avoid the loss of cultural and biological diversity? What transformations are needed to retain this diversity while fostering the resilience-based stewardship needed to sustain the Earth System?

- *Navigating transitions in human population and consumption that place increasing pressure on the planet's natural resources*: The absolute increase in human population and its demand for food, energy, and other natural resources are projected to be greater in the next four decades than at any time in the history or likely future of our planet. What steps can be initiated *now* that will reduce this pressure on planetary resources and speed the transition to a more sustainable future path?

- *Using urbanization as an opportunity to enhance regional resilience*: Rapid movement of people from rural areas to urban areas provides opportunities to reshape patterns and pathways of using ecosystem services. How can rapid urbanization be used to enhance resilience and the generation of ecosystem services to meet human needs (not desires) and to rebuild rural diversity and support social–ecological sustainability?

We suggest that resilience-based ecosystem stewardship provides a framework to effectively address these and other urgent issues that face society today and in the future. Like all complex adaptive systems, resilience-based ecosystem stewardship is always a work in progress that will inevitably adapt and transform as new insights and conditions emerge.

Review Questions

1. Describe, for a social–ecological system of your choice, specific changes in policies and institutions that would reduce its vulnerability to economic decline or expected climatic changes. What practical steps would facilitate these changes? How can these be initiated *now*?

2. Describe, for a social–ecological system of your choice, specific changes in policies and institutions that would enhance the adaptive capacity and resilience to economic decline or expected climatic changes. What practical steps would facilitate these changes? How can these be initiated *now*?

3. Describe, for a social–ecological system of your choice, the unintended regime shifts or transformations that might plausibly occur within the next 20–40 years. What are the costs and benefits to different stakeholders of these large potential changes? What practical steps can be initiated *now* to reduce the likelihood of these changes?

4. Describe, for a social–ecological system of your choice, plausible transformations that might be actively navigated to enhance opportunities or to avoid major social–ecological problems that currently exist or are likely to occur. Propose a pragmatic set of actions that would make this transformation likely to occur, if the appropriate window of opportunity presented itself. How might this window of opportunity be created?

Abbreviations

AD	Anno Domini (years after the birth of Christ)	EU	European Union
AIDS	Acquired immunodeficiency syndrome	FAO	Food and Agriculture Organization
AMA	Adaptive Management Area	FARMCs	Fishery and Aquatic Resource Management Councils
B	Billion	FERTIMEX	Mexican Fertilizer Company
BES	Baltimore Ecosystem Study LTER program	FDI	Foreign direct investment
CAFSAC	Canadian Atlantic Fisheries Scientific Advisory Committee	GBRMP	Great Barrier Reef Marine Park
CAM	Crassulacian acid metabolism	GCM	General circulation model; global climate model
CAMPFIRE	Communal Areas Management Program for Indigenous Resources	GDP	Gross domestic product
		GECs	Global environmental changes
		GEF	Global Environment Facility
		GIS	Geographic information system
CAP	Central Arizona-Phoenix LTER program	GNP	Gross national product
		GSCs	Global social changes
CBO	Community-based organization	HIV	Human immunodeficiency virus
CCAMLR	Convention on the Conservation of Antarctic Marine Living Resources	IBM	International Business Machine
		IPAT	Generalization relating human impact, population, affluence, and technology
CFC	Chlorofluorocarbon		
CH	Camp household	IPCC	Intergovernmental Panel on Climate Change
CH$_4$	Methane		
CITES	Convention on International Trade in Endangered Species	IPM	Integrated pest management
		IT	Information technology
CMA	Catchment management agency	ITQ	Individual transferable quota
CO$_2$	Carbon dioxide	IUCN	International Union for Conservation of Nature
CRP	Conservation reserve program		
CV	Coefficient of variation	IUU	Illegal, unreported, and unregulated (fishing)
DDE	Breakdown product from an insecticide		
		IVQ	Individual vessel quota
DDP	Dryland development paradigm	KNP	Kruger National Park
DDT	An insecticide	LDC	Least developed countries
EEZ	Exclusive economic zone	LTER	Long-Term Ecological Research program
ENSO	El Niño-Southern Oscillation		

M	Million	ppm	Parts per million
MDGs	UN Millennium Development Goals	REIT	Real Estate Investment Trust
		Tg	Teragrams
MEA	Millennium Ecosystem Assessment	TIMO	Timber Investment Management Organization
MPA	Marine protected area	TPC	Threshold of probable concern
MSY	Maximum sustained yield	TRI	Toxic release inventory
N_2	Di-nitrogen, the predominant form of nitrogen in the atmosphere	TURF	Territorial Use Rights Fishery
		UK	United Kingdom
		UN	United Nations
NAFO	Northwest Atlantic Fisheries Organization	UNDP	United Nations Development Programme
NAFTA	North American Free Trade Agreement	UNEP	United Nations Environment Programme
NGO	Nongovernmental organization	UNESCO	United Nations Educational, Scientific and Cultural Organization
NTFP	Non-timber forest product		
NTT	Nusa Tenggara Timor in Indonesia	UNFCCC	United Nations Framework Convention on Climate Change
NWFP	Northwest Forest Plan	US	United States
NYC	New York City	USDA	US Department of Agriculture
ODA	Official development assistance	USSR	Union of Soviet Socialist Republics
OECD	Organization for Economic Cooperation and Development		
		VH	Village household
OSY	Optimum sustained yield	WCED	World Commission on Environment and Development
PAM	Plant-available moisture		
PAN	Plant-available nutrients	WDF	Washington Department of Fisheries
PCB	Polychlorinated biphenyl (class of compounds frequently used as pesticides)		
		WRI	World Resources Institute

Glossary

Aboriginal. See indigenous. The first group to occupy a region. In Canada and Australia used synonymously with indigenous.

Active adaptive management. Intentional manipulation of the system to test its response and apply understanding to future decisions.

Actor group. A group of people often with different skills (e.g. knowledge carriers, entrepreneurs) engaged in an activity such as ecosystem stewardship or poverty alleviation.

Adaptability. See adaptive capacity.

Adaptation. Adjustment to a change in environment. Defined by biologists and anthropologists as a genetic change in a population. Anthropologists also refer to adaptation as a social, economic or cultural adjustment to a change in the physical or social environment.

Adaptive capacity. Capacity of human actors, both individuals and groups, to respond to, create, and shape variability and change in the state of the system. Synonymous with adaptability.

Adaptive co-management. Resource management that seeks social-ecological sustainability through a multi-scale and collaborative process of intentionally learning from experience.

Adaptive cycles. Cycles of system disruption and renewal.

Adaptive governance. Active experimentation in governance with institutional and political frameworks designed to adapt to changing relationships between society and ecosystems in ways that sustain ecosystem services; expands the focus from adaptive management of ecosystems to address the broader social contexts that enable ecosystem-based management.

Adaptive learning. Process in which one or more groups (1) carefully and regularly observe social-ecological conditions, (2) draw on those observations to improve understanding of the system's behavior, (3) evaluate the implications of emergent conditions and the various options for actions, and (4) respond in ways that support the resilience of the social-ecological system.

Adaptive management. Resource management approach based on the science of learning by doing. See also active and passive adaptive management.

Afforestation. Planting of forests on previously unforested lands.

Agency capture. Condition in which a special interest group establishes a controlling relationship with an agency, which works almost exclusively on the interest group's behalf.

Agistment. Practice in which private properties that have too many animals for the available forage in a particular year will transport stock to other properties where the imbalance is in reverse, and the senders pay the receivers.

Albedo. Reflectance of incoming solar (shortwave) radiation.

Alternative stable states. Alternative system states, each of which is plausible in a particular environment.

Amplifying feedback. Feedback that augments changes in process rates and tend to destabilize the system. It occurs when two interacting components cause one another to change in the same direction (both components increase or both decrease). Synonymous with positive feedback.

Anion. Negatively charged ion.

Anthropocene. New planetary epoch beginning with the advent of industrialization characterized by global processes that are strongly shaped by humanity.

Aquaculture. Fish and shellfish cultured in confined systems.

Arable land. Land under temporary crops, temporary meadows for mowing or pasture, land under market and kitchen gardens and land temporarily fallow (less than five years). It does not include land that is potentially cultivable.

Aridity index. Ratio of precipitation to potential evapotranspiration.

Artisanal fishery. Small-scale mixed commercial and subsistence fishery, predominantly in coastal areas.

Assisted migration. Movement of organisms by people to a more favorable climate because their natural migration is too slow to enable them to adapt to climatic change.

Backcasting. The process of identifying societal goals and working backwards to explore how to arrive at them. Often used with simulation models and scenario analysis.

Bennett's law. Generalization that the caloric intake of households is dominated by starchy staples at low levels of income, but is characterized by a diversified diet of fruits, vegetables, and animal products with income growth.

Benthic. Bottom-dwelling organisms of aquatic ecosystems.

Biodiversity. Number and relative abundance of organisms in an area.

Bioeconomic equilibrium. Equilibrium harvest level dictated by harvest effort and profitability.

Blue water. Liquid water in rivers, lakes, reservoirs, and groundwater aquifers.

Bonding network. Network of individuals contributing to group cohesion, often based on long-term familiar relationships.

Bottom-up effects. The impact on a system of activities by low levels in a hierarchy, e.g., effects of plants on their consumers or stakeholders on government.

Boundary organizations. Organizations that facilitate the transfer of knowledge between groups or social processes, for example, between science and policy.

Bounded rationality. Model of human decision making reflecting how limitations in understanding and information affect choice; this contrasts with a purely economic rational approach.

Bridging network. Network that links individuals and groups across scales to access to resources such as information and expertise.

Bridging organization. Group, such as a board, council or other organization, that communicates information and coordinates collaborations among local stakeholders and actors across several organizational levels or cultural systems.

Brousse tigré. Banded landscape patterns.

Built capital. The physical means of production beyond that which occurs in nature (e.g., tools, clothing, shelter, dams, and factories). Synonymous with manufactured capital.

Bundles of services. Groups of ecosystem services that co-occur because of tight linkages through ecosystem processes.

Bureaucratization. Process by which, over time, an organization becomes increasingly dependent upon formal rules and procedures.

Bycatch. Species that are unintentionally caught in the process of fishing for other species.

Cap-and-trade. A market-based approach for controlling use of a common-pool resource. The cap is the maximum tolerable level of resource use and is set by regulation. Marketable credits issued by the regulatory authority can be bought or sold to allow resource conservation by some users to balance resource use by others.

Capital. Assets or productive base of a social-ecological system, i.e., its human, manufactured, and natural assets.

Carrying capacity. Maximum quantity of an organism that the environment can support.

Catchability coefficient. Fishing mortality rate generated by one unit of fishing effort.

Cation. Positively charged ion.

Chaotic behavior. Behavior that is unpredictable and depends primarily on initial conditions and/or the nature of the perturbation.

CitiStat-type programs. Urban programs that provide accurate and timely intelligence, develop effective tactics and strategies, rapidly deploy resources, and facilitate follow-up and assessments.

Citizen science. A type of knowledge production involving non-professional volunteers who perform monitoring and research-related tasks, often to track changes and/or answer real-world problems.

Civil society. All voluntary civic activities, social organizations, and institutions that are the basis for a functioning society and its collective action.

Clay. Fine soil mineral particles.

Coefficient of variation. Standard deviation divided by the mean.

Cognitive maps. The collection of beliefs, experiences, and information that a person uses to orient to one's environment. Can refer to the mental models used to perceive, contextualize, simplify, and make sense of otherwise complex problems.

Collective action. Cooperation of individuals to pursue a common goal.

Co-management. Sharing of power and responsibility between resource user communities and state agencies to manage resources.

Command-and-control. Decision making by upper management of an organization to direct individuals' behavior with the goal of maintaining a constant output of some ecosystem good or service.

Common-pool resources. Shared resources that are subtractable and from which it is costly to exclude people's use. Formerly termed common-property resources.

Common-property resources. See common-pool resources.

Community-based wildlife management program. Program with high levels of community involvement in resource management, where explicit incentives for conservation encourage stewardship behavior and protect threatened species; a strategy for promoting both conservation and local economic development.

Community quota. A property right held in common by a community to harvest a resource; it is issued through legal means and is usually transferable only among members of a specified organization, but not at speculative prices.

Complex adaptive system. System whose components interact in ways that cause the system to adjust (i.e., "adapt") in response to changes in conditions.

Complex problem. Problem with many potential solutions that are quite different in execution, and rankable in quality of outcome.

Congestible resources. Resources that decrease in quality when the number of resource users reaches a threshold, for reasons other than ecological productivity (e.g., enjoyment of wilderness).

Connector. Individual who knows lots of people in terms of both numbers and kinds of people, in particular the diversity of acquaintances.

Conservation by utilization. Ecosystem management policy, which acknowledges that landowners should benefit from ecosystem services.

Conservation phase. Phase of an adaptive cycle during which interactions among components of the system become more specialized and complex.

Context-dependent. Actions framed by an overall context, for example a legal framework, norms and rules, a historical trajectory or a cultural belief system.

Coping. Short-term adjustment by individuals or groups to minimize the impacts of hazards or stresses.

Coupled human-environment system. See social-ecological system.

Crisis. Time when a group of people or a society perceives that some components of the present system are dysfunctional.

Critical ecosystem services. Ecosystem services that (1) society depends on or values; (2) are undergoing (or are vulnerable to) rapid change; and/or (3) have no technological or off-site substitutes.

Cross-scale linkages. Processes and networks that connect the dynamics of a system to events that occur at other times and places.

Cross-scale surprises. Surprises that occur when there are cross-scale interactions, such as when local variables coalesce to generate an unanticipated regional or global pattern, or when a process exhibits contagion (as with fire, insect outbreak, and disease).

Cultural ecology. Study of the relationship of a given society and its natural environment, with a focus on changing social and ecological conditions.

Cultural heritage. Stories, legends, and memories of past cultural ties to the environment.

Cultural identity. A sense of membership by an individual or group to a culture.

Cultural services. Non-material benefits that society receives from ecosystems (e.g., cultural identity, recreation, and aesthetic, spiritual, and religious benefits).

Cumulative processes. Processes that occur locally in different parts of the world, but over time the effect aggregates to produce regional-to-global consequences.

Cyanobacteria. Nitrogen-fixing bluegreen algae.

Decision-support tools. Tools that integrate information to assess the state of knowledge, assess management actions, direct future monitoring and research, and explore the implications of alternative futures.

Decomposer. Organism that uses dead organic matter as a source of energy.

Decomposition. Chemical breakdown of dead organic matter by soil organisms.

De facto. Used in practice, often referred to with respect to property rights.

Deforestation. Conversion of forests to a nonforested ecosystem type, frequently agriculture.

Degradation. Deterioration of a system to a less desirable state as a result of failure to actively adapt or transform.

De jure. Formal, based in a legal convention.

Denitrification. Conversion of nitrate to gaseous forms (N_2, NO, and N_2O).

Depensatory decline. Population decline that accelerates as the stock size declines.

Desalinization. Conversion of salt water into freshwater.

Desertification. Soil degradation that occurs in drylands that is triggered by drought, reduced vegetation cover, over-grazing, or their interactions.

Deterministic. Governed by and predictable in terms of a set of laws or rules.

Development and empowerment organizations. Organizations that work with communities and other stakeholder groups to build local capacity to address current and future problems, and improve communication internally and with other groups. Commonly associated with economic development.

Directional change. Change with a persistent trend over time.

Disciplinary. Belonging to a field of study that shares a common perspective of the world.

Discount rate. The estimated value of goods and services that can be obtained in the future, relative to the value of the same goods and services that are available today.

Discourse. Formal or orderly expression of thought on a subject, including what is discussed, how it is discussed, and which aspects are deemed legitimate or are marginalized.

Disequilibrial. Perpetually buffeted away from any equilibrium.

Disturbance. Relatively discrete event in time and space that alters ecosystem structure and causes changes in resource availability or physical environment.

Disturbance regime. Characteristic severity, frequency, type, size, timing, and intensity of a set of disturbances in an ecosystem.

Double-loop learning. Feedback process in decision making by which practitioners question the assumptions, knowledge, and learning models that underlie goals and strategies before taking further action.

Drip irrigation. Irrigation method that minimizes the use of water and fertilizer by allowing water to drip slowly to the roots of plants, either onto the soil surface or directly into the root zone.

Dryland development paradigm (DDP). See Drylands syndrome.

Drylands syndrome. Suite of key attributes that characterize most drylands of the world: unpredictability, resource scarcity, sparse populations, remoteness and 'distant voice.' Synonymous with dryland development paradigm.

Dustbowl. Region of the central US where wind erosion removed massive amounts of soil during the extended drought of the 1930s.

Dynamics of denial. Perspective in which a group of people, for example fishermen, argue that scientific assessments must be wrong because they are still catching plenty of fish (and if there is a decline, recovery will occur naturally due to favorable environmental circumstances).

Earth System. Planet Earth as a social-ecological system.

Economic rent. Income gained relative to the minimum income necessary to make an activity economically viable. The economic rent in a sales transaction is the difference between the payment actually received and the second-best price the owner could otherwise get for using land, labor or capital in another activity.

Ecosystem. All of the organisms (plants, microbes, and animals—including people) and the physical components (atmosphere, soil, water, etc.) with which they interact.

Ecosystem goods. See provisioning services.

Ecosystem management. Management paradigm that emphasizes practices for multiple ecosystem services by capitalizing on, sustaining, and enhancing ecosystem processes.

Ecosystem services. Benefits that people receive from ecosystems, including supporting, provisioning, regulating, and cultural services.

Ecotone. Edge or transition zone between two ecological communities.

Effect diversity. The diversity of organisms with respect to their effects on ecosystem processes.

Emergent properties. Property of a complex system that emerges from interactions of subunits and cannot be understood or predicted by studying individual subunits.

Engel's law. Generalization that, as incomes grow, the proportion of household income spent on food in the aggregate declines.

Engineering fishery policies. Increase the safe catch curve through engineered habitat management and artificial propagation (hatchery) systems.

Entitlements. Sets of alternative benefits that people can access, depending on their rights and opportunities, sometimes guaranteed through law. It also refers, in a more casual sense to someone's belief of deserving some particular reward or benefit.

Environmental event. Physical events occurring in ecosystems such as floods and droughts.

Environmental injustice. Uneven burden of environmental hazards among different social groups.

Equilibrium. Condition of a system that remains unchanged over time because of a balance among opposing forces.

Equity. Fairness.

Erosion. Transport of soil particles by wind or water from one place to another.

Essential resource. Resource for which no substitute exists.

Eutrophication. Nutrient enrichment (typically nitrogen and phosphorus pollution of aquatic systems).

Excludable resources. Resources from which potential users can be excluded at low cost.

Exclusive economic zone (EEZ). Marine waters within 200 miles of a nation's coast.

Exogenous factor. Factor external to the system being examined, which therefore is not incorporated into management practices.

Expert knowledge. Knowledge of a system or phenomenon based on extensive experience, which can include formal study, by individuals.

Exposure. Nature and degree to which the system experiences environmental or socio-political stress.

Extensification. Increased agricultural production achieved through ecosystem conversion to agricultural lands.

Evolution. Changes in organisms as a result of genetic responses to past events.

Fast variable. Variable that responds sensitively to daily, seasonal, and interannual variation in exogenous or endogenous conditions.

Feedback. See Amplifying feedback and stabilizing feedback.

Fishery collapse. 90% or greater reduction in catch from peak historic levels.

Fixed escapement rule. Allow the same fixed number of fish to spawn each year.

Fixed exploitation rate. Harvest the same fixed percentage of the stock each year, leaving the rest to spawn.

Food security. Ability of a group of people at all times to achieve physical and economic access to the food needed to lead a productive and healthy life.

Forecasting. Projection of future conditions and their social-ecological consequences based on extrapolation of recent trends.

Forest certification. Procedure for assessing forest management practices against standards for sustainability so purchasers can support sustainable management.

Formal institutions. Formal sets of written rules typically recognized and enforced by governments such as constitutions, laws, and legally based conventions.

Formal knowledge. "Scientific knowledge" learned from books or through formal educational systems.

Fossil groundwater. Groundwater that accumulated under a different climate regime and is no longer being renewed at a significant rate.

Framing. Defining a problem or issue in a context that conveys its value to the public and to decision makers.

Free-rider problem. Situation that occurs when individuals or groups of actors do not assume their fair share of responsibilities while consuming more than their share of the resources.

Fugitive resources. Resources that move across a range of jurisdictions and can have many user groups.

Functional redundancy. Diversity *within* a functional type.

Functional silos. Fragmentation of responsibilities and authorities among individuals or agencies. Synonymous with stove pipes.

Functional types. Groups of organisms that exert similar effects on social-ecological systems (effect functional type) or which show similar response (response functional type) to an environmental change (e.g., evergreen trees, algal-eating fish).

Genuine investment. Increase in the productive base (total capital or inclusive wealth) of a social-ecological system.

Global Environment Facility (GEF). Multilateral funding mechanism administered jointly by UNEP, UNDP, and the World Bank and dedicated to providing funds needed to help developing countries address large-scale environmental problems like climate change.

Globalization. Global interconnectedness and interdependence (e.g., of economy or culture).

Goal displacement. Condition in which the survival of the organization assumes greater priority than efforts to meet its stated mission.

Governance. Pattern of interaction among actors, their sometimes conflicting objectives, and the instruments chosen to steer social and environmental processes within a particular policy area.

Great acceleration. The explosion of the human enterprise in terms of both population and economic activities after the Second World War.

Green revolution. Design and dissemination of high-yielding seed varieties for the major cereal crops requiring intensive application of fertilizers and pesticides.

Green water. Moisture in the soil that supports evapotranspiration from all non-irrigated vegetation, including rain-fed crops, pastures, timber, and terrestrial natural vegetation.

Growth phase. Phase of the adaptive cycle during which environmental resources are incorporated into living organisms, and policies become regularized.

Habitat/species management area. Protected area managed mainly for conservation through management intervention.

Hierarchical system. A vertically organized system, often in the shape of a pyramid, with each row of objects linked to objects directly beneath it, for example, an army or a computer file system.

Horizontal interplay. Interaction among institutions at the same level of social organization or across space.

Horizontal linkages. Linkages that occur at the same level of social organization and across spatial scales, e.g., in treaties among countries.

Human agency. Capacity of people to make individual and collective choices and to impose their choices on the world.

Human capital. Capacity of people to accomplish their goals given their skills at hand.

Human ecology. See cultural ecology.

Hydroponics. Method of growing plants using mineral nutrient solutions instead of soil.

Iconic species. Species that symbolize important nature-based societal values.

Inclusive wealth. Total capital (natural, manufactured, human, and social) that constitutes the productive base available to society.

Income elasticity of demand. Growth in the demand for a good in response to an increase in income of people demanding that good.

Indecision as rational choice. Failure of regulators to make decisions when faced with convincing evidence from stakeholders of detrimental economic impacts and uncertain scientific evidence of degradation of the resource being regulated.

Indicator. A measure of one aspect of a system used to communicate its state or direction of change.

Indigenous. Group of people who have long inhabited a geographic region, often as the original or earliest known inhabitants. It generally refers to a group that is ethnically distinct from a group that colonized a region in historical times. In Canada the preferred term is aboriginal.

Individual transferable quota (ITQ). Quotas issued by government that privatize the access and withdrawal rights to a specific marine resource.

Industrial roundwood. Sawlogs and pulpwood (and the resulting chips, particles, and wood residues). Synonymous with timber.

Informal institution. Unwritten rules such as sanctions, taboos, customs, and traditions.

Institution. Rules in use that create enduring regularities of human action in situations structured by rules, norms, and shared strategies.

Institutional fit. Match of characteristics of institutions with the dimensions of a social-ecological system in which they are embedded.

Institutional interplay. Interactions among institutions.

Institutional learning. See social learning.

Integrated pest management (IPM). Multidimensional approach for managing agricultural pests by development of new host plant resistance and limited spraying of chemicals.

Intensification. Increased agricultural production achieved through increased inputs of fertilizers, pesticides, irrigation, and technology to enhance yield per unit land area.

Interdisciplinary. Integration between two or more academic disciplines to define problems, share methods, and create and answer common questions.

Interplay. Interactions among human agency, organizations, or institutions.

Investment. Increase in the quantity of an asset times its value.

IPAT law. Rule of thumb stating that the impact (I) of any human activity is equal to the product of the human population size (P), affluence of the population (A), and the technology employed in production (T).

Keystone species. Species that have disproportionately large effects on ecosystems, typically because they alter critical slow variables.

Knowledge system. Culturally defined way of knowing.

Landscape function. Capacity of a landscape to regulate nutrients and water, concentrating them in fertile, vegetated patches where soil biota maintain nutrient cycles and water infiltration and where vegetation cover impedes surface water flow, retains nutrients and seeds, maintains infiltration and protects against erosion.

Learning networks. See communities of practice.

Legacies. Stored past experiences of the dynamics of social-ecological systems.

Life-support system. Supporting ecosystem services that give rise to the provisioning, regulating, and cultural ecosystem services for society.

Limited entry. Policy requiring permits in zones of use.

Litter. Layer of dead leaves on the soil surface.

Livelihood. Strategy undertaken by individuals or social groups to create or maintain a living.

Local knowledge. A dynamic system of place-based observations, interpretations, and local preferences that inform people's use of and relationship with their environment and with other people. It may include a mix of social, ecological, and practical knowledge and involve a belief component. Overlaps with traditional ecological knowledge.

Local surprises. Surprises that occur locally and are created by a narrow breadth of experience with a particular system, either temporally or spatially.

Managed resource protected area. Protected area managed mainly for the sustainable use of natural ecosystems.

Manufactured capital. See built capital.

Maven. An altruistic individual with social skills who serves as information broker, sharing and trading what (s)he knows.

Maximum sustained yield. Policy that seeks to maximize the harvest of forests, fish, and wildlife to meet current needs, without reducing the potential to continue these yields in the future. Often optimistic about current and future potential yields.

Mental map. See cognitive map.

Metapopulations. Populations of a species that consist of partially isolated subpopulations.

Mitigation. Reduction in the exposure of a system to a stress or hazard.

Mobile links. Biological or physical processes that link patches on a landscape or seascape.

Muddling through. Process of non-strategic trial-and-error decision making in which solutions to problems are sought by organizations without the benefits of careful reflection.

Multidisciplinary. Perspectives from multiple disciplines on how the world works, with each discipline working within its own disciplinary framework.

Multilevel governance. Governance that takes place across several levels of institutions.

Multiple-use management. Management that explicitly seeks to sustain a broad array of ecosystem services.

Multi-stakeholder body. Group that is often convened by policy makers to scope issues, seek solutions to problems, achieve broader public participation, and foster public consensus.

Mycorrhizae. Symbiotic relationship between fungi and plant roots leading to an exchange of fungal nutrients for plant carbohydrates.

National monument. Protected area managed mainly for conservation of specific natural features.

National park. Protected area managed mainly for ecosystem protection and recreation.

Natural capital. Nonrenewable and renewable natural resources that support the production of goods and services on which society depends.

Natural enemies. Pathogens, predators, and parasitoids of agricultural pests.

Negative feedback. See stabilizing feedback.

Neo-liberalism. Political movement that espouses economic liberalism (i.e., minimal governmental interference) as a means of promoting economic development and securing political liberty.

Nitrogen fixation. Conversion of gaseous nitrogen (N_2) to ammonium by organisms (biological nitrogen fixation) or industrially using fossil fuels as an energy source (industrial nitrogen fixation).

Nonpoint pollution. Nutrients or toxins scoured by runoff from agricultural or urban lands.

Norms. Rules for social behavior.

Normative concepts. Concepts with a values orientation.

Ontogenetic habitat shifts. Habitat shifts that occur as fish develop from juvenile to adult.

Open access. Situation in which potential users are not excluded from using a resource.

Open systems. System characterized by flows of materials, organisms, and information into and out of the system.

Opportunity costs. Potential benefits that are forgone as a result of a particular choice or action.

Organization. Social collective with membership and resources, which functions as a component of broader social networks.

Overland flow. Movement of water across the soil surface.

Panarchy. Mosaics of nested subsystems that are at different stages of their adaptive cycles, with moments of interaction across scales.

Passive adaptive management. Learning through intensive examination of historic cause-effect relationships.

Path dependence. Effects of historical legacies on the future trajectory of a system, or, more narrowly, the co-evolution of institutions and social-ecological conditions in a particular historical context.

Pelagic. Water column-dwelling.

Piscivore. Organisms that eat fish.

Planktivore. Organisms that eat plankton.

Permanent crops. Land cultivated with crops that occupy the land for long periods and need not be replanted after each harvest, such as fruit trees, cocoa, coffee, and rubber.

Permanent pastures. Land used permanently (five years or more) for herbaceous forage crops, either cultivated or growing wild (wild prairie or grazing land).

Pluralism. Affirmation and acceptance of a diversity of perspectives, mental models, or knowledge systems.

Policy community. Group that shares an interest in one or more specific issues and collaborates to change policy.

Polycentric governance. The organization of small, medium, and large-scale democratic units that each exercise independence to make and enforce rules within a scope of authority for a specified geographical area.

Polycentric institutions. Complex array of interacting institutions with overlapping and varying objectives, levels of authority, and strengths of linkages.

Positive feedback. See amplifying feedback.

Poverty. Pronounced deprivation of well-being.

Poverty trap. Situation characterized by persistent poverty, reflecting a loss of options to develop or deal with change.

Power. Having influence over others through various resources, including formal authority, threat, charisma, money, and information; intentionally imposed or achieved through passive or disruptive behavior.

Precautionary fishery policies. Stock assessments and harvest regulations that are deliberately chosen to be conservative enough to assure that the safe harvest curve remains high over time.

Precautionary principle. If an action or policy might cause severe harm, but the outcome is uncertain, then the burden of proof falls on those who advocate taking the action.

Pressure group. Group that lobbies to bring about institutional change.

Property rights. The relationship between a resource user, a community, a society, and the rights, rules and responsibilities that govern access to and use of resources. The three classes of property rights are private, communal, and state property.

Prospective actions. Actions taken today because people believe those actions will influence the capacity of future residents and governance networks to meet short-term and long-term challenges and opportunities.

Protected landscape/seascape. Protected area managed mainly for landscape/seascape conservation and recreation.

Provisioning services. Products of ecosystems that are directly harvested by society (e.g., fresh water, food, fiber, fuelwood, biochemicals, and genetic resources). Synonymous with **ecosystem goods** or renewable resources.

Public consultation processes. Strategy for linking across scales, commonly used by government management agencies to inform decision making about the interests and concerns of local communities.

Punctuated equilibrium. Temporal pattern in which long periods of stability and incremental change are separated by abrupt, non-incremental, large-scale changes.

Pure public goods. Goods that are not subtractable and nonexcludable.

Redundancy. Diversity of functionally similar components (e.g., species or institutions) to provide multiple means of accomplishing the same ends, in the event that some components disappear.

Reflexive behavior. Human behavior that allows planning for the future by taking into account the likely consequences of actions.

Reforestation. Regrowth or planting of forests on previously forested lands.

Regime shift. Abrupt large-scale transition to a new state or stability domain characterized by very different structure and feedbacks.

Regulating services. Regulation by ecosystems of processes that extend beyond their boundaries (e.g., regulation of climate, water quantity and quality, disease, and pollination).

Release phase. Phase of an adaptive cycle that radically and rapidly reduces the structural complexity of a system.

Renewable resource. Resource whose rate of extraction does not exceed the replenishment rate.

Renewal phase. Phase of an adaptive cycle in which the system reorganizes through the development of stabilizing feedbacks that tend to sustain properties over time.

Rent. See economic rent.

Rent seeking. Situation that occurs when individuals or organizations glean the benefits of a transaction without contributing to, and in some cases subtracting from, the welfare of society.

Resilience. Capacity of a social-ecological system to absorb a spectrum of shocks or perturbations and to sustain and develop its fundamental function, structure, identity, and feedbacks as a result of recovery or reorganization in a new context.

Resilience-based ecosystem stewardship. A suite of approaches whose goal is to sustain social-ecological systems, based on reducing vulnerability and enhancing adaptive capacity, resilience, and transformability. Its goals are to respond to and shape change in social-ecological systems in order to sustain the supply and opportunities for use of ecosystem services by society.

Resilience learning. Form of social learning that fosters society's capacity to be prepared for the long term by enhancing adaptive capacity to deal with change.

Resource regime. Cluster of institutions governing management of a particular area of interest or resource.

Response diversity. The diversity of responses to environmental change among organisms contributing to the same ecosystem function.

Rigidity trap. Situation in which people and institutions try to resist change and persist with their current management and governance system despite a clear recognition that change is essential.

Roving bandits. People who range widely to exploit resources without regard to established institutions.

Rules-in-use. Practiced institutions (vs. espoused rules).

Runoff. Water that moves as overland flow or groundwater from a terrestrial to an aquatic system. Gives rise to blue water.

Salesman. Individual with the social skills to persuade people unconvinced of what they are hearing.

Salinization. Accumulation of salts in soils or freshwaters to the point that productivity declines.

Salmonid. Fish belonging to the salmon family.

Scenario. A plausible, simplified, synthetic description of how the future of a system might develop, based on a coherent and internally consistent set of assumptions about key driving forces and relationships among key variables.

Sector. A division of society, often associated with an economic activity (e.g., the business sector).

Self-efficacy. Sense of having an effect on events through one's efforts, having the power to make a difference.

Self-organization. The development of system structure as a result of stabilizing feedbacks among system components.

Sense of place. Self-identification with a particular location or region.

Shadow network. Group that is indirectly involved in decision making and supportive to the process.

Shifting baseline syndrome. Depletion of a stock that occurs so gradually that current levels are accepted as normal, and no policies are implemented to prevent further depletion.

Shifting cultivation. See swidden agriculture.

Single-loop learning. Feedback process that adjusts actions to meet identified management goals (e.g., modifies harvest rate to conform to specified catch limits) but does not evaluate basic assumptions and approaches.

Single-species management. Management to maintain the abundance or productivity of a single species.

Six degrees of separation. Hypothesis that everyone on Earth has a maximum of six linkages that separate him/her from any other person.

Slash-and-burn agriculture. See swidden agriculture.

Slow variables. Variables that strongly influence social-ecological systems but remain relatively constant over years-to-decades.

Social capital. Capacity of groups of people to act collectively to solve problems.

Social-ecological governance. Collective coordination of efforts to define and achieve societal goals related to human-environment interactions.

Social-ecological processes. Feedbacks and interconnections among components of a social-ecological system.

Social-ecological system. System with interacting and interdependent physical, biological, and social components, emphasizing the 'humans-in-nature' perspective.

Social learning. Process by which groups assess social-ecological conditions and respond in ways that meet objectives.

Social memory. Memory of past experiences that is retained by groups, providing a legacy of knowing how to do things under different circumstances.

Social movement. A process in which disaggregated groups operate over broad geographic scales, seeking to advance a particular ideological perspective or objective.

Social network. Linkages that establish relations among individuals and organizations (and their institutions) across time and space.

Soil. Mixture of small mineral and organic particles that retain the water and nutrients required for growth of terrestrial plants.

Special interest group. Organization that advocates for policies that serve its interests at local-to-regional scales.

Species diversity. Number of species, adjusted for relative abundance.

Stability. Tendency of the system to maintain the same properties over time.

Stabilizing feedback. Feedback that tends to reduce fluctuations in process rates, although, if extreme, can induce chaotic fluctuations. A stabilizing feedback occurs when two interacting components cause one another to change in opposite directions. Synonymous with negative feedback.

Stakeholders. Individuals and organizations, including government, affected by policy decisions.

Staple. Crops that form the basis of the traditional diet of a region.

Steady state. Condition of a system in which there is no *net* change in system structure or functioning over a particular time period.

Steady-state mosaic. Landscape in which different stands are at different successional stages, but there is no net change in landscape composition over time.

Stewardship. See resilience-based ecosystem stewardship.

Stove pipes. See functional silos.

Strict nature reserve. Protected area managed mainly for science.

Substitution. Replacement of one form of capital input by another.

Subtractable. One person's use of the resource reduces the availability of that resource for use by others.

Succession. Directional change in ecosystem properties resulting from biologically driven changes in resource supply.

Supporting services. Fundamental ecological processes that sustain ecosystem functioning (maintenance of soil fertility, cycling of essential elements, biological diversity, and cycles of disturbance and renewal).

Surprise. An unexpected and unimagined occurrence. See cross-scale surprises, local surprises, and true-novelty surprises.

Sustainability. Use of the environment and resources to meet the needs of the present without compromising the ability of future generations to meet their own needs. Maintenance of the productive base (total capital) over time.

Sustainable development. Development that seeks to improve human well-being, while at the same time sustaining the natural resource base and opportunities on which future generations depend.

Sustainable management. Management to sustain the functional properties of social-ecological systems that are important to society.

Sustained yield. Production of a biological resource (e.g., timber or fish) under management procedures that ensure replacement of the part harvested by regrowth or reproduction before the next harvest occurs.

Swidden agriculture. Agricultural system involving cycles of forest clearing, growing of crops, and regrowth of forests in small patches. Synonymous with slash-and-burn agriculture or shifting cultivation.

Synergy. Ecosystem services or societal benefits that co-occur with other services and benefits (e.g., aesthetic value and carbon sequestration provided by natural forests).

Systemic processes. Processes for which changes anywhere on the planet rapidly affect the Earth System at the global scale.

Temporal tradeoffs. Tradeoffs between short-term benefits and long-term capacity of ecosystems to provide services to future generations. Synonymous with intergenerational tradeoffs.

Theory of weak ties. Theory that critical information is most commonly received from those outside one's stronger social network.

Threshold. Critical level of one or more drivers or state variables that, when crossed, triggers an abrupt change (regime shift) in the system.

Threshold of probable concern (TPC). Upper and lower levels in selected indicators of management goals that together define current views about acceptable heterogeneity in conditions of ecosystems.

Timber. See industrial roundwood.

Tipping point. Threshold for transformation from an old to a new system controlled by different critical slow variables and feedbacks.

Top-down effects. Effects of predators (ecology) or large-scale organizations and institutions on a group of actors.

Tradeoffs. Ecosystem services or societal benefits that can be obtained only at the expense of other services and benefits (e.g., clearcut logging and old-growth conservation of the same forest).

Traditional knowledge. A cumulative body of knowledge observations, understanding, practice, and belief, evolving by adaptive processes and handed down through generations by cultural transmission, about the relationship of living beings (including humans) with one another and with their environment. Typically viewed as held by indigenous people or a group unique to a society. Overlaps with local knowledge.

Tragedy of the commons. Outcome in which the self-interest of resource consumers and poorly defined property rights result in significant degradation of common-pool resources, sometimes referred to the tragedy of open access.

Transaction costs. Costs associated with search, communication, negotiation, monitoring, coordination and enforcement of rules.

Transdisciplinary. Integration that transcends traditional disciplines to formulate problems in new ways.

Transformability. Capacity to re-conceptualize and create a fundamentally new system with different characteristics.

Transformation. Fundamental change in a social-ecological system that results in different control variables defining the state of the system, new ways of making a living, and often changes in scales of critical feedbacks. Transformations can be purposefully navigated or unintended.

Transformative learning. Learning that re-conceptualizes the system as fundamentally distinct through processes of reflection and engagement in a way that supports transformational change.

Transhumance. Seasonal movements of pastoralists.

Transpiration. Evaporation of water from cell surfaces inside leaves that supports plant production. Synonymous with green water.

Triple-loop learning. Learning that redefines norms and protocols as a basis for changes in governance.

Trophic cascade. Changes in species composition driven by changes in top predators that affect lower trophic levels, primary producers, bacteria and nutrient cycles.

True-novelty surprise. Never-before-experienced phenomena for which strict pre-adaptation is impossible.

Turnover time. Amount of time that it takes to replace the pool of a material in an ecosystem, if the pool size is steady over time.

Type 1 error. Acceptance of a proposition that turns out to be false.

Type 2 error. Failure to reject a false proposition.

Upwelling. Movement of deep nutrient-rich water to the surface.

Utility. Capacity of individuals or society to meet its own needs.

Vertical interplay. Interaction among institutions in multilevel governance systems.

Vertical linkages. Linkages that occur as part of inter-organizational relationships, such as in the transactions between local- and national-level management systems operating in a given region.

Virtual water. Volume of freshwater needed to produce a specific product or service.

Vulnerability. Degree to which a system is likely to experience harm due to exposure to a specified hazard or stress.

Vulnerable. Likely to change in state in response to a stress or stressor.

Wallace's Line. Zoographic boundary between Asian and Australasian faunas.

Watershed. A lake or stream and all the lands that drain into it.

Weathering. Breakdown of rocks to form soil particles.

Well-being. Quality of life; basic material needs for a good life, freedom and choice, good social relations, and personal security. Also, the present value of future utility.

Wicked problem. Problem that is so complex that each attempted solution creates new problems for other segments of society or other times and places.

Wilderness area. Protected area managed mainly for wilderness protection.

Window of opportunity. Short periods that offer possibilities for large-scale change.

Worldview. Framework by which a group interprets events and interacts with its social-ecological system.

References

Abel, N.M. 1997. Mis-measurement of the productivity and sustainability of African communal rangelands: A case study and some principles from Botswana. *Ecological Economics* 23: 113–133.

Abel, N.M., and P. Blaikie. 1986. Elephants, people, parks and development: The case of the Luangwa Valley, Zambia. *Environmental Management* 10:735–751.

Abel, N.M., D.H.M. Cumming, and J.M. Anderies. 2006. Collapse and reorganization in social-ecological systems: Questions, some ideas, and policy implications. *Ecology and Society* 11(1):17. http://www.ecologyandsociety.org/vol11/iss1/art17/.

Acheson, J.M. 1989. Management of common-property resources. Pages 351–475 *in* S. Plattner, editor. *Economic Anthropology*. Stanford University Press, Stanford.

Adams, L.W. 1994. *Urban Wildlife Habitats: A Landscape Perspective*. University of Minnesota Press, Minneapolis.

Addams, L., D. Battisti, E. McCullough, J.L. Minjares, and G. Schoups. in press. Water resource management in the Yaqui Valley. *in* P.A. Matson, W.P. Falcon, editors. *Agriculture, Development and the Environment in the Yaqui Valley (Mexico)*. Island Press, Washington.

Adger, W.N. 2006. Vulnerability. *Global Environmental Change* 16:268–281.

Adger, W.N., N.W. Arnell, and E.L. Tompkins. 2005. Successful adaptation to climate change across scales. *Global Environmental Change* 15:77–86.

Agar, N. 2001. *Life's Intrinsic Value: Science, Ethics and Nature*. Columbia University Press, New York.

Agardy, T., J. Alder, P. Dayton, S. Curran, A. Kitchingman, et al. 2005. Coastal systems. Pages 513–549 *in* R. Hassan, R. Scholes, and N. Ash, editors. *Ecosystems and Human Well-Being: Current State and Trends. Millennium Ecosystem Assessment*. Island Press, Washington.

Agrawal, A. 2002. Commons resources and institutional sustainability. Pages 41–86 *in* NRC (National Research Council); E. Ostrom, T. Dietz, N. Dolšak, P.C. Stern, S. Stovich et al., editors. *The Drama of the Commons*. National Academy Press, Washington.

Agrawal, A., A. Chhatre, and R. Hardin. 2008. Changing governance of the world's forests. *Science* 320:1460–1462.

Alimaev, I.I., and R.H.J. Behnke. 2008. Ideology, land tenure and livestock mobility in Kazakhstan. Pages 151–178 *in* K.A. Galvin, R.S. Reid, R.H.J. Behnke, and N.T. Hobbs, editors. *Fragmentation in Semi-arid and Arid Landscapes: Consequences for Human and Natural Systems*. Springer, Dordrecht.

Allan, C., and A. Curtis. 2005. Nipped in the bud: Why regional-scale adaptive management is not blooming. *Environmental Management* 36:414–425.

Allan, J.A. 1998. Virtual water: A strategic resource. Global solutions to regional deficits. *Ground Water* 36:545–546.

Allison, E.H., and F. Ellis. 2001. The livelihoods approach and management of small-scale fisheries. *Marine Policy* 25:377–388.

Allison, H.E., and R.J. Hobbs. 2004. Resilience, adaptive capacity, and the "Lock-in Trap" of the Western Australian agricultural region. *Ecology and Society* 9(1):3. http://www.ecologyandsociety.org/vol9/iss1/art3/.

Alverson, D.L., M.H. Freeberg, S.A. Murawski and J.G. Pope. 1994. A global assessment of fisheries bycatch and discards. *FAP Fisheries Technical Paper* 339, 235 pp.

AMAP (Arctic Monitoring and Assessment Programme). 2003. *AMAP Assessment 2002: Human Health in the Arctic*. AMAP, Oslo.

Amelung, T., and M. Diehl. 1992. *Deforestation of Tropical Rain Forests: Economic Causes and Impact on Development*. J.C.B. Mohr, Tübingen, Germany.

Anderies, J.M., M.A. Janssen, and E. Ostrom. 2004. A framework to analyze the robustness of social-ecological systems from an institutional perspective. *Ecology and Society* 9(1):18. http://www.ecologyandsociety.org/vol9/iss1/art18/.

Anderson, E.N. 1996. Learning from the land otter: Religious representation of traditional resource management. Pages 54–72 *in* E.N. Anderson, editor. *Ecologies of the Heart: Emotions, Beliefs, and the Environment*. Oxford University Press, Oxford.

Andersson, E., S. Barthel, and K. Ahrne. 2007. Measuring social-ecological dynamics behind the generation of ecosystem services. *Ecological Applications* 17:1267–1278.

Andow, D.A. 1991. Vegetational diversity and arthropod population response. *Annual Review of Entomology* 36:561–586.

Angelsen, A., and D. Kaimowitz. 1999. Rethinking the causes of deforestation: Lessons from economic models. *World Bank Research Observer* 14:73–98.

Argyris, C. 1992. *On Organizational Learning*. Blackwell Business, Oxford.

Argyris, C., and D.A. Schöen. 1978. *Organizational Learning: A Theory of Action Perspective*. Addison-Wesley Publishing Company, Reading.

Armitage, D., F. Berkes, and N. Doubleday, editors. 2007. *Adaptive Co-Management: Collaboration, Learning, and Multi-Level Governance*. University of British Columbia Press, Vancouver.

Arrow, K., L. Goulder, P. Dasgupta, G. Daily, P. Ehrlich, et al. 2004. Are we consuming too much? *Journal of Economic Perspectives* 18:147–172.

Atkinson, B.A., G.A. Rose, E. Murphy, and C.A. Bishop. 1997. Distribution change and abundance of northern cod (*Gadus morhua*): 1981–1993. *Canadian Journal of Fisheries and Aquatic Sciences* 54(Suppl. 1):132–138.

Attorre, F., A. Stanisci, and F. Bruno. 1997. The urban woods of Rome. *Plant Biosystems* 131:113–135.

Baker, L.A., D. Hope, Y. Xu, J. Edmonds, and L. Lauver. 2001. Nitrogen balance for the Central Arizona-Phoenix Ecosystem. *Ecosystems* 4:582–602.

Bala, G., K. Caldeira, M. Wickett, T.J. Phillips, D.B. Lobell, et al. 2007. Combined climate and carbon-cycle effects of large-scale deforestation. *Proceedings of the National Academy of Sciences* 104:6550–6555.

Baland, J.-M., and J.-P. Platteau. 1996. *Halting Degradation of Natural Resources: Is there a Role for Rural Communities?* Clarendon Press, Oxford.

Baliga, S., and E. Maskin. 2003. Mechanism design for the environment. Pages 305–324 *in* K.-G. Mäler and J.R. Vincent, editors. *Handbook of Environmental Economics*. Elsevier, Amsterdam.

Balirwa, J.S., C.A. Chapman, L.J. Chapman, I.G. Cowx, K. Geheb, et al. 2003. Biodiversity and fishery sustainability in the Lake Victoria basin: An unexpected marriage? *BioScience* 53:703–715.

Band, L.E., M.L. Cadenasso, S. Grimmond, and J.M. Grove. 2006. Heterogeneity in urban ecosystems: Pattern and process. Pages 257–278 *in* G. Lovett, C.G. Jones, M.G. Turner, and K.C. Weathers, editors. *Ecosystem Function in Heterogeneous Landscapes*. Springer-Verlag, New York.

Bandura, A. 1982. Self-efficacy mechanisms in human agency. *American Psychologist* 37:122–147.

Banerjee, A.V., and E. Duflo. 2007. The economic lives of the poor. *Journal of Economic Perspectives* 21:141–167.

Barbier, E.B. 1987. The concept of sustainable economic development. *Environmental Conservation* 14:101–110.

Barbier, E.B., and J.R. Thompson. 1998. The value of water: Floodplain versus large-scale irrigation benefits in northern Nigeria. *Ambio* 27:434–440.

Bardhan, P., and D. Mookerjee, editors. 2006. *Decentralization and Local Governance in Developing Countries*. MIT Press, Cambridge, MA.

Bates, R.H. 1997. *Open-Economy Politics: The Political Economy of the World Coffee Trade*. Princeton University Press, Princeton.

Batty, M. 2006. Rank clocks. *Nature* 444:592–596.

Batty, M. 2008. The size, scale, and shape of cities. *Science* 319:769–771.

Bawa, K.S., and S. Dayanandan. 1997. Socioeconomic factors and tropical deforestation. *Nature* 386:562–63.

Bawa, K.S., R. Seidler, and P.H. Raven. 2004. Reconciling conservation paradigms. *Conservation Biology* 18:859–860.

Bayliss, K., and T. McKinley. 2007. Providing basic utilities in sub-Saharan Africa. *Environment* 49(3):24–32.

Befu, H. 1980. The political ecology of fishing in Japan: Techno-environmental impact of industrialization in the Inland Sea. *Research in Economic Anthropology* 3:323–392.

Behnke, R.H.J. 2008. The drivers of fragmentation in arid and semi-arid landscapes. Pages 305–340 *in* K.A. Galvin, R.S. Reid, R.H.J. Behnke, and N.T. Hobbs, editors. *Fragmentation in Semi-arid and Arid Landscapes: Consequences for Human and Natural Systems*. Springer, Dordrecht.

Bellwood, D., T.P. Hughes, C. Folke, and M. Nyström. 2004. Confronting the coral reef crisis. *Nature* 429:827–833.

Bengtsson, J., P. Angelstam, T. Elmqvist, U. Emanuelsson, C. Folke, et al. 2003. Reserves, resilience, and dynamic landscapes. *Ambio* 32: 389–396.

Benjamin, C., R. Orozco, R.F.M. Sales, and S.T. Jayme. 2003. Civil Society Participation in the Fisheries and Aquatic Resource Management Council (FARMC) and the National Agriculture and Fisheries Council (NAFC): The Case of Cavite and Bataan. Research conducted by the Philippine Rural Reconstruction Movement (PRRM) for the NGO for Fishery Reform (NFR).

Bennett, E.M., S.R. Carpenter, and N.F. Caraco. 2001. Human impact on erodable phosphorus and eutrophication: A global perspective. *BioScience* 51:227–234.

Bennett, E.M., S.R. Carpenter, G.D. Peterson, G.S. Cumming, M. Zurek, et al. 2003. Why global scenarios need ecology. *Frontiers in Ecology and the Environment* 1:322–329.

Bennett, E.M., T. Reed-Andersen, J.N. Houser, J.R. Gabriel, and S.R. Carpenter. 1999. A phosphorus budget for the Lake Mendota watershed. *Ecosystems* 2:69–75.

Berger, J.R. 2001. African elephant, human economies, and international law: Bridging the great rift for East and Southern Africa. *Georgetown International Environmental Law Review* 13:417–462.

Berkes, F. 1988. Environmental philosophy of the Cree people of James Bay. Pages 7–21 *in* M. Freeman and L. Carbyn, editors. *Traditional Knowledge and Renewable Resource Management in Northern Regions*. Boreal Institute for Northern Studies, Edmonton.

Berkes, F., editor. 1989. *Common Property Resources: Ecology and Community-Based Sustainable Development*. Belhaven Press, London.

Berkes, F. 1995. Indigenous knowledge and resource management systems: A native Canadian case study from James Bay. Pages 99–109 *in* S. Hanna and M. Munasinghe, editors. *Property Rights in a Social and Ecological Context: Case Studies and Design Applications*. Beijer International Institute of Ecological Economics and the World Bank, Washington.

Berkes, F. 1998. Indigenous knowledge and resource management systems in the Canadian subarctic. Pages 98–128 *in* F. Berkes and C. Folke, editors. *Linking Social and Ecological Systems: Management Practices and Social Mechanisms for Building Resilience*. Cambridge University Press, Cambridge.

Berkes, F. 2002. Cross-scale institutional linkages: Perspectives from the bottom up. Pages 293–321 *in* NRC (National Research Council); E. Ostrom, T. Dietz, N. Dolšak, P.C. Stern, S. Stovich, et al., editors. *The Drama of the Commons*. National Academy Press, Washington.

Berkes, F. 2004. Rethinking community-based conservation. *Conservation Biology* 18: 621–630.

Berkes, F. 2007. Community-based conservation in a globalized world. *Proceedings of the National Academy of Sciences* 104:15188–15193.

Berkes, F. 2008. *Sacred Ecology: Traditional Ecological Knowledge and Resource Management*. 2nd Edition. Taylor and Francis, Philadelphia.

Berkes, F., N. Bankes, M. Marschke, D. Armitage, and D. Clark. 2005. Cross-scale institutions: Building resilience in the Canadian North. Pages 225–247 *in* F. Berkes, R. Huebert, H. Fast, M. Manseau, and A. Diduck, editors. *Breaking Ice: Renewable Resource and Ocean Management in the Canadian North*. University of Calgary Press, Calgary.

Berkes, F., J. Colding, and C. Folke. 2000. Rediscovery of traditional ecological knowledge as adaptive management. *Ecological Applications* 10:1251–1262.

Berkes, F., J. Colding, and C. Folke, editors. 2003. *Navigating Social-Ecological Systems: Building Resilience for Complexity and Change*. Cambridge University Press, Cambridge.

Berkes, F., and I.J. Davidson-Hunt. 2006. Biodiversity, traditional management systems, and cultural landscapes: Examples from the boreal forest of Canada. *International Social Science Journal* 187:35–47.

Berkes, F., D. Feeny, B.J. McCay, and J.M. Acheson. 1989. The benefits of the commons. *Nature* 340:91–93.

Berkes, F., and C. Folke. 1998a. Linking social and ecological systems for resilience and sustainability. Pages 1–25 *in* F. Berkes and C. Folke, editors. *Linking Social and Ecological Systems: Management Practices and Social Mechanisms for Building Resilience*. Cambridge University Press, Cambridge.

Berkes, F., and C. Folke, editors. 1998b. *Linking Social and Ecological Systems: Management Practices and Social Mechanisms for Building Resilience*. Cambridge University Press, Cambridge.

Berkes, F., and C. Folke. 2002. Back to the future: Ecosystem dynamics and local knowledge. Pages 121–146 *in* L.H. Gunderson and C.S. Holling, editors. *Panarchy: Understanding Transformations in Human and Natural Systems*. Island Press, Washington.

Berkes, F., T.P. Hughes, R.S. Steneck, J.A. Wilson, D.R. Bellwood, et al. 2006. Globalization, roving bandits, and marine resources. *Science* 311:1557–1558.

Berkowitz, A.R., C.H. Nilon, and K.S. Holweg, editors. 2003. *Understanding Urban Ecosystems: A New Frontier for Science and Education*. Springer-Verlag, New York.

Bernstein, S.F. 2002. *The Compromise of Liberal Environmentalism*. Columbia University Press, New York.

Berry, W. 2005. Renewing husbandry: After mechanization, can modern agriculture reclaim its soul? *Crop Science* 45:1103–1106.

Bhagwat, S., C. Kushalappa, P. Williams, and N. Brown. 2005. The role of informal protected areas in maintaining biodiversity in the Western Ghats of India. *Ecology and Society* 10 (1):8. http://www.ecologyandsociety.org/vol10/iss1/art8/.

Biermann, F. 2007. Earth system governance as a crosscutting theme of global change research. *Global Environmental Change* 17:326–337.

Biggs, H.C., and K.H. Rogers. 2003. An adaptive system to link science, monitoring, and management in practice. Pages 59–80 *in* J.T. du Toit, K.H. Rogers, and H.C. Biggs, editors. *The Kruger Experience: Ecology and Management of Savanna Heterogeneity*. Island Press, Washington.

Binswanger, H. 1990. The policy response in agriculture. Pages 231–258 *in* Proceedings of the World Bank Annual Conference on Development Economics 1989. World Bank, Washington.

Birkeland, P.W. 1999. *Soils and Geomorphology*. 3rd Edition. Oxford University Press, New York.

Bodin, Ö., and J. Norberg. 2005. Information network topologies for enhanced local adaptive management. *Environmental Management* 35:175–193.

Bolund, P., and S. Hunhammar. 1999. Ecosystem services in urban areas. *Ecological Economics* 29:293–301.

Bonfil, R., G. Munro, U.R. Sumaila, H. Valtysson, M. Wright, et al. 1998. Impacts of distant water fleets: An ecological, economic and social assessment. Pages 11–111 *in The Footprint of Distant Water Fleets on World Fisheries: Endangered Seas Campaign*, WWF International, Godalming, UK.

Boone, C. 2002. An assessment and explanation of environmental inequity in Baltimore. *Urban Geography* 23:581–595.

Borgerhoff Mulder, M., and P. Coppolillo. 2005. *Conservation: Linking Ecology, Economics, and Culture*. Princeton University Press, Princeton, NJ.

Borgstrom, S.T., T. Elmquist, P. Angelstrom, and C. Alfsen-Norodom. 2006. Scale mismatches in management of urban landscapes. *Ecology and Society* 11(2):16. http://www.ecologyandsociety.org/vol11/iss2/art16/.

Bormann, B.T., and A.R. Kiester. 2004. Options forestry: Acting on uncertainty. *Journal of Forestry* 102:22–27.

Bormann, F.H., and G.E. Likens. 1967. Nutrient cycling. *Science* 155:424–429.

Bormann, F.H., and G.E. Likens. 1979. *Patterns and Processes in a Forested Ecosystem*. Springer-Verlag, New York.

Boudouris, K., and K. Kalimtzis, editors. 1999. *Philosophy and Ecology*. Iona Publications, Athens.

Bowles, S., S. Durlauf, and K. Hoff, editors. 2006. *Poverty Traps*. Princeton University Press, Princeton.

Boyden, S., S. Millar, K. Newcombe, and B. O'Neill. 1981. *The Ecology of A City and Its People: The Case of Hong Kong*. Australian National University Press, Canberra.

Bradshaw, A.D. 1983. The reconstruction of ecosystems. *Journal of Ecology* 20:1–17.

Brady, N.C., and R.R. Weil. 2001. *The Nature and Properties of Soils*. 13th Edition. Prentice Hall, Upper Saddle River, NJ.

Brand, S. 2006. City Planet. strategy + business. http://www.strategy-business.com/press/16635507/16606109.

Brock, W.A. 2006. Tipping points, abrupt opinion changes, and punctuated policy change. Pages 47–77 *in* R. Repetto, editor. *Punctuated Equilibrium and the Dynamics of U.S. Environmental Policy*. Yale University Press, New Haven.

Brodie, J., K. Fabricius, G. De'ath, and K. Okaji. 2005. Are increased nutrients responsible for more outbreaks of crown-of-thorns starfish? *Marine Pollution Bulletin* 51:266–278.

Bromley, D.W. 1989. *Economic Interests and Institutions: The Conceptual Foundations of Public Policy*. Basil Blackwell, New York.

Bromley, D.W., editor. 1992. *Making the Commons Work: Theory Practice and Policy.* Institute for Contemporary Studies, San Francisco.

Brouzes, R.J.P., A.J. Liem, and V.A. Naish. 1978. *Protocol for Fish Tainting Bioassays.* CPAR Report. No. 775–1. Environmental Protection Service, Ottawa.

Brown, K. 2002. Innovations for conservation and development. *The Geographical Journal* 168: 6–17.

Brown, K. 2003. Integrating conservation and development: A case of institutional misfit. *Frontiers in Ecology and the Environment* 1:479–487.

Brunner, R.D., T.A. Steelman, L. Coe-Juell, C.M. Cromley, C.M. Edwards, et al., editors. 2005. *Adaptive Governance: Integrating Science, Policy, and Decision Making.* Columbia University Press, New York.

Burch, W.R. Jr., and D.R. DeLuca. 1984. *Measuring the Social Impact of Natural Resource Policies.* New Mexico University Press, Albuquerque.

Burke, E.J., S.J. Brown, and N. Christidis. 2006. Modeling the recent evolution of global drought and projections for the twenty-first century with the Hadley Centre climate model. *Journal of Hydrometeorology* 7:1113–1125.

Burke, M., M. Mastrandrea, W.P. Falcon, and R. Naylor. 2008. *Ending hunger by assumption.* Program on Food Security and the Environment (under review).

BurnSilver, S.B. 2009. Pathways of change and continuity: Maasai livelihoods in Amboseli, Kajaido District, Kenya. Pages 161–208 *in* K. Homewood, P. Kristjanson, and P. Chevevix Trench, editors. *Staying Maasai?* Livelihoods, Conservation and Development in East African Rangelands. Springer, New York.

Burton, I., R.W. Kates, and G.F. White. 1993. *The Environment as Hazard.* 2nd Edition. Guilford, New York.

Burton, P.J., C. Messier, G.F. Weetman, E.E. Prepas, W.L. Adamowicz, et al. 2003. The current state of boreal forestry and the drive for change. Pages 1–40 *in* P.J. Burton, C. Messier, D.W. Smith, and W.L. Adamowicz, editors. *Towards Sustainable Management of the Boreal Forest.* National Research Council of Canada, Ottawa.

Busch, D.E., and J.C. Trexler. 2002. *Monitoring Ecosystems: Interdisciplinary Approaches for Evaluation Ecoregional Initiatives.* Island Press, Washington.

Butzer, K.W. 1980. Adaptation to global environmental change. *Professional Geographer* 32: 269–278.

Caddy, J.F., and T. Surette. 2005. In retrospect the assumption of sustainability for Atlantic fisheries has proved an illusion. *Reviews in Fish Biology and Fisheries* 15:313–337.

Cadenasso, M.L., S.T.A. Pickett, and J.M. Grove. 2006. Dimensions of ecosystem complexity: Heterogeneity, connectivity, and history. *Ecological Complexity* 3:1–12.

Campbell, R.A., B.D. Mapstone, and A.D.M. Smith. 2001. Evaluating large-scale experimental designs for management of coral trout on the Great Barrier Reef. *Ecological Applications* 11: 1763–1777.

Campbell, B.D., D.M. Stafford Smith, and GCTE Pastures and Rangelands Network members. 2000. A synthesis of recent global change research on pasture and rangeland production: Reduced uncertainties and their management implications. *Agriculture, Ecosystems and Environment* 82: 39–55.

Canadell, J.G., C. Le Quéré, M.R. Raupach, C.B. Field, E.T. Buitehuls, et al. 2007. Contributions to accelerating atmospheric CO_2 growth from economic activity, carbon intensity, and efficiency of natural sinks. *Proceedings of the National Academy of Sciences* 104:10288–10293.

Carpenter, S.R. 2003. *Regime Shifts in Lake Ecosystems: Pattern and Variation.* Ecology Institute, Oldendorf/Luhe, Germany.

Carpenter, S.R., E.M. Bennett, and G.D. Peterson. 2006a. Scenarios for ecosystem services: An overview. *Ecology and Society* 11(1):29. http://www.ecologyandsociety.org/vol11/iss1/art29/.

Carpenter, S.R., D. Bolgrien, R.C. Lathrop, C.A. Stow, T. Reed, et al. 1998a. Ecological and economic analysis of lake eutrophication by nonpoint pollution. *Australian Journal of Ecology* 23:68–79.

Carpenter, S.R., and W.A. Brock. 2004. Spatial complexity, resilience and policy diversity: Fishing on lake-rich landscapes. *Ecology and Society* 9(1):8. http://www.ecologyandsociety.org/vol9/iss1/art8/.

Carpenter, S.R., W.A. Brock, and D. Ludwig. 2002. Collapse, learning, and renewal. Pages 173–193 *in* L.H. Gunderson and C.S. Holling, editors. *Panarchy: Understanding Transformations in Human and Natural Systems.* Island Press, Washington.

Carpenter, S.R., N.F. Caraco, D.L. Correll, R.W. Howarth, A.N. Sharpley, and V.H. Smith. 1998b. Non-point pollution of surface waters with phosphorus and nitrogen. *Ecological Applications* 8:559–568.

Carpenter, S.R., and L.H. Gunderson. 2001. Coping with collapse: Ecological and social dynamics in ecosystem management. *BioScience* 51:451–457.

Carpenter, S.R., and J.F. Kitchell. 1993. *The Trophic Cascade in Lakes*. Cambridge University Press, Cambridge.

Carpenter, S.R., and R.C. Lathrop. 1999. Lake restoration: Capabilities and needs. *Hydrobiologia* 395/396:19–28.

Carpenter, S.R., R.C. Lathrop, P. Nowak, E.M. Bennett, T. Reed, et al. 2006b. The ongoing experiment: Restoration of Lake Mendota and its watershed. Pages 236–256 *in* J.J. Magnuson, T.K. Kratz, and B.J. Benson, editors. *Long-Term Dynamics of Lakes in the Landscape*. Oxford University Press, London.

Carpenter, S.R., D. Ludwig, and W.A. Brock. 1999. Management of eutrophication for lakes subject to potentially irreversible change. *Ecological Applications* 9:751–771.

Carpenter, S.R., and M.G. Turner. 2000. Hares and tortoises: Interactions of fast and slow variables in ecosystems. *Ecosystems* 3:495–497.

Carr, C. 2008. *Kalashnikov Culture: Small Arms Proliferation and Irregular Warfare*. Greenwood, Portsmouth, NH.

Carson, R. 1962. *Silent Spring*. Crest, New York.

Cassman, K.G. 1999. Ecological intensification of cereal production systems: Yield potential, soil quality, and precision agriculture. *Proceedings of the National Academy of Sciences* 96:5952–5959.

Cassman, K.G., A. Dobermann, D.T. Walters, and H. Yang. 2003. Meeting cereal demand while protecting natural resources and improving environmental quality. *Annual Review of Environment and Resources* 28:316–351.

Cassman, K.G., V. Eidman, and E. Simpson. 2006. Convergence of agriculture and energy: Implications for research and policy. Council for Agricultural Science and Technology (CAST) Commentary, 12 pp.

Cassman, K.G., and A.J. Liska. 2007. Food and fuel for all: Realistic or foolish? *Biofuels, Bioproducts, and Biorefining* 1(1):18–23.

Cassman, K.G., S. Wood, P.S. Choo, H.D. Cooper, C. Devendra, et al. 2005. Cultivated systems. Pages 745–794 *in* R. Hassan, R.J. Scholes, and N. Ash, editors. *Ecosystems and Human Well-Being: Current State and Trends. Millennium Ecosystem Assessment*. Island Press, Washington.

Chambers, R., and G. Conway. 1991. *Sustainable Rural Livelihoods: Practical Concepts for the 21st Century*. IDS, Brighton.

Chambers, S. 1998. Short- and Long-Term Effects of Clearing Native Vegetation for Agricultural Purposes. PhD. Dissertation. Flinders University of South Australia, Adelaide.

Chan, S. 2007. Considering the Urban Planet 2050. New York Times City Blog. Posted December 4, 2007, 6:11 pm. http://cityroom.blogs.nytimes.com/2007/12/04/considering-the-urban-planet-of-2050/?scp=1-b&sq=&st=nyt.

Chapin, F.S., III, and K. Danell. 2001. Boreal forest. Pages 101–120 *in* F.S. Chapin, III, O.E. Sala, and E. Huber-Sannwald, editors. *Global Biodiversity in a Changing Environment: Scenarios for the 21st Century*. Springer-Verlag, New York.

Chapin, F.S., III, K. Danell, T. Elmqvist, C. Folke, and N.L. Fresco. 2007. Managing climate change impacts to enhance the resilience and sustainability of Fennoscandian forests. *Ambio* 36:528–533.

Chapin, F.S., III, M. Hoel, S.R. Carpenter, J. Lubchenco, B.Walker, et al. 2006a. Building resilience and adaptation to manage arctic change. *Ambio* 35:198–202.

Chapin, F.S., III, A.L. Lovecraft, E.S. Zavaleta, J. Nelson, M.D. Robards, et al. 2006b. Policy strategies to address sustainability of Alaskan boreal forests in response to a directionally changing climate. *Proceedings of the National Academy of Sciences* 103:16637–16643.

Chapin, F.S., III, P.A. Matson, and H.A. Mooney. 2002. *Principles of Terrestrial Ecosystem Ecology*. Springer-Verlag, New York.

Chapin, F.S., III, J.T. Randerson, A.D. McGuire, J.A. Foley, and C.B. Field. 2008. Changing feedbacks in the earth-climate system. *Frontiers in Ecology and the Environment* 6:313–320.

Chapin, F.S., III, and G.R. Shaver. 1985. Individualistic growth response of tundra plant species to environmental manipulations in the field. *Ecology* 66:564–576.

Chapin, F.S., III, M.S. Torn, and M. Tateno. 1996. Principles of ecosystem sustainability. *American Naturalist* 148:1016–1037.

Chapin, F.S., III, E.S. Zavaleta, V.T. Eviner, R.L. Naylor, P.M. Vitousek, et al. 2000. Consequences of changing biotic diversity. *Nature* 405:234–242.

Chase-Dunn, C. 2000. Guatemala in the global system. *Journal of Interamerican Studies and World Affairs* 42:109–126.

Chen, S., and M. Ravallion. 2007. Absolute poverty measures for the developing world, 1981–2004. *Proceedings of the National Academy of Sciences* 104:16757–16762.

Chevalier, J., and D. Buckles. 1999. Conflict management: A heterocultural perspective. Pages 13–41 *in* D. Buckles, editor. *Cultivating Peace: Conflict and Collaboration in Natural Resource Management*. International Development Research Centre, Ottawa.

Choi, J.S., K.T. Frank, W.C. Leggett, and K. Drinkwater. 2004. Transition to an alternate state in a continental shelf ecosystem. *Canadian Journal of Fisheries and Aquatic Sciences* 61: 505–510.

Choi, Y.D. 2007. Restoration ecology to the future: A call for a new paradigm. *Restoration Ecology* 15:351–353.

Christensen, N.L., A.M. Bartuska, J.H. Brown, S. Carpenter, C. D'Antonio, et al. 1996. The report of the Ecological Society of America committee on the scientific basis for ecosystem management. *Ecological Applications* 6:665–691.

Christensen, V., S. Guenette, J.J. Heymans, C.J. Walters, R. Watson, et al. 2003. Hundred-year decline of North Atlantic predatory fishes. *Fish and Fisheries* 4(1):1–24.

Christy, F.T. 1982. *Territorial Use Rights in Marine Fisheries: Definitions and Conditions*. Food and Agriculture Association Fisheries Technical Paper 227. Food and Agriculture Association, Rome.

Christy, F.T. 1997. Economic waste in fisheries: Impediments to change and conditions for improvement. Pages 28–39 *in* E.K. Pikitch, D.D. Huppert and M.P. Sissenwine, editors. *Global Trends: Fisheries Management*. American Fisheries Society, Bethesda, MD.

CIA. 2007. *The World Factbook*. Central Intelligence Agency, Washington.

Cilliers, S.S., and G.J. Bredenkamp. 1999. Analysis of the spontaneous vegetation intensively managed open spaces in the Potchefstroom Municipal Area, North West Province, South Africa. *South African Journal of Botany* 65:59–68.

Cinner, J., M.J. Marnane, T.R. McClanahan, and G.R. Almany. 2005. Periodic closures as adaptive coral reef management in the Indo-Pacific. *Ecology and Society* 11(1):31. http://www.ecologyandsociety.org/vol11/iss1/art31/.

Cissel, J.H., F.J. Swanson, and P.J. Weisberg. 1999. Landscape management using historical fire regimes: Blue River, Oregon. *Ecological Applications* 9:1217–1231.

Clallam County Department of Community Development. 2004. State of the Waters of Clallam County: A Report on the Health of Our Streams & Watersheds. http://www.clallam.net/streamkeepers/html/state_of_the_waters.htm.

Clark, C.W. 1985. *Bioeconomics Modeling and Fishery Management*. Wiley Interscience, New York.

Clark, C.W. 2007. *The Worldwide Crisis in Fisheries: Economic Models and Human Behavior*. Cambridge University Press, Cambridge.

Clark, W.C., and N.M. Dickson. 2003. Sustainability science: The emerging research program. *Proceedings of the National Academy of Sciences* 100:8059–8061.

Clay, G. 1973. *Close Up: How to Read the American City*. Praeger Publishers, New York.

Cohen, F. 1986. *Treaties on Trial: The Continuing Controversy over Northwest Indian Fishing Rights*. University of Washington Press, Seattle.

Cohen, M.D., J.G. March, and J.P. Olsen. 1972. A garbage can model of organizational choice. *Administrative Science Quarterly* 17:1–25.

Colding, J. 2007. Ecological land-use complementation' for building resilience in urban ecosystems. *Landscape and Urban Planning* 81: 46–55.

Colding, J., J. Lundberg, and C. Folke. 2006. Incorporating green-area user groups in urban ecosystem management. *Ambio* 35:237–244.

Cole, J.J., Y.T. Prairie, N.F. Caraco, W.H. McDowell, L.J. Tranvik, et al. 2007. Plumbing the global carbon cycle: Integrating inland waters into the terrestrial carbon budget. *Ecosystems* 10: 172–184.

Cole, M.M. 1982. The influence of soils, geomorphology and geology on the distribution of plant communities in savanna ecosystems. Pages 145–174 *in* B.J. Huntley and B.H. Walker, editors. *Ecology of Tropical Savannas*. Springer-Verlag, Berlin.

Coleman, J. 1990. *Foundations of Social Theory*. Harvard University Press, Cambridge, MA.

Collier, P. 2007. *The Bottom Billion: Why the Poorest Countries Are Failing and What Can Be Done About It*. Oxford University Press, Oxford.

Collins, S., S.M. Swinton, C.W. Anderson, B.J. Benson, T.L. Gragson, et al. 2007. *Integrated Science for Society and the Environment: A strategic research initiative*. LTER Network Office, Albuquerque, NM.

Contreras-Hermosilla, A. 2002. *Law Compliance in the Forestry Sector: An Overview*. The World Bank, Washington.

Conway, G. 1997. *The Doubly Green Revolution: Food for All in the 21st Century*. Cornell University Press, Ithaca.

Cooke, G.D., E.B. Welch, S. Peterson, and S.A. Nichols. 2005. *Restoration and Management of Lakes and Reservoirs*. 3rd Edition. CRC Press, Boca Raton, FL.

Corcoran, P.J. 2006. Opposing views of the "Ecosystem Approach" to fisheries management. *Conservation Biology* 20:617–619.

Cordell, J. editor. 1989. *A Sea of Small Boats*. Cultural Survival, Inc., Cambridge, MA.

Cork, S.J., G.D. Peterson, E.M. Bennett, G. Petschel-Held, and M. Zurek. 2006. Synthesis of the storylines. *Ecology and Society* 11(2):11. http://www.ecologyandsociety.org/vol11/iss2/art11/.

Costa, M.H., A. Botta, and J.A. Cardille. 2003. Effects of large-scale changes in land cover on the discharge of the Tocantins River, Southeastern Amazonia. *Journal of Hydrology* 283:206–217.

Costa, M.H., and J.A. Foley. 1999. Trends in the hydrological cycle of the Amazon basin. *Journal of Geophysical Research* 104:14189–14198.

Costanza, R., editor. 1991. *Ecological Economics: The Science and Management of Sustainability.* Columbia University Press, New York.

Costanza, R., and H. Daly. 1992. Natural capital and sustainable development. *Conservation Biology* 6:37–46.

Costanza, R., R. d'Arge, R. de Groot, S. Farber, M. Grasso, et al. 1997. The value of the world's ecosystem services and natural capital. *Nature* 387:253–260.

Costanza, R., and C. Folke. 1996. The structure and function of ecological systems in relation to property-rights regimes. Pages 13–34 *in* S. Hanna, C. Folke, K.-G. Mäler, and Å. Jansson, editors. *Rights to Nature: Ecological, Economic, Cultural, and Political Principles of Institutions for the Environment.* Island Press, Washington.

Couzin, J. 2008. Living in the danger zone. *Science* 319:748–749.

Crona, B. 2006. Supporting and enhancing development of heterogeneous ecological knowledge among resource users in a Kenyan seascape. *Ecology and Society* 11(1):32 http://ecologyandsociety.org/vol11/iss1/art32/.

Cronon, W. 1991. *Nature's Metropolis: Chicago and the Great West.* W.W. Norton & Co., New York.

Crutzen, P. 2002. Geology of Mankind – The Anthropocene. *Nature* 415:23.

Cutter, S.L. 1996. Vulnerability to environmental hazards. *Progress in Human Geography* 20:529–539.

Daily, G.C., editor. 1997. *Nature's Services: Societal Dependence on Natural Ecosystems.* Island Press, Washington.

D'Antonio, C.M., and P.M. Vitousek. 1992. Biological invasions by exotic grasses, the grass-fire cycle, and global change. *Annual Review of Ecology and Systematics* 23:63–87.

Dasgupta, P. 2001. *Human Well-Being and the Natural Environment.* Oxford University Press, Oxford.

Dasgupta, P., and K.-G. Mäler. 2000. Net national product, wealth, and social well-being. *Environment and Development Economics* 5:69–93.

Davidson, R. 2006. No fixed address: Nomads and the fate of the planet. *Quarterly Essay* 24:1–53.

Davidson-Hunt, I.J., P. Jack, E. Mandamin, and B. Wapioke. 2005. Iskatewizaagegan (Shoal Lake) plant knowledge: An Anishinaabe (Ojibway) ethnobotany of northwestern Ontario. *Journal of Ethnobiology* 25:189–227.

Davis, M. 2006. *Planet of Slums.* Verso, London.

DeAngelis, D.L., and W.M. Post. 1991. Positive feedback and ecosystem organization. Pages 155–178 *in* M. Higashi and T.P. Burns, editors. *Theoretical Studies of Ecosystems: The Network Perspective.* Cambridge University Press, Cambridge.

Degnbol, P. 2003. Science and the user perspective: The gap co-management must address. Pages 31–50 *in* D.C. Wilson, J.R. Nielsen, and P. Degnbol, editors. *The Fisheries Co-Management Experience: Accomplishments, Challenges and Prospects.* Kluwer, Dordrecht.

de Groot, R., P.S. Ramakrishnan, A. van de Berg, T. Kulenthran, and S. Muller, 2005. Cultural and amenity services. Pages 455–476 *in* R. Hassan, R. Scholes, and N. Ash, editors. *Ecosystems and Human Well-Being: Current State and Trends. Millennium Ecosystem Assessment.* Island Press, Washington.

Dercon, S. 2004. Growth and shocks: Evidence from rural Ethiopia. *Journal of Development Economics* 74:309–329.

Deutsch, B. 2007. *A Threat So Big, Academics Try to Collaborate.* The New York Times, New York.

Deutsch, L., S. Gräslund, C. Folke, M. Troell, M. Huitric, et al. 2007. Feeding aquaculture growth through globalization: Exploitation of marine ecosystems for fishmeal. *Global Environmental Change* 17:238–249.

Deutsch, L., A. Jansson, M. Troell, P. Ronnback, C. Folke, et al. 2000. The "ecological footprint": Communicating human dependence on nature's work. *Ecological Economics* 32: 351–355.

deYoung, B., R.M. Peterman, A.R. Dobell, E. Pinkerton, Y. Breton, et al. 1999. Canadian marine fisheries in a changing and uncertain world. *Canadian Special Publication of Fisheries and Aquatic Sciences* 129:199.

Diamond, J. 1997. *Guns, Germs and Steel.* W.W. Norton & Co. New York.

Diamond, J. 2005. *Collapse: How Societies Choose or Fail to Succeed.* Viking, New York.

Díaz, S., J. Fargione, F.S. Chapin, III, and D. Tilman. 2006. Biodiversity loss threatens human well-being. *Plant Library of Science (PLoS)* 4: 1300–1305.

Diener, E., and M.E.P. Seligman. 2004. Beyond money: Toward an economy of well-being. *Psychological Science in the Public Interest* 5:1–31.

Dietz, T., E. Ostrom, and P.C. Stern. 2003. The struggle to govern the commons. *Science* 302:1907–1912.

Dirzo, R., and P.H. Raven. 2003. Global state of biodiversity and loss. *Annual Review of the Environment and Resources* 28:137–167.

D'Odorico, P., and A. Porporato, editors. 2006. *Dryland Ecohydrology*. Springer, Dordrecht.

Douglas, M. 1986. *How Institutions Think*. Syracuse University Press, Syracuse.

Dow, K. 2000. Social dimensions of gradients in urban ecosystems. *Urban Ecosystems* 4: 255–275.

Downing, J.A., Y.T. Prairie, J.J. Cole, C.M. Duarte, L.J. Tranvik, et al. 2006. The global abundance and size distribution of lakes, ponds and impoundments. *Limnology and Oceanography* 51: 2388–2397.

Driscoll, C.T., G.B. Lawrence, A.J. Bulger, T.J. Butler, C.S. Cronan, et al. 2001. Acidic deposition in the northeastern United States: Sources and inputs, ecosystem effects and management strategies. *BioScience* 51:180–198.

Duit, A., and V. Galaz. 2008. Governance and complexity: Emerging issues for governance theory. *Governance* 21:311–335.

Duncan, O.D. 1961. From social system to ecosystem. *Sociological Inquiry* 31:140–149.

Duncan, O.D. 1964. Social organization and the ecosystem. Pages 37–82 in R.E.L. Faris, editor. *Handbook of Modern Sociology*. Rand McNally & Co., Chicago.

du Toit, J.T., K.H. Rogers, and H.C. Biggs, editors. 2003. *The Kruger Experience: Ecology and Management of Savanna Heterogeneity*. Island Press, Washington.

Dyson, M., G. Bergkamp, and J. Scanlon, editors. 2007. *Flow: The Essentials of Environmental Flows*. IUCN, Gland, Switzerland.

Eagle, J., R. Naylor, and W. Smith. 2004. Why farm salmon out-compete fishery salmon. *Marine Policy* 28:269–270.

Easterlin, R.A. 2001. Income and happiness: Towards a unified theory. *Economic Journal* 111:465–484.

Easterling, W.E., P.K. Aggarwal, P. Batima, K.M. Brander, L. Erda, et al. 2007. Food, fibre and forest products. Pages 273–313 in M.L. Parry, O.F. Canziani, J.P. Palutikof, P.J. van der Linden, and C.E. Hanson, editors. *Climate Change 2007: Impacts, Adaptation and Vulnerability. Contribution of Working Group II to the Fourth Assessment Report of the Intergovernmental Panel on Climate Change*. Cambridge University Press. Cambridge.

Ebbin, S. 1998. Emerging Cooperative Institutions for Fisheries Management: Equity and Empowerment of Indigenous Peoples of Washington and Alaska. PhD. dissertation. Yale University, New Haven.

Ehrlich, P.R., and J.P. Holdren. 1974. Impact of population growth. *Science* 171:1212–1217.

EIA. 2006. *International Energy Annual 2004*. International Energy Administration, Washington.

Ellis, E.C., and N. Ramankutty. 2008. Putting people on the map: Anthropogenic biomes of the world. *Frontiers in Ecology and the Environment* 6: 439–447.

Ellis, F. 1998. Household strategies and rural diversification. *Journal of Developmental Studies* 35: 1–38.

Ellis, J. 1994. Climate variability and complex ecosystem dynamics: Implications for pastoral development. Pages 37–46 in I. Scoones, editor. *Living with Uncertainty*. Intermediate Technology Publishers, London.

Ellison, A.M., M.S. Bank, B.D. Clinton, E.A. Colburn, K. Elliott, et al. 2005. Loss of foundation species: Consequences for the structure and dynamics of forested ecosystems. *Frontiers in Ecology and the Environment* 3:479–486.

Elmqvist, T., J. Colding, S. Barthel, S. Borgstrom, A. Duit, et al. 2004. The dynamics of social-ecological systems in urban landscapes: Stockholm and the National Urban Park, Sweden. *Annals of New York Academy of Sciences* 1023:308–322.

Elmqvist, T., C. Folke, M. Nyström, G. Peterson, J. Bengtsson, et al. 2003. Response diversity, ecosystem change, and resilience. *Frontiers in Ecology and the Environment* 1:488–494.

Elmqvist, T., M. Pyykonen, M. Tengo, F. Rakotondrasoa, E. Rabakonandrinina, et al. 2007. Patterns of loss and regeneration of tropical dry forest in Madagascar: The social institutional context. *Plant Library of Science (PLoS)* 2:e402:doi:410.1371/journal.pone.0000402.

Enfors, E. and L. Gordon. 2007. Analysing resilience in dryland agro-ecosystems: A case study of the Makanya catchment in Tanzania over the past 50 years. *Land Degradation & Development* 18:680–696.

Engerman, S., and K.L. Sokoloff. 2006. The persistence of poverty in the Americas: The role of institutions. Pages 43–78 in S. Bowles, S. Durlauf, and K. Hoff, editors. *Poverty Traps*. Princeton University Press, Princeton.

Epstein, M.J. 2008. *Making Sustainability Work: Best Practices in Managing and Measuring Corporate Social, Environmental, and Economic Impacts.* Greenleaf Publishing Ltd, Sheffield, UK.

Evans, L.T. 1998. *Feeding the Ten Billion: Plants and Population Growth.* Cambridge University Press. Cambridge.

Ewel, J.J. 1999. Natural systems as models for the design of sustainable systems of land use. *Agroforestry Systems* 45:1–21.

Fabricius, C., C. Folke, G. Cundill, and L. Schultz. 2007. Powerless spectators, coping actors, and adaptive co-managers: A synthesis of the role of communities in ecosystem management. *Ecology and Society* 12(1):29. http://www.ecologyandsociety.org/vol12/iss1/art29/.

Falcon, W.P. 1991. Whither food aid? Pages 237–246 in C.P. Timmer, editor. *Agriculture and the State: Growth, Employment, and Poverty in Developing Countries.* Cornell University Press. Ithaca.

Falcon, W.P., and R.L. Naylor. 2005. Rethinking food security for the 21st century. *American Journal of Agricultural Economics* 87:1113–1127.

Falkenmark, M., and J. Rockström. 2004. *Balancing Water for Humans and Nature: The New Approach in Ecohydrology.* Earthscan, London.

FAO (Food and Agriculture Organization). 1995. *Code of Conduct for Responsible Fisheries.* FAO, Rome.

FAO (Food and Agriculture Organization). 2004. *The State of Food Insecurity in the World. Monitoring Progress towards the World Food Summit and the Millennium Development Goals.* FAO. Rome.

FAO (Food and Agriculture Organization). 2006. *The State of the World Fisheries and Aquaculture.* United Nations FAO. Rome.

FAOSTAT. 2007. *FAO Statistical Database.* FAO, Rome.

Farber, S., C. Costanza, D.L. Childers, J. Erickson, K. Gross, et al. 2006. Linking ecology and economics for ecosystem management: A services-based approach with illustrations from LTER sites. *BioScience* 56:117–129.

Fargher, J.D., B.M. Howard, D.G. Burnside, and M.H. Andrew. 2003. The economy of Australian rangelands: Myth or mystery? *Rangeland Journal* 25:140–156.

Felson, A.J., and S.T.A. Pickett. 2005. Designed experiments: New approaches to studying urban ecosystems. *Frontiers in Ecology and the Environment* 10:549–586.

FEMAT (Forest Ecosystem Management Assessment Team). 1993. Forest ecosystem management: An ecological, economic, and social assessment. USDA Forest Service: U.S. Department of Commerce, National Oceanic and Atmospheric Administration, National Marine Fisheries Service; USDI Bureau of Land Management, Fish and Wildlife Service, National Park Service, Environmental Protection Agency, Portland, OR.

Fernandez, M., and J.C. Castilla. 2005. Marine conservation in Chile: Historical perspective, lessons, and challenges. *Conservation Biology* 19: 1752–1762.

Fernández, R., E.R.M. Archer, A.J. Ash, H. Dowlatabadi, P.H.Y. Hiernaux, et al. 2002. Degradation and recovery in socio-ecological systems: A view from the household/farm level. Pages 297–323 in J.F. Reynolds and D.M. Stafford Smith, editors. *Global Desertification: Do Humans Cause Deserts?* Dahlem University Press, Berlin.

Fernholz, K. 2007. *TIMOs & REITs: What, why, & how they might impact sustainable forestry.* Dovetail Partners. http://www.dovetailinc.org/reportView.php?action=displayReport&reportID=78.

Field, C.B., D.B. Lobell, H.A. Peters, and N.R. Chiariello. 2007. Feedbacks of terrestrial ecosystems to climate change. *Annual Review of Environment and Resources* 32:1–29.

Field, D.R., P.R. Voss, T.K. Kuczenski, R.B. Hammer, and V.C. Radeloff. 2003. Reaffirming social landscape analysis in landscape ecology: A conceptual framework. *Society & Natural Resources* 16:349–361.

Finlayson, A.C., and B.J. McCay. 1998. Crossing the threshold of ecosystem resilience: The commercial extinction of Northern cod. Pages 311–337 in F. Berkes and C. Folke, editors. *Linking Social and Ecological Systems: Management Practices and Social Mechanisms for Building Resilience.* Cambridge University Press, Cambridge.

Finlayson, C.M., R. D'Cruz, N. Aladin, D.R. Barker, G. Beltram, et al. 2005. Inland water systems. Pages 551–583 in R. Hassan, R. Scholes, and N. Ash, editors. *Ecosystems and Human Well-Being: Current State and Trends. Millennium Ecosystem Assessment.* Island Press, Washington.

Fischer, F. 1993. Citizen participation and the democratization of policy expertise: From theoretical inquiry to practical cases. *Policy Sciences* 26:165–187.

Fischer, F. 2000a. *Citizens, Experts, and the Environment: The Politics of Local Knowledge.* Duke University Press, Durham, NC.

Fischer, J. 2000b. Participatory research in ecological fieldwork: A Nicaraguan study. Pages 41–54 *in* B. Neis, and L. Felt, editors. *Finding Our Sea Legs*. Institute of Social and Economic Research, St. Johns, Canada.

Fischer, J., A.D. Manning, W. Steffen, D.B. Rose, K. Daniell, et al. 2007. Mind the sustainability gap. *Trends in Ecology and Evolution* 22:621–624.

Foley, J.A., M.T. Coe, M. Scheffer, and G. Wang. 2003a. Regime shifts in the Sahara and Sahel: Interactions between ecological and climatic systems in Northern Africa. *Ecosystems* 6: 524–539.

Foley, J.A., M.H. Costa, C. Delire, N. Ramankutty, and P. Snyder. 2003b. Green surprise? How terrestrial ecosystems could affect earth's climate. *Frontiers of Ecology and the Environment* 1:38–44.

Foley, J.A., R. DeFries, G.P. Asner, C. Barford, G. Bonan, et al. 2005. Global consequences of land use. *Science* 309:570–574.

Foley, J.A., J.E. Kutzbach, M.T. Coe, and S. Levis. 1994. Feedbacks between climate and boreal forests during the Holocene epoch. *Nature* 371:52–54.

Folke, C. 2003. Freshwater for resilience: A shift in thinking. *Philosophical Transactions of the Royal Society of London, Series B* 358:2027–2036.

Folke, C. 2006. Resilience: The emergence of a perspective for social-ecological systems analysis. *Global Environmental Change* 16:253–267.

Folke, C., S.R. Carpenter, T. Elmqvist, L. Gunderson, C.S. Holling and B. Walker. 2002. Resilience and sustainable development: Building adaptive capacity in a world of transformations. *Ambio* 31:437–440.

Folke, C., S.R. Carpenter, B.H. Walker, M. Scheffer, T. Elmqvist. 2004. Regime shifts, resilience and biodiversity in ecosystem management. *Annual Review in Ecology Evolution and Systematics* 35:557–581.

Folke, C., J. Colding, and F. Berkes. 2003. Synthesis: Building resilience and adaptive capacity in social-ecological systems. Pages 352–387 *in* F. Berkes, J. Colding, and C. Folke, editors. *Navigating Social-Ecological Systems: Building Resilience for Complexity and Change*. Cambridge University Press, Cambridge.

Folke, C., T. Hahn, P. Olsson, and J. Norberg. 2005. Adaptive governance of social-ecological systems. *Annual Review of Environment and Resources* 30:441–473.

Folke, C., M. Hammer, R. Costanza, and A. Jansson. 1994. Investing in natural capital: Why, what, and how? Pages 1–20 *in* A. Jansson, M. Hammer, C. Folke, and R. Costanza, editors. *Investing in Natural Capital*. Island Press, Washington.

Folke, C., A. Jansson, J. Larsson, and R. Costanza. 1997. Ecosystem appropriation by cities. *Ambio* 26:167–172.

Folke, C., J.L. Prichard, F. Berkes, J. Colding, and U. Svedin. 1998. *The Problem of Fit Between Ecosystems and Institutions*. Institutional Dimensions of Global Environmental Change Program, Bonn, Germany.

Folke, C., J.L. Prichard, F. Berkes, J. Colding, and U. Svedin. 2007. The Problem of Fit between Ecosystems and Institutions: Ten Years Later. *Ecology and Society* 12(1):30. http://www.ecologyandsociety.org/vol12/iss1/art30/.

Ford, J.D., and B. Smit. 2004. A framework for assessing the vulnerability of communities in the Canadian Arctic to risks associated with climate change. *Arctic* 57:389–400.

Fore, L.S., K. Paulsen, and K. O'Laughlin. 2001. Assessing the performance of volunteers in monitoring streams. *Freshwater Biology* 46: 109–123.

Foster, D.R., and J.D. Aber. 2004. *Forests in Time: The Environmental Consequences of 1,000 Years of Change in New England*. Yale University Press, New Haven.

Foster, S.S.D., and P J. Chilton. 2003. Groundwater: The processes and global significance of aquifer degradation. *Philosophical Transactions of the Royal Society of London, Series B* 358:1957–1972.

Fowler, C., and T. Hodgkin. 2004. Plant genetic resources for food and agriculture: Assessing global availability. *Annual Review of Environment and Resources* 29:143–179.

Fox, J., J. Krummel, S. Yarnasarn, M. Ekasingh, and N. Podger. 1995. Land-use and landscape dynamics in Northern Thailand: Assessing change in three upland watersheds. *Ambio* 24:328–34.

Fox, J.J. 1991. Managing the ecology of rice production in Indonesia. Pages 61–84 *in* J. Hardjono, editor. *Indonesia: Resources, Ecology, and Environment*. Oxford University Press. Oxford.

Francis, R., and R. Shotton. 1997. "Risk" in fisheries management. *Canadian Journal of Fisheries and Aquatic Sciences* 54:1699–1715.

Frank, K.T., B. Petrie, J.S. Choi, and W.C. Leggett. 2005. Trophic cascades in a formerly cod-dominated ecosystem. *Science* 308:1621–1623.

Freeman, M.M.R., and U.P. Kreuter, editors. 1994. *Elephants and Whales: Resources for Whom?* Gordon and Breach Publishers, Basel, Switzerland.

Friedlingstein, P., K.C. Prentice, I.Y. Fung, J.G. John, and G.P. Brasseur. 1995. Carbon-biosphere-climate interactions in the last glacial maximum climate. *Journal of Geophysical Research* 100:7203–7221.

Frost, P., E. Medina, J.-C. Menaut, O. Solbrig, M. Swift, et al. 1986. *Response of Savannas to Stress and Disturbance: A Proposal for a Collaborative Programme of Research*. International Union of Biological Sciences, Special Issue 10. Paris.

Funke, N., K. Nortje, K. Findlater, M. Burns, A. Turton, et al. 2007. Redressing inequality: South Africa's new water policy. *Environment* 49:10–23.

Galaz, V., P. Olsson, T. Hahn, C. Folke, and U. Svedin. 2008. The problem of fit among biophysical systems, environmental and resource regimes, and broader governance systems: Insights and emerging challenges. Pages 147–186 *in* O.R. Young, L.A. King, and H. Schröder, editors. *Institutions and Environmental Change: Principal Findings, Applications, and Research Frontiers*. MIT Press, Cambridge, MA.

Gallant, A.L., E.F. Binnian, J.M. Omernik, and M.B. Shasby. 1995. *Ecoregions of Alaska*. USGS Professional Paper 1567. USGS, Washington. http://agdcftp1.wr.usgs.gov/pub/projects/fhm/ecoreg.gif.

Gallopin, G.C. 2006. Linkages between vulnerability, resilience, and adaptive capacity. *Global Environmental Change* 16:293–303.

Galloway, J., M. Burke, E. Bradford, W. Falcon, J. Gaskell, et al. 2007. International trade in meat: The tip of the pork chop. *Ambio* 36:622–629.

Galvin, K.A., R.S. Reid, R.H.J. Behnke, and N.T. Hobbs, editors. 2008. *Fragmentation in Semi-Arid and Arid Landscapes: Consequences for Human and Natural Systems*. Springer, Dordrecht.

Garreau, J. 1991. *Edge City: Life on the New Frontier*. Doubleday, New York.

Geertz, C. 1963. *Agricultural Involution: The Process of Ecological Change in Indonesia*. University of California Press. Berkeley.

Gelcich, S., G. Edwards-Jones, M.J. Kaiser, and J.C. Castilla. 2006. Co-management policy can reduce resilience in traditionally managed marine ecosystems. *Ecosystems* 9:951–966.

Getz, W.M., L. Fortmann, D. Cumming, J. du Toit, J. Hilty, et al. 1999. Sustaining natural and human capital: Villagers and scientists. *Science* 283: 1855–1856.

Geist, H.J., and E.F. Lambin. 2004. Dynamic causal patterns of desertification. *BioScience* 53:817–829.

Ghassemi, F., A.J. Jakeman, and H.A. Nix. 1995. *Salinisation of Land and Water Resources: Human Causes, Extent, Management, and Case Studies*. CAB International. Wallingford, UK.

Giller, K., and C. Palm. 2004. Cropping systems: Slash-and-burn cropping systems of the tropics.

Pages 262–366 *in* R.M. Goodman, editors. *Encyclopedia of Plant and Crop Science*. Marcel Dekker, New York.

Gillson, L., and K.I. Duffin. 2007. Thresholds of potential concern as benchmarks in the management of African savannahs. *Philosophical Transactions of the Royal Society of London, Series B* 362:309–319.

Ginn, W.J. 2005. *Investing in Nature: Case Studies of Land Conservation in Collaboration with Business*. Island Press, Washington.

Gladwell, M. 2000. *The Tipping Point: How Little Things Can Make a Big Difference*. Abacus, London.

Gleick, P.H. 1996. Basic water requirements for human activities: Meeting basic needs. *Water International* 21:83–92.

Gleick, P.H. 2000. The World's Water 2000–2001. Island Press, Washington.

Gleick, P.H. 2002. *The World's Water: The Biennial Report on Freshwater Resources 2002–2003*. Island Press, Washington.

Gleick, P.H., H. Cooley, D. Katz, E. Lee, J. Morrison, et al. 2006. *The Worlds Water: The Biennial Report on Freshwater Resources 2006–2007*. Island Press, Washington.

Goodwin, N.B., A. Grant, A.L. Perry, N.K. Dulvy, and J.D. Reynolds. 2006. Life history correlates of density-dependent recruitment in marine fishes. *Canadian Journal of Fisheries and Aquatic Sciences* 63:494–509.

Gordon, L.J., G.D. Peterson, and E.M. Bennett. 2008. Agricultural modifications of hydrological flows create ecological surprises. *Trends in Ecology and Evolution* 23:211–219.

Gordon, M.R., J.C. Mueller, and C.C. Walden. 1980. Effect of biotreatment on fish tainting propensity of bleach kraft whole mill effluent. *Transactions of the Technical Association of Pulp and Paper Industry* 6:TR2-TR8.

Gorgens, A.H.M., and B.W. van Wilgen. 2004. Invasive alien plants and water resources in South Africa: Current understanding, predictive ability and research challenges. *South African Journal of Science* 100:27–33.

Gottdiener, M., and R. Hutchinson. 2001. *The New Urban Sociology*. 2nd Edition. McGraw-Hill Higher Education, New York.

Granovetter, M.D. 2004. The impact of social structures on economic development. *Journal of Economic Perspectives* 19:33–50.

Gray, B. 1989. *Collaborating: Finding Common Ground for Multiple Problems*. Jossey-Bass, San Francisco.

Gray, B., and D.J. Wood. 1991. Collaborative alliances: Moving from practice to theory. *Journal of Applied Behavioral Sciences* 27:3–22.

Griffith, B., D.C. Douglas, N.E. Walsh, D.D. Young, T.R. McCabe, et al. 2002. The Porcupine caribou herd. Pages 8–37 *in* D.C. Douglas, P.E. Reynolds, and E.B. Rhode, editors. *Arctic Refuge Coastal Plain Terrestrial Wildlife Research Summaries.* U.S. Geological Survey, Biological Resources Division, Biological Sciences Report USGS/BRD BSR-2002–0001, Alaska.

Grimm, N., J.M. Grove, S.T.A. Pickett, and C.L. Redman. 2000. Integrated approaches to long-term studies of urban ecological systems. *Bio-Science* 50:571–584.

Grimm, N.B., S.H. Faeth, N.E. Golubiewski, C.L. Redman, J. Wu, et al. 2008. Global change and the ecology of cities. *Science* 319:756–760.

Groffman, P., N.L. Law, K.T. Belt, L.E. Band, and G.T. Fisher. 2004. Nitrogen fluxes and retention in urban watershed ecosystems. *Ecosystems* 7: 393–403.

Groffman, P.M., D.J. Bain, L.E. Band, K.T. Belt, G.S. Brush, et al. 2003. Down by the riverside: Urban riparian ecology. *Frontiers in Ecology and the Environment* 1:315–321.

Groffman, P.M., J.S. Baron, T. Blett, A.J. Gold, M.A. Palmer, et al. 2006. Ecological thresholds: The key to successful environmental management or an important concept with no practical application? *Ecosystems* 9:1–13.

Grove, J.M. 1995. Excuse me, Could I speak to the property owner, please? *Common Property Resources Digest* 35:7–8.

Grove, J.M., and W.R. Burch. 1997. A social ecology approach to urban ecosystems and landscape analysis. *Urban Ecosystems* 1:185–199.

Grumbine, R.E. 1994. What is ecosystem management? *Conservation Biology* 8:27–38.

Gunderson, L.H. 1999. Resilience, flexibility and adaptive management: Antidotes for spurious certitude? *Conservation Ecology* 3(1):7. http://www.consecol.org/vol3/iss1/art7/.

Gunderson, L.H. 2003. Adaptive dancing: Interactions between social resilience and ecological crises. Pages 33–52 *in* F. Berkes, J. Colding, and C. Folke, editors. *Navigating Social–Ecological Systems: Building Resilience for Complexity and Change.* Cambridge University Press, Cambridge.

Gunderson, L.H., S.R. Carpenter, C. Folke, P. Olsson, and G.D. Peterson. 2006. Water RATs (resilience, adaptability, and transformability) in lake and wetland social-ecological systems. *Ecology and Society* 11(1):16. http://www.ecologyandsociety.org/vol11/iss1/art16/.

Gunderson, L.H., and C.S. Holling, editors. 2002. *Panarchy: Understanding Transformations in Human and Natural Systems.* Island Press, Washington.

Gunderson, L.H., C.S. Holling, and S.S. Light. 1995. *Barriers and Bridges to the Renewal of Ecosystems and Institutions.* Columbia University Press, New York.

Gunderson, L.H., and L. Pritchard, editors. 2002. *Resilience and the Behavior of Large-Scale Ecosystems.* Island Press, Washington.

Gunter, J. 2004. *The Community Forestry Guidebook: Tools and Techniques for Communities in British Columbia.* FORREX, Kamloops, Canada.

Hahn, T., P. Olsson, C. Folke, and K. Johansson. 2006. Trust-building, knowledge generation and organizational innovations: The role of a bridging organization for adaptive co-management of a wetland landscape around Kristianstad, Sweden. *Human Ecology* 34:573–592.

Hanna, S.S., C. Folke, and K.-G. Maler, editors. 1996. *Rights to Nature: Ecological, Economic, Cultural, and Political Principles of Institutions for the Environment.* Island Press, Washington, DC.

Hannah, L., G. Midgley, T. Lovejoy, W.J. Bond, M. Bush, et al. 2002. Conservation of bioidversity in a changing climate. *Conservation Biology* 16: 264–268.

Hara, M., and J. Nielsen. 2003. Experiences with fisheries co-management in Africa. Pages 81–97 *in* D.C. Wilson, C. Douglas, J.R. Nielsen, and P. Degnbol, editors. *The Fisheries Co-Management Experience: Accomplishments, Challenges and Prospects.* Kluwer, Dordrecht.

Hardin, G. 1968. Tragedy of the Commons. *Science* 162:1243–1248.

Harley, S.J., R.A. Myers, and A. Dunn. 2001. Is catch-per-effort proportional to abundance? *Canadian Journal of Fisheries and Aquatic Sciences* 58:1760–1772.

Harmon, M.E., W.K. Ferrell, and J.F. Franklin. 1990. Effects on carbon storage of conversion of old-growth forests to young forests. *Science* 247: 699–702.

Hawley, A.H. 1950. *Human Ecology: A Theory of Community Structure.* Ronald Press, New York.

Haynes, R.W., B.T. Bormann, D.C. Lee, and J.R. Martin, editors. 2006. *Northwest Forest Plan – The First 10 Years (1994–2003): Synthesis of Monitoring and Research Results. General Technical Report PNW-GTR-651.* U.S. Department of Agriculture, Forest Service, Pacific Northwest Research Station, Portland.

Heal, G. 2000. *Nature and the Marketplace: Capturing the Value of Ecosystem Services*. Island Press, Washington.

Heaslip, R. 2008. Monitoring salmon aquaculture waste: The contribution of First Nations' rights, knowledge, and practices in British Columbia. *Marine Policy* 32:988–996.

Held, D., A. McGrew, D. Goldblatt, and J. Perraton. 1999. *Global Transformations: Politics, Economics, and Culture*. Stanford University Press, Stanford.

Hennessey, T., and M. Healey. 2000. Ludwig's ratchet and the collapse of New England groundfish stocks. *Coastal Management* 28:187–213.

Hewitt, K. 1997. *Regions of Risk: A Geographical Introduction to Disasters*. Longman, Harlow, UK.

Hibbard, K.A., P.J. Crutzen, E.F. Lambin, D. Liverman, N.J. Mantua, et al. 2006. Decadal interactions of humans and the environment. Pages 341–375 *in* R. Costanza, L. Graumlich, and W. Steffen, editors. *Integrated History and Future of People on Earth. Dahlem Workshop Report 96*. MIT Press, Cambridge, MA.

Hiernaux, P., and M.D. Turner. 2002. The influence of farmer and postoralist management practices on desertification processes in the Sahel. Pages 135–148 *in* J.F. Reynolds and D.M. Stafford Smith, editors. *Global Desertification: Do Humans Cause Deserts?* Dahlem University Press, Berlin.

Hilborn, R. 1992. Hatcheries and the future of salmon in the northwest. *Fisheries* 17:5–8.

Hilborn, R., and D. Eggers. 2000. A review of the hatchery programs for pink salmon in Prince William Sound. *Transactions of the American Fisheries Society* 129:333–350.

Hilborn, R., R.H. Parrish, and C.J. Walters. 2006. *Peer Review, California Marine Life Protection Act (MLPA) Science Advice and MPA Network Proposals*. California Fisheries Coalition, Sacramento, CA.

Himmel, M.E., S. Ding, D.K. Johnson, W.S. Adney, M.R. Nimlos, J.W. Brady, and T.D. Foust. 2007. Biomass recalcitrance: Engineering plants and enzymes for biofuels production. *Science* 315: 804–807.

Hoekstra, A.Y., and A.K. Chapagain. 2007. Water footprints of nations: Water use by people as a function of their consumption pattern. *Water Resources Management* 21:35–48.

Holling, C.S. 1973. Resilience and stability of ecological systems. *Annual Review of Ecology and Systematics* 4:1–23.

Holling, C.S., editor. 1978. *Adaptive Environmental Assessment and Management*. John Wiley & Sons, New York.

Holling, C.S. 1986. Resilience of ecosystems: Local surprise and global change. Pages 292–317 *in* W.C. Clark and R.E. Munn, editors. *Sustainable Development and the Biosphere*. Cambridge University Press, Cambridge.

Holling, C.S. 1995. What barriers? What bridges? Pages 3–34 *in* L.H. Gunderson, C.S. Holling, and S.S. Light, editors. 1995. *Barriers and Bridges to the Renewal of Ecosystems and Institutions*. Columbia University Press, New York.

Holling, C.S. 1998. Two cultures of ecology. *Conservation Ecology* 2(2):4. http://www.consecol.org/vol2/iss2/art4/.

Holling, C.S., F. Berkes, and C. Folke. 1998. Science, sustainability, and resource management. Pages 342–362 *in* F. Berkes, and C. Folke, editors. *Linking Social and Ecological Systems: Management Practices and Social Mechanisms for Building Resilience*. Cambridge University Press, Cambridge.

Holling, C.S., and L.H. Gunderson. 2002. Resilience and adaptive cycles. Pages 25–62 *in* L.H. Gunderson and C S. Holling, editors. *Panarchy: Understanding Transformations in Human and Natural Systems*. Island Press, Washington.

Holling, C.S., L.H. Gunderson, and D. Ludwig. 2002a. In quest of a theory of adaptive change. Pages 3–22 *in* L.H. Gunderson and C.S. Holling, editors. *Panarchy: Understanding Transformations in Human and Natural Systems*. Island Press, Washington.

Holling, C.S., L.H. Gunderson, and G.D. Peterson. 2002b. Sustainability and panarchies. Pages 63–102 *in* L.H. Gunderson and C.S. Holling, editors. *Panarchy: Understanding Transformations in Human and Natural Systems*. Island Press, Washington.

Holling, C.S., and G.K. Meffe. 1996. Command and control and the pathology of natural resource management. *Conservation Biology* 10:328–337.

Holling, C.S., and S. Sanderson. 1996. Dynamics of (dis)harmony in ecological and social systems. Pages 57–85 *in* S. Hanna, C. Folke, K.-G. Mäler, and Å. Jansson, editors. *Rights to Nature: Ecological, Economic, Cultural, and Political Principles of Institutions for the Environment*. Island Press, Washington.

Hostetler, M. 1999. Scale, birds, and human decisions: A potential for integrative research in urban ecosystems. *Landscape and Urban Planning* 45:15–19.

Hough, M. 1984. *City Form and Natural Process: Towards a New Urban Vernacular*. Van Nostrand Reinhold Company, New York.

Hughes, T.P., D.R. Bellwood, C. Folke, L.J. McCook, and J.M. Pandolfi. 2007. No-take areas: Herbivory and coral reef resilience. *Trends in Ecology and Evolution* 22:1–3.

Huitric, M. 2005. Lobster and conch fisheries of Belize: A history of sequential exploitation. *Ecology and Society* 10(1):21. http://www.ecologyandsociety.org/vol10/iss1/art21/.

Hulme, D., and M. Murphree, editors. 2001. *African Wildlife and Livelihoods: The Promise and Performance of African Conservation*. James Currey Publishers, Oxford.

Hutton, G., and L. Haller. 2004. *Evaluation of the Costs and Benefits of Water and Sanitation Improvements at the Global Level*. World Health Organization, Geneva.

Imhoff, D. 2007. *Food Fight: A Citizen's Guide to a Food and Farm Bill*. University of California Press. Berkeley.

Imhoff, M.L., L. Bounoua, T. Ricketts, C. Loucks, R. Harriss, and W.T. Lawrence. 2004. Global patterns in human consumption of net primary production. *Nature* 429:870–873.

IPCC (Intergovernmental Panel on Climate Change). 2007a. *Climate Change 2007: The Physical Science Basis, Contribution of Working Group I to the Fourth Assessment Report of the Intergovernmental Panel on Climate Change*. Cambridge University Press, Cambridge.

IPCC (Intergovernmental Panel on Climate Change). 2007b. *Climate Change 2007: Impacts, Adaptation and Vulnerability, Contribution of Working Group II to the Fourth Assessment Report of the Intergovernmental Panel on Climate Change*. Cambridge University Press, Cambridge.

IFPRI (International Food Policy Research Institute). 2002. *Reaching Sustainable Food Security for All by 2020: Getting the Priorities and Responsibilities Right*. IFPRI, Washington.

IUCN (International Union for the Conservation of Nature). 1991. *Caring for the Earth: A Strategy for Sustainable Living*. World Conservation Union, United Nations Environment Programme, World Wildlife Fund for Nature, Gland, Switzerland.

Ives, A.R., K. Gross, and J.L. Klug. 1999. Stability and variability in competitive communities. *Science* 286:542–544.

Jackson, J.B.C. 2001. What was natural in the coastal oceans? *Proceedings of the National Academy of Sciences* 98:5411–5418.

Jackson, J.B.C., M.X. Kirby, W.H. Berger, K.A. Bjorndal, L.V. Botsford, et al. 2001. Historical overfishing and the recent collapse of coastal ecosystems. *Science* 293:629–638.

Jacobs, J. 1961. *The Death and Life of Great American Cities*. Vintage Books, New York.

James, C.D., J. Landsberg, and S.R. Morton. 1999. Provision of watering points in the Australian arid zone: A review of effects on biota. *Journal of Arid Environments* 41:87–121.

Janssen, M.A., T.A. Kohler, and M. Scheffer. 2003. Sunk-cost effects and vulnerability to collapse in ancient societies. *Current Anthropology* 44: 722–728.

Janssen, M.A., and M. Scheffer. 2004. Overexploitation of renewable resources by ancient societies and the role of sunk-cost effects. *Ecology and Society* 9(1):6. http://www.ecologyandsociety.org/vol9/iss1/art6/.

Janssen, M.A., M.L. Schoon, W. Ke, and K. Borner. 2006. Scholarly networks on resilience, vulnerability and adaptation within the human dimensions of global environmental change. *Global Environmental Change* 16:240–252.

Jansson, Å., C. Folke, J. Rockström, L. Gordon, and M. Falkemark. 1999. Linking freshwater flows and ecosystem services appropriated by people: The case of the Baltic Sea Drainage Basin. *Ecosystems* 2:351–366.

Jarvis, A., A. Lane, and R. Hijmans. 2008. The effect of climate change on crop wild relatives. *Agriculture, Ecosystems and Environment* 126:13–23.

Jenkins, J.C., and R. Riemann. 2003. What does non-forest land contribute to the global C balance? Pages 173–179 *in* R. McRoberts, G.A. Reams, P.A. Van Deusen, and J.W. Moser, editors. *Proceedings, Third Annual FIA Science Symposium*. USDA Forest Service General Technical Report NC-230. St. Paul, MN.

Jentoft, S. 1993. *Dangling Lines: The Fisheries Crisis and the Future of Coastal Communities: The Norwegian Experience*. Institute for Social and Economic Studies, St. Johns, Canada.

Jiang, H. 2002. Culture, ecology, and Nature's changing balance: Sandification on Mu Us Sand Land, Inner Mongolia, China. Pages 181–196 *in* J.F. Reynolds and D.M. Stafford Smith, editors. *Global Desertification: Do Humans Cause Deserts?* Dahlem University Press, Berlin.

Johannes, R. 1978. Traditional marine conservation methods in Oceania and their demise. *Annual Review of Ecological Systems* 9:349–364.

Johannes, R. 1981. *Words of the Lagoon: Fishing and Marine Lore in the Palau District of Micronesia.* University of California Press, Berkeley.

Johnson, E.A. 1992. *Fire and Vegetation Dynamics: Studies from the North American Boreal Forest.* Cambridge University Press, Cambridge.

Johnson, S. 2001. *Emergence: The Connected Lives of Ants, Brains, Cities, and Software.* Scribner, New York.

Johnson, S. 2006. *The Ghost Map: The Story of London's Most Terrifying Epidemic – and How It Changed Science, Cities, and the Modern World.* Riverhead Books, New York.

Jones, C.G., J.H. Lawton, and M. Shachak. 1994. Organisms as ecosystem engineers. *Oikos* 69: 373–386.

Justice, M. 2007. *Volunteering in Fish-Habitat Rehabilitation Projects in British Columbia.* MRM Research Report No. 435. (thesis) School of Resource and Environmental Management, Simon Fraser University, Vancouver.

Kahneman, D., and A. Tversky. 1979. Prospect theory: An analysis of decisions under risk. *Econometrica* 47:263–291.

Kaldor, M. 2003. *Global Civil Society: An Answer to War.* Polity Press, Cambridge.

Kanter, R. 1983. *The Change Masters: Corporate Entrepreneurs at Work.* Simon and Schuster, New York.

Kaplan, S., and R. Kaplan. 1989. *The Experience of Nature: A Phychological Perspective.* Cambridge University Press, New York.

Kasperson, R.E., K. Dow, E.R.M. Archer, D. Caceres, T.E. Downing, et al. 2005. Vulnerable peoples and places. Pages 143–164 *in* R. Hassan, R. Scholes, and N. Ash, editors. *Ecosystems and Human Well-Being: Current State and Trends. Millennium Ecosystem Assessment.* Island Press, Washington.

Katz, B., and J. Bradley. 1999. Divided we sprawl. *Atlantic Monthly* 284:26–42.

Kaufman, L. 1992. Catastrophic change in species-rich freshwater ecosystems: The lessons of Lake Victoria. *BioScience* 42:846–858.

Keane, J. 2003. *Global Civil Society.* Cambridge University Press, Cambridge.

Keen, I. 2004. *Aboriginal Economy and Society: Australia at the Threshold of Colonisation.* Oxford University Press, South Melbourne.

Keene, E. 2002. *Beyond the Anarchical Society: Grotius, Colonialism and World Order.* Cambridge University Press, Cambridge.

Keesing, J.K., R.H. Bradbury, L.M. DeVantier, M.J. Riddle, and G. De'ath. 1992. Geological evidence for recurring outbreaks of the crown-of-thorns starfish: A reassessment from an ecological perspective. *Coral Reefs* 11:79–85.

Kennedy, V.S., and L.L. Breisch. 1983. Sixteen decades of political mismanagement of the oyster fishery in Maryland's Chesapeake Bay. *Journal of Environmental Management* 16:153–171.

Kent, M., R.A. Stevens, and L. Zhang. 1999. Urban plant ecology patterns and processes: A case study of the flora of the City of Plymouth, Devon, UK. *Journal of Biogeography* 26: 1281–1298.

Khagram, S. 2004. *Dams and Development: Transnational Struggles for Water and Power.* Cornell University Press. Ithaca.

Khagram, S., W.C. Clark, and D.F. Raad. 2003. From the environment and human security to sustainable security and development. *Journal of Human Development* 4:289–313.

Kidder, T. 1981. *The Soul of a New Machine.* Aron Books, New York.

King, J.M., R.E. Tharme, and M.S. de Villers. 2000. *Environmental Flow Assessments for Rivers: Manual for the Building Block Methodology.* Water Research Commission, Pretoria, South Africa.

Kinzig, A., D. Starrett, K. Arrow, B. Bolin, P. Dasgupta, et al. 2003. Coping with uncertainty: A call for a new science-policy forum. *Ambio* 32: 330–335.

Kinzig, A.P., P. Ryan, M. Etienne, T. Elmqvist, H.E. Allison, et al. 2006. Resilience and regime shifts: Assessing cascading effects. *Ecology and Society* 11(1):20. http://www.ecologyandsociety.org/vol11/iss1/art20/.

Kirwan, J. 2008. Using indigenous fire practices to manage coastal wetlands. Paper read to the Conference of the Society for Applied Anthropology, Memphis, Tennessee.

Kitchell, J.F., editor. 1992. *Food Web Management: A Case Study of Lake Mendota.* Springer-Verlag, New York.

Kocher, T.D. 2004. Adaptive evolution and explosive speciation: The cichlid fish model. *Nature Reviews Genetics* 5:288–298.

Koehane, R.O., and E. Ostrom, editors. 1995. *Local Commons and Global Interdependence.* Sage Press, London.

Kofinas, G.P. 1998. The Costs of Power Sharing: Community Involvement in Canadian Porcupine Caribou Co-Management. PhD Dissertation, University of British Columbia, Vancouver.

Kofinas, G.P. 2005. Hunters and researchers at the co-management interface: Emergent dilemmas and the problem of legitimacy in power sharing. *Anthropologica* 47:179–196.

Kofinas, G.P., and Communities of Aklavik, Arctic Village, Old Crow, and Fort McPherson. 2002. Community contributions to ecological monitoring: Knowledge co-production in the U.S.-Canada Arctic Borderlands. Pages 54–91 *in* I. Krupnik and D. Jolly, editors. *The Earth Is Faster Now: Indigenous Observations of Arctic Environmental Change*. Arctic Research Consortium of the United States, Fairbanks, Alaska.

Kofinas, G.P., and J.R. Griggs. 1996. Collaboration and the B.C. Round Table: Analysis of a "better way" of deciding. *Environments: A Journal of Interdisciplinary Studies* 23:17–40.

Kofinas, G.P., S.J. Herman, and C.L. Meek. 2007. Novel problems require novel solutions: Innovation as an outcome of adaptive co-management. Pages 249–267 *in* D. Armitage, F. Berkes, and N. Doubleday, editors. *Adaptive Co-Management: Collaboration, Learning, and Multi-Level Governance*. University of British Columbia Press, Vancouver.

Kofinas, G., P. Lyver, D. Russell, R. White, A. Nelson, et al. 2003. Towards a protocol for community monitoring of caribou body condition. *Rangifer* 14:43–52.

Kokou, K., and N. Sokpon. 2006. Les forêts sacrées du culoir du Dahomey. *Bois et Forêts des Tropiques* 288:15–23.

Kolar, C., and D.M. Lodge. 2000. Freshwater nonindigenous species: Interactions with other global changes. Pages 3–30 *in* H.A. Mooney and R.J. Hobbs, editors. *Invasive Species in a Changing World*. Island Press, Washington.

Kolata, G. 2000. *Flu: The Story of the Great Influenza Pandemic of 1918 and the Search for the Virus that Caused It*. Macmillan, London.

Kolstad, C. 1999. *Environmental Economics*. Oxford University Press. Oxford.

Krauss, M.E. 1982. *Native Languages of Alaska (Map)*. Alaska Native Language Center, University of Alaska, Fairbanks. http://www.uaf.edu/anlc/.

Krupnik, I., and D. Jolly, editors. 2002. *The Earth is Faster Now: Indigenous Observations of Arctic Environmental Change*. Arctic Research Consortium of the United States, Fairbanks, Alaska.

Kundzewicz, Z.W., and H J. Schellnhuber. 2004. Floods in the IPCC perspective. *Natural Hazards* 31:111–128.

Kursar, T.A., C.C. Caballero-George, T.L. Capson, L. Cubilla-Rios, W.H. Gerwick, et al. 2006. Securing economic benefits and promoting conservation through bioprospecting. *BioScience* 56:1005–1012.

Lach, D., P. List, B. Steel, and B. Shindler. 2003. Advocacy and credibility of ecological scientists in resource decision making: A regional study. *BioScience* 53:171–179.

Lambin, E.F. 2007. *The Middle Path: Avoiding Environmental Catastrophe*. University of Chicago Press, Chicago.

Lambin, E.F., B.L. Turner, H.J. Geist, S.B. Agbola, A. Angelsen, et al. 2001. The causes of land-use and land-cover change: Moving beyond the myths. *Global Environmental Change* 11:261–269.

Landsberg, J., C.D. James, S.R. Morton, W.J. Müller, and J. Stol. 2003. Abundance and composition of plant species along grazing gradients in Australian rangelands. *Journal of Applied Ecology* 40:1008–1024.

Lane, M.B., and G. McDonald. 2002. Towards a general model of forest management through time: Evidence from Australia, USA, and Canada. *Land Use Policy* 19:193–206.

Langdon, S. 1999. *Communities and Quotas. Alternatives in the North Pacific Fisheries*. Presentation to the Pacific Marine States Fisheries Commission. Semiahmoo, Washington.

Langdon, S. 2008. The community quota program in the Gulf of Alaska: A vehicle for Alaska Native village sustainability. Pages 155–194 *in* M.E. Lowe and C. Carothers, editors. *Enclosing the Fisheries: People, Places, and Power*. American Fisheries Society. Bethesda, MD.

Lansing, S. 2006. *Perfect Order: Recognizing Complexity in Bali*. Princeton University Press, Princeton.

Larson, A.M., and J.C. Ribot. 2007. The poverty of forestry policy: Double standards on an uneven playing field. *Sustainability Science* 2:189–204.

Larsson, S., and K. Danell. 2001. Science and the management of boreal forest biodiversity. *Scandinavian Journal of Forest Research Supplement* 3:5–9.

Lathrop, R.C. 1992a. Nutrient loadings, lake nutrients and water clarity. Pages 69–96 *in* J.F. Kitchell, editor. *Food Web Management: A Case Study of Lake Mendota*. Springer-Verlag, New York.

Lathrop, R.C. 1992b. Decline in zoobenthos densities in the profundal sediments of Lake Mendota (Wisconsin, USA). *Hydrobiologia* 235/236: 353–361.

Lathrop, R.C., B.M. Johnson, T.B. Johnson, M.T. Vogelsang, S.R. Carpenter, et al. 2002. Stocking piscivores to improve fishing and water clarity: A synthesis of the Lake Mendota biomanipulation project. *Freshwater Biology* 47:2410–2424.

Lebel, L., N.H. Tri, A. Saengnoree, S. Pasong, U. Buatama, et al. 2002. Industrial transformation and shrimp aquaculture in Thailand and Vietnam: Pathways to ecological, social, and economic sustainability? *Ambio* 31:311–323.

Lee, K.N. 1993. *Compass and Gyroscope: Integrating Science and Politics for the Environment.* Island Press, Washington.

Lee, K.N. 1999. Appraising adaptive management. *Conservation Ecology* 3(2):3. http://www.consecol.org/vol3/iss2/art3/.

Lee, R.B. 1979. *The !Kung San: Men, Women and Work in a Foraging Society.* Cambridge University Press, Cambridge.

Lee, R., and I. DeVore, editors. 1968. *Man the Hunter.* Aldine, Chicago.

Lee, S., and C. Webster. 2006. Enclosure of the urban commons. *GeoJournal* 66:27–42.

Leemans, R., and A. Kleidon. 2002. Regional and global assessment of the dimensions of desertification. Pages 215–231 *in* J.F. Reynolds and D.M. Stafford Smith, editors. *Global Desertification: Do Humans Cause Deserts?* Dahlem University Press, Berlin.

Leopold, A. 1949. *A Sand County Almanac.* Oxford University Press, Oxford.

Levin, S.A. 1998. Ecosystems and the biosphere as complex adaptive systems. *Ecosystems* 1:431–436.

Levin, S.A. 1999. *Fragile Dominion: Complexity and the Commons.* Perseus Books, Reading, MA.

Levy, M., S. Babu, K. Hamilton, V. Rhoe, A. Catenazzi, et al. 2005. Ecosystem conditions and human well-being. Pages 123–142 *in* R. Hassan, R. Scholes, and N. Ash, editors. *Ecosystems and Human Well-Being: Current State and Trends. Millennium Ecosystem Assessment.* Island Press, Washington.

Lewis, H.T. 1989. Ecological and technological knowledge of fire: Aborigines versus park managers in Northern Australia. *American Anthropologist* 91:940–61.

Lewis, J. 2002. Agrarian change and privitization of *ejido* land in Northern Mexico. *Journal of Agrarian Affairs* 3:402–420.

Lewis, H.T., and T.A. Ferguson. 1988. Yards, corridors and mosaics: How to burn a boreal forest. *Human Ecology* 16:57–77.

Liebman, M., and C.P. Staver. 2001. Crop diversification for weed management. Pages 322–374 *in* M. Liebman, C.L. Mohler, and C.P. Staver, editors. *Ecological Management of Agricultural Weeds.* Cambridge University Press, Cambridge.

Likens, G.E., F.H. Bormann, R.S. Pierce, J.S. Eaton, and N.M. Johnson. 1977. *Biogeochemistry of a Forested Ecosystem.* Springer-Verlag, New York.

Lindblom, C.E. 1959. The science of "muddling through". *Public Administration Review* 19:70–88.

Lindblom, C.E. 1972. Still muddling, not yet through. *Public Administration Review* 39:517–526.

Lindgren, E., and R. Gustafson. 2001. Tick-borne encephalitis in Sweden and climate change. *Lancet* 358:16–18.

Lindgren, E., L. Tälleklint, and T. Polfeldt. 2000. The impact of climate change on the northern latitude limit and population density of the disease-transmitting European tick *Ixodes ricinus. Environmental Health Perspectives* 108:119–123.

Lobell, D., M. Burke, C. Tebaldi, M. Mastrandrea, W. Falcon, et al. 2008. Prioritizing climate adaptation needs for food security in 2030. *Science* 319:607–610.

Lobell, D.B., and J.I. Ortiz-Monasterio. 2008. Satellite monitoring of yield responses to irrigation practices across thousands of fields. *Agronomy Journal* 100:1005–1012.

Longhurst, A. 2006. The sustainability myth. *Fisheries Research* 81:107–112.

Loucks, L.A. 2005. The Evolution of the Area 19 Snow Crab Co-Management Agreement: Understanding the Inter-relationship Between Transaction Costs, Credible Commitment, and Collective Action. PhD. dissertation. Simon Fraser University, Vancouver.

Lovelock, J.E. 1979. *Gaia: A New Look at Life on Earth.* Oxford University Press, Oxford.

Low, B., E. Ostrom, C. Simon, and J. Wilson. 2003. Redundancy and diversity: Do they influence optimal management? Pages 83–114 *in* F. Berkes, J. Colding, and C. Folke, editors. *Navigating Social-Ecological Systems: Building Resilience for Complexity and Change.* Cambridge University Press, Cambridge.

Ludwig, D., W.A. Brock, and S.R. Carpenter. 2005. Uncertainty in discount models and environmental accounting. *Ecology and Society* 10(2):13. http://www.ecologyandsociety.org/vol10/iss2/art13/.

Ludwig, D., R. Hilborn, and C. Walters. 1993. Uncertainty, resource exploitation, and conservation: Lessons from history. *Science* 260:17, 36.

Ludwig, D., M. Mangel, and B. Haddad. 2001. Ecology, conservation, and public policy. *Annual Review of Ecology and Systematics* 32:481–517.

Ludwig, J.A., D.J. Tongway, G.N. Bastin, and C.D. James. 2004. Monitoring ecological indicators of rangeland functional integrity and their relation to biodiversity at local to regional scales. *Austral Ecology* 29:108–120.

Ludwig, J.A., D.J. Tongway, and S.G. Marsden. 1999. Stripes, strands or stipples: Modelling the influence of three landscape banding patterns on resource capture and productivity in semi-arid woodlands, Australia. *CATENA* 37:257–273.

Ludwig, J.A., B.P. Wilcox, D.D. Breshears, D.J. Tongway, and A.C. Imeson. 2005. Vegetation patches and runoff-erosion as interacting ecohydrological processes in semiarid landscapes. *Ecology* 86:288–297.

Lukes, S. 2002. *Power: A Radial View*. Palgrave Macmillan, New York.

Luthar, S.S., and D. Cicchetti. 2000. The construct of resilience: Implications for interventions and social policies. *Development and Psychopathology* 12:857–885.

MacCall, A. 1990. *Dynamic Geography of Fish Populations*. University of Washington Press, Seattle.

Machlis, G.E., J.E. Force, and W.R. Burch, Jr. 1997. The human ecosystem. Part I: The human ecosystem as an organizing concept in ecosystem management. *Society and Natural Resources* 10: 347–367.

Magadlela, D., and N. Mdzeke. 2004. Social benefits in the Working for Water programme as a public works initiative. *South African Journal of Science* 100:94–96.

Magnuson, J.J., T.K. Kratz, and B.J. Benson, editors. 2006. *Long-Term Dynamics of Lakes in the Landscape*. Oxford University Press, London.

Magnuson, J.J., and R.C. Lathrop. 1992. Historical changes in the fish community. Pages 193–232 in J.F. Kitchell, editor. *Food Web Management*. Springer-Verlag, New York.

Mahler, J. 1997. Influences of organizational culture on learning in public agencies. *Journal of Public Administration Research and Theory* 7:519–540.

Makse, H.A., S. Havlin, and H.E. Stanley. 1995. Modeling urban growth patterns. *Nature* 377:608–612.

Malthus, T.R. 1798. *An Essay on the Principle of Population*. Johnson, London.

Manning, R. 2002. Agriculture versus biodiversity: Will market solutions suffice? *Conservation in Practice* 3:18–27.

Manning, R. 2004. *Against the Grain: How Agriculture Has Hijacked Civilization*. North Point Press, New York.

Mansuri, G., and V. Rao. 2004. Community-based (and –driven) development: A critical review. *The World Bank Research Observer* 19(1):1–39.

Martin, P.S. and R.G. Klein, editors. 1984. *Quaternary Extinctions: A Prehistoric Revolution*. University of Arizona Press, Tucson.

Maskrey, A. 1989. *Disaster Mitigation: A Community-based Approach*. Oxfam, Oxford.

Maslow, A.H. 1943. A theory of human motivation. *Psychological Review* 50:370–396.

Matson, P., A. Luers, K. Seto, and R. Naylor. 2005. People, land use, and environment in the Yaqui Valley, Sonora, Mexico. Pages 238–264 in B. Entwisle, P.C. Stern, editors. *Population, Land Use, and Environment: Research Directions*. National Resource Council of the National Academies. The National Academies Press, Washington, D.C.

Matson, P.A., R.L. Naylor, and I. Ortiz-Monasterio. 1998. Integration of environmental, agronomic, and economic aspects of fertilizer management. *Science* 280:112–115.

Matson, P.A., W.J. Parton, A.G. Power, and M.J. Swift. 1997. Agricultural intensification and ecosystem properties. *Science* 227:504–509.

McAllister, R.R.J., I.J. Gordon, M.A. Janssen, and N. Abel. 2006. Pastoralists' responses to variation of rangeland resources in time and space. *Ecological Applications* 16:572–583.

McCarthy, J.J., M.L. Martello, R. Corell, N.E. Selin, S. Fox, et al. 2005. Climate change in the context of multiple stressors and resilience. Pages 945–988 in ACIA, editor. *Arctic Climate Impact Assessment*. Cambridge University Press.

McCay, B.J. 2002. Emergence of institutions for the commons: Contexts, situations, and events. Pages 361–402 in NRC (National Research Council); E. Ostrom, T. Dietz, N. Dolšak, P.C. Stern, S. Stovich, et al., editors. *The Drama of the Commons*. National Academy Press, Washington.

McCay, B.J., and J. Acheson, editors. 1987. *The Question of the Commons: The Culture and Ecology of Communal Resources*. University of Arizona Press, Tucson.

McGinnis, M.D. editor. 1999. *Polycentric Governance and Development: Readings from the Workshop in Political Theory and Policy Analysis*. University of Michigan Press, Ann Arbor.

McGoodwin, J. 1994. "Nowadays, nobody has any respect": The demise of folk management in a rural Mexican fishery. Pages 43–54 in C.L. Dyer and J.R. McGoodwin, editors. *Folk Management in the World's Fisheries. Lessons for Modern Fisheries Management*. University Press of Colorado, Niwot, CO.

McGrath, B., V. Marshall, M.L. Cadenasso, J.M. Grove, S.T.A. Pickett, et al., editors. 2008. *Designing Patch Dynamics*. Columbia University, New York.

McKeon, G., W. Hall, B. Henry, G. Stone, and I. Watson, editors. 2004. *Pasture Degradation and Recovery in Australia's Rangelands: Learn-*

ing from History. Queensland Department of Natural Resources, Mines and Energy, Brisbane, Australia.

McLachlan, J.S., J. Hellmann, and M. Schwartz. 2007. A framework for debate of assisted migration in an era of climate change. *Conservation Biology* 21:297–302.

McNaughton, S.J. 1977. Diversity and stability of ecological communities: A comment on the role of empiricism in ecology. *American Naturalist* 111:515–525.

McNeill, J. 2000. *Something New Under the Sun: An Environmental History of the Twentieth Century*. Penguin Press, London.

MEA (Millennium Ecosystem Assessment). 2005a. *Ecosystems and Human Well-being: Current Status and Trends*. Island Press, Washington.

MEA (Millennium Ecosystem Assessment). 2005b. *Ecosystems and Human Well-being: Desertification Synthesis*. World Resources Institute, Washington.

MEA (Millennium Ecosystem Assessment). 2005c. *Ecosystems and Human Well Being: Scenarios*. Island Press, Washington.

MEA (Millennium Ecosystem Assessment). 2005d. *Ecosystems and Human Well-being: Synthesis*. Island Press, Washington, D.C.

Meadows, D.H, J. Randers, D.L. Meadows, and W.W. Behrens. 1972. *The Limits to Growth: A Report for the Club of Rome's Project on the Predicament of Mankind*. Universe Books, New York.

Mears, L.A. 1981. *The New Rice Economy of Indonesia*. Gadja Mada University Press, Yogyakarta, Indonesia.

Melosi, M.V. 2000. *The Sanitary City: Urban Infrastructure in America from Colonial Times to the Present*. Johns Hopkins University Press, Baltimore.

Memmott, J. 1997. The structure of a plant-pollinator food web. *Ecology Letters* 2:276–280.

Middleton, N.J., and D.S.G. Thomas, editors. 1997. *World Atlas of Desertification*. 2nd Edition. U.N. Environment Programme, Edward Arnold, New York.

Millar, C.I., N.L. Stephenson, and S.L. Stephens. 2007. Climate change and forests of the future: Managing in the face of uncertainty. *Ecological Applications* 17:2145–2151.

The Millennium Development Goals Report. 2007. United Nations. New York, NY. http://www.un.org/millenniumgoals/pdf/mdg2007.pdf.

Mintzberg, H., and F. Westley. 1992. Cycles of organizational change. *Strategic Management Journal* 13:39–59.

Mitchell, C.E., and A.G. Power. 2003. Release of invasive plants from fungal and viral pathogens. *Nature* 421:625–627.

Mollenhoff, D. 2004. *Madison - A History of the Formative Years*. University of Wisconsin Press, Madison.

Morrow, P., and C. Hensel. 1992. Hidden dissension: Minority-majority relationships and the use of contested terminology. *Arctic Anthropology* 29(1):38–53.

Mullon, C., P. Freon, and P. Cury. 2005. The dynamics of collapse in world fisheries. *Fish and Fisheries* 6:111–120.

Münch, R., and N.J. Smesler. 1987. Relating the micro and macro. Pages 356–387 *in* J.C Alexander, B. Giesen, R. Munch, and N.J. Smesler, editors. *The Micro-Macro Link*. University of California Press, Berkeley.

Mwangi, E. 2006. The footprints of history: Path dependence in the transformation of property rights in Kenya's Maasailand. *Journal of Institutional Economics* 2:157–180.

Myers, N.A. 1993. Tropical forests: The main deforestation fronts. *Environmental Conservation* 20: 9–16.

Myers, N.A., R.A. Mittermeier, C.G. Mittermeier, G.A.B. da Fonseca, and J. Kent. 2000. Biodiversity hotspots for conservation priorities. *Nature* 403:853–858

Myers, R.A. 1998. When do environment-recruit correlations work? *Reviews in Fish Biology and Fisheries* 8:285–305.

Myers, R.A., K.G. Bowen, and N.J. Barrowman. 1999. Maximum reproductive rate of fish at low population sizes. *Canadian Journal of Fisheries and Aquatic Sciences* 56:2404–2419.

Myers, R.A., and B. Worm. 2003. Rapid worldwide depletion of predatory fish communiies. *Nature* 423:283.

Naeem, S. 1998. Species redundancy and ecosystem reliability. *Conservation Biology* 12:39–45.

Nagendra, H. 2007. Drivers of reforestation in human-dominated forests. *Proceedings of the National Academy of Sciences* 104:15218–15223.

Naiman, R.J., H. Décamps, and M.E. McClain. 2005. *Riparia*. Elsevier Academic Press, Amsterdam.

Narayan, D., R. Chambers, M.K. Shah, and P. Petesch. 2000. *Voices of the Poor: Crying Out for Change*. Oxford University Press, New York.

Naylor, R.L. 1991. The rural labor market in Indonesia. Pages 58–98 *in* S. Pearson, W. Falcon, P. Heytens, E. Monke, and R. Naylor, editors. *Rice Policies in Indonesia*. Cornell University Press. Ithaca.

Naylor, R.L. 1994. Culture and agriculture: Employment practices affecting women in Java's rice economy. *Economic Development and Cultural Change* 42:509–535.

Naylor, R.L. 2000. Agriculture and global change. Pages 462–475 in G. Ernst, editor. *Earth Systems: Processes and Issues.* Cambridge University Press, Cambridge.

Naylor, R.L., D.S. Battisti, D.J. Vimont, W.P. Falcon, and M.B. Burke, 2007a. Assessing risks of climate variability and climate change for Indonesian rice agriculture. *Proceedings of the National Academy of Sciences* 104:7752–7757.

Naylor, R.L., and M. Burke. 2005. Aquaculture and ocean resources: Raising tigers of the sea. *Annual Review of Environment and Resources* 30: 185–218.

Naylor, R.L., and P.R. Ehrlich. 1997. Natural pest control services and agriculture. Pages 151–174 in G.C. Daily, editor. *Nature's Services: Societal Dependence on Natural Ecosystems.* Island Press, Washington.

Naylor, R.L., and W.P. Falcon. In press. The role of policy in agricultural transition. in P.A. Matson, R.L. Naylor, I. Ortiz-Monasterio, and W.P. Falcon, editors. *Agriculture, Development and the Environment in the Yaqui Valley (Mexico).* Island Press, Washington.

Naylor, R.L., W. Falcon, and C. Fowler. 2007b. The conservation of global crop genetic resources in the face of climate change. Summary report from Plant Genetic Resources in the Face of Climate Change Conference, Bellagio, Italy, September 3–7, 2007. http://fse.stanford.edu/.

Naylor, R.L., W.P Falcon, R.M. Goodman, M.M. Jahn, T. Sengooba, et al. 2004. Biotechnology in the developing world: A case for increased investments in orphan crops. *Food Policy* 29(1): 15–44.

Naylor, R.L., R.J. Goldburg, J.H. Primavera, N. Kautsky, M.C.M. Beveridge, et al. 2000. Effect of aquaculture on world fish supplies. *Nature* 405:1017–1024.

Naylor, R.L., A. Liska, M. Burke, W. Falcon, J. Gaskell, et al. 2007c. The ripple effect: Biofuels, food security and the environment. *Environment* 49(9):30–43.

Naylor, R.L., H. Steinfeld, W. Falcon, J. Galloway, V. Smil, et al. 2005. Losing the links between livestock and land. *Science* 310:1621–1622.

Neis, B., and L. Felt. 2000. *Finding Our Sea Legs: Linking Fishery People and Their Knowledge with Science and Management.* Institute for Social and Economic Research, St. Johns, Canada.

Nelson, D.R., W.N. Adger, and K. Brown. 2007. Adaptation to environmental change: Contributions of a resilience framework. *Annual Review of Environment and Resources* 32:395–419.

Neuwirth, R. 2006. *Shadow Cities: A Billion Squatters: An Urban New World.* Routledge, New York.

Nicholls, N., and K.K. Wong. 1990. Dependence of rainfall variability on mean rainfall, latitude, and the southern oscillation. *Journal of Climate* 3: 163–170.

Niemela, J. 1999. Ecology and urban planning. *Biodiversity and Planning* 8:118–131.

NOAA (National Oceanic and Atmospheric Administration). 2004. Population Trends Along the Coastal United States: 1980–2008. U.S. Department of Commerce, Washington.

Norberg, J., J. Wilson, B.H. Walker, and E. Ostrom. 2008. Diversity and resilience of social-ecological systems. Pages 46–79 in J. Norberg and G.S. Cumming, editors. *Complexity Theory for a Sustainable Future.* Columbia University Press, New York.

Norgaard, R.B. 1989. The case for methodological pluralism. *Ecological Economics* 1:35–57.

North, D.C. 1990. *Institutions, Institutional Change and Economic Performance.* Cambridge University Press, New York.

North, D.C. 1991. Institutions. *Journal of Economic Perspectives* 5:97–112.

Northridge, M.E., E.D. Sclar, and P. Biswas. 2003. Sorting out the connection between the built environment and health: A conceptual framework for navigating pathways and planning healthy cities. *Journal of Urban Health: Bulletin of the New York Academy of Medicine* 80:556–568.

Nowak, D.J. 1994. Atmospheric carbon dioxide reduction by Chicago's urban forest. Pages 83–94 in E.G. McPherson, editor. *Chicago's Urban Forest Ecosystem: Results of the Chicago Urban Climate Project.* Northeastern Research Station, USDA Forest Service, Radnor, PA.

NRC (National Research Council). 1996. *The Bering Sea Ecosystem.* National Academy Press, Washington.

NRC (National Research Council). 1999. *Sharing the Fish: Toward a National Policy on Individual Fishing Quotas.* NRC (Committee on Ecosystem Management and Sustaining Marine Fisheries, Ocean Studies Board, Commission on Geosciences, Environment, and Resources). National Academy Press, Washington.

NRC (National Research Council). 2000. *Watershed Management for Potable Water Supply: Assessing the New York City Strategy.* National Academy Press, Washington.

NRC (National Research Council); E. Ostrom, T. Dietz, N. Dolšak, P.C. Stern, S. Stovich, et al., editors. 2002. *The Drama of the Commons.* National Academy Press, Washington.

NRC (National Research Council). 2003. *Decline of the Stellar Sea Lion in Alaskan Waters: Untangling Food Webs and Fishing Nets.* National Academy Press, Washington.

Nyström, M. 2006. Redundancy and response diversity of functional groups: Implications for the resilience of coral reefs. *Ambio* 35:30–35.

Nyström, M., and C. Folke. 2001. Spatial resilience of coral reefs. *Ecosystems* 4:406–417.

Odum, E.P. 1989. *Ecology and Our Endangered Life-Support Systems.* Sinauer Associates, Sunderland, MA.

OECD (Organization for Economic Cooperation and Development). 1997. *Towards Sustainable Fisheries: Economic Aspects of the Management of Living Marine Resources.* Organization for Economic Cooperation and Development, Paris.

O'Flaherty, R., I.J. Davidson-Hunt, and M. Manseau. 2008. Indigenous knowledge and values in planning for sustainable forestry: Pikangikum First Nation and the Whitefeather Forest Initiative. *Ecology and Society* 13(1):6. http://www.ecologyandsociety.org/vol13/iss1/art6/

Oki, T., and S. Kanae. 2006. Global hydrological cycles and world water resources. *Science* 313:1068–1072.

Oldfield, F., and W. Steffen. 2004. The earth system. Page 7 *in* W. Steffen, A. Sanderson, P. Tyson, J. Jäger, P. Matson, et al., editors. *Global Change and the Earth System: A Planet Under Pressure.* IGBP Global Change Series. Springer-Verlag, New York.

Olsen, J.P. 2001. Garbage cans, new institutionalism and the study of politics. *American Political Science Review* 95:191–198.

Olsson, P., C. Folke, and F. Berkes. 2004a. Adaptive co-management for building resilience in social-ecological systems. *Environmental Management* 34:75–90.

Olsson, P., C. Folke, and T. Hahn. 2004b. Social-ecological transformation for ecosystem management: The development of adaptive co-management of a wetland landscape in southern Sweden. *Ecology and Society* 9(4):2. www.ecologyandsociety.org/vol9/iss4/art2/.

Olsson, P., C. Folke, and T.P. Hughes. 2008. Navigating the transition to ecosystem-based man-agement of the Great Barrier Reef, Australia. *Proceeding of the National Academy of Sciences* 105:9489–9494.

Olsson, P., L.H. Gunderson, S.R. Carpenter, P. Ryan, L. Lebel, et al. 2006. Shooting the rapids: Navigating transitions to adaptive governance of social-ecological systems. *Ecology and Society* 11(1):18. http://www.ecologyandsociety.org/vol11/iss1/art18/.

Oppenheimer, S. 2004. *Out of Eden: The Peopling of the World.* Constable, London.

Ostfeld, R.S., and F. Keesing. 2000. Biodiversity and disease risk: The case of Lyme disease. *Conservation Biology* 14:722–728.

Ostrom, E. 1990. *Governing the Commons: The Evolution of Institutions for Collective Action.* Cambridge University Press, Cambridge.

Ostrom, E. 1999a. Coping with the tragedies of the commons. *Annual Review of Political Science* 2:493–535.

Ostrom, E. 1999b. Institutional rational choice: An assessment of the institutional analysis and development framework. Pages 35–71 *in* P.A. Sabatier, editor. *Theories of the Policy Processes.* Westview Press, Boulder.

Ostrom, E., 2005. *Understanding Institutional Diversity.* Princeton University Press, Princeton.

Ostrom, E., M.A. Janssen, and J.M. Anderies. 2007. Going beyond panaceas. *Proceeding of the National Academy of Sciences* 104:15176–15178.

Otterman, J., A. Manes, S. Rubin, P. Alpert, and D.O. Starr. 1990. An increase of early rains in southern Israel following land-use change. *Boundary-Layer Meteorology* 53:333–351.

Page, S.E. 2007. Uncertainty, difficulty, and complexity. *Journal of Theoretical Politics* 20:115–149.

Palm, C.A., S.A. Vosti, P.A. Sanchez, and P.J. Ericksen. 2005. *Slash-and-Burn Agriculture: The Search for Alternatives.* Columbia University Press. New York.

Paloheimo, J.E., and L. Dickie. 1964. Abundance and fishing success. *Rapports et Proces-verbaux des Reunions, Council International pour l'Exploration de la Mer* 155:152–63.

Parry, M., C. Rosenzweig, A. Iglesias, M. Livermore, and G. Fischer. 2004. Effects of climate change on global food production under SRES emissions and socio-economic scenarios. *Global Environmental Change* 14:53–67.

Patz, J.A., U.E.C. Confalonieri, F.P. Amerasinghe, K.B. Chua, P. Daszak, et al. 2005. Human health: Ecosystem regulation of infectious diseases. Pages 391–415 *in* R. Hassan, R. Scholes, and N. Ash, editors. *Ecosystems and Human Well-Being: Current*

State and Trends. Millennium Ecosystem Assessment. Island Press, Washington.

Patz, J.A., P. Daszak, G.M. Tabor, A.A. Aguirre, and M. Pearl. 2004. Unhealthy landscapes: Policy recommendations on land use change and infectious disease emergence. *Environmental Health Perspectives* 101:1092–1098.

Pauly, D. 1995. Anecdotes and the shifting baseline syndrome. *Trends in Ecology & Evolution* 10:430.

Pauly, D., J. Alder, A. Bakun, S. Heileman, K.-H. Kock, et al. 2005. Marine fisheries systems. Pages 477–511 *in* R. Hassan, R. Scholes, and N. Ash, editors. *Ecosystems and Human Well-Being: Current State and Trends. Millennium Ecosystem Assessment*. Island Press, Washington.

Pellow, D.N. 2000. Environmental inequality formation: Toward a theory of environmental injustice. *American Behavioral Scientist* 43:581–601.

Perera, A.J., L.J. Buse, and M.G. Weber. 2004. *Emulating Natural Forest Landscape Disturbances: Concepts and Applications*. Columbia University Press, New York.

Peterson, G.D., G.S. Cumming, and S.R. Carpenter. 2003. Scenario planning: A tool for conservation in an uncertain world. *Conservation Biology* 17:358–366.

Peterson, J.H. 1994. Sustainable wildlife use for community development in Zimbabwe. Pages 99–111 *in* M.M.R. Freeman and U.P. Kreuter, editors. *Elephants and Whales: Resources for Whom?* Gordon and Breach Publishers, Basel.

Petit, J.R., J. Jouzel, D. Raynaud, N.I. Barkov, J.M. Barnola, et al. 1999. Climate and atmospheric history of the past 420,000 years from the Vostok ice core, Antarctica. *Nature* 399:429–436.

Petschel-Held, G., A. Block, M. Cassel-Gintz, J. Kropp, M.K.B. Ludeke, et al. 1999. Syndromes of global change: A qualitative modeling approach to assist global environmental management. *Environmental Modeling and Assessment* 4: 295–314.

Pfeffer, J. 1981. *Power in Organizations*. Harvard Business, Stanford.

Pickett, S.T.A., W.R. Burch, and S. Dalton. 1997a. Integrated urban ecosystem research. *Urban Ecosystems* 1:183–184.

Pickett, S.T.A., W.R. Burch, S. Dalton, T. Foresman, J.M. Grove, et al. 1997b. A conceptual framework for the study of human ecosystems in urban areas. *Urban Ecosystems* 1:185–199.

Pickett, S.T.A., and M. Cadenasso. 2007. Linking ecological and built components of urban mosaics: An open cycle of ecological design. *Ecology Future Directions* 1:1–5.

Pickett, S.T.A., and M. Cadenasso. 2008. Patch dynamics as a conceptual tool to link ecology and design. Pages 94–103 *in* B. McGrath, V. Marshall, M.L. Cadenasso, J.M. Grove, S.T.A. Pickett, et al., editors. *Designing Patch Dynamics*. Columbia University, New York.

Pickett, S.T.A., M.L. Cadenasso, and J.M. Grove. 2004. Resilient cities: Meaning, models and metaphor for integrating the ecological, socio-economic, and planning realms. *Landscape and Urban Planning* 69:369–384.

Pickett, S.T.A., M.L. Cadenasso, J.M. Grove, C.H. Nilon, R.V. Pouyat, et al. 2001. Urban ecological systems: Linking terrestrial ecological, physical, and socioeconomic components of metropolitan areas. *Annual Review of Ecology and Systematics* 32:127–157.

Pickett, S.T.A., P. Groffman, M.L. Cadenasso, J.M. Grove, L.E. Band, et al. 2008. Beyond urban legends: An emerging framework of urban ecology as illustrated by the Baltimore Ecosystem Study. *BioScience* 58:139–150.

Pickett, S.T.A., and P.S. White. 1985. *The Ecology of Natural Disturbance as Patch Dynamics*. Academic Press, New York.

Pimentel, D., L. Lach, R. Nuniga, and D. Morrison. 2000. Environmental and economic costs of non-indigenous species in the United States. *BioScience* 50:53–65.

Pinkerton, E.W. 1987. The fishing-dependent community. Pages 293–325 *in* P. Marchak, N. Guppy, and J. McMullan, editors. *Uncommon Property: The Fishing and Fish Processing Industries of British Columbia*. University of British Columbia Press, Vancouver.

Pinkerton, E.W. 1989a. Attaining better fisheries management through co-management: Prospects, problems, and propositions. Pages 3–33 *in* E.W. Pinkerton, editor. *Co-operative Management of Local Fisheries: New Directions for Improved Management and Community Development*. University of British Columbia Press, Vancouver.

Pinkerton, E.W., editor. 1989b. *Co-operative Management of Local Fisheries: New Directions for Improved Management and Community Development*. University of British Columbia Press, Vancouver.

Pinkerton, E.W. 1991. Locally based water quality planning: Contributions to fish habitat protection. *Canadian Journal of Fisheries and Aquatic Sciences* 48:1326–1333.

Pinkerton, E.W. 1992. Translating legal rights into management practice: Overcoming barriers to the exercise of co-management. *Human Organization* 51:330–341.

Pinkerton, E.W. 1993. Analyzing co-management efforts as social movements: The Tin-Wis coalition and the drive for forest practice legislation in British Columbia. *Alternatives* 19(3):33–38.

Pinkerton, E.W. 2003. Toward specificity in complexity: Understanding co-management from a social science perspective. Pages 61–77 *in* D.C. Wilson, J.R. Nielsen, and P. Degnbol, editors. *The Fisheries Co-Management Experience: Accomplishments, Challenges and Prospects.* Kluwer, Dordrecht.

Pinkerton, E. and D. Edwards. 2009. The elephant in the room: The hidden costs of leasing Individual Transferable Fishing Quotas. *Marine Policy* 33 (in press).

Pinkerton, E.W., and L. John. 2008. Creating local management legitimacy. *Marine Policy* 32: 680–691.

Pinkerton, E.W., and M. Weinstein. 1995. *Fisheries that Work: Sustainability through Community-Based Management.* The David Suzuki Foundation, Vancouver.

Plummer, R., and D. Fennel. 2007. Exploring co-management theory: Prospects for sociobiology and reciprocal altruism. *Journal of Environmental Management* 85:944–955.

Poff, N.L., J.D. Allan, M.A. Palmer, D.D. Hart, B. Richter, et al. 2003. River flows and water wars: Emerging science for environmental decision making. *Frontiers in Ecology and the Environment* 1:298–306.

Pomeroy, R.S. 2003. The government as a partner in co-management. Pages 247–261 *in* D.C.Wilson., J.R. Nielsen, and P. Degnbol, editors. *The Fisheries Co-Management Experience: Accomplishments, Challenges and Prospects.* Kluwer, Dordrecht.

Pomeroy, R.S., and F. Berkes. 1997. Two to tango: The role of government in fisheries co-management. *Marine Policy* 21:465–480.

Pomeroy, R.S., and K. Viswanathan. 2003. Experiences with fisheries co-management in Southeast Asia and Bangladesh. Pages 99–117 *in* D.C. Wilson, J.R. Nielsen, and P. Degnbol, editors. *The Fisheries Co-Management Experience: Accomplishments, Challenges and Prospects.* Kluwer, Dordrecht.

Portney, P., and J. Weyant, editors. 1999. *Discounting and Intergenerational Equity.* Resources for the Future, Washington.

Posey, D.A., editor. 1999. *Cultural and Spiritual Values of Biodiversity.* UNEP and Intermediate Technology Publications, Nairobi.

Post, J.R., M. Sullivan, S.P. Cox, N.P. Lester, C.J. Walters, et al. 2002. Canada's recreational fisheries: The invisible collapse? *Fisheries* 27:6–17.

Postel, S.L. 1999. *Pillar of Salt: Can the Irrigation Miracle Last?* W.W. Norton, New York.

Postel, S.L. 2001. Growing more food with less water. *Scientific American* 284(2):46–49.

Postel, S.L. 2005. *Liquid Assets.* Worldwatch Institute, Washington.

Postel, S.L., and B. Richter. 2003. *Rivers for Life: Managing Water for People and Nature.* Island Press, Washington.

Postel, S.L., G.C. Daily, and P.R. Ehrlich. 1996. Human appropriation of renewable fresh water. *Science* 271:785–788.

Power, A.G. 1999. Linking ecological sustainability and world food needs. *Environment, Development, and Sustainability* 1:185–196.

Power, A.G., and A.S. Flecker. 1996. The role of biodiversity in tropical managed ecosystems. Pages 173–194 *in* G.H. Orians, R. Dirzo, and J.H. Cushman, editors. *Biodiversity and Ecosystem Processes in Tropical Forests.* Springer-Verlag, New York.

Pretty, J. 2002. *Agriculture: Reconnecting People, Land and Nature.* Earthscan Publications Ltd., London.

Pringle, H.J.R., and K.L. Tinley. 2003. Are we overlooking critical geomorphic determinants of landscape change in Australian rangelands? *Ecological Management and Restoration* 4:180–186.

Pringle, R.M. 2005a. The Nile perch in Lake Victoria: Local responses and adaptations. *Africa* 75:510–538.

Pringle, R.M. 2005b. The origins of the Nile Perch in Lake Victoria. *BioScience* 55:780–787.

Putnam, R.D. 2000. *Bowling Alone: The Collapse and Revival of American Community.* Simon & Schuster, New York.

Pyne, S.J. 1997. *World Fire: The Culture of Fire on Earth.* University of Washington Press, Seattle.

Quinn, J. 1985. Managing innovation: Controlled chaos. *Harvard Business Review*, May-June 63: 73–84.

Rabalais, N.N., R.E. Turner, and W.J. Wiseman, Jr. 2002. Gulf of Mexico hypoxia, A.K.A. "The dead zone". *Annual Review of Ecology and Systematics* 33:235–263.

Radeloff, V.C., R.B. Hammer, S.I. Stewart, J.S. Fried, S.S. Holcomb, et al. 2005. The wildland-urban interface in the United States. *Ecological Applications* 15:799–805.

Rahel, F. 2000. Homogenization of fish faunas across the United States. *Science* 288:854–856.

Raivio, S., E. Normark, B. Pettersson, and P. Salpakivi-Salomaa. 2001. Science and the management of boreal forest biodiversity: Forest industries' views. *Scandinavian Journal of Forest Research Supplement* 3:99–104.

Ramakrishnan, P.S. 1992. *Shifting Agriculture and Sustainable Development: An Interdisciplinary Study from North-Eastern India*. Parthenon Publishing Group, Park Ridge, NJ.

Ramakrishnan, P.S., K.G. Saxena, and U.M. Chandrashekara, editors. 1998. *Conserving the Sacred for Biodiversity Management*. Oxford and IBH, New Delhi.

Rahmstorf, S., A. Cazenave, J.A. Church, J.E. Hansen, R.F. Keeling, et al. 2007. Recent climate observations compared to projections. *Science* 316:709.

Ranjan, R., and V.P. Upadhyay. 1999. Ecological problems due to shifting cultivation. *Current Science* 77:1246–50.

Rappaport, R.A. 1967. *Pigs for the Ancestors*. Yale University Press, New Haven.

Raupach, M.R., G. Marland, G., P. Ciais, P., C. Le Quere, J.G. Canadell, et al. 2007. Global and regional drivers of accelerating CO_2 emissions. *Proceeding of the National Academy of Sciences* 104:10288–10293.

Reardon, T., and C.P Timmer. 2007. Transformation of markets for agricultural output in developing countries since 1950: How has the thinking changed? Pages 2807–2855 *in* R. Evenson and P. Pingali, editors. *Handbook of Agricultural Economics*, Volume 3. Elsevier Press, Amsterdam.

Redman, C.L. 1999. *Human Impact on Ancient Environments*. University of Arizona Press, Tucson.

Regier, H.A., and G.L. Baskerville.1986. Sustainable redevelopment of regional ecosystems degraded by exploitive development. Pages 75–101 *in* W.C. Clark, and R.E. Munn, editors. *Sustainable Development of the Biosphere*. Cambridge University Press, London.

Renault, D., and W.W. Wallender. 2000. Nutritional water productivity and diets. *Agricultural Water Management* 45:275–296.

Repetto, R., editor. 2006. *Punctuated Equilibrium and the Dynamics of U.S. Environmental Policy*. Yale University Press, New Haven.

Rerkasem, B., editor. 1996. *Montane Mainland Southeast Asia in Transition Chiang Mai University Consortium*. Chiang Mai, Thailand.

Reynolds, J.E., and D.F. Greboval. 1988. *Socioeconomic Effects of the Evolution of the Nile Perch Fisheries in Lake Victoria: A Review*. FAO, Rome, Italy.

Reynolds, J.F., and D.M. Stafford Smith, editors. 2002. *Global Desertification: Do Humans Cause Deserts?* Dahlem University Press, Berlin.

Reynolds, J.F., D.M. Stafford Smith, E.F. Lambin, B.L. Turner, II, M. Mortimore, et al. 2007. Global desertification: Building a science for dryland development. *Science* 316:847–851.

Ricketts, T.H., G.C. Daily, P.R. Ehrlich, and C.D. Michener. 2004. Economic value of tropical forest to coffee production. *Proceedings of the National Academy of Sciences* 101:12579–12582.

Ridgwell, A.J., and A.J. Watson. 2002. Feedback between aeolian dust, climate, and atmospheric CO_2 in glacial time. *Paleoceanography* 17:1059, doi:10.1029/2001PA000729, 2002

Rietkerk, M., S.C. Dekker, P.C. de Ruiter, and J. van de Koppel. 2004. Self-organized patchiness and catastrophic shifts in ecosystems. *Science* 305:2004–2009.

Robertson, G.P., and S.M. Swinton. 2005. Reconciling agricultural productivity and environmental integrity: A grand challenge for agriculture. *Frontiers in Ecology and the Environment* 3:38–46.

Robinson, J.B., editor. 1996. *Life in 2030: Exploring a Sustainable Future for Canada*. University of British Columbia Press, Vancouver.

Robinson, J.G., and E.L. Bennett, editors. 2000. *Hunting for Sustainability in Tropical Forests*. Columbia University Press, New York.

Rockström, J., L. Gordon, C. Folke, M. Falkenmark, and M. Engwall. 1999. Linkages among water vapor flows, food production, and terrestrial ecosystem services. *Conservation Ecology* 3:5. http://www.consecol.org/vol3/iss2/art5/.

Rose, G.A., B. deYoung, D.W. Kulka, S.V. Goddard, and G.L. Fletcher. 2000. Distribution shifts and overfishing the northern cod (*Gadus morhua*): A view from the ocean. *Canadian Journal of Fisheries and Aquatic Sciences* 57:644–663.

Ross, M.L. 2008. Blood barrels: Why oil wealth fuels conflict. *Foreign Affairs* 87:2–8.

Rothschild, B.J., J.S. Ault, P. Goulletquer, and M. Heral. 1994. Decline of the Chesapeake Bay oyster population: A century of habitat destruction and overfishing.*Marine Ecology Progress Series* 111:29–39.

Ruthenberg, H., J.D. MacArthur, H.D. Zandstra, and M.P. Collinson, 1980. *Farming Systems in the Tropics*. 3rd Edition. Clarendon, Oxford.

Ruttan, V.W., and Y. Hyami. 1984. Induced innovation model of agricultural development. Pages 97–114 *in* C.K. Eicher and J.M. Staatz, editors. *Agricultural Development in the Third World*. Johns Hopkins University Press, Baltimore.

Safriel, U., Z. Adeel, D. Niemeijer, J. Puigdefabregas, R. White, et al. 2005. Dryland Systems. Pages 623–662 *in* R. Hassan, R. Scholes, and N. Ash, editors. *Ecosystems and Human Well-Being: Current*

State and Trends. Millennium Ecosystem Assessment. Island Press, Washington.

Sainsbury, K.J. 1988. The ecological basis of multi-species fisheries and management of a demersal fishery in Tropical Australia. Pages 349–382 *in* J. Gulland, editor. *Fish Population Dynamics.* John Wiley & Sons, New York.

Sainsbury, K.J., R.A. Campbell, R. Lindholm, and A.W. Whitelaw. 1997. Experimental management of an Australian multispecies fishery: Examining the possibility of trawl-induced habitat modification. Pages 107–112 *in* E.K. Pikitch, D.D. Huppert, and M.P. Sissenwine, editors. *Global Trends in Fisheries Management.* American Fisheries Society, Bethesda, MD.

Sampson, R.N., N. Bystriakova, S. Brown, P. Gonzalez, L.C. Irland, et al. 2005. Timber, fuel, and fiber. Pages 585–621 *in* R. Hassan, R. Scholes, and N. Ash, editors. *Ecosystems and Human Well-Being: Current State and Trends. Millennium Ecosystem Assessment.* Island Press, Washington.

Sandford, S. 1983. *Management of Pastoral Development in the Third World.* John Wiley, London.

Sarkar, S., R.L. Pressey, D.P. Faith, C.R. Margules, T. Fuller, et al. 2006. Biodiversity conservation planning tools: Present status and challenges for the future. *Annual Review of Environment and Resources* 31:123–159.

Sass, G.G., J.F. Kitchell, S.R. Carpenter, T.R. Hrabik, A.E. Marburg, et al. 2006. Fish community and food web responses to a whole-lake removal of coarse woody habitat. *Fisheries* 31:321–330.

Schaaf, T., X. Zhao, and G. Keil. 1995. *Towards a Sustainable City: Methods of Urban Ecological Planning and Its Application in Tianjin, China.* Urban System Consult GmbH, Berlin.

Scheffer, M., S.R. Carpenter, J. Foley, C. Folke, and B. Walker. 2001. Catastrophic shifts in ecosystems. *Nature* 413:591–596.

Scheffer, M., and E.H. van Nes. 2004. Self-organized similarity: The evolutionary emergence of groups of similar species. *Proceedings of the National Academy of Sciences* 103:6230–6235.

Scheffer, M., and F.R. Westley. 2007. The evolutionary basis of rigidity: Locks in cells, minds, and society. *Ecology and Society* 12(2):36. http://www.ecologyandsociety.org/vol12/iss2/art36/.

Schelling, T.C. 1978. *Micromotives and Macrobehavior.* W.W. Norton, New York.

Schellnhuber, H.-J. 2002. Coping with Earth System complexity and irregularity. Pages 151–156 *in* W. Steffen, J. Jaeger, D.J. Carson, and C. Bradshaw, Clare, editors. *Challenges of a Changing Earth.* Springer-Verlag, Berlin.

Schlager, E., and E. Ostrom. 1993. Property rights regimes and coastal fisheries: An empirical analysis. Pages 13–41 *in* T.L. Anderson and R.T. Simmons, editors. *The Political Economy of Customs and Culture: Informal Solutions to the Commons Problem.* Rowman and Littlefield Publishers, Lantham, MD.

Schlesinger, W.H. 1997. *Biogeochemistry: An Analysis of Global Change.* 2nd Edition. Academic Press, San Diego.

Schmidhuber, J. 2007. *Biofuels: An Emerging Threat to Europe's Food Security?* Notre-Europe. aris, France. http://www.notre-europe.eu/uploads/tx_publication/Policypaper-Schmidhuber-EN.pdf.

Schnore, L.F. 1958. Social morphology and human ecology. *American Journal of Sociology* 63:620–624, 629–634.

Schoennagel, T., T.T. Veblen, and W.H. Romme. 2004. The interaction of fire, fuels, and climate across Rocky Mountain forests. *BioScience* 54:661–676.

Schreiber, D. 2003. Salmon farming and salmon people: Identity and environment in the Leggatt Inquiry. *American Indian Culture and Research Journal* 27(4):79–103.

Schreiber, D. 2006. First Nations, consultation, and the rule of law: Salmon farming and colonialism in British Columbia. *American Indian Culture and Research Journal* 30(4):19–40.

Schreiber, E.S.G., A.R. Bearlin, S.J. Nicol, and C.R. Todd. 2004. Adaptive management: A synthesis of current understanding and effective application. *Ecological Management and Restoration* 5:177–182.

Schultz, L., C. Folke, and P. Olsson. 2007. Enhancing ecosystem management through social-ecological inventories: Lessons from Kristianstads Vattenrike, Sweden. *Environmental Conservation* 34:140–152.

Schulze, E.-D., C. Wirth, and M. Heimann. 2000. Climate change: Managing forests after Kyoto. *Science* 289:2058–2059.

Scoones, I. 1995. Exploiting heterogeneity: Habitat use by cattle in dryland Zimbabwe. *Journal of Arid Environments* 29:221–237.

Scott, W.R. 2000. *Institutions and Organizations.* Sage Press, Thousand Oaks, CA.

Selby, M.J. 1993. *Hillslope Materials and Processes.* 2nd Edition. Oxford University Press, Oxford.

Sen, A.K. 1981. *Poverty and Famines: An Essay on Entitlement and Deprivation.* Clarendon, Oxford.

Senge, P.M. 1990. *The Fifth Discipline: The Art and Practice of the Learning Organization.* Doubleday/Currency, New York.

Shah, T., D. Molden, R. Sakthivadivel, and D. Seckler. 2000. *The Global Groundwater Situation: Overview of Opportunities and Challenges*. International Water Management Institute, Colombo, Sri Lanka.

Shane, G.D. 2005. *Recombinant Urbanism: Conceptual Modeling in Architecture, Urban Design, and City Theory*. John Wiley & Sons Ltd., Chichester, UK.

Shertzer, K.W., and M.H. Prager. 2007. Delay in fishery management: Diminished yield, longer rebuilding, and increased probability of stock collapse. *ICES Journal of Marine Science* 64:149–159.

Shindler, B.A., and L.A. Cramer. 1999. Shifting public values for forest management: Making sense of wicked problems. *Western Journal of Applied Forestry* 14:28–34.

Shumway, D.L., and G.G. Chandwick. 1971. Influence of kraft mill effluent on the flavor of salmon flesh. *Water Research* 5:997–1003.

Shvidenko, A., D.V. Barber, R. Persson, P. Gonzalez, R. Hassan, et al. 2005. Forest and woodland systems. Pages 585–621 in R. Hassan, R. Scholes, and N. Ash, editors. *Ecosystems and Human Well-Being: Current State and Trends. Millennium Ecosystem Assessment*. Island Press, Washington.

Singer, P. 1993. *Practical Ethics*. Cambridge University Press, Cambridge.

Singleton, S. 1998. *Constructing Cooperation: The Evolution of Institutions of Co-Management*. University of Michigan Press, Ann Arbor.

Smil, V. 2000. *Feeding the World: A Challenge for the 21st Century*. MIT Press, Cambridge, MA.

Smil, V. 2002. Eating meat: Evolution, patterns, and consequences. *Population and Development Review* 28:599–639.

Smit, B., and O. Pilifosova. 2003. From adaptation to adaptive capacity and vulnerability reduction. Pages 9–28 in J.B. Smith, R.J.T. Klein, and S. Huq, editors. *Climate Change, Adaptive Capacity and Development*. Imperial College Press, London.

Smit, B., and J. Wandel. 2006. Adaptation, adaptive capacity and vulnerability. *Global Environmental Change* 16:282–292.

Smith, B.D. 1998. *The Emergence of Agriculture*. Scientific American Library, New York.

Solow, R.M. 1991. Sustainability: An economist's perspective. Eighteenth J. Seward Johnson Lecture, Marine Policy Center, Woods Hole Oceanographic Institution, Woods Holes, MA. Pages 179–187 in R. Dorfman and N.S. Dorfman, editors. *Economics of the Environment: Selected Readings*. W.W. Norton, New York.

Songorwa, A.N., T. Buhrs, and K.F.D. Hughey. 2000. Community-based wildlife management in Africa: A critical assessment of the literature. *Natural Resources Journal* 40:603–643.

Spirn, A.W. 1984. *The Granite Garden: Urban Nature and Human Design*. Basic Books, Inc., New York.

Stafford Smith, D.M. 2003. Linking environments, decision-making and policy in handling climatic variability. Pages 131–151 in L. Botterill and M. Fisher, editors. *Beyond Drought in Australia: People, Policy and Perspectives*. CSIRO Publishing, Melbourne.

Stafford Smith, M. 2008. The 'desert syndrome:' Causally-linked factors that characterise outback Australia. *The Rangeland Journal* 30:3–14.

Stafford Smith, D.M., G.M. McKeon, I.W. Watson, B.K. Henry, G.S. Stone, et al. 2007. Learning from episodes of degradation and recovery in variable Australian rangelands. *Proceedings of the National Academy of Sciences* 104:20690–20695.

Stankey, G.H., B.T. Bormann, C. Ryan, B. Shindler, V. Sturtevant, et al. 2003. Adaptive management and the Northwest Forest Plan: Rhetoric and reality. *Journal of Forestry* 101:40–46.

Starfield, A.M., K.A. Smith, and A.L. Bleloch. 1990. *How to Model it: Problem Solving for the Computer Age*. McGraw Hill Inc., New York.

Stearns, F., and T. Montag, editors. 1974. *The Urban Ecosystem: A Holistic Approach*. Dowden, Hutchinson & Ross, Inc., Stroudsburg, PA.

Steffen, W.L., A. Sanderson, P.D. Tyson, J. Jäger, P.A. Matson, et al. 2004. *Global Change and the Earth System: A Planet Under Pressure*. Springer-Verlag, New York.

Steinfeld, H., P. Gerber, T. Wassenaar, V. Castel, M. Rosales, et al. 2006. *Livestock's Long Shadow: Environmental Issues and Options*. U.N. Food and Agricultural Organization, Rome, Italy.

Steinfeld, H., and T. Wassenaar. 2007. The role of livestock production in carbon and nitrogen cycles. *Annual Review of Environment and Resources* 32:271–294.

Stephens, S.L. 1998. Evaluation of the effects of silviculture and fuels treatments on potential fire behavior in Sierra Nevada mixed-conifer forests. *Forest Ecology and Management* 105: 21–35.

Stern, N.H. 2007. *The Economics of Climate Change: The Stern Review*. Cambridge University Press, Cambridge.

Stevens, S., editor. 1997. *Conservation Through Cultural Survival: Indigenous Peoples and Protected Areas*. Island Press, Washington.

Stiassny, M.L., and A. Meyer. 1999. Cichlids of the rift lakes. *Scientific American* 280:64–69.

Stiglitz, J.E. 2002. *Globalization and Its Discontents.* W.W. Norton, New York.

Stiglitz, J.E. 2006. *Making Globalization Work.* W.W. Norton, New York.

Stoffle, B.W., D. Halmo, R. Stoffle, and G. Burpee. 1994. Folk management and conservation ethics among small-scale fishers of Buen Hombre, Dominican Republic. Pages 115–138 *in* C.L. Dyer and J.R. McGoodwin, editors. *Folk Management in the World's Fisheries: Lessons for Modern Fisheries Management.* University Press of Colorado, Niwot, CO.

Straussfogel, D. 1997. World-systems theory: Toward a heuristic and pedagogic conceptual tool. *Economic Geography* 73:118–130.

Suding, K.N., S. Lavorel, F.S. Chapin, III, J.H.C. Cornelissen, S. Diaz, et al. 2008. Scaling environmental change through the community level: A trait-based response-and-effect framework for plants. *Global Change Biology* 14:1125–1140.

Sukopp, H., S. Hejny, and I. Kowarik, editors. 1990. *Urban Ecology: Plants and Plant Communities in Urban Environments.* SPB Academic Publishing, The Hague.

Swain, D.P., and Sinclair, A.F. 1994. Fish distribution and catchability: What is the appropriate measure of distribution? *Canadian Journal of Fisheries and Aquatic Sciences* 51:1046–1054.

Swanson, F.J. 2004. Roles of scientists in forestry policy and management: Views from the Pacific Northwest. Pages 112–126 *in* K. Arabas and J. Bowersox, editors. *Forest Futures: Science, Politics, and Policy for the Next Century.* Rowman & Littlefield Publishers, Inc., Lanham, MD.

Swanson, F.J., C. Goodrich, and K.D. Moore. 2008. Bridging boundaries: Scientists, creative writers, and the long view of the forest. *Frontiers in Ecology and the Environment* 6:499–504.

Swezey, S., and R. Heizer. 1977. Ritual management of salmonid fish resources in California. *Journal of California Anthropology* 4(1):6–29.

Szaro, R.C., N.C. Johnson, W.T. Sexton, and A.J. Malk, editors. 1999. *Ecological Stewardship: A Common Reference for Ecosystem Management.* Elsevier Science Ltd, Oxford.

Tainter, J. 1988. *The Collapse of Complex Societies.* Cambridge University Press, Cambridge.

Tansley, A.G. 1935. The use and abuse of vegetational concepts and terms. *Ecology* 16:284–307.

Taylor, B.R., editor. 2005. *Encyclopedia of Religion and Nature.* Thoemmes Continuum, London.

Taylor, D.E. 2000. The rise of the environmental justice paradigm: Injustice framing and the social construction of environmental discourses. *American Behavioral Scientist* 43:508–580.

Taylor, L.H., S.M. Latham, and M.E.J. Woolhouse. 2001. Risk factors for human disease emergence. *Philosophical Transactions of the Royal Society of London, Series B* 356:983–989.

Taylor, M., and S. Singleton. 1993. The communal resource: Transaction costs and the solution of collective action problems. *Politics and Society* 21:195–214.

Thalenberg, E. 1998. Fisheries Beyond the Crisis. Canadian Broadcasting Company documentary film. 55 minutes. Box 500 Station A, Toronto, Ontario M5W 1E6.

Thom, R.M. 2000. Adaptive management of coastal ecosystem restoration projects. *Ecological Engineering* 15:365–372.

Thomas, S.J. 1994. Seeking equity in common property wildlife in Zimbabwe. Pages 129–142 *in* M.M.R. Freeman and U.P. Kreuter, editors. *Elephants and Whales: Resources for Whom?* Gordon and Breach Publishers, Basel.

Thompson, G.E., and F.R. Steiner, editors. 1997. *Ecological Design and Planning.* John Wiley & Sons, Inc., New York.

Tietenberg, T. 2002. The tradable permits approach to protecting the commons: What have we learned?" Pages 197–232 *in* NRC (National Research Council); E. Ostrom, T. Dietz, N. Dolšak, P.C. Stern, S. Stovich, et al., editors. *The Drama of the Commons.* National Academy Press, Washington.

Tiffin, M., and M. Mortimore. 1994. Malthus controverted: The role of capital and technology in growth and environment recovery in Kenya. *World Development* 22:997–1010.

Tilman, D., K. Cassman, P. Matson, R. Naylor, and S. Polasky. 2002. Agricultural sustainability and intensive production practices. *Nature* 418: 671–677.

Tilman, D., J. Hill, and C. Lehman. 2006. Carbon-negative biofuels from low-input high-diversity grassland biomass. *Science* 314:1598–1600.

Timmer, C.P. 1975. The political economy of rice in Asia: Indonesia. *Food Research Institute Studies* 14:197–231.

Timmer, C.P. 2005. *Operationalizing Pro-poor Growth: A Case Study for the World Bank on Indonesia.* The World Bank, Washington.

Timmer, C.P., W. Falcon, and S. Pearson. 1983. *Food Policy Analysis.* The Johns Hopkins University Press, Baltimore.

Timmer, V., and C. Juma. 2005. Biodiversity conservation and poverty reduction come together in the tropics: Lessons from the Equator Initiative. *Environment* 47(4):24–47.

Toepfer, K. 2005. From the Desk of Klaus Toepfer, United Nations Under-Secretary-General and Executive Director, UNEP. Our Planet.

Tongway, D.J., and J.A. Ludwig. 1997. The conservation of water and nutrients within landscapes. Pages 13–22 in J.A. Ludwig, D.J. Tongway, D.O. Freudenberger, J.C. Noble, and K.C. Hodgkinson, editors. *Landscape Ecology Function and Management: Principles from Australia's Rangelands.* CSIRO Publishing, Melbourne.

Troy, A., and J.M. Grove. 2008. Property values, parks, and crime: A hedonic analysis in Balitmore, MD. *Landscape and Urban Planning* 87: 233–245.

Turner, B.L., II, W.C. Clark, R.W. Kates, J.F. Richards, J.T. Mathews, et al., editors. 1990. *The Earth as Transformed by Human Action: Global and Regional Changes in the Biosphere over the Past 300 Years.* Cambridge University Press, Cambridge.

Turner, B.L., II, R.E. Kasperson, P.A. Matson, J.J. McCarthy, R.W. Corell, et al. 2003. A framework for vulnerability analysis in sustainability science. *Proceedings of the National Academy of Sciences* 100:8074–8079.

Turner, B.L., II, and S.R. McCandless. 2004. How humankind came to rival nature: A brief history of the human-environment condition and the lessons learned. Pages 227–243 in W.C. Clark, P. Crutzen, and H.-J. Schellnhuber, editors. *Earth System Analysis for Sustainability.* MIT Press, Cambridge, MA.

Turner, M.G., R.H. Gardner, and R.V. O'Neill. 2001. *Landscape Ecology in Theory and Practice: Pattern and Process.* Springer-Verlag, New York.

Turner, N.J., I.J. Davidson-Hunt, and M. O'Flaherty. 2003. Living on the edge: Ecological and cultural edges as sources of diversity for social-ecological resilience. *Human Ecology* 31:439–463.

Turner, N.J., M.B. Ignace, and R. Ignace. 2000. Traditional ecological knowledge and wisdom of aboriginal peoples in British Columbia. *Ecological Applications* 10:1275–1287.

Tversky, A., and D. Kahneman. 1974. Judgment under uncertainty: Heuristics and biases. *Science* 185:1124–1131

Ulrich, R.S. 1983. Aesthetic and affective response to natural environment. Pages 85–125 in I. Altman and J.F. Wohlwill, editors. *Human Behavior and Environment: Advances in Theory and Research.* Plenum Press, New York.

UN (United Nations). 2003. *Water for People: Water for Life. UN World Water Development Report.* United Nations World Water Assessment Program. UNESCO, Paris.

UN (United Nations). 2007a. *The Millennium Development Goals Report.* United Nations, New York.

UN (United Nations). 2007b. Press Release on The Millennium Development Goals Report. 2 July

UN (United Nations). 2008. Population Statistics. http://unstats.un.org/unsd/demographic/products/vitstats/. Accessed January 15, 2008.

UNDP (United Nations Development Programme). 2003. *Human Development Report 2003: Millennium Development Goals: A Compact among Nations to End Human Poverty.* Oxford University Press, New York.

UNDP (United Nations Development Programme). 2008. United Nations Development Programme. The Equator Initiative. http://www.undp.org/equatorinitiative.htm.

USDA and USDI (US Department of Agriculture and US Department of Interior). 1994. *Record of Decision for Amendments to Forest Service and Bureau of Land Management Planning Documents within the Range of the Northern Spotted Owl.* USDA Forest Service Regional Office, Portland, OR.

Valiela, I. 1995. *Marine Ecological Processes.* 2nd Edition. Springer-Verlag, New York.

van der Brugge, R., and R. van Raak. 2007. Facing the adaptive management challenge: Insights from transition management. *Ecology and Society* 12(2):33. http://www.ecologyandsociety.org/vol12/iss2/art33/.

Vannote, R.I., G.W. Minshall, K.W. Cummings, J.R. Sedell, and C.E. Cushing. 1980. The river continuum concept. *Canadian Journal of Fisheries and Aquatic Sciences* 37:120–137.

Vetter, S. 2005. Rangelands at equilibrium and non-equilibrium: Recent developments in the debate. *Journal of Arid Environments* 62: 321–341.

Vincent, G., and Y. Bergeron. 1985. Weed synecology and dynamics in urban environment. *Urban Ecology* 9:161–175.

Vitousek, P.M. 2004. *Nutrient Cycling and Limitation: Hawai'i as a Model System.* Princeton University Press, Princeton.

Vitousek, P.M., J.D. Aber, R.W. Howarth, G.E. Likens, P.A. Matson, et al. 1997. Human alteration of the global nitrogen cycle: Sources and consequences. *Ecological Applications* 7:737–750.

Vitousek, P.M., P.R. Ehrlich, A.H. Ehrlich, and P.A. Matson. 1986. Human appropriation of the products of photosynthesis. *BioScience* 36: 368–373.

Vitousek, P.M., H.A. Mooney, J. Lubchenco, and J.M. Melillo. 1997. Human domination of Earth's ecosystems. *Science* 277:494–499.

Vörösmarty, C.J., C. Leveque, C. Revenga, R. Bos, C. Caudill, et al. 2005. Fresh water. Pages 165–207 *in* R. Hassan, R. Scholes, and N. Ash, editors. *Ecosystems and Human Well-Being: Current State and Trends. Millennium Ecosystem Assessment*. Island Press, Washington.

Walbran, P.D., R.A. Henderson, J.W. Faithful, H.A. Polach, R.J. Sparks, et al. 1989. Crown-of-thorns starfish outbreaks on the Great Barrier Reef: A geological perspective based upon the sediment record. *Coral Reefs* 8:67–78.

Walker, B.H. 1993. Rangeland ecology: Understanding and managing change. *Ambio* 22:80–87.

Walker, B.H. 1995. Conserving biological diversity through ecosystem resilience. *Conservation Biology* 9:747–752.

Walker, B.H., L.H. Gunderson, A.P. Kinzig, C. Folke, S.R. Carpenter, et al. 2006. A handful of heuristics and some propositions for understanding resilience in social-ecological systems. *Ecology and Society* 11(1):13. http://www.ecologyandsociety.org/vol11/iss1/art13/.

Walker, B.H., C.S. Holling, S.R. Carpenter, and A.P. Kinzig. 2004. Resilience, adaptability and transformability in social–ecological systems. *Ecology and Society* 9(2):5 http://www.ecologyandsociety.org/vol9/iss2/art5/.

Walker, B.H., A. Kinzig, and J. Langridge. 1999. Plant attribute diversity, resilience, and ecosystem function: The nature and significance of dominant and minor species. *Ecosystems* 2:95–113.

Walker, B.H., and J.L. Langridge. 1997. Predicting savanna vegetation structure on the basis of plant available moisture (PAM) and plant available nutrients (PAN): A case study from Australia. *Journal of Biogeography* 24:813–25

Walker, B.H., and J.A. Meyers. 2004. Thresholds in ecological and social–ecological systems: A developing database. *Ecology and Society* 9(2):3. http://www.ecologyandsociety.org/vol9/iss2/art3/.

Walker, B.H., and D. Salt. 2006. *Resilience Thinking: Sustaining Ecosystems and People in a Changing World*. Island Press, Washington.

Walker, B.H., D.M. Stafford Smith, N. Abel, and J. Langridge. 2002. A framework for the determinants of degradation in arid ecosystems. Pages 75–94 *in* J.F. Reynolds and D.M. Stafford Smith,

editors. *Global Desertification: Do Humans Cause Deserts?* Dahlem University Press, Berlin.

Walter, E., M. M'Gonigle, and C. McKay. 2000. Fishing around the law: The Pacific salmon management system as a "structural infringement" of aboriginal rights. *McGill Law Journal* 45: 263–314.

Walters, C.J. 1986. *Adaptive Management of Renewable Resources*. Blackburn Press, Caldwell, NJ.

Walters, C.J. 1997. Challenges in adaptive management of riparian and coastal ecosystems. *Conservation Ecology* 1(2):1. http://www.consecol.org/vol1/iss2/art1/.

Walters, C.J. 2007. Is adaptive management helping to solve fisheries problems? *Ambio* 36:304–307.

Walters, C.J., V. Christensen, and D. Pauly. 1999. ECOSPACE: Prediction of mesoscale spatial patterns in trophic relationships of exploited ecosystems. *Ecosystems* 2:539–544.

Walters, C.J., R. Goruk, and D. Radford. 1993. Rivers Inlet Sockeye salmon: An experiment in adaptive management. *North American Journal of Fisheries Management* 13:253–262.

Walters, C.J., and R. Hilborn. 1976. Adaptive control of fishing systems. *Journal of the Fisheries Research Board of Canada* 33:145–159.

Walters, C.J., and C.S. Holling. 1990. Large-scale management experiments and learning by doing. *Ecology* 71:2060–2068.

Walters, C.J., and S.J.D. Martell. 2004. *Fisheries Ecology and Management*. Princeton University Press, Princeton.

Walters, C.J., and McGuire, J.J. 1996. Lessons for stock assessment from the northern cod collapse. *Reviews in Fish Biology and Fisheries* 6: 125–137.

Wapner, P. 1997. Governance in global civil society. Pages 65–84 *in* O.R. Young, editor. *Global Governance: Drawing Insights from the Environmental Experience*. MIT Press, Cambridge, MA.

Warren-Rhodes, K., and A. Koenig. 2001. Escalating trends in the urban metabolism of Hong Kong: 1971–1997. *Ambio* 30:429–438.

Wasserman, S., and K. Faust. 1994. *Social Network Analysis: Methods and Applications*. Cambridge University Press, Cambridge.

Watts, D.J. 2003. *Six Degrees: The Science of a Connected Age*. Vintage, London.

WCD (World Commission on Dams). 2000. *Dams and Development: A New Framework for Decision-Making*. Earthscan, London.

WCED (World Commission on Environment and Development). 1987. *Our Common Future*. Oxford University Press, Oxford.

Weber, J.R., and C.S. Word. 2001. The communication process as evaluative context: What non-scientists hear when scientists speak? *BioScience* 51:487–495.

Weinstein, M. 2006. Not "either-or": A presentation about traditional knowledge and regulation in relation to salmon farm management in the Broughton Archipelago. Legislative Assembly of British Columbia, Special Committee on Sustainable Aquaculture. June 26. 9 pp.

Weinstein, M. 2007. Five years of progress? Understanding changes to clam beaches in the Broughton. Presentation to the Workshop on Clam Bed Research in the Broughton Archipelago, Malaspina University-College, Institute for Coastal Research, In partnership with BC Ministry of Agriculture and Lands, BC Ministry of Environment, and Fisheries & Oceans Canada. April 12.

Westley, F. 1995. Governing design: The management of social systems and ecosystems management. Pages 391–427 *in* L.H. Gunderson, C.S. Holling, and S. Light, editors. *Barriers and Bridges to the Renewal of Ecosystems and Institutions*. Columbia University Press, New York.

Westley, F. 2002. The devil in the dynamics: Adaptive management on the front lines. Pages 333–360 *in* L.H. Gunderson and C.S. Holling, editors. *Panarchy: Understanding Transformations in Human and Natural Systems*. Island Press, Washington.

Westley, F., S.R. Carpenter, W.A. Brock, C.S. Holling and L.H. Gunderson. 2002. Why systems of people and nature are not just social and ecological systems. Pages 103–119 *in* L.H. Gunderson and C.S. Holling, editors. *Panarchy: Understanding Transformations in Human and Natural Systems*. Island Press, Washington.

Westoby, M., B. Walker, and I. Noy-Meir. 1989. Opportunistic management for rangelands not at equilibrium. *Journal of Range Management* 42:266–274.

White, W.B., G. McKeon, and J. Syktus. 2003. Australian drought: The interference of multispectral global standing modes and travelling waves. *International Journal of Climatology* 23: 631–662.

Wibe, S., and T. Jones, editors. 1992. *Forests: Market and Intervention Failures*. Earthscan Publications Ltd., London.

Wilson, D.C., J.R. Nielsen, and P. Degnbol, editors. 2003. *The Fisheries Co-Management Experience: Accomplishments, Challenges,* and Prospects. Kluwer Academic Publishers, Dordrecht.

Wilson, E.O. 1998. *Consilience: The Unity of Knowledge*. Alfred A. Knopf, New York.

Wilson, J. 2002. Scientific uncertainty, complex systems, and the design of common-pool institutions. Pages 327–360 *in* NRC (National Research Council); E. Ostrom, T. Dietz, N. Dolšak, P.C. Stern, S. Stovich et al., editors. 2002. *The Drama of the commons*. National Academy Press, Washington.

Wilson, J.A., J. Acheson, M. Metcalf, and P. Kleban. 1994. Chaos, complexity, and community management of fisheries. *Marine Policy* 19:291–305.

Williamson, O.E. 1993. Transaction cost economics and organizational theory. *Industrial and Corporate Change* 2:107–155.

Wolf, A.T. 1998. Conflict and cooperation along international waterways. *Water Policy* 1: 251–265.

Wollenberg, E., D. Edmunds, and L. Buc. 2000. *Scenarios as a tool for adaptive forest management*. Center for International Forestry Research, Bogor, Indonesia. http://www.cifor.cgiar.org/ publications/pdf_files/Other/SCENARIO.pdf.

Wondolleck, J.M., and S.L. Yaffee. 2000. *Making Collaboration Work: Lessons from Innovation in Natural Resource Management*. Island Press, Washington.

World Bank. 2004. *World Development Report*. World Bank. Washington.

World Bank. 2007. *World Development Indicators*. [online database] URL: http://devdata. world-bank.org/dataonline/. Accessed June 1, 2007.

Worm, B., E.B. Barbier, N. Beaumont, J.E. Duffy, C. Folke, et al. 2006. Impacts of biodiversity loss on ocean ecosystem services. *Science* 314:787–790.

WRI (World Resources Institute). 2005. *World Resources 2005: The Wealth of the Poor - Managing Ecosystems to Fight Poverty*. World Resources Institute, Washington.

Wright, M.M., and R.C. Brown. 2007. Comparative economics of biorefineries based on the biochemical and thermochemical platforms. *Biofuels, Bioproducts and Biorefining* 1(1):49–56.

Xue, Y.K., H.M.H. Juang, W.P. Li, S. Prince, R. DeFries, et al. 2004. Role of land surface processes in monsoon development: East Asia and West Africa. *Journal of Geophysical Research-Atmospheres* 109:D03105, doi:10.1029/2003JD003556.

Yaffee, S.L. 1997. Why environmental policy nightmares recur. *Conservation Biology* 11:328–337.

Yodzis, P. 2001. Must top predators be culled for the sake of fisheries? *Trends in Ecology and Evolution* 16:78–84.

Young, M.D. 1985. The influence of farm size on vegetation condition in an arid area. *Journal of Environmental Management* 21:193–203.

Young, O.R. 1982. *Resource Regimes, Natural Resources and Social Institutions*. University of California Press, Berkeley.

Young, O.R. 1994. The problem of scale in human/environment relationships. *Journal of Theoretical Politics* 6:429–447.

Young, O.R. 2002a. Institutional interplay: The environmental consequence of cross-scale interactions Pages 263–292 *in* NRC (National Research Council); E. Ostrom, T. Dietz, N. Dolšak, P.C. Stern, S. Stovich, et al., editors. *The Drama of the Commons*. National Academy Press, Washington.

Young, O.R. 2002b. *The Institutional Dimensions of Environmental Change: Fit, Interplay, and Scale*. MIT Press, Cambridge, MA.

Young, O.R. 2007. Rights, rules, and common pools or the perils of studying human/environment relations. *Natural Resources Journal* 47:1–16.

Young, O.R., F. Berkhout, G.C. Gallopin, M.A. Janssen, E. Ostrom, and S. van der Leeuw. 2006. The globalization of socio-ecological systems: An agenda for scientific research. *Global Environmental Change* 16:304–316.

Young, O.R., L.A. King, and H. Schroeder, editors. 2008. *Institutions and Environmental Change: Principal Findings, Applications, and Research Frontiers*. MIT Press, Cambridge

Zhu, Y., H. Chen, J. Fan, Y. Wang, Y. Li, J. Chen, J. Fan, et al. 2000. Genetic diversity and disease control in rice. *Nature* 406:718–722.

Zimov, S.A., V.I. Chuprynin, A.P. Oreshko, F.S. Chapin, III, J.F. Reynolds, et al. 1995. Steppe-tundra transition: An herbivore-driven biome shift at the end of the Pleistocene. *American Naturalist* 146:765–794.

Zwanenburg, K.C.T., D. Bowen, A. Bundy, K. Drinkwater, K. Frank, et al. 2002. Decadal changes in the Scotian Shelf large marine ecosystem. Pages 105–150 *in* K. Sherman and H.R. Skjoldal, editors. *Large Marine Ecosystems of the North Atlantic*. Elsevier, Amsterdam.

Index